Introduction to

INDIAN KNOWLEDGE SYSTEM
Concepts and Applications

B. MAHADEVAN
Professor
Indian Institute of Management Bangalore

VINAYAK RAJAT BHAT
Associate Professor
Center for Indian Knowledge System
Chanakya University
Bengaluru

NAGENDRA PAVANA R.N.
Assistant Professor
School of Vedic Knowledge Systems
Chinmaya Vishwa Vidyapeeth
Ernakulam

PHI Learning Private Limited
Delhi-110092
2026

*In fond memory of **Shri Asoke K. Ghosh** (October 1942 – February 2024), Founder Chairman and Managing Director of PHI Learning, whose vision endlessly inspires.*

The Legacy Continues....

Published by Pushpita Ghosh, PHI Learning Private Limited, Rimjhim House, 111, Patparganj Industrial Estate, Delhi-110092 and Printed by Guruji Printers, 426, Ground Floor, Pocket–C, IFC Paper Market, Ghazipur, New Delhi-110096.

₹795.00

INTRODUCTION TO INDIAN KNOWLEDGE SYSTEM: Concepts and Applications
B. Mahadevan, Vinayak Rajat Bhat and Nagendra Pavana R.N.

ISBN-978-93-91818-20-3 (Print Book)
ISBN-978-93-91818-21-0 (e-Book)

TO

The Great Sages and Saints of Ancient India and the lineage of Gurus who selflessly contributed to the IKS knowledge repository

|| न हि ज्ञानेन सदृशं पवित्रमिह विद्यते ||

There is nothing as purifying as knowledge

Bhagavad-Gita (4.38)

The development of background material for preparation of the textbook was through a project funded by the **Ministry of Education, Government of India**.

Contents in Brief

Contents

Part 4
Humanities and Social Sciences in IKS

प्रो. अनिल डी. सहस्रबुद्धे
अध्यक्ष
Prof. Anil D. Sahasrabudhe
Chairman

सत्यमेव जयते

आज़ादी का अमृत महोत्सव

अखिल भारतीय तकनीकी शिक्षा परिषद्
(भारत सरकार का एक सांविधिक निकाय)
(शिक्षा मंत्रालय, भारत सरकार)
नेल्सन मंडेला मार्ग, वसंत कुंज, नई दिल्ली–110070
दूरभाष : 011–26131498
ई–मेल : chairman@aicte-india.org

ALL INDIA COUNCIL FOR TECHNICAL EDUCATION
(A STATUTORY BODY OF THE GOVT. OF INDIA)
(Ministry of Education, Govt. of India)
Nelson Mandela Marg, Vasant Kunj, New Delhi-110070
Phone : 011-26131498
E-mail : chairman@aicte-india.org

Foreword
by Anil Sahasrabudhe

The formal Indian education system inherited from British policymakers famously known as Macaulian system has kept the Ancient Indian knowledge heritage out of the reach of the budding young students, citing reasons of lack of rigor and scientific value. This is arguably an incorrect perspective. However, truth is otherwise. Ours is one of the only continuing, surviving ancient civilization with huge repository of knowledge created by the forefathers, which can provide great value for any society. It enables the current generation to understand the thought processes and frameworks, analyse the received wisdom in a contemporary context and provide new opportunities to assimilate the accrued wisdom and synthesize new knowledge.

Therefore, All India Council for Technical Education (AICTE), the apex body for technical education spanning the Engineering and Management Education in India while revising the curricula in 2018, introduced a mandatory non-credit course on Indian Knowledge Systems (IKS) along with courses on constitution of India and environment science. While there are several books on the latter courses, unfortunately, there are no textbooks currently available for the subject that will help students understand the various components of IKS. Hence a textbook for the course on Indian Knowledge Systems (IKS) is a timely and valuable contribution to the education system in the country and I am happy to note that this book has been written to address the requirements of this course.

The book has detailed coverage of IKS topics, and the authors have taken the right approach to balancing the concepts with the applications. I notice extensive endnotes for each chapter that points to the depth of research the authors have carried out in culling out relevant ideas to the students. The chapters on Number Systems, Mathematics, and Astronomy truly showcase the scientific rigour and heritage that was existing in the country. Similarly, the chapters on Metal Working and other engineering applications will be true eye openers for the young engineering students. Another interesting feature of this book is the discover IKS exercises at the end of the chapters. This will provide further opportunities for the students to experientially discover the multi-faceted nature of IKS.

Although the authors claim that the book was mainly intended for the IKS course in Engineering institutions, I believe that the book can serve as a useful textbook for any undergraduate or postgraduate degree programs in the country wanting to introduce an elective course on IKS. Incidentally, the NEP 2020 does refer to Indian Knowledge Systems couple of times. The book will therefore be a valuable asset in the context of the New Education Policy (NEP 2020).

I hope the book will motivate professionals and IKS researchers to deep dive into each chapter and write a separate textbook based on each chapter.

(Anil Sahasrabudhe)

Foreword
by Subhash Kak

It is generally recognized that historical accounts of science in India, the world's third-largest economy on a purchasing power parity basis, are woefully inadequate and need to see beyond the colonial lens. Therefore, I am pleased that this book *Introduction to Indian Knowledge System – Concepts and Applications* written by B. Mahadevan, Vinayak Rajat Bhat, and R.N. Nagendra Pavana will provide the students with an excellent introduction to the astonishing breadth and depth of the Indian scientific and knowledge tradition.

Ṣā'id al-Andalusī, writing in Toledo, Spain in 1068 in the *Tabaqāt al-'Umam*, comparing science in different parts of the world declared that India was the most advanced nation: "The first nation to have cultivated science is India. India is known for the wisdom of its people." India appears to have remained ahead of others in science until about the seventeenth century.

India's sciences are based on fundamental principles, axioms, logical inference, and empirical observations and were generally written down in texts called śāstras and sūtras. For example, the Śulba-sūtra geometry includes the so-called "Pythagoras theorem" several centuries prior to its later discovery in Greece; there is also Pāṇini's astonishing grammar that describes the Sanskrit language in 4000 algebraic rules with a structure that is now compared to a computer program. The logic of Navya Nyāya is equivalent to mathematical logic, which is the foundation of analysis by modern machines.

Kaṇāda's physics has its laws of motion, and it speaks of nine classes of substances, some of which are non-atomic and some atomic. Every substance was taken to be composed of four different kinds of atoms, two of which had mass and two did not. A thousand or more years after Kaṇāda, Āryabhaṭa postulated that earth rotated and advanced the basic idea of relativity of motion. It is fascinating that Kaṇāda's ideas were communicated by Swami Vivekananda to Nikola Tesla in connection with the possibility of conversion of mass into energy.

It is generally believed that the discovery of infinite series and calculus by Newton and Leibniz heralded the Scientific Revolution that was to change the world. But new research has shown that the Kerala School of Mathematics had already developed calculus over two centuries prior. Some historians suggest that this advanced astronomical knowledge from Kerala went abroad via the Jesuits and sparked its further development in Europe.

The ancient Āyurveda texts include the notion of germs and inoculation and postulate mind-body connection, which has become an important area of contemporary research. Āyurveda uses tripartite analysis (three *doṣas*, for example), and since modern logic shows that ternary logic is more powerful than binary logic, this fact may prove to be an advantage in the further development of medicine.

There are indirect ways that Indian ideas led to scientific advances. Mendeleev was inspired by the two-dimensional structure of the Sanskrit alphabet to propose a similar two-dimensional structure of chemical elements. Erwin Schrödinger, a founder of quantum theory, credited ideas in the Upanishads for the key notion of superposition that was to bring about the quantum revolution in physics that has transformed modern science. A Vedantic vision guided Jagadish Chandra Bose in his pathbreaking discoveries in several fields, and he is considered the true father of radio science which, as we know, heralded a new age.

Introduction to Indian Knowledge System – Concepts and Applications provides a grand view of the Indian knowledge tradition, and it goes back to the Vedic texts for its overarching unity. The book covers various topics across linguistics, mathematics, astronomy, engineering, town planning, and architecture; it also shows the application of these ideas to health, psychology, and management. Apart from the careful study of motion, vibratory phenomena, acoustics, Indian sages studied the transforming power of fire. It was known that lightning bolts had electricity. Indian technology, which was quite advanced, was used in mechanical devices, engineering, art, and civil projects.

The phenomenon of consciousness is the frontier from the perspectives of biology and physics; it is also central to answering whether machines will become conscious. The Upanishads speak of two kinds of science: one about things and concepts and their mutual relationships (*aparā vidyā*); the other about consciousness (*parā vidyā*). The book will also serve as an excellent foundation for those who wish to investigate how the Indian tradition of consciousness science can be harnessed for further scientific advance and the betterment of the world.

Professor Subhash Kak
Regents Professor
School of Electrical and Computer Engineering
Oklahoma State University–Stillwater
USA

Foreword
by S. Sadagopan

I have gone through this book in detail and it has been an *enriching, fulfilling* and *satisfying* experience. The authors have done an outstanding job covering a whole gamut of things, yet with sufficient depth, to avoid a superficial survey.

The authors have addressed the contributions to Foundational knowledge, Science, Engineering & Technology, and Humanities & Social Sciences through a structured classification. **IKS** (Indian Knowledge System) have evolved over centuries, in fact, over millenniums. It has a wide range, and the well-known among them being **Astronomy**, including planetary movements, solar-centric world, shape and diameter of the Earth; **Health and Well-being**, including plants & herbs, surgical procedures; **Mathematics and Computing**, including the discovery of zero, decimal system of numerals, and approximation algorithms for computation of Pi; **Languages and Linguistics**, including Panini's Sanskrit grammar; **Metallurgy,** including steel-making and zinc-smelting; and **Public Administration**, including Good Governance and Taxation.

Thanks to the Royal patronage over centuries, there was a system where scholars could study IKS throughout their lives without worrying about their daily livelihood. This led to several Schools of thought, often expounded by several principal works, notably, *Brahmasutra Bhashya* by **Sankaracharya**, *Sri Bhashya* by **Ramanujacharya**, *Gita Bhashya* by **Madhwacharya**, *Mulamadhyamakarika* by **Nagarjuna**, and *Pramana Samucchaya* by **Dignaga.** Several cities and towns could house thousands of scholars leading to Centres of learning, for example, **Kasi** in the North and **Kanchi** in the South. Institutions of learning, including **Nalanda** and **Takshashila,** paved the way for graduating generations of scholars.

Particular mention must be made of the many texts in IKS that are referred to in the footnotes at the end of the chapters. For most people, IKS works stop with *Ramayana, Mahabharata* and *Srimad Bhagavatam*; Patanjali's *Yogasutras* on **Yoga**, Sayana's *Sayana Bhashya* on **Vedas,** Bharata's *Natyasastra* on **Dance**, Kautilya's *Arthasastra* on **Economics**, Vishnu Sharma's *Pancha Tantra* **Stories,** Sushruta's *Sushruta Samhita* on **Medicine and Surgery,** and more recently Panini's *Ashtadhyayi* on **Sanskrit Grammar** that caught the attention of Computer Scientists. I particularly like the references to the less known and equally impactful works like *Rasa Ratna Samuccaya* that details complex **Metallurgy**, *Aryabhatiya* and Varahamihira's *Brihat Samhita* on **Astronomy**, *Shulba Sutras* on **Geometry,** *Lilavati* of Bhaskaracharya a treatise on **Mathematics**, Gautama's *Nyaya Sutra* on **Logic**, Kanada's *Vaiseshika Sutras* on **Ontology**, Narada's *Shilpasastra* on **Architecture** and **Civil Engineering**, Mayamuni's *Mayamata* (**Architecture**), Bhoja's *Yuktikalpataru* (**Ship-building**), and, Kshemasarma's *Kshemakutuhalam* on **Dietetics and Well-being**.

Most of these works were set in the Sanskrit language widespread across the country. With the higher education system switching completely to the English language over the past two centuries (more so after Indian Independence in 1947 CE), IKS became inaccessible to most Indians. There has been sporadic interest in the IKS, particularly from Western scholars, in specific areas, thanks to pioneering efforts of individuals; for example,

- Swami Vivekananda and BKS Iyengar – Indian Philosophical Systems and Yoga
- Pandit Ravi Shankar and MS Subbalakshmi – North Indian and South Indian Classical Music

Yet, for most University students in India today, IKS is not in their consideration set!

It is against this backdrop that the study of IKS (Indian Knowledge System) is an idea whose time has come. I am delighted that AICTE & UGC are making efforts to support scholars and textbook creators in the IKS domain.

I had a personal brush with reality when I was looking for an authentic and widely available text of Rig Veda that I realized that the Indian tradition was largely "oral". I ended up creating the first-ever digital edition of Rig Veda through the book *Rig Veda Samhita*, SAKSI, 1998.

This book by Professor Mahadevan et al. is a timely release, particularly in the Post Covid-19 World, when India plays a leading role in the science of discovering vaccines and managing distribution and logistics. It makes IKS accessible to a much larger set of readers, notably University students. The book is interspersed with illustrations and context setting "action points". It provides ample opportunities for further exploration through the pointers for further study that include several authentic works and Websites. I am sure the additional material provided through the companion website will make the book a place that young readers will visit multiple times.

Happy reading!

Professor Sadagopan
Chairman, BoG, IIITDM-Kancheepuram
Founder Director, IIIT-Bangalore

Preface

India is a nation with a long civilizational history with recorded history, cultural artefacts, and evidence pointing to more than five millennia of existence. Not surprisingly, such a society would have discovered enormous knowledge cutting across various dimensions of human life and existence. Despite a continuous onslaught of foreign invasions for more than a millennium, the knowledge practices have more or less remained intact and have been passed on from generation to generation 'orally'. There have also been attempts to record the knowledge and practices in written form in the last millennium. However, during the last 200 years, this knowledge has been relegated to the background and the knowledge paradigm of the West has been imposed on the Indian society. Even after Independence, no concrete efforts were made to introduce indigenous knowledge in the educational curriculum formally.

However, knowing the thinking patterns and the knowledge repository created by the forefathers provides great value for any society. It enables the current generation to understand the thought processes and frameworks, and synthesize new knowledge. Arguably, no society can hope to flourish by simply severing itself from its past and embracing alternative ideas alien to the indigenous thoughts and practices. After some time, the need to revisit will be strongly felt, and the Indian society is currently going through such a phase.

The policymakers and the government have taken cognizance of this and have taken several steps to address this requirement. One of the steps is to introduce a course on the Indian Knowledge System (IKS) in higher educational institutions. Unfortunately, we do not have a textbook to teach this course. This textbook responds to this growing need felt by the Indian society at large and is primarily intended to facilitate offering a one-semester or two-semester course on IKS to undergraduate and graduate-level students. This textbook is a culmination of our efforts to offer a two-semester course on IKS in Chinmaya Vishwavidyapeeth, Ernakulam.

ORGANIZATION OF THE BOOK

This book seeks to fill in the gap for offering a required course on IKS, recently mandated by AICTE. The proposed textbook seeks to introduce the epistemology and ontology of IKS to the Engineering and Science students in a way they can relate, appreciate and explore further should there be a keen interest in the matter. Moreover, the New Education Policy (NEP) has also provided a clear trajectory for imparting IKS in the higher education curriculum, necessitating a book of this kind in several higher education institutions in the country in the days to come.

The book presents the topics in a two-part study of IKS. The first is to develop an overall understanding of some key components of IKS. The second aspect is the application of IKS for some gainful use. Specific areas of applications can be studied in a focused manner by drawing the relevant portions of the IKS showcasing the potential for practical application either in theoretical advancement of concepts or practical use in addressing some issues that we may face either as an individual or at an institutional level. The book presents these aspects in four parts:

- Part I: Indian Knowledge System: An Introduction
- Part II: Foundational Concepts relevant for Science, Engineering, and Technology Applications
- Part III: Specific Concepts related to Science, Engineering, and Technology
- Part IV: Humanities and Social Sciences Applications

The details of the chapters and the topics discussed in the chapter are schematically presented below.

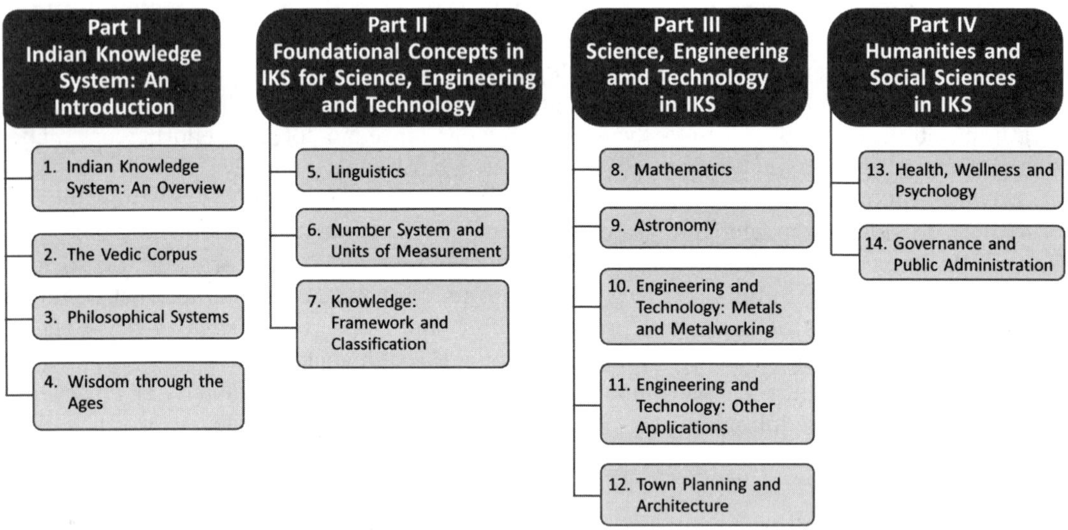

Although the book has been primarily developed for use by the engineering institutions, the structure and the contents lend itself easily to address the requirement in other University systems (Liberal Arts, Medicine, Science and Management) for such a book. Chapter 1 of the book provides these suggestions.

UNIQUE FEATURES

One approach taken to present IKS has been to lay clear emphasis merely on the 'content' of IKS. Therefore, efforts were made to present the 'what' of IKS rather than the 'why' or 'how' of IKS. The application, implications, and practical relevance of IKS are often left for self-introspection by the students. Another approach is to merely showcase the practical relevance of IKS. However, to impress that IKS has application potential, the material is presented with generous and sweeping extrapolation and innovative interpretations. We have taken a middle path and have made special efforts to present IKS in a contextually relevant fashion by delicately balancing the 'why' or 'how' of IKS and the 'what' of IKS. Moreover, the textbook follows International best practices by incorporating several pedagogical features that will make learning effective and enjoyable for the students.

Chapter summaries, review questions, and exercises at the end of the chapters help students check their progress in comprehending the subject matter. Besides this, Multiple choice questions are also available at the companion website (https://www.phindia.com/introduction_to_indian_knowledge_system) for this textbook.

However, following features are also provided to make the process enjoyable for both the teacher and the student:

IKS in Action: This feature enables the students to develop an early appreciation of the subject matter being discussed and provides a context where the ideas discussed in the chapter have application.

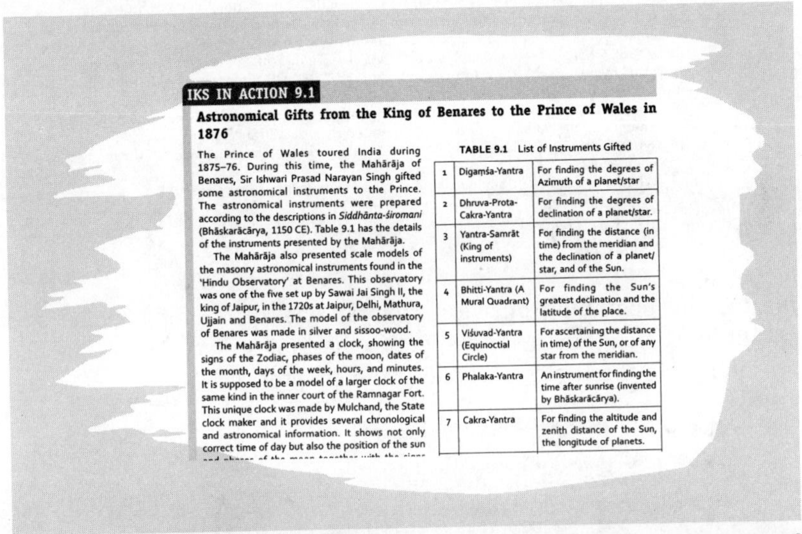

IKS IN ACTION 9.1

Astronomical Gifts from the King of Benares to the Prince of Wales in 1876

The Prince of Wales toured India during 1875–76. During this time, the Mahārāja of Benares, Sir Ishwari Prasad Narayan Singh gifted some astronomical instruments to the Prince. The astronomical instruments were prepared according to the descriptions in *Siddhānta-śiromani* (Bhāskarācārya, 1150 CE). Table 9.1 has the details of the instruments presented by the Mahārāja.

The Mahārāja also presented scale models of the masonry astronomical instruments found in the 'Hindu Observatory' at Benares. This observatory was one of the five set up by Sawai Jai Singh II, the king of Jaipur, in the 1720s at Jaipur, Delhi, Mathura, Ujjain and Benares. The model of the observatory of Benares was made in silver and sissoo-wood.

The Mahārāja presented a clock, showing the signs of the Zodiac, phases of the moon, dates of the month, days of the week, hours, and minutes. It is supposed to be a model of a larger clock of the same kind in the inner court of the Ramnagar Fort. This unique clock was made by Mulchand, the State clock maker and it provides several chronological and astronomical information. It shows not only correct time of day but also the position of the sun

TABLE 9.1 List of Instruments Gifted

1	Digaṃśa-Yantra	For finding the degrees of Azimuth of a planet/star
2	Dhruva-Prota-Cakra-Yantra	For finding the degrees of declination of a planet/star.
3	Yantra-Samrāt (King of instruments)	For finding the distance (in time) from the meridian and the declination of a planet/star, and of the Sun.
4	Bhitti-Yantra (A Mural Quadrant)	For finding the Sun's greatest declination and the latitude of the place.
5	Viśuvad-Yantra (Equinoctial Circle)	For ascertaining the distance in time) of the Sun, or of any star from the meridian.
6	Phalaka-Yantra	An instrument for finding the time after sunrise (invented by Bhāskarācārya).
7	Cakra-Yantra	For finding the altitude and zenith distance of the Sun, the longitude of planets.

Rich Illustrations: The chapters are dotted with several illustrations by way of figures and tables that help concisely summarise complex concepts and facilitate easy understanding and retention by the students.

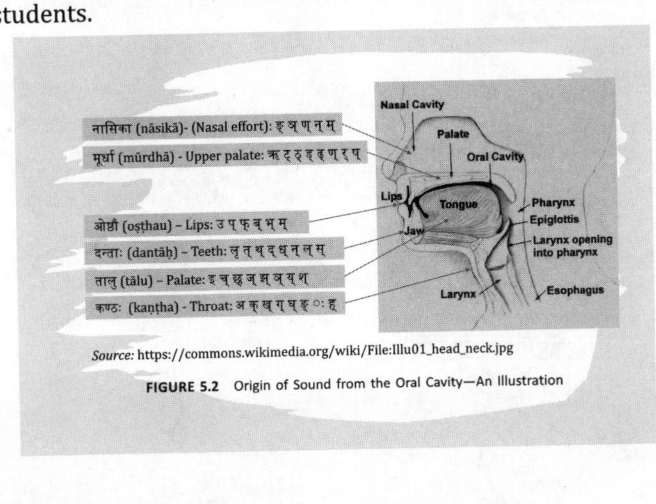

Source: https://commons.wikimedia.org/wiki/File:Illu01_head_neck.jpg

FIGURE 5.2 Origin of Sound from the Oral Cavity—An Illustration

Discover IKS: Every chapter has an end-of-the-chapter feature that points to useful videos on some of the concepts discussed in the chapter. The videos are actual illustrations, existing artefacts, or expert opinions. Students are encouraged to go through the video and prepare a report of their understanding of the ideas presented. Students can develop applied learning, appreciation of concepts, and their relevance to practice.

DISCOVER IKS

1. India's contribution to mathematics has been a subject that has attracted considerable attention in recent times. Several attempts have been made to inquire into this issue. Watch this video produced by BBC, titled, *History of Indian Mathematics, Part-1* by pursuing the link: https://youtu.be/pElvQdcaGXE. After watching the video carefully, prepare a write-up to answer the following questions:
   (a) What are the main contributions of ancient Indians to the number system?
   (b) What is the importance of the number '0'? How does it contribute to science?
   (c) What are the contributions of Brahmagupta in use of numbers in mathematics?
2. It is often mentioned that India's contribution to the world of mathematics is the number zero. This leaves an impression there is not much other than this contribution. On the contrary, there are significant contributions by ancient Indians in a number of areas of mathematics that has led to the development of mathematical thought and applications in a significant way in the West. Watch the talk by Prof. C.K. Raju on the topic, "Not just the Zero – India's gifts of mathematics and science to the world", by pursuing the link: https://youtu.be/22uZ3D5AgaE. After watching the video, prepare a three-page note that covers answers to the following questions:
   (a) Was India's contribution to Mathematics merely the number zero or something more? Enumerate the other salient contributions and their significance.
   (b) What are the salient aspects of Mādhava's sine table (kaṭapayādi)? What is the importance of sine table?
   (c) What is the value of traditional knowledge in mathematics? Do we need it?

Opening Vignette: The beginning of every chapter has an opening vignette, which helps develop a clear set of expectations with respect to the topics discussed in the chapter. It also generates early interest in the topics covered in the chapter.

LEARNING OUTCOMES

After finishing this chapter, you will be able to:

- Develop familiarity with the Science, Engineering and Technology (S & T) heritage of ancient and medieval India
- Understand the role of archaeological, and other evidence in assessing the S & T heritage of India
- Understand about ancient Indian pursuits in the area of metal and metalworking and idol making

This is a gold coin of the Gupta king Samudragupta (330–376 CE) kept at the British Museum. Samudragupta, with halo, standing facing left, wearing cap, decorated coat and trousers and earrings, holding a spear in left hand and making an offering with right hand over altar. In left field is a Garuda standard with ribbons and crescent above.

Stamping such coins with intricate details requires knowledge of mining, extraction of Gold and further metal forming techniques such as die casting.

Source: https://commons.wikimedia.org/wiki/File:SamudraguptaCoin.jpg

Endnotes: Material for the textbook has been drawn from several original sources and research papers. To establish authenticity and enable the students and the teachers to access the original sources of information, these have been listed at the end of every chapter.

ENDNOTES

1. For a good account of this refer to **Dharampal (2021).** "Indian Science and Technology in the eighteenth century: Some contemporary European Accounts", Dharampal Classic Series 2, Rashtrotthaana Sahitya, Bengaluru.
2. See for example this extract from Chamaka Praśna — ... हिरंण्यं च॒ मे ऽयंश्च मे॒ सीसं॑ च मे॒ त्रपु॑श्च मे श्या॒मं च॑ मे लो॒हं च॑ मे॒ ... hiráṇyaṃ ca̱ me 'yáśca me̱ sīsaṃ́ ca me̱ trapúśca me śyā̱mamṃ cá me lo̱haṃ cá me̱ ...
3. तिग्मं चिदेम महि वर्षो अस्य भसदश्वो न यमसान आसा । विजेहमानः परशुर्न जिह्वां द्रविर्न द्रावयति दारु धक्षत् ॥ tigmaṃ cidema mahi varṣo asya bhasadaśvo na yamasāna āsā | vijehamānaḥ paraśurna jihvāṃ dravirna drāvayati dāru dhakṣat || Ṛgveda Saṃhitā 6.3.4
4. तद्यथा लवणेन सुवर्णं सन्दध्यात्सुवर्णेन रजतं रजतेन त्रपु त्रपुणा सीसं सीसेन लोहं लोहेन दारु दारु चर्मणा ॥ tadyathā lavaṇena suvarṇaṃ sandadhyātsuvarṇena rajataṃ rajatena trapu trapuṇā sīsaṃ sīsena lohaṃ lohena dāru dāru carmaṇā || **Swami Gambhirananda (2003).** "Chāndogya Upaniṣad", Advaita Ashrama, Kolkata, p. 308.
5. लोहितेन स्वधितिना मिथुनं कर्णयोः कृधि । अकर्तामश्विना लक्ष्म तदस्तु प्रजया बहु ॥ lohitena svadhitinā mithunaṃ karṇayoḥ kṛdhi | akartāmaśvinā lakṣma tadastu prajayā bahu || Atharvaveda 6.141.2
6. को अस्मिन्नापो व्यदधात्विषूवृतः पुरूवृतः सिन्धुसृत्याय जाताः । तीव्रा अरुणा लोहिनीस्ताम्रधूम्रा ऊर्ध्वा अवाचीः पुरुषे तिरश्चीः ॥ ko asminnāpo vyadadhātviṣūvṛtaḥ purūvṛtaḥ sindhusṛtyāya jātāḥ | tīvrā aruṇā lohinīstāmradhūmrā ūrdhvā avācīḥ puruṣe tiraścīḥ || Atharvaveda, 10.2.11
7. For a detailed report on this, please see, **Srinivasan, K.R. (1958).** "The Pallava Architecture of South India", *Ancient India*, **14,** pp. 114–138.
8. For more details on this and issues addressed in this section see, **Deva, K. (1959).** "Temples of Khajuraho in Central India", *Ancient India*, **15,** pp. 43–65.
9. For more details on this and the other issues addressed in this section see, **Singh, P.K., Dey, P., Jaina, S.K. and Mujumdar, P.P. (2020).** "Hydrology and water resources management in ancient India", *Hydrol. Earth Syst. Sci.*, **24**, pp. 4691–4707.
10. https://en.wikipedia.org/wiki/Stepwell Last accessed on Oct. 1, 2021.
11. **Jain—Neubauer, J. (1999).** "The stepwells of Gujarat ", *India International Centre Quarterly*, **26**(2), pp. 75–80.

Suggested Readings: A list of additional readings has been provided at the end of every chapter to help the students pursue further studies in the topics covered in the chapter. It also helps the teachers prepare effectively the topics covered in the chapter.

SUGGESTED READINGS

Datta, B. and Singh, A.N. (1962). *History of Hindu Mathematics: Parts I and II*, Asia Publishing House, Mumbai.

Datta, B. (1932). *Ancient Hindu Geometry: The Science of Sulbas*, Calcutta University Press, Reprinted Cosmos Pug., New Delhi, 1993.

Divakaran, P.P. (2018). *The Mathematics of India: Concepts, Methods, Connections.* Springer (Hindustan Book Agency), New Delhi.

Joseph, G.G. (1990). *The Crest of the Peacock: Non-European Roots of Mathematics*, Penguin.

Joseph, G.G. (2009). *A Passage to Infinity: Medieval Indian Mathematics from Kerala and Its impact.* Sage, New Delhi.

Joseph, G.G. (2016). *Indian Mathematics Engaging with the World from Ancient to Modern Times*, World Scientific, London.

Kedarnath (1938). *Chandaḥ śāstra of Piṅgala with Commentary Mṛtasañjīvanī of Halāyudha Bhaṭṭa*, 3rd ed., Mumbai, 1938.

Kolachana, A., Mahesh K. and Ramasubramanian, K. (2019). "Hindu mathematics in the seventh century as found in Bhāskara I's commentary on the *Āryabhaṭīya* (IV)", In: Kolachana A., Mahesh K., Ramasubramanian K. (Eds), Studies in Indian Mathematics and Astronomy, *Sources and Studies in the History of Mathematics and Physical Sciences*, Springer, Singapore.

Madhusudana Sastri, (1994). *Vṛttaratnākara of Kedāra with Commentaries Nārāyaṇī and Setu*", Chaukhamba, Varanasi.

Parameswaran, S. (1998). *The Golden Age of Indian Mathematics*, Swadeshi Science Movement, Kerala.

Plofker, K. (1963). *Mathematics in India*, Princeton University Press, New Jersey, USA.

Ramasubramanaian, K. (2015). "Gaṇitānada": Selected works of Radha Charan Gupta on History of Mathematics, *Indian Society for History of Mathematics*, New Delhi.

Sarasvati Amma, T.A. (1999). *Geometry in Ancient and Medieval India*, Motilal Banarsidass, 2nd ed., New Delhi.

Sen, S.N. and Bag, A.K. (1983). "The Sulbasutras", INSA, New Delhi.

Our work in developing a textbook for IKS is perhaps the first of its kind. IKS is a generic phrase that covers practically everything pertaining to India. For a nation with more than 5000 years of recorded history, defining what constitutes IKS is itself a huge challenge. Literature, oral knowledge, societal practices, and knowledge assets in all Indian languages will technically fall under the ambit of IKS. We have taken a certain view on what constitutes IKS to make it manageable for a university-level course. We are aware that alternative views can be taken on this issue. We believe that our work will encourage other authors to bring more textbooks taking alternative perspectives on IKS.

B. Mahadevan
Vinayak Rajat Bhat
Nagendra Pavana R.N.

Acknowledgements

The need for developing a textbook on Indian Knowledge System is increasingly felt in the Indian society at large and the Higher Education institutions in particular. When we approached the then Honourable Union Minister for Education, Shri Ramesh Pokhriyal, with our proposal to develop a textbook, he readily agreed to fund and directed AICTE to coordinate the same. We are grateful to the Honourable Minister for this gesture and support to textbook development.

We are grateful to Prof. Anil Sahasrabudhe, Chairman, AICTE, for providing us with the required funding, encouragement, and moral support. He has constantly been monitoring the progress of the book and ensuring the timely completion of the book project. This has gone a long way in enabling us to focus on the project unhindered. We are also thankful to Professor Dilip Malkhede, Advisor – I, Policy and Academic Planning Bureau, AICTE, for his constant support and encouragement during the development of this textbook. Dr. Shamasundar was one of our co-investigators in the textbook development project, which preceded the textbook development. He was instrumental in initiating this entire exercise, contributing to certain parts of the textbook.

We are thankful to Swami Advayananda ji of Chinmaya International Foundation, Ernakulam, for his encouragement. Developing a course outline for the Indian Knowledge System during the formative stages of Chinmaya Vishwavidyapeeth (CVV) was Swamiji's idea. It culminated in this textbook, and we are grateful to him. We are also thankful to the Indian Institute of Management Bangalore (IIMB) and CVV for their support to us in many ways to complete this project.

A project of this magnitude is beyond our normal means, given the vastness of the subject matter to be covered. While we were self-motivated and focused on the task, a group of experts in IKS could play a significant role in ensuring we do not miss out on the essentials. It also required someone to closely monitor, go through the material, and offer critical comments for improving it. We were indeed fortunate to have a team of high-profile and accomplished academic colleagues in the area of the Indian knowledge system, who, despite their busy schedules, agreed to serve in our expert committee. The expert committee consists of the following members:

1. Professor Michel Danino, Visiting Professor, Humanities and Social Sciences, IIT, Gandhinagar.
2. Professor R N Iyengar, Distinguished Professor and Director, Centre for Ancient History and Culture, Jain University, Bangalore.
3. Professor Subhash C Kak, Regents Professor, School of Electrical and Computer Engineering, Oklahoma State University–Stillwater, USA.

4. Professor K Ramasubramanian, Humanities and Social Sciences, IIT Bombay.
5. Vidwan Chamu Krishna Shastry, Sanskrit Promotion Foundation, New Delhi.
6. Professor M.D. Srinivas, Centre for Policy Studies, Chennai.

We are grateful to every one of them for their support. Needless to say, we were greatly benefited from their advice, support, views, and suggestions on the material that we were developing. While all improvements are due to them, any errors and omissions are ours.

Professor M S Sriram of K V Sarma Research Foundation, Chennai, gave critical inputs for the chapter on Astronomy. We gratefully acknowledge the support rendered by Prof Sriram. When we sent our manuscripts to several subject matter experts, they were kind enough to go through them and patiently offer valuable comments and suggestions. Although it is impossible to name every one of them, we do want to mention a few names: Professors Gauri Mahulikar, Amba Kulkarni, Shrinivasa Varakhedi, R Venkataraghavan, B. Narsing Rao, Srinath Mohandas, Ashok Nene, Manoj Gundanna, Rama Jayasundar, P.L.T. Girija, and Mala Sinha. We are grateful to them.

We also acknowledge the support rendered by Dr. Poonam Purohit, Senior Research Associate, who put together diverse material into a clean format. She played a key role both administratively and academically in supporting the project. Namrata Praveen and the Library Staff at IIMB and CVV supported us by procuring the required material and archiving them in the library. We are thankful for their support. Ms. Shivani Garg, Senior Editor at PHI Learning, played a very important role in bringing out this book. Due to her constant attention to the project, we could publish the book faster. Our special thanks are due to her and her editorial team, who did a wonderful job of bringing out this book. We also thank Ms. Babita Misra, Editorial Coordinator PHI Learning for her support to this project.

The support of our family members enabled us to stay afloat and push us to our limits repeatedly. Without their support, we would have abandoned the project long back. We are thankful for their moral support, and we were encouraged when they steadfastly supported us throughout the project.

Finally, in the best traditions of India, we believe in the maxim of Gita–"दैवं चैवात्र पञ्चमम्" (daivaṃ caivātra pañcamam). Several forces have played an invisible role in the development of this textbook, and we realize that we are merely instruments in giving it a shape and structure.

B. Mahadevan
Vinayak Rajat Bhat
Nagendra Pavana R.N.

Guide to Transliteration Symbols Used in the Book

All Sanskrit words used in this book are presented using roman scripts as transliteration. The transliteration schema used in this book is based on International Alphabet of Sanskrit Transliteration (**IAST**). IAST is a transliteration scheme that allows a lossless romanization of scripts as employed by the Sanskrit language. The details are given below.

Vowels

अ	a	उ	u	ऌ	ḷ	ओ	o
आ	ā	ऊ	ū	ॡ	ḹ	औ	au
इ	i	ऋ	ṛ	ए	e	ं/अं	ṃ/aṃ
ई	ī	ॠ	ṝ	ऐ	ai	ः/अः	ḥ/aḥ

Consonants

क	ka	च	ca	ट	ṭa	त	ta	प	pa
ख	kha	छ	cha	ठ	ṭha	थ	tha	फ	pha
ग	ga	ज	ja	ड	ḍa	द	da	ब	ba
घ	gha	झ	jha	ढ	ḍha	ध	dha	भ	bha
ङ	ṅa	ञ	ña	ण	ṇa	न	na	म	ma

य	ya	र	ra	ल	la	व	va	श	śa
		ष	ṣa	स	sa	ह	ha		

Others

क्ष	kṣa	त्र	tra	ज्ञ	jña	श्र	śra	ऽ	‘

Part 1

INDIAN KNOWLEDGE SYSTEM
AN INTRODUCTION

CHAPTER

1 Indian Knowledge System: An Overview

LEARNING OUTCOMES

After finishing this chapter, you will be able to:

- Understand and appreciate the importance of ancient knowledge to a society
- Understand the term 'Indian Knowledge System' (IKS)
- Familiarise with the key components of the IKS
- Develop some appreciation of IKS historicity

Built by Raja Raja Chola I. The construction was completed in 1010. This is one of the earliest granite temples in the world. Around 60,000 tons of granite is said to be used to build the temple. It has one of the tallest vimānas (temple tower) and its kumbham (the structure on the top) weighs approximately 80 tons.

Source: https://upload.wikimedia.org/wikipedia/commons/7/7d/Brihadeeswarar_Temple_Full_View.jpg

IKS IN ACTION 1.1

Do We Need Indian Knowledge System?

There is a question ruminating over our minds, "Why should we bother so much about Indian Knowledge System (IKS)?" IKS is not about merely knowing about some ancestral knowledge. If we closely follow the emerging patenting regime and the economic power arising out of a knowledge society, it becomes clear that the issue merits serious attention. It is about protecting received wisdom, economic security, and national pride.

Efforts by an international coalition of environmentalists to get the US patents on products of the neem tree cancelled did not fructify. We all know that we use neem leaves to keep away insects from our kitchen garden. Every farmer in India knew from time immemorial that it was a good pesticide among other things. Yet, a few years ago, a US company was awarded a patent for neem as a pesticide. The company claimed it had developed an agent that would make the active pesticide agent in neem last for more than the normal two weeks. Although our scientists had been tinkering around with research on neem for years, they had not applied for this specific process and the battle was lost. Nor we were able to bring our vast knowledge system to argue against the case. The patenting of traditional remedies from developing countries became a global issue after patents were granted for neem.

On the other hand, after a legal battle for more than a year India's Council of Scientific and Industrial Research (CSIR), has successfully forced the US Patent and Trademark Office (PTO) to revoke a contentious patent it granted two years ago to researchers in the United States on the use of powdered turmeric (Curcuma longa) for wound healing. The turmeric patent was granted in 1995 to two researchers, Soman K. Das and Harihar Kohli of the University of Mississippi Medical Center. Their six patent claims covered the oral and topical use of turmeric powder to heal surgical wounds and ulcers.

Turmeric is a native Indian plant, and Indians have been using it for centuries for wound healing. With the support of several documents including books on home remedies and Āyurvedic texts, CSIR was able to argue the case. Das and Kohli contested CSIR's objections, but the patent office rejected all their claims. The patenting of indigenous knowledge by foreign corporations is a cultural threat to countries like India as well as an economic one. The case of turmeric is a perfect example since it plays such an extensive role in India's culinary and health practices, among its other uses.

As the turmeric patent case makes it evident, the current patent system seems to allow biopiracy. Patents on Neem, Amla, Jar Amla, Anar, Salai, Dudhi, Gulmendhi, Bagbherenda, Karela, Rangoon-ki-bel, Erand, Vilayetishisham, and Chamkura all need to be revoked based on the logic that these are part of Indian indigenous knowledge and 'prior art'.

India's fight for the turmeric patent was necessary to uphold 'national pride' and to dispel unfounded fears that India was incapable of protecting its traditional knowledge base. The then CSIR's director, Dr. Mashelkar highlighted the importance of documenting traditional knowledge, to provide evidence of prior knowledge. Our scientists and technologists need to wake up and focus their efforts on building scientific data on many of these traditional knowledge treasures that we possess and work towards getting them patented. Knowledge of IKS becomes critical in this journey.

Source: Jayaraman, K.S., US patent office withdraws patent on Indian herb. Nature 389, 6 (1997). https://doi.org/10.1038/37838 and several other reports available on the Internet.

India is a country with a long-surviving civilisational history and practice known to mankind. While the modern western scholars date the civilisation to at least 5000–8000 years, the indigenous sources and belief systems in India date the civilisation to a very ancient period, almost time immemorial. Notwithstanding these differences, a country with such a long history ought to have accumulated some knowledge over its long period of existence. There is an impression in contemporary society that all knowledge that we benefit from has originated from the West. Since the western civilisation is of relatively recent origin compared to Indian or other civilisations such as the Chinese and the Egyptian, this also implies that all knowledge is of recent origin. This idea is counterintuitive and illogical.

Human beings are inherently knowledge generating in nature endowed with unique capabilities. By using the power of discrimination, reasoning, and rational thinking, human beings constantly process the newly acquired knowledge. Therefore, it is not surprising that a rich repository of knowledge accumulated in the Indian subcontinent and manifested in terms of traditions and practices. One or two examples help us understand this aspect. Indians were good in steel making until the 17th century. The Indian 'wootz' steel was used to manufacture what was famously known as 'Damascus blades' and despite several attempts by the metallurgists in the past, it was not possible to replicate the properties of the wootz steel. Indians' contributions in the fields of Number Systems, Mathematics, and Astronomy in the first millennia of CE contributed to several other developments. These ideas were percolating into the West via the Arabic countries, and they ought to have influenced the scientific developments in the West beginning from 15th century CE. Unfortunately, in our current educational system, we do not have an inkling of the nature of the contributions made by the Indians. This raises several questions. Where has all this knowledge gone today? Have we lost this knowledge totally? Is it of no use or interest to us today? Is there a sudden loss of continuity? What has caused this?

As many of us are aware, the ancient knowledge in India was preserved and transmitted 'orally' until a few centuries back. There was an uninterrupted lineage of 'Guru–Śiṣya' that took responsibility for the preservation and transmission of knowledge down the generations. Quite often, the teacher-student was a father-son combination and a group of related family members. These people formed a clan, who preserved the knowledge, practiced it by making a living, and transmitted it to their offsprings. The use of print media in recent history and the palm leaf scripts earlier have served to formally capture this oral knowledge and store it.

- Indians were extraordinary in steel making until the 17th century. The Indian 'wootz' steel was used to manufacture what was famously known as 'Damascus blades'.
- Due to major changes in the educational system in India introduced about 200 years back, there was a rather abrupt end to the process of knowledge transmission.

Unfortunately, due to major changes in the educational system introduced in India about 200 years back, there was an abrupt end to this process of knowledge transmission and the continuity is mostly lost. The newly introduced educational system demanded the society acquire only such knowledge as made available through the educational system. Those who aligned themselves to the new educational system were assured of jobs and salaries by the ruling class. Arguably, it would have taken about 50 years for most of the population to abandon old ways of doing things and come 'on board' the new system that promises economic prosperity. Once this transition happened, the oral transmission dwindled dramatically, created a sudden void and loss of continuity, thereby confining the knowledge to whatever was available in palm leaf manuscripts and other archives and personal collections.

We have continued with the British system of education in independent India, by keeping the ancient knowledge repository out of consideration. The erstwhile British policies on education excluded the ancient Indian knowledge citing reasons of lack of rigour and scientific values. This textbook is an effort to bring snippets of the Indian knowledge by providing a fresh relook at the corpus and culling out relevant portions that may generate renewed interest in the subject and motivate several to engage in a study of the knowledge repository of interest.

1.1 IMPORTANCE OF ANCIENT KNOWLEDGE

Ancient knowledge is the accrued knowledge over several generations and preserved in formal and informal means. Formal means include documented knowledge and informal means include shared values and practices through oral traditions. Sadly, as explained above, ancient Indian knowledge has been relegated to millions of palm manuscripts lying scattered all over the country and it is gathering dust. While several scholars are engaged in the process of bringing the hidden knowledge out of these manuscripts by researching and republishing such works, it does not match the scale required to make a meaningful impact. It is a herculean proposition to uncover the knowledge and bring it to the attention of modern society. On the other hand, the oral traditions continue in some rural pockets and are at the threat of getting extinct for want of patronage. The question in front of us is, "Does any society need to preserve, protect and pass on the ancient knowledge to the future generations?"

- If the underlying knowledge systems are abruptly withdrawn from society, the cultural practices will be rudely jolted.
- Ancient knowledge provides a head start to a society to march on the highway of innovation and new knowledge creation.

The thinking patterns and the repository of knowledge created by the forefathers in any society enable the current generation to understand the thought processes and frameworks of the previous generations. It will allow them to analyse the received wisdom in a contemporary context and identify new opportunities to assimilate the accrued wisdom and synthesize new knowledge. Therefore, keeping the current generation in the dark about the contributions of the ancestors is an inefficient, and a short-sighted option for society. Ancient knowledge serves multiple roles for society. Figure 1.1 schematically captures these.

FIGURE 1.1 Importance of Ancient Knowledge

Identity

The quintessential value that ancient knowledge brings to society is the identity it provides to fellow members of the society. Essentially it defines the context for several aspects of the day-to-day living of every individual. The social practices and norms have continuity as most of them are transmitted from generation to generation through practices and supporting knowledge repositories. Therefore, preserving this knowledge and baton passing them on to the next generation is an important step for contemporary society. In the absence of this continuity, individuals lose their conviction on several living practices. They lose their ability to 'meaning making' of much of the knowledge. Eventually it challenges one's own identity and that of the society.

Culture

Culture has several dimensions. In a direct sense, it is the manifestation of human intellectual achievement regarded collectively by society over time. From a social perspective, culture is nothing but the set of ideas, customs, and behaviour of society. In other words, culture provides a sense of identity at a societal level by providing a common medium for communication and the transaction of ideas. The prevailing knowledge and the literary traditions play a significant role in shaping the culture of the society. If the underlying knowledge systems are abruptly withdrawn from society, the cultural practices will be rudely jolted. It may create distortions and discontinuities in societal progress.

Received Wisdom

Knowledge and innovation are in a continuum. Innovation and new knowledge creation in any society is 'path-dependent'. What it essentially means is that the road travelled so far determines the future path. Without the continuity of thoughts, it is very difficult to make further progress in terms of new ideas. The other equally important issue is the risk of reinventing the wheel. When the benefit of prior knowledge and the thought process is lost by society, it will lead to reinventing the wheel, making innovation and new knowledge creation inefficient. In this context, ancient knowledge plays the valuable role of 'received wisdom' and provides a head start to a society to march on the highway of innovation and new knowledge creation.

Economic Value

One of the compelling arguments in support of the ancient knowledge systems is the huge potential it offers from an economic value standpoint. The emerging world order puts greater emphasis on knowledge society. The prevailing 'military power' will give way for 'knowledge power' and such nations who demonstrate the superiority of knowledge traditions are bound to lead the rest of the world. Transforming knowledge into economic value has been fully formalized with the global intellectual property rights regulations and patent laws. Therefore, the ancient knowledge system will be beneficial to a country like India (see IKS in Action 1.1 at the beginning of the chapter for an illustration of this idea).

- The prevailing 'military power' will give way for 'knowledge power' and such nations who demonstrate the superiority of knowledge traditions are bound to lead the rest of the world.
- Unless we preserve the ancient knowledge, we will not be able to prevent the spillover of our economic value arising out of our ancient knowledge.

Let us see an example to understand this aspect. The US patent and trademark office granted patent rights on knowledge of the usage of pigeon pea extracts for treating diabetes, hypoglycemia, obesity, and arthero-sclerotic cardiovascular disease (clogged arteries) to Insmed Inc, based in Richmond in Virginia. The company claimed its novelty in the invention of pigeon pea extracts for treating these diseases. In the patent applications, Insmed acknowledged only a handful of uses of pigeon peas in traditional medicines by citing some references of journal articles that appeared in 1957 and 1968 that describe the effects of pigeon pea and its extracts on blood sugar. The patent application did not include references to the traditional use of pigeon peas in the treatment of the mentioned diseases.

Pigeon pea (botanical name *Cajanus cajan*) is commonly known as arhar or red gram in India. There are several instances of the use of pigeon pea extracts in traditional medicines in India. A study of plant medicines by researchers in the department of pharmacology at the All-India Institute of Medical Sciences (AIIMS) tested pigeon pea extracts as they are used to treat diabetes in Ayurvedic medicines. The scientists at the Council of Scientific and Industrial Research (CSIR) observed that there is a need to gather strong evidence from our traditional texts to challenge such patent rights. Unless we preserve and be aware of the ancient knowledge, we will not be able to prevent the spill over of our economic value arising out of our ancient knowledge[1].

1.2 DEFINING INDIAN KNOWLEDGE SYSTEM

Indian Knowledge Systems (IKS) is a generic phrase that covers practically everything about India. For a nation with more than 5000 years of recorded history, abundant cultural and archaeological artifacts, literature, and social and community practices defining what constitutes Indian Knowledge is itself a huge challenge. Literature, cultural and social practices, historical evidence, and other such knowledge assets available in all Indian languages, dialects, and geographical regions will all technically fall under the ambit of IKS. The other aspect of the issue is the time dimension. Knowledge is continuously synthesized by any society. Knowledge assets available in India from the pre-historic times to the current day will all qualify to be part of the IKS. Therefore, it requires an unambiguous scope for defining IKS for this book.

Arguably, IKS can evoke different meanings to different stakeholders. The term IKS has three words in it. To better understand what we mean by IKS in the context of this book, we shall analyse each of these words separately.

Indian

By this term, we mean the indigenous sources of knowledge generated by the Indian society. The current political formation called 'India' is of recent origin and it alone does not qualify to be called 'Indian'. The term 'Indian' points to the undivided Indian subcontinent (Akhaṇḍa Bhārata). We mean the geographical area spanning from Burma on the east to modern-day Afghanistan in the west and Himalayas in the North to the Indian Ocean in the south. This region has common cultural, literary, and social practices, and has witnessed a continuous exchange of people, and ideas among them throughout the history of undivided India. Despite several political formations and princely states ruling this entire region for the last several hundred years until the consolidation begun from the 16th century CE, the society was unified under the common umbrella of social practices. Cāṇakya could get educated in Takṣaśilā in the western part of the sub-continent and be instrumental in establishing a powerful Mauryan empire with Pāṭalīputra as the capital in the Eastern part. Similarly, Pāṇini, a Sanskrit Grammarian from Gāndhāra in the North-Western corner of undivided India (now in Pakistan) could influence the thinking of people in the entire country on the Sanskrit language.

A second aspect to this is only such knowledge synthesized, codified, and made available by the 'Indians' is considered Indian knowledge. This implies that they ought to have been part of the Indian subcontinent, born and lived there, and are part of the knowledge system in an integral fashion. This is especially important because India witnessed several foreign travellers who visited its universities, stayed for some time and wrote about the country, the knowledge, and cultural practices. These have significantly contributed to the export of this knowledge to the west and other parts of the world. For example, some reports have extensively studied the role of such authors in taking mathematical thinking to the west via the Arab world[2]. These are considered as 'about IKS' rather than IKS itself.

Knowledge

The second component of IKS is the 'knowledge', which is always tacit. It primarily arises in the form of the wisdom of the knowledge seekers. It is obtained by the insights gained by personal experiences with life situations, facing problems, and coming up with means of solving them. At other times, one obtains knowledge by means of intense observation of events, experimentation, conjecturing, and analysis. Knowledge may or may not be converted to a literary format. The tacit knowledge can be preserved and transmitted through an oral tradition without loss. India has a rich tradition of folklore practices even to date, that belongs to this category. While both these forms of knowledge are equally important and valuable it is impossible to formally study knowledge transmitted through oral traditions. Therefore, by 'knowledge', we mean in this book, a formal repository of knowledge available in literary sources.

The tacit knowledge gained by a seeker is eventually transmitted systematically in the form of some 'explicit' knowledge. This happens by way of proposing a new theory, framework, or literary work. Furthermore, knowledge pervades all three domains: spiritual, religious, and others addressing social and day-today issues. We can summarise the term 'knowledge' as that emanating from the wisdom and insights arising out of deep experiences, observation, experimentation, and analysis and validated, improved, and augmented further.

System

By 'System' in IKS, we mean a structured methodology and a classification scheme to access the available corpus of knowledge. By its inherent nature, knowledge could be accessed in any manner depending on the interest, purpose, and capacity of the seeker. For an uninitiated, this vastness could throw a challenge as the seeker may be clueless as to where to begin and how to proceed. Therefore, the available knowledge needs to be collected, grouped, and arranged logically. Codification and classification of the available knowledge using a definite framework would constitute one dimension of the word 'System' in IKS. The other important requirement is the interconnection between the part of the knowledge in the classification framework. The framework used to represent knowledge should also provide some logical relationships between the different parts of the proposed framework. This helps easy understanding of the overall contribution of the knowledge and how the different components of the knowledge complement each other. We take up this issue for discussion in the next section and present a systematic approach to classifying IKS for this book.

1.3 THE IKS CORPUS – A CLASSIFICATION FRAMEWORK

There are many ways to define and identify what constitutes IKS. For example, one approach is to merely pick the important topics representative of the knowledge corpus such as the Vedas,

Yoga, Vāstu, Śilpa Śāstras, Āyurveda, Buddhism, and Jainism to define IKS. Another approach is to select phrases such as Indian Psychology, Indian Arts, Dance, and Architecture and put together related works into it to construct IKS. These examples bring out the components of IKS. However, whether they will qualify to be a good framework for IKS or not depends on their ability to meet the requirements of a classification framework for IKS.

The usefulness of a classification framework depends on three factors: completeness, compactness, and inter-connectedness.

- *Completeness* ensures that all important components of the IKS are included in the proposed framework. If significant omissions are found in the classification, it makes the IKS non-exhaustive and non-representative.
- *Compactness* indicates the efficacy of the grouping of various topics in IKS in a congruent and logical fashion. This makes the representation simple, concise, and easy to understand and remember.
- *Inter-connectedness* brings logical relationships among the various sub-classifications. The classification framework will identify how the different components are logically connected. Thus, it presents a unified picture of the entire knowledge.

Closer scrutiny of the IKS knowledge repository provides us the following details:

(a) As already noted, the knowledge is available in both formal literary sources and informal non-literary sources.

(b) Among the literary sources, we can broadly identify three categories. One of the major sources is the Vedic and allied literature, which we shall designate as Sanātana-dharma literature, presented mainly in the Sanskrit language. This comprises the religious and philosophical part consisting of the Vedic and allied corpus, which forms the core and a good repository of other literature spanning areas such as sciences, architecture, and aesthetics. The second major source is the literature on other dharmic traditions. The third group is a large repository of knowledge in other Indian languages and dialectics.

(c) The non-literary source is predominantly available through a rich set of oral traditions found throughout the country.

Figure 1.2 pictorially presents the classification scheme based on the above observations. Let us see some details of these categories.

Sanātana-dharma – Core Literature

This comprises a vast repository of knowledge starting with the Vedas, known as Śruti. Although the Vedic corpus is oral in nature and is still transmitted using oral methods, these have been later systematically documented in written form. The Vedas are considered foundational by the Indian society and several important literary works were developed later, which substantially added to the Vedic corpus by facilitating better understanding and implementation of the ideas presented in the Vedas. This literature owed its allegiance to the Vedas and extracted their cardinal assumptions and principles from the Vedas. For example, six schools of philosophical thought, known as Darśanas developed their basic assumptions from the Vedas while stating their prescriptions. The Vedic and allied repository has several sub-components and divisions and is best understood from a classification methodology adopted, which we will see in the Section 1.4.

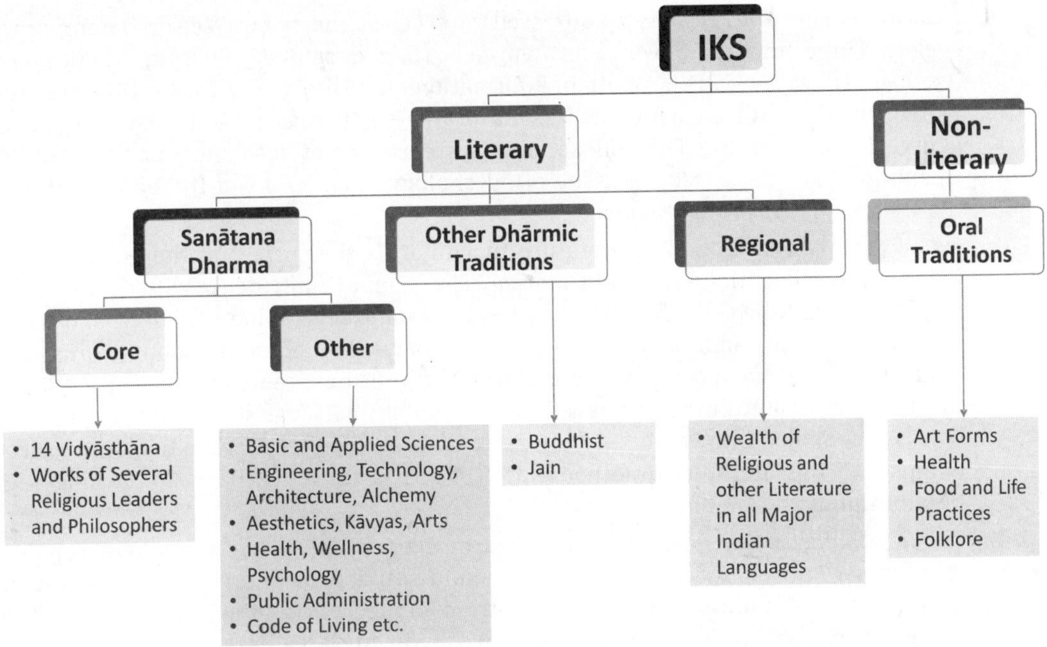

FIGURE 1.2 The IKS Corpus – A Classification Framework

Sanātana-dharma – Other Literature

The other literature consists of works that addressed key issues of day-to-day life such as health, wellness, science, engineering, technology that aided societal progress and development, and aesthetics and art forms. Although they were primarily addressing a variety of issues, they still owed their allegiance to Sanātana-dharma and acknowledged the core assumptions laid out in the Vedas as the ultimate source for valid knowledge and drew relevant ideas wherever it applied to their work. For example, the work on astronomy of Āryabhaṭa recognizes the ideas of what constitutes a year, and the notion of four yugas from the Vedic corpus while proceeding with specific discussions on the mathematical aspects.

This literature flourished from the first millennia in the BCE and was continuously augmented by multiple works. Several areas were covered in this category, and the following are prominent among them:

- *Basic and applied sciences* (Mathematics, Astronomy, Plant Sciences). A series of studies were carried out continuously from the beginning of CE. Later works improved, expanded, and added new components to the existing knowledge repository as evident from the literary sources. We discuss these aspects in Chapters 6, 8, and 9 of the book in some detail.
- *Engineering and Technology* (Metalworking Technology, Shipbuilding, Dams and Watershed Management, Alchemy, Cosmetics, Perfumes, Dyes, Town Planning, and Architecture). The literature presents both evidence of these as well as specific ideas and techniques developed in many of these topics. Chapters 10, 11, and 12 look at these aspects of IKS.

- *Health, Wellness, and Psychology* addressed the crucial issue that we face in contemporary society. Three important works on Āyurveda (Caraka-saṃhitā, Suśruta-saṃhitā, and Aṣṭāṇga-hṛdaya) provide a wealth of information on health and wellness. Other related works dealing with alchemy such as Rasaratna-samuccaya provide information on āyurvedic formulations. The philosophical systems such as Yoga and Sāṃkhya and the Upaniṣads have discussed the issue of psychology. Chapter 13 of the book discusses the issues related to this theme.
- *Nīti-śāstras* is a collection of literature that informs the society of the good code of living through poetic verses and stories. The sāmānya-nīti deals with elements of good living, and the role of ethics and morality in life. Several life situations and wide-ranging topics are addressed through stories, parables, and short poetic works. The famous pañcatantra and the works of Bhartṛhari are some representative examples of this category of literature. An extensive collection of such ideas scattered in the various works is compiled into what is known as subhāṣitas. Chapter 4 of the book introduces glimpses of this literature. Another aspect of nīti-śāstra is the Rāja-nīti, dealing with public administration and governance.
- *Public administration* deals with the idea of governance of state and public policy measures required for administration. Manu-smṛti provides rich information on governance and administration. Arthaśāstra compiled during the 3rd century BCE is a seminal work and it triggered further works in the area. Notable among them is the Nīti-sāra of Kāmandaka. Chapter 14 of the book takes up this issue for discussion.
- *Aesthetics, Kāvyas, and Performing Arts* is another area with rich contributions. The Sanskrit language is the vehicle through which the entire knowledge corpus of the Sanātana-Dharma (both the core and the other) is presented. Linguistics and phonetics of the Sanskrit language is a fundamental work that sets the stage for rich literature development. Chapter 5 of the book introduces some concepts related to the Sanskrit language. Works of great poets such as Kālidāsa, Daṇḍin, and Bāṇabhaṭṭa, works such as Kāmasūtra of Vātsyāyana, and Nāṭyaśāstra of Bharata are some of the examples.

Other Dharmic Traditions

Other dharmic traditions have stayed out of the Vedic framework but have immensely contributed to IKS in the religious, philosophical, and other domains. Two of them, the Buddhist and the Jain literature are noteworthy, and they have contributed right from 500 BCE to IKS. While the religious and philosophical part of the literature is based on the respective tenets of the school of thought, other literature has applications in areas of science, technology, and other areas.

Buddhist literature has dealt with the religious concepts in its canonical texts. However, there are several Buddhist works in which many issues such as mathematical concepts, maritime activities and alchemy are also discussed. The work of Nāgārjuna, Rasaratnākara in the 1st century CE is an early contribution to alchemy. The Jain sacred literature consists of canonical texts. They considered mathematics as an integral part and have dedicated 'Gaṇitānuyoga', a portion of their literature to mathematics[3]. Tattvārtha-sūtra, composed by Umāsvāti during 2nd–3rd century CE is an important Jain literature. Some popular Jain texts dealing with mathematics include Anuyogadvāra-sūtra, Vyavahāra-sūtra, and Sūrya-prajñapti. Mahāvīrācārya's work, Gaṇita-sāra-saṃgraha (850 CE) is one of the important contributions to the development of mathematics in India.

Regional Literature

The separation of regional languages in this figure is only for convenience. Ideally, it could be included under Sanātana-dharma. Indian subcontinent has a rich and diverse mix of cultural and linguistic variations. The 8th schedule of the Indian constitution has listed 22 languages of the country. In each of these languages, there is a huge corpus of religious, philosophical, and other literature. For example, Tamil literature has several contributions in the Sangam period (first millennium BCE). Several of the works in the regional literature have drawn from the Sanskrit resources and have either explained them in detail in the chosen regional language or extrapolated them further with some more ideas. The new literature created broadly follows the Sanātana-dhārmic literature and utilises the basic framework laid in the Sanātana-dharma literature. The sheer volume and vastness of the regional literature introduces constraints in drawing substantially from this corpus for the present book.

- The Buddhist and the Jain literature have contributed significantly to IKS right from 500 BCE.
- The 8th schedule of the Indian constitution has listed 22 languages of the country and in each of these there is a huge corpus of sacred and other literature.

Oral Traditions

The diversity of cultural practices and regional preferences have paved the way for oral traditions to preserve and transmit knowledge across generations. These have been primarily in the form of folklore artistic endeavours, skilful jobs, food and life practices, and health. The 64 Kalās mentioned in the IKS literature are mostly skill-based and artistic chores that are orally transmitted.

Sanskrit has been the dominant language for transacting knowledge for a long time in India. Therefore, for the purpose of the book, the main sources of knowledge to discuss various aspects of IKS are drawn from the Sanskrit literature. As we have already seen, the Sanātana-dharma literature, the Jain and some of the Buddhist literature are presented using Sanskrit as the medium of language. The choice of a Sanskrit-based knowledge repository does not imply that similar knowledge was not available in other regional languages in the country. However, as noted earlier, the process becomes complex and unwieldy to present all these in a single book. The other aspect for inclusion is that the knowledge shall be quoted, cross-referenced, and acknowledged by the indigenous people in the domain. This provides internal consistency and validation of the knowledge by the indigenous society.

The other issue that merits attention is, "how recent a history we must include in the definition?" The culture of new knowledge creation is an unhindered process in the country until the early 19th century. However, beginning the 16th century CE, there has been a wave of invasions in the country, introducing newer dimensions and priorities in society. Therefore, there is a greater focus to preserve the existing knowledge repository. Further, beginning the 16th century the ascend of the Western knowledge systems and scientific discoveries had its influence on the native knowledge practices also. Therefore, the knowledge sources dated up to the 16th century CE are mainly considered for the purpose of this book.

1.4 CATURDAŚA-VIDYĀSTHĀNA

A classification framework for the Sanātana-dharma literature as we have defined in the previous section is available within the resources itself[4]. In this framework, the literature is systematically organised under 14 major divisions. Therefore, it is referred to as "Caturdaśa-Vidyāsthāna" (Caturdaśa in Sanskrit means fourteen). The components of the 14-part knowledge contain the four Vedas (and their Upa-Vedas), the six Vedāṅgas, Purāṇas, the Dharma-śāstras, Nyāya in its detailed form and Mīmāṃsā (both Pūrva and the Uttara portions). The Mīmāṃsā and the Nyāya (in its expanded form) together constitute the six darśanas. A pictorial representation of the above classification is available in Figure 1.3.

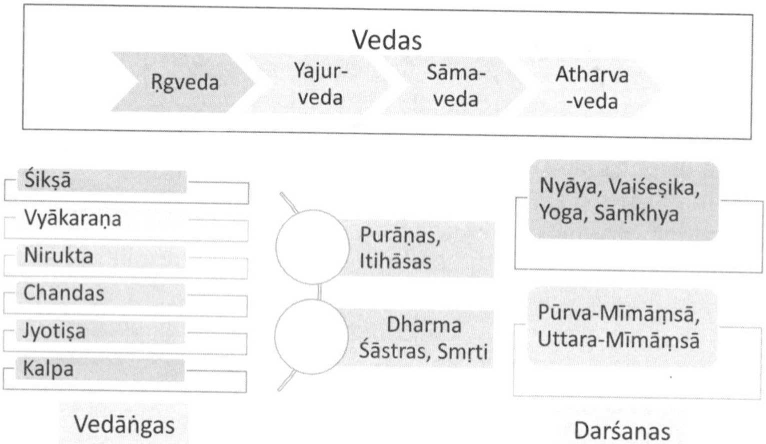

FIGURE 1.3 A Pictorial Representation of Caturdaśa-Vidyāsthāna

Vedas

The Vedas are the primordial source of knowledge in the scheme of the Sanātana-dharma literature. They are also known as Śruti as they are not authored by anyone but heard by the seers of the Vedic lore. The mantras that were revealed to them were later orally passed down the generations through a lineage of 'Guru–Śiṣya'. The other sources of knowledge are considered secondary to the Vedas. The primary purpose of the other knowledge is to expand the ideas contained in the Vedas so that it has practical applications and relevance. The other purpose is also to clarify the Vedic intent by way of stories, explanatory notes, and operational guidelines for daily life. The Upa-Vedas are typically associated with the Vedas on account of the material being found in the respective Vedas. In this sense, they are integral to the Vedas.

- The Vedas are the primordial source of knowledge in the scheme of the Sanātana-dharma literature.
- Purāṇas contains a rich repository of ideas that seek to explain various aspects of the Vedic thoughts, several socio-cultural ideas and practices for living.

Since the primordial source of the Sanātana-dharma literature is the Vedas, it provides the overall defining framework for living. The larger issue of meaning and purpose of life as stated in the Vedas need to be understood clearly. Therefore, there is a need to expand the tenets contained in this framework. Detailed explanations in terms of 'how-to' aspects of the ideas expressed in the Vedas is also required. Moreover, whenever conflicting situations emerge in the applications of the principles, we need to know how to resolve them. Some of the other components of caturdaśa-vidyāsthāna serve to address these requirements in the context of the Vedic repository.

Vedāṅgas

To benefit fully from the Vedas some complementary tools and skills are required. These help to understand the exact meaning and intent of what is presented in the Vedas and follow them based on specific instructions provided. These are collectively referred to as Vedāṅgas. Chapter 2 of the book has more details on the Vedas and the Vedāṅgas.

Darśanas

It is a natural quest for everyone to understand three forces that operate and interact with one another: an individual (Jīva), the Universe (Jagat), and a larger force governing the other two (variously referred to as Īśvara, Brahman etc.). Establishing the connection between these three becomes an issue of philosophical thinking. Darśana essentially means a philosophical thought or view. There are six schools of philosophy in the Sanātana-dharma literature and other schools outside the realm of this. These are discussed in some detail in Chapter 3 of the book.

Purāṇas and Itihāsas

Purāṇas contains a rich repository of ideas that seek to explain various aspects of the Vedic thoughts using detailed stories and anecdotes. They present several socio-cultural ideas and practices for living. Furthermore, they address some of the common issues that mankind faces and provide answers using the overarching framework of the Vedas. The stories in the Purāṇas relate to pre-historic events and the subject matter discussed follows a set pattern[5]. Itihāsas, on the other hand, relate to historical events that have taken place which can be associated with specific timelines. As we know, Mahābhārata and Rāmāyaṇa are two well-known itihāsas. In some sense, this literature represents the wisdom that we have accrued through the ages. Chapter 4 of the book discusses these aspects in some detail.

Dharma-śāstras and Smṛtis

Let us think for a moment about what we practice in our modern-day work life. For example, if we need to be part of a company as an employee, we are supposed to know the rules and norms of the organisation, the do's and don'ts, expectations on the part of the employer and the co-workers, social etiquettes, our limits and entitlements and the consequences of wrong actions. If this clarity is not there, there will be chaos and we will end up with unproductive work. If this is the situation in an office, such norms and rules are essential in a society where several entities have complex interactions among them, both in structured and unstructured ways. In other words, a guide to lead a dhārmic life based on the principles of the Veda is required.

The dhārmic principles engrained in the Vedic corpus are presented in multiple formats in our knowledge traditions. Smṛtis are rule books with specific operating guidelines of how to put the dhārmic principles into action and what are the consequences of not doing so. Itihāsas demonstrate dhārmic principles 'in action', through case studies and real-life situations. Nīti-śāstras and Subhāṣitas are pearls of wisdom articulated by learned people in the society, clearly showing the value of adhering to the dhārmic principles in life. A discussion of some of these topics is available in Chapter 4 of the book.

IKS IN ACTION 1.2

Caturdaśa-Vidyāsthāna and the Income Tax Act, 1961

The value of the various components defined in the Caturdaśa-Vidyāsthāna framework is best understood by relating it to a modern-day concept. For the sake of simplicity let us take the Income Tax Act (IT Act) 1961 of the Government of India. This act is the basis for all matters pertaining to the issue of tax incidence, in the hands of those who have earned income, collection of tax, refund, and penalties and legal action, etc.

The bare act will be very concisely stated, and it may at most run to 30 pages. The bare act will merely contain statements of sections and sub-sections laying out the schema for the income tax matters in the country. This by itself cannot be put to direct use as there will be several questions in terms of implementing it in practice. Therefore, another book, *A Guide to Income Tax Act 1961* will be published. This book may be about 500 pages. In this book, details will be provided as to how to operationalize the provisions enlisted in the bare act. Detailed guidelines, step-by-step procedures, sample calculations, and do's and don'ts will be provided to improve our clarity and understanding of the IT Act 1961. In this manner, it will enable us to put the IT Act to practical use. Even as greater details about the Act are provided in this book, the basic tenets laid out in the act will not be violated in letter and spirit.

However, even after providing such an elaborate guideline for putting IT Act 1961 to practice, in actual implementation numerous situations would have emerged. This provides an opportunity to check our understanding of the Act, its intentions, and our ability to adhere to it correctly. The judiciary system often steps in by providing the needed clarifications by constantly interpreting the Act.

A vast accumulated repository of such judgments delivered adds to the corpus of knowledge about IT Act 1961. By knowing these situations, our ability to adhere to the IT Act 1961 will only become better as we could avoid several traps by not reinventing the wheel. Therefore, another book will be published, *A Book of Case Laws in IT Act 1961*. This book will have more than 10,000 pages and will keep growing as more and more case laws are handled over time. This new book will inform us of the consequences of adhering or otherwise and ways of adhering to the provisions of the IT Act better.

The Caturdaśa-Vidyāsthāna framework that we have discussed in this chapter exactly resembles this situation. The Vedas are the equivalent of the IT Act 1961. The Smṛtis is equivalent of the guidebook to IT Act 1961. The Itihāsas, Nīti-śāstras, and Subhāṣitas play the role equivalent to the book on the case laws in IT Act 1961. Similar to the IT Act example, the Smṛtis and other Dharma-śāstras will not violate the Vedas in letter and spirit. They will merely facilitate the process of adhering to the tenets enjoined in the Vedas by those who follow it.

1.5 HISTORICITY OF IKS

Dating of the Indian literature is a major challenge for today's researchers, primarily on account of several reasons. The Indian knowledge repository pre-dates the western civilisational knowledge repository by several millennia. The western knowledge sources originated mostly in the common era (several of them in the last millennia). On the other hand, several works constituting the Indian knowledge belong to the BCE and attributed to 500 BCE or before. The method of representing knowledge, storing, and archiving ought to be very different between these two eras. Applying our contemporary methods of dating the knowledge to such old knowledge sources can pose serious limitations.

The other related challenge is that as we already mentioned, most of the Indian knowledge repositories in the BCE were oral. The earliest available sources in the form of temple inscriptions and palm leaf manuscripts are often considered by contemporary researchers as reliable. In this manner, the dating of the Indian knowledge becomes conservative, approximate, and much later than what ought to have been its original date. Western knowledge repository began in an era well established with written literature and therefore using the parameters that work well for such knowledge repository to others who preceded them may not yield the correct picture. This is perhaps the reason that Western scholars and independent Indology researchers have been able to accurately date the Indian contributions in the common era (post 100 CE for example) but not those belonging to the BCE. Sometimes, some stone inscriptions and archaeological artifacts help the process of resolving some of these confusions.

- Dating of the Indian literature is a major challenge.
- A useful source of data to fix the date of the Indian knowledge is the astronomical references found in the texts.

Another useful source of data to fix the date of the Indian knowledge is the astronomical references found in the texts. This could be one of the possible sources and so far, we have not been able to cover much ground. Fortunately, in an era of information technology-driven research, some novel methods help us to date the literature using the wealth of astronomical data that we have in the Indian knowledge repository. A case in point is the dating of the Śatapatha-brāhmaṇa using the astronomy software Skymap Pro. There is a mention of the fact that Kṛttikā stars never deviate from the east in Śatapatha-brāhmaṇa. The Skymap Pro software can plot a night sky in any place in the Universe between 4000 BCE and 8000 CE. Using this software, some studies suggest that the Śatapatha-brāhmaṇa ought to have been written sometime around 3000 BCE.[6]

Despite these limitations and constraints, an effort has been made to present the available information to get some idea of the historicity of the IKS. Table 1.1 presents a sample of some important works on IKS separated by some periods. As evident from the table, we can broadly divide the IKS into three time periods:

Before 3,000 BCE

In modern parlance, this era is categorised as the dark ages. This is indeed a dark age for Western civilisation since no evidence of any organised knowledge repository (either oral or written) is available. However, in India considerable amount of work had already been done and the resultant knowledge was orally shared among generations of people living. The main contributions include the Vedas, Purāṇas, and the Itihāsas. Despite several attempts to date these resources, as new evidence (such as the discovery of new underwater archaeological artifacts in Dvārakā), and methods to date the knowledge becomes available (such as new planetarium software) the earlier estimations are proved to be erroneous. Culturally, in India, it is believed that these texts belong to antiquity and any attempt to date them will be futile.

- Earlier estimates of time periods for Vedas, Purāṇas, and the Itihāsas prove to be erroneous as new evidence and methods to date the knowledge becomes available.
- In an oral tradition, the entire knowledge is to be committed to one's memory. Therefore, it needs to be concise, and specific.

3,000 BCE to 500 CE

This period roughly starts with the beginning of recorded human history as per the western context and extending as far as 500 CE, which coincides roughly with the fall of the roman empire. As evident from the table, while the rest of the World was picking up the elementary skills of organised living, the Indian counterpart had been actively creating many new and useful knowledge for the society. Significant contributions were made in the areas of Linguistics, Literature, Health and wellness, Mathematics, and code of living. In modern parlance, this is an indication of the vibrancy of society and its receptiveness to new ideas and methods of improving the quality of life.

TABLE 1.1 A Sample List of the IKS Repository

Sl. No.	*Name of the Work*	*Keyword 1*	*Keyword 2*
	Before 3,000 BCE		
1	Vedas	Dharma (Code of Living)	Several Other Topics
2	Purāṇas*		
3	Mahābhārata, Rāmāyaṇa		
	3,000 BCE to 500 CE		
1	Vedāṅga-jyotiṣa	Astronomy	
2	Manu-smṛti	Public Administration	Dharma (Code of Living)
3	Śulba-sūtras	Mathematics	Dharma (Code of Living)
4	Suśruta-saṃhitā	Health	Wellness
5	Aṣṭādhyāyī, Nirukta	Linguistics	Grammar
6	Nāṭyaśāstra	Art Forms	Dance, Theatre
7	Buddhist Texts	Philosophy	Mathematics
8	Nyāya and Vaiśeṣika Sūtras	Logic, Epistemology	Knowledge Framework
9	Jaina Mathematical Works	Mathematics	
10	Arthaśāstra	Public Administration	Finance, Foreign Policy
11	Chandaḥ-śāstra	Metrical Pattern, Prosody	Binary Maths Ideas
12	Yoga-sūtras	Control of Mind	Philosophy
13	Kāmasūtra	Art Forms	Dharma (Code of Living)
14	Mahā-bhāṣya	Sanskrit Language	Grammar
15	Rasaratnākara	Alchemy	
16	Caraka-saṃhitā	Health	Wellness
17	Sāṃkhya-darśana	Philosophy	Psychology
18	Amarakośa	Linguistics	Lexicography
19	Sūrya-siddhānta	Astronomy	Mathematics
20	Bṛhat-saṃhitā	Astronomy, Mathematics	Several Other Topics
	500 CE to 1,800 CE		
1	Āryabhaṭīya, Ārya-siddhānta	Astronomy	Mathematics
2	Pañca-siddhāntikā	Astronomy	
3	Mayamata	Architecture	
4	Brāhmasphuṭa-siddhānta	Astronomy	Mathematics
5	Mānasāra	Architecture	Town Planning
6	Āryabhatīya-bhaṣya, Mahābhāskarīya	Astronomy	Mathematics
7	Nārada-śilpa-śāstra	Architecture	Iconography
8	Gaṇita-sāra-saṅgraha	Mathematics	
9	Siddhānta-śekhara	Astronomy	

Sl. No.	Name of the Work	Keyword 1	Keyword 2
10	Yukti-kalpataru	Shipbuilding	Several Other Topics
11	Samarāṅgaṇa-sūtradhāra	Architecture	Several Other Topics
12	Siddhānta-śiromaṇi	Astronomy	Mathematics
13	Kāśyapa-śilpa-śāstra	Temple Architecture	Iconography
14	Aṣṭāṅga-hṛdaya, Rasaratna-samuccaya	Alchemy	Health, Wellness
15	Kerala School of Mathematics	Mathematics	Astronomy
16	Graha-lāghava	Astronomy	

* As per modern researchers, these are variously dated much later.

500 CE to 1,800 CE

During this era, Indians have made significant strides in the area of mathematics, astronomy, philosophy, and spirituality. With strong foundations in mathematics, several allied areas have also grown, notable among them are architecture and technology. It is no wonder that some of the best temple complexes, rust-free iron pillars in the open ground, and musical pillars in several temples have withstood the onslaught of time and bear testimony to these skills even today. In this period the supremacy of the Indians in the areas of astronomy and mathematics continued to be strengthened as there were continuous contributions on several aspects. Several contributions have also been made in other areas such as alchemy, metalworking, etc.

1.6 SOME UNIQUE ASPECTS OF IKS

1.6.1 Nuances of an Oral Tradition

IKS is by and large an oral tradition by its very nature as we have already seen. If knowledge needs to be transmitted orally, it requires a few things. In a written tradition there is scope for being elaborate. Specific ideas can be taken up for a detailed discussion running to several pages. For example, one can have a separate discussion on the philosophical ideas in a certain work, another can analyse from a perspective of aesthetics and grammatical structure, a third can dwell on religious aspects, etc. However, in an oral tradition, the entire knowledge is to be transmitted orally and committed to one's memory. Therefore, it needs to be concise, and specific. Moreover, it may be difficult to have one treatise for one subject, say mathematics, and another for spirituality, etc. Therefore, we will find that in several works the issues discussed are many. The Nyāya, though being a philosophical school of thought discusses other subjects such as Logic and argumentation and valid means of knowledge. The Purāṇas and Itihāsas, for example, are encyclopaedic in nature. Issues discussed include cosmology, cosmogeny, politics, public administration, aesthetics, morals of life, and so on. Even in a work on mathematics, we may find apart from mathematical concepts, poetry, and philosophy among other things.

To generalise the above discussion, the pattern of IKS literature shows that a single work can provide simultaneously ideas on three streams seamlessly: Spiritual, Religious, and Secular. Therefore, the charm and power of IKS lie in its multi-dimensional perspective. Let us take the case of Bhagavadgītā as an example to understand this aspect prevalent in IKS literature. A true devotee of Lord Krishna may want to read Gītā as it is a matter of religion to him. On the other hand, a spiritual seeker may view Gītā as a spiritual text. However, there is a third aspect to Gītā, which many of us maybe not aware of. This is the 'secular' perspective. By this we mean a set of ideas that help us conduct our life sensibly from day-to-day, working perspective. Let us look at specific examples from Gītā to bring clarity to the idea.

Religious vs Material Dimensions

Verses 7 and 8 in Chapter 4 of the Gītā quintessentially brings the *Avatāra-puruṣa* dimension. Whenever there is a deterioration of dharma the God takes one more incarnation (*Avatāra*) to uphold the dharma. The incarnation of God is to protect the good people, destroy the evil ones and restore dharma in society once again. That is how the *Avatāra Puruṣa* concept manifests in terms of the context, and purpose of the incarnation.

The above verses could be interpreted from a 'secular' perspective to convey an idea often taught in several engineering and management schools. *Stability and long-term sustainability of the system happens because there are* ***regenerative points****. When the system attains disequilibrium and shows signs of being unstable, measures have to be taken to restore the equilibrium in the system. If the regenerative points are not there, the system will become unstable and eventually perish.*

This is a typical 'systems engineering' idea according to which there are regenerative points in the system. The regenerative points indicated in the above translation corresponds to the incarnation idea in the original verse. One can relate this idea even to some well-known concepts in Economics and Management. The demand-supply equilibrium, pricing decisions in alternative market structures, how organisations continue to root out bad CEOs or Managers over time, the mechanisms to prevent opportunistic behaviours, in the long run, could all be explained by this idea expressed through these verses.

Spiritual vs Material Dimensions

In Chapter 2 of Gītā, Krishna brings into focus the notion of time. Let us consider verse 22 in the chapter[7]. The meaning of this verse is as follows. *Just as a person discard worn pieces of cloth and takes new ones, the ātman also discards old bodies and acquires new ones.* This verse explains the idea of a chain of birth and death events taking the analogy of a shirt. There is a spiritual angle to it as true seekers of knowledge will deeply contemplate it. However, if we reflect on this verse furthermore, we can explore other interpretations. One such interpretation is, *To be successful and sustainable organisations need to continuously engage themselves in discarding old ideas (mindset!) and embrace new ones. This is the fundamental building block of innovation and creating competitive advantage.*

- In IKS a single work can provide simultaneously ideas on three streams: Spiritual, Religious, and Secular.
- A large number of works in IKS are in verses set to a metrical structure irrespective of whether the subject matter is literature, mathematics, or engineering.

When we interpret the verse in this manner, it reminds us of the recent work in economics on creative destruction and innovation. The biggest challenge in organisations is mindset inertia. This puts realistic limits to creating better organisations over time. In this verse, the need for discarding old ideas and mindset is emphasized by describing the process of death. Another example can be found in a paper[8] that shows how a verse in Chapter 3 of Gītā (3.27) indeed connects to some of the issues related to cybernetics and control theory.

1.6.2 Typical Presentation Style – Sūtras, Encryptions

A related aspect to the above, which makes IKS unique is the use of specific structural aspects to make it a concise piece of work. Notable among them are the following:

- A large number of them are in verses set to a metrical structure. This is used irrespective of whether the work is original or a commentary on another work. It is also independent of whether the subject matter is literature, mathematics, or engineering. Since the work is in prosody, it requires the author to use minimum words and syllables that conform to the metrical structure. For example, one of the approximations for the value of π is given in a verse as follows:

  चतुरधिकं शतमष्टगुणं द्वाषष्टिस्तथा सहस्राणाम् ।
  अयुतद्वयविष्कम्भस्यासन्नो वृत्तपरिणाहः॥

  caturadhikaṃ śatamaṣṭaguṇaṃ dvāṣaṣṭistathā sahasrāṇām |
  ayutadvayaviṣkambhasyāsanno vṛttapariṇāhaḥ ||

  This verse computes the value: $\pi = \frac{\text{Circumference}}{\text{Diameter}} = \frac{62832}{20000} = 3.1416$

- Typically, mnemonics (sūtras) are employed to convey the message. A mnemonic is a memory mechanism and a learning technique that facilitates information retention or retrieval in human memory. For example, Piṅgala in his work on Chandaḥ-śāstra dating back to the 2nd century BCE defined eight groups of binary numbers each of word length 3 (equivalent to what is now known in computer science as De Bruijn sequence). This can be represented using a simple mnemonic: यमाता-राजभान-सलगम् (yamātā-rājabhāna-salagam). Chapter 6 has more details on this. We discuss more details of sūtras in Chapter 5 of the book when we analyse the Sanskrit language and syntax.
- Several innovative methods are used to make the message concise and amenable to a metrical presentation. A good example is to use encryptions to represent an idea. In Chapter 6 we discuss two methods, the Kaṭapayādi system, and the Bhūta-saṃkhyā system to represent numbers in unique ways so that these could be easily remembered and incorporated in a verse while discussing mathematical operations, numbers, and results. Another shining example is the Āryabhaṭan system for number representation, with which he could represent the entire sine table (of differences) values in a verse (a couplet)[9].

1.7 ORGANISATION OF THE BOOK

A proper understanding of IKS and its relevance to contemporary society requires a two-part study of IKS. The first is to develop an overall understanding of some key components of IKS. This is important before we explore the nature of applications. Therefore, Part I of the book provides a quick introduction to the key components of IKS. The second aspect is the application of IKS for some gainful use. Specific areas of applications can be studied in a focused manner by drawing the relevant portions of the IKS showcasing the potential for practical application either in theoretical advancement of concepts or practical use in addressing some issues that we may face either as an individual or at an institutional level. The rest of the book presents these aspects in three parts:

- Part II: Foundational concepts relevant for Science, Engineering, and Technology applications.

- Part III: Specific concepts related to Science, Engineering, and Technology.
- Part IV: Humanities and Social Sciences applications.

The details of the chapters and the topics discussed in the chapter are schematically presented in Figure 1.4.

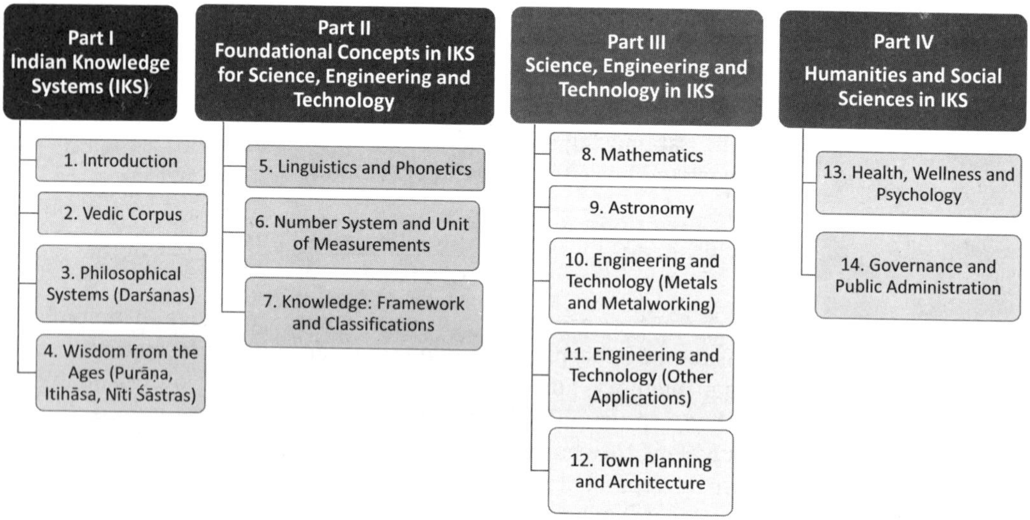

FIGURE 1.4 Organisation of the Chapters in the Book

Textbook use for Alternative Streams

The selection of topics and the material included in the chapters are done with a broader audience in mind. Three streams of programs have been identified as possible users of the textbook: Engineering, Liberal Arts, Commerce, and Science. While the first two parts of the book help develop an overall understanding of IKS, the remaining chapters in the textbook can be selectively used as a curriculum for each of the streams based on the relevance of these topics to the respective streams. We provide in Figure 1.5 a suggestive set of chapters for each of the streams identified above.

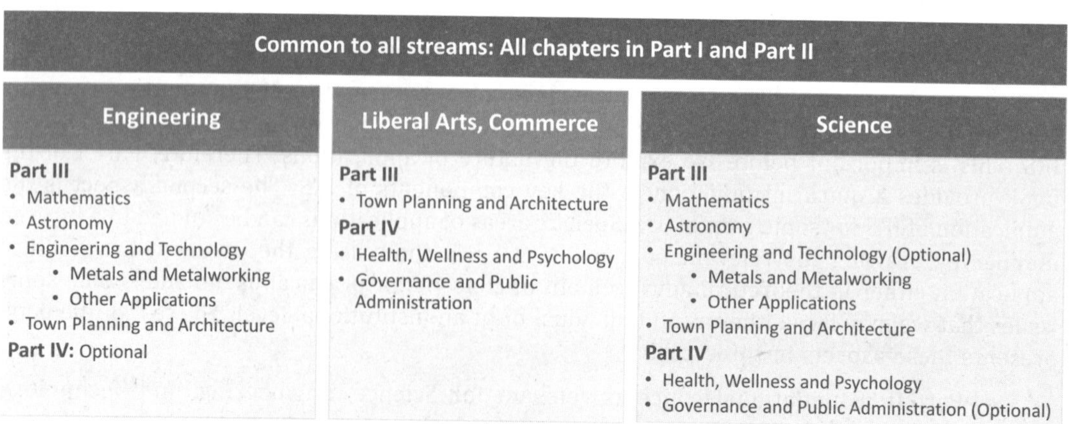

FIGURE 1.5 Suggested Chapters for Different Streams

SUMMARY

- The ancient knowledge in India was preserved and transmitted 'orally' until a few centuries back. There was a rather abrupt end to this process of knowledge transmission.
- It is very important to know the thinking patterns and the repository of knowledge created by the forefathers as it provides numerous values to society.
- Transforming knowledge into economic value has been fully formalized with the intellectual property rights regulations and patent laws. This is one area where the ancient knowledge system will benefit a country like India.
- Knowledge is a systematic body of literature emanating from the wisdom and insights arising out of deep experiences, observation, experimentation, and analysis and validated and augmented over time.
- The knowledge is available in both formal literary sources and informal non-literary sources.
- Among the literary sources, we can broadly identify three categories: Sanātana-Dharma literature, presented mainly in the Sanskrit language, literature on other dharmic traditions, and a large repository of knowledge in other Indian languages and dialectics.
- Buddhist and the Jain literature have significantly contributed to IKS from 500 BCE. The Jain sacred literature consists of canonical texts. 'Gaṇitānuyoga', a portion of their literature is dedicated to mathematics.
- In each of the regional languages, there is a huge corpus of sacred and other literature.
- A classification framework for Sanātana-Dharma literature, referred to as 'Caturdaśa-Vidyāsthāna, has organised the knowledge repository under 14 major divisions.
- The Vedas are the primordial source of knowledge in the scheme of IKS. Vedāṅgas provide complementary tools and skills to fully appreciate the content and also benefit from the Vedas.
- Normal methods used to date literature are grossly inadequate and misleading in the case of IKS.
- IKS literature shows that a single work can provide simultaneously ideas on three streams seamlessly: Spiritual, Religious, and Material.
- On account of the oral transmission, IKS literature is fundamentally a concise piece of work. To make it concise, specific structural aspects are deployed.

REVIEW QUESTIONS

1. What is the current status of IKS in India? Briefly explain the reasons for its current status.
2. Do you think ancient knowledge is useful for a society? Support your answer with suitable arguments.
3. Define the term 'Indian Knowledge System'. Briefly explain what you understand by each word in this term.
4. Outline the broad classification of the IKS repository. How do the categories in the classification framework differ from one another?
5. What do you understand by the term 'Caturdaśa-vidyāsthāna'? Briefly explain the various components of this framework.
6. Comment on the statement, "The components of 'Caturdaśa-vidyāsthāna' are all related to the Vedas".
7. Briefly state the salient features of IKS when viewed from a historicity point. What are the areas of key contributions?

8. Indian knowledge repository pre-dated several of the Western works in several areas. Do you agree with the statement? Prepare a note either supporting the statement or otherwise.
9. Comment on the statement, "The oral tradition employed by ancient Indians necessitated use of some unique methods to represent the knowledge".

DISCOVER IKS

1. There is a general trend in the Indian psyche to follow many of the ideas and practices of the west. While inter-mingling of culture and practices is welcome, there is a need to also understand the distinctive features of the Indian society. Rajiv Malhotra discusses this issue in the video: https://www.youtube.com/watch?v=2WYZtS_LLog. After watching the video prepare a two-page note to answer the following questions:
   (a) How is the Indian mind and practices different from the West?
   (b) How has the Western thought influenced the Indian thinking? How has this percolated into current thinking?
   (c) Can you identify a few areas in which the Indian society is distinct from the West?
2. Indians had several indigenous practices in science and technology. Some of them have been recorded in literary works while several others are transmitted through an oral tradition. Watch this video, in which Anupam Mishra illustrates several water harvesting practices of the people of Rajasthan living close to the desert, with scanty annual rainfall: https://www.ted.com/talks/anupam_mishra_the_ancient_ingenuity_of_water_harvesting?language=en or https://www.youtube.com/watch?v=eJCTAXb_BWs. After watching the video, prepare a two-page note to answer the following questions:
   (a) Briefly outline the annual rainfall in the region and the status of their water availability.
   (b) What are the methods adopted for water conservation and harvesting? Have they been effective in meeting the water requirements of the region?
   (c) How do the modern systems of water conservation and harvesting compare with this? Do you have any advice for modern-day water conservationists?
3. Western researchers consider the time between Sindhu-Sarasvati civilisation (3rd millennium in BCE), and 500 BCE as 'dark ages', as they have no information about India. It does not mean people suddenly forgot to read and write. Watch this interview with S. R. Rao, Archaeologist on excavation of Dvārakā: https://www.youtube.com/watch?v=R1PGp7706HY. After watching the video, prepare a two-page note to answer the following questions:
   (a) What is the significance of the excavation of Dvārakā for Indian civilisation?
   (b) What is the relationship between Sindhu-Sarasvati civilisation and Dvārakā?
   (c) What are the contents of the excavated portion of underwater city, Dvārakā?

SUGGESTED READINGS

"US Firm Bags Patent On Arhar Extracts", Financial Express, August 3, 2003. https://www.financialexpress.com/archive/us-firm-bags-patent-on-arhar-extracts/89352/ Last accessed on May 1, 2020.

Beer, S. (1994). "May the Whole Earth be Happy: Lokasamastat Sukhino Bhavantu", *Interfaces,* **24**(4), pp. 83–93.

Danino, M. (2010). *The Lost River - On the Trail of Saraswati*, Penguin Books, New Delhi.

Datta, B. and Singh, A.N. (1962). *History of Hindu Mathematics: Parts I and II*, Asia Publishing House, Mumbai.

Joseph, G.G. (2016). "Indian Mathematics: Engaging with the World from Ancient to Modern Times", World Scientific, pp. 97–114.

Kak, S. (2010). "India", Chapter 6 in Heritage Sites of Astronomy and Archeoastronomy in the context of the UNESCO World Heritage Convention: A Thematic Study, Clive Ruggles and Michel Cotte (Eds.), Published by ICOMOS, pp. 99–112.

Lal, B.B. (1997). *The Earliest Civilization of South Asia*, Aryan Books International, New Delhi.

Mukerji, R. (1977). "Notes on Ancient Hindu Shipping" in R. G. Bhandarkar Commemoration Volume, Bharatiya Publishing House, Delhi, pp. 447–455.

Pusalker, A.D. (1955). *Studies in the Epics and Purāṇas*, Bharatiya Vidya Bhavan, Mumbai.

Raju, C.K. (2007). "Cultural Foundations of Indian Mathematics", Centre for Studies in Civilization, New Delhi.

Rao, S.S. (2006). "Indigenous Knowledge Organization: An Indian Scenario", *International Journal of Information Management,* **26**, pp. 224–233.

Ray, P. (1956). "History of Chemistry in Ancient and Medieval India", Indian Chemical Society, Kolkata.

Sampad and Vijay (2011). "The Wonder that is Sanskrit", Sri Aurobindo Society, Puducherry.

Sarasvati Amma, T.A. (1999). *Geometry in Ancient and Medieval India*, Motilal Banarsidass, 2nd ed., New Delhi.

Sen, S.N. and Shukla, K.S. (2000). *History of Astronomy in India*, Indian National Science Academy, 2nd ed., New Delhi.

Subbarayappa, B.V. (1982). "Glimpses of Science and Technology in Ancient and Mediaeval India", *Endeavour, New Series*, **6**(4), pp. 177–182.

ENDNOTES

1. For more details on this see, https://www.financialexpress.com/archive/us-firm-bags-patent-on-arhar-extracts/89352/ Last accessed on May 1, 2020.
2. For more details, see, **Ifrah, G. (2005).** "The Universal History of Numbers II", Penguin Books, pp. 1–24.
3. For some details see Chapter 4: From Zero to Infinity: Mathematics in Jain and Buddhist Literature in **Joseph, G.G. (2016).** "Indian Mathematics: Engaging with the World from Ancient to Modern Times", World Scientific, pp. 97–114.
4. In the Śivamahāpurāṇa vāyavīya saṃhitā the 25th verse in Chapter 1 brings this idea, अङ्गानि वेदाश्चत्वारो मीमांसा न्यायविस्तरः । पुराणं धर्मशास्त्रं च विद्या ह्येताश्चतुर्दश ॥ aṅgāni vedāścatvāro mīmāṃsā nyāyavistaraḥ | purāṇaṃ dharmaśāstraṃ ca vidyā hyetāścaturdaśa || One can see the same verse quoted in Vāyu Purāṇa, Mahābhāratam and Yājñavalkya Smṛti also. To explain this more clearly, the Viṣṇu Purāṇa has classified this into 18 parts, as evident from these two ślokas, अङ्गानि वेदाश्चत्वारो मीमांसा न्यायविस्तरः । पुराणं धर्मशास्त्रं च विद्या ह्येताश्चतुर्दश ॥ 3.6.28 आयुर्वेदो धनुर्वेदो गान्धर्वश्चेत्यनुक्रमात् । अर्थशास्त्रं परं तस्मात् विद्या ह्यष्टादश स्मृता: ॥ 3.6.29 . aṅgāni vedāścatvāro mīmāṃsā nyāyavistaraḥ | purāṇaṃ dharmaśāstraṃ ca vidyā hyetāścaturdaśa ॥ 3.6.28 āyurvedo dhanurvedo gāndharvaścetyanukramāt | arthaśāstraṃ paraṃ tasmāt vidyā hyaṣṭādaśa smṛtāḥ || 3.6.29 These can be referred in the book Published by Gitapress, Gorakhapur.
5. Purāṇas are characterised by certain 'Lakṣaṇas'. These are discussed in Chapter 4 of the book.

6. For full details on this see, **Narahari Achar, B.N. (2000).** "On the Astronomical Basis on the Date of Satapatha Brahmana: A Re-examination of Dikshit's Theory", *Indian Journal of History of Science*, **35**(1), pp. 1–19.
7. वासांसि जीर्णानि यथा विहाय नवानि गृह्णाति नरोऽपराणि । तथा शरीराणि विहाय जीर्णान्यन्यानि संयाति नवानि देही ॥ 2.22. vāsāṃsi jīrṇāni yathā vihāya navāni gṛhṇāti naro'parāṇi | tathā śarīrāṇi vihāya jīrṇānyanyāni saṃyāti navāni dehī || 2.22. For details see, **Swami Chinmayananda (2002).** "The Holy Geeta", Central Chinmaya Mission Trust, Mumbai.
8. For details, see **Beer, S. (1994).** "May the Whole Earth be Happy: Lokasamastāḥ Sukhino Bhavantu", *Interfaces*, **24**(4), pp. 83–93.
9. Āryabhaṭa gives the sine difference table using his system of representation of numbers in the following verse:

   मखि भकि फखि धखि णखि ञखि ङखि हस्झ स्ककि किष्ग श्घकि किघ्व ।

   घ्लकि किग्र हक्य धकि किच स्ग झश ङ्व क्ल प्त फ छ कलार्धज्याः ॥

   makhi bhaki phakhi dhakhi ṇakhi ñakhi ṅakhi hasjha skaki kiṣga śghaki kighva |

   ghlaki kigra hakya dhaki kica sga jhaśa ṅva kla pta pha cha kalārdhajyāḥ ||

   These represent the following numbers which are nothing but Rsine differences: 225, 224, 222, 219, 215, 210, 205, 199, 191, 183, 174, 164, 154, 143, 131, 119, 106, 93, 79, 65, 51, 37, 22, 7. For more details, see **Kripa Kumar Shankar (1976).** "Āryabhaṭīya of Āryabhaṭa", Indian National Science Academy, New Delhi. Chapter 1, Gītikā Section, Verse 12, p. 29.

CHAPTER

2 The Vedic Corpus

LEARNING OUTCOMES

After finishing this chapter, you will be able to:

- Get introduced to the overall organisation of the Vedic Corpus
- Familiarise with the four Vedas and their sub-components
- Recognise the multi-faceted nature of the knowledge contained in the Vedic Corpus
- Appreciate the role of the Vedāṅgas and develop a basic familiarity of the Vedāṅgas

The figure given below is a brick alter identified as Śyena-Citi excavated from an ancient site at Purola in Uttarakhand. The remains of pottery are assignable to 1st century BCE–2nd century CE. Vedic people lived a life in which Yajña was central to their life practices. A Yajña is performed in an altar, such as the one shown here.

Source: https://www.asidehraduncircle.in/uttarkashi.html

IKS IN ACTION 2.1

Yajña and Project Management

In order to understand the various divisions of the Veda better, we need to inquire into some of the governing principles behind the living style of the Vedic people. The life of the Vedic people in some ways revolved around Yajñas. Every aspect of life and celebration was linked to Yajña, where the devatās were invoked and offerings made. Agni was the carrier of the offerings to the intended devatā.

These were performed to request for material blessings, wealth, health and overall evolution of an individual. They were also performed as an expression of gratefulness for the bounty showered on oneself and the society at large. A large corpus of the Vedic knowledge provided intricate details for performing the yajña related rituals. Therefore, there was a very evolved structure and methodology to the performance of yajña. The way the yajña was done provides a good insight into the project management skills that they brought into the act.

Certain types of yajñas require a team of 18 people for managing the entire set of activities. The yajamāna and his wife are the first two required because they are the underlying cause for the yajña to happen. They are perhaps seeking specific favours from the devatā or are expressing their sense of gratitude to the devatās. In order to perform the yajña 16 more people consisting of four groups of four each are required.

The first group is called 'Hotṛ-gaṇa', consisting of four people, a head and three assistants, who are experts in the Ṛgveda. They invoke the intended devatā by uttering relevant hymns from the Ṛgveda.

The second group of four is called 'Adhvaryu-gaṇa', whose job is to indeed perform the rituals and make offerings. They are experts in the Yajurveda and use the relevant mantras and prescribed practices from Yajurveda. The third group of four experts from Sāmaveda is called 'Udgātṛ-gaṇa'. They select the relevant mantras from Sāmaveda and sing in praise of the invoked devatā.

The fourth group is called 'Brahma-gaṇa' and are experts from Atharvaveda. They play the role of overall supervision of the yajña, ensuring that the rituals happen as per plan and prescribed methods. They intervene and rectify any deviations and address issues emerging during the conduct of the ritual.

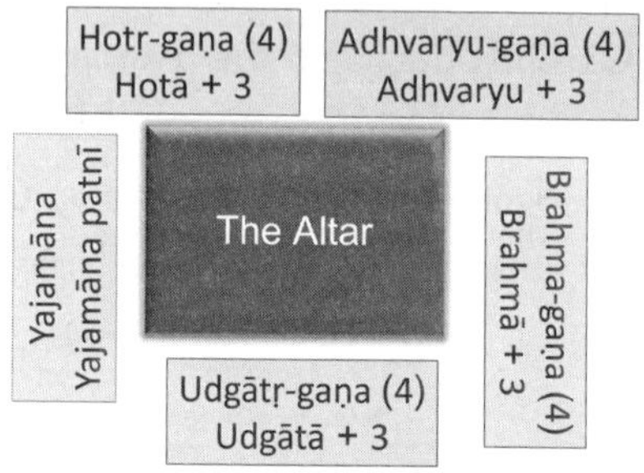

Ṛgveda: – Hotṛ-gaṇa; Yajurveda – Adhvaryu-gaṇa
Sāmaveda – Udgātṛ-gaṇa; Atharvaveda – Brahma-gaṇa

FIGURE 2.1 Organisation of a Yajña

Yajñas differ in size and scale, some of them running to several days involving huge outlay. Therefore, a high level of organisation and management skills need to be put in place. This explains why the Vedic people have exhibited such project management skills in performing a yajña, which was central to their living.

2.1 INTRODUCTION TO VEDAS

The word 'Veda', is generally derived from the Sanskrit root विद् (vid - to know). The word Veda can be derived from five verbal roots[1]. These mean to exist, to know, to discriminate, to obtain, and to make known. Veda indicates a vast body of knowledge concerning the eternal spiritual values and principles and practices for gainful and happy living revealed to the ṛṣis through their deep meditation. The Hindu religious tradition has accorded the Vedas the highest place in its canonical literature and are revered as the basic scriptures. The Vedas are not merely considered as scriptures but as the fountainhead of Indian culture and human civilisation. It is believed in the Indian tradition that the Veda is the poetry of Gods and it neither fades nor becomes stale by the passing of time. The actual dating of the Vedas has been a subject matter for considerable debate. The latest date attributed by many modern researchers falls a few millennia before the common era.

Vedas are a unique repository of knowledge that distinguishes itself from other forms of knowledge. Normally, we attribute knowledge to a person. However, the Vedic corpus is generally understood as a non-human source (known as **Apauruṣeya**). Vedas are revealed to the ṛṣis from time to time in their state of deep contemplation. That is why it is apauruṣeya. Despite this, we will find that specific Vedic mantras are associated with certain ṛṣis. The ṛṣis are associated with certain mantras, not as authors but as those who indeed discovered the mantra through their deep concentration and focus. In this sense, the ṛṣis of the Vedic lore are 'seers' of the mantras and not the authors.

- Veda is transmitted orally and preserved intact from time immemorial.
- Vedas have a well-developed system of phonetics that has ensured its preservation.
- Recognized by UNESCO as heritage.

The Vedic repository is preserved and transmitted through an oral tradition. Hence, the Vedas are referred to as śruti, 'what is heard'. From an operational perspective, the Vedic knowledge was transmitted in an unbroken line of transmission from teacher to student that was formalized early on. It has survived several thousand years on account of scientific methods of oral rendering. This has been possible on account of a well-developed system of phonetics, known as **Śikṣā**, one of the vedāṅgas. This ensured an impeccable textual transmission superior to the classical texts of such antiquity. This is perhaps the reason for the UNESCO recognising the Vedas as a heritage for preservation.

The seers concluded that the purpose of human life is to realise oneself, rid oneself of limitations and constraints, and experience bliss by knowing the absolute truth. Given this objective, they proposed a path for the evolution of an individual. During the early stage of one's life, each one of us is supposed to engage in the world of activities, contribute actively, and enjoy a secured living. Towards this end, the seers proposed the **Karma-kāṇḍa**. The karma-kāṇḍa nudges an individual to obtain a purity of mind as we engage with the world. Since the Vedic living was centrally focused on Yajña, several instructions and operating frameworks pertaining to the Yajña form part of the karma-kāṇḍa.

As individuals progress, there is a need for single pointedness of the mind through contemplation and focus. In the Vedic corpus, the **Upāsanā-kāṇḍa** provides these ideas for an individual. The word Upāsanā literally means sitting near (Upa + āsanā). It is a method of developing a certain attitude to worship. In simple terms, Upāsanā refers to a state of concentration where 'whatever is meditated upon' is completely identified and absorbed with self. It provides a systematic method and an opportunity to harness the inherently wandering mind towards a defined goal through contemplation. Through systematic practices prescribed in the upāsanā-kāṇḍa, an individual will be able to develop calmness of the mind and mental equipoise. Several meditation practices are prescribed in the Vedas as part of the upāsanā. Upāsanā ideas are indeed very valuable in the modern world. A purified and sharpened mind becomes a storehouse of energy, and it enables an individual to perform very effectively both in his professional life as well as in his meditative practices. At a later stage of life, this is an important requirement for every individual. The other part of the Veda addresses the issue of self-realisation. The **Jñāna-kāṇḍa** opens the mind of an individual to its limits and facilitates much deeper inquiry about oneself.

English language-based work on Vedas in India is very minimal. The western efforts seem to have taken a main place in the modern-day English language-based sources on Vedas and their interpretations. The Western approach to the study of the Vedic corpus has primarily been in extracting the archival material and publishing them afresh, with English commentaries. In the Western approach, they often view the entire exercise as purely intellectual and linguistic in nature. Since most of them would have had very little opportunity to experience and imbibe native traditions, the work could ignore or overlook traditional and cultural dimensions that are required to present the knowledge in an appropriate context and perspective. Moreover, the Western works generally ascribe to the Aryan invasion theory and use that as one of the main lenses to study the Vedic scriptures. One needs to be aware of these aspects while dealing with such works.

2.2 THE FOUR VEDAS

As per the tradition it is held that the Vedas were originally three and they together were called the Trayī Vidyā. Ṛk is typically a hymn and is distinct from Yajus, which is a sacrificial formula. Certain Ṛks were, set to singing and they are called the Sāmas. These Sāmas were sung at sacrifices or at the time of extracting the Soma juice. The fourth Veda is the Atharvaveda, recognised later and it contains some hymns as old as the Ṛgvedic hymns, while others are evidently of later date in terms of the structure, style of the language and matter. Although the Vedas existed for a long time there was a need to organise them systematically so that the available knowledge is put to correct use by society. The credit to organise the Vedic repository in the manner we understand it today goes to Kṛṣṇa-Dvaipāyana popularly known as Vyāsa. Vyāsa organised the Vedic corpus into four major divisions: Ṛgveda, Yajurveda, Sāmaveda and Atharvaveda. To ensure that the knowledge is passed down to future generations without interruption, he taught these divisions to four of his primary disciples and made each of them responsible to primarily propagate one of the four. Paila was associated with Ṛgveda, Vaiśampāyana Yajurveda, Jaimini Sāmaveda and Sumantu Atharvaveda.

- Ṛgveda is the oldest, largest and the main repository.
- Ṛgveda should be studied by one who wants to understand Indian literature and spiritual culture.
- Lofty and interesting set of ideas found in Ṛgveda makes it contextually relevant.

Ṛgveda

The Ṛgveda represents the earliest sacred book of India. It is the oldest and biggest amongst all the four Vedas[2]. All the features of classical Sanskrit poetry can be traced to the Ṛgveda. In the Ṛgveda we find the origins of the religious and philosophical development of the most ancient society. Thus, both for its poetry and its religious and philosophical importance, the Ṛgveda should be studied by one who wants to understand Indian literature and spiritual culture. The Ṛgveda priest is known and Hotṛ (see the opening box for a description of the roles of the priests in the conduct of a yajña), who employs the mantras to sing the praise of devatā invoked during the ritual. The Ṛgvedic hymns are various and not always prayers addressed to the god to whom a sacrifice is being offered. The Ṛgvedic verses are essentially the utterances of the Vedic sages on several topics in the form of poetry.

The inherent curiosity and quest for new knowledge of ancient Indians are quite evident from the varied theme and character of the sūktas in the Ṛgveda. These provide a rich repository of creative thinking, opening our understanding to several aspects of life and their inter-connections. The origin of the Universe, for example, is a question that has captured the attention of today's scientists. However, there are several sūktas in the Ṛgveda which has taken up this theme. The Nāsadīya-sūkta (RV10.129) which speculates on the origin of the Universe has attracted several commentaries both in Indian darśanas and in Western philology. The other sūktas that inquired into the origin of the Universe include Hiraṇyagarbha-sūkta (RV10.121) and Puruṣa-sūkta (RV10.90). The lofty and interesting set of ideas that one finds in Ṛgveda makes it special and contextually relevant. It promotes a high sense of unity in diversity by proclaiming that the truth is one but learned ones articulate it in different ways (*ekam sat viprāḥ bahudā vadanti*, RV1.164.46). Rich philosophical ideas expounded in Vedānta literature have their seeds in the Ṛgveda mantras. These sūktas are set in a highly mystic and poetical form that requires correct understanding to derive the full benefit of the intended message.

Yajurveda

Yajurveda confines itself to the major issue of conducting the sacrifices. The word Yajurveda is derived from the root *Yaj, m*eaning, the worship associated with sacrifice. This Veda mainly focuses on yajña and a list of various yajñas is found in this Veda. The mantras in Yajurveda are mostly in prose form although a small fraction is in the metrical form, among these many are borrowed directly from the Ṛgveda. The mantras in Yajurveda are referred to as yajus[3]. The Adhvaryu-priest who is mainly charged with the performance of sacrifices makes use of the yajus. The Yajurveda is essentially a guidebook for the Adhvaryu priest who had to do practically all ritualistic works in a sacrifice. The Adhvaryu priest needs to perform a variety of tasks including the selection of a plot of land for the sacrificial altar, offering oblations to the sacred fires with relevant mantras for the devatā. Though the major topic of Yajurveda is Yajña, many other topics are discussed in it. These include human anatomy, metals, constellation, seasons, numbers and geometry, grains, and yogic insights.

- Yajurveda mainly focuses on Yajña and a list of various yajñas are found in this Veda.
- Yajurveda is in two major branches: Kṛṣṇa-Yajurveda and Śukla-Yajurveda.

The Yajurveda is divided into two branches: the Kṛṣṇa (Black) and the Śukla (White). The distinguishing aspect is that the Kṛṣṇa-Yajurveda is more ancient than the Śukla-Yajurveda. Till the time of Sage Yājñavalkya, Yajurveda was a single scripture. Sage Yājñavalkya learned Yajurveda from his guru Vaiśampāyana. Later, because of some misunderstanding between them, Yājñavalkya is said to have learned the new Veda which is known as Śukla-Yajurveda and the earlier one is known as Kṛṣṇa-Yajurveda. Yājñavalkya transferred this knowledge to fifteen of his disciples. The śākhās of Śukla-Yajuveda are named after these disciples.

Sāmaveda

The word Sāmaveda is derived from the Sanskrit root, 'Sāma' indicating 'to please, pacify or satisfy'. Essentially, it refers to the singing of Ṛgveda mantras. The mantras in Sāmaveda are typically referred to as 'Sāma'. It is a Ṛgveda mantra set to music. Sāmaveda currently has three branches viz. Kauthuma, Raṇāyanīya, and Jaiminīya. However, there are references in Mahābhāṣya of Patañjali, Śrīmad-Bhāgavata-Mahāpurāṇa, and other sources which suggest that there were 1,000 branches of Sāmaveda, indicating different traditions and versatile ways of singing the mantras. In a yajña, Sāmaveda is used to please the devatās by singing mantras after making the offering.

Sāmaveda is divided into two parts: Pūrvārcikam and Uttarārcikam, consisting of a total of 1,875 mantras[4]. Out of these, except 75 mantras, the rest are taken from the Ṛgveda saṁhitā. There are more than 150 seers associated with Sāmaveda. Unlike the other three Vedas, the mantras of the Sāmaveda, are related to musical scales, similar to the seven scales of classical music. Therefore, in some ways, the origin of Indian classical music lies in the Sāmaveda.

Atharvaveda

The etymology of the word 'Atharvan' brings out the multi-faceted nature and characteristics of this Veda. It means one which brings wellness, seen by sage Atharvan and one with no falsehood or movement. As already mentioned, it is generally believed that the Atharvaveda is a later addition to the original set of the three Vedas (Ṛg–Yajur–Sāma), chronologically speaking. The Atharvaveda priest is known as Brahman, whose main job is overall coordination and monitoring of the Vedic ritual. Before starting any activity in the yajña, Brahman's permission is sought. When there are deviations or changes, the Brahman steps in and makes the necessary amendments. In other words, the Atharvaveda priest plays the crucial role of quality control and compliance when rituals are performed. Viewed from this perspective, the Atharvaveda priest must be a knower of all the other three Vedas to flawlessly execute this task of overall coordination and quality control.

Originally, nine śākhās of this Veda are known to have existed, but only two are extant: Paippalāda and Śaunakīya. Of the two, it is only the latter that is available in a complete form. The hierarchy of the arrangement of mantras in Atharvaveda is quite similar to what we see in Ṛgveda. At the highest level, the Atharvaveda-saṁhitā is divided into four books. There are 20 kāṇḍas or books in all. Except for Books 15 and 16, the text is in poem form deploying a diversity of Vedic metres. Each kāṇḍa is again subdivided into sūktas or hymns, and the sūktas into mantras. There are 6,077 mantras, in 736 sūktas. About a sixth of the Atharvaveda texts adapts verses from the Ṛgveda. In particular, the last kāṇḍa, i.e., the 20th, has borrowed heavily from the Ṛgveda-saṁhitā.

Messages in the Vedas

Vedas are the quintessential wisdom that forms the foundation for the Sanātana-dharma and is considered as the ultimate reference for every aspect of living in India. The subject matter covered in the Vedic repository is vast. The details about performing several rituals to propitiate the devatās form a major component, however, several other issues are addressed. These include, for example, inquiring into the origin of the Universe, human beings' intricate relationship with nature, reflecting on some observed celestial happenings leading to astronomical insights, marriage, health and wellness, and larger questions such as our purpose in life, and many methods of inquiring into these subjects. There is a rigor with which several aspects of life are inquired into.

- Atharvaveda has details on diseases and their cure.
- Prayers for prosperity and peace in the Vedas invariably included all the living organisms in the Universe, not just the mankind.

Prayers for prosperity and peace in the Vedas invariably included all living organisms in the Universe, not just mankind. For example, in the Śānti-sūkta in the Atharvaveda, the prayer is to bless both the two-legged and four-legged creatures with peace and prosperity. In certain other mantras, mention of eight-legged creatures and nine-legged creatures is also indicated. Similarly, peace and prosperity are prayed for the Earth, the Interspace, and Space above. Such is the vastness of appeal and degree of inclusivity in the thinking of the Vedic people.

Our ability to succeed in our endeavours require a unified vision of what needs to be achieved, the oneness of thought and purpose. Therefore, developing a mutual understanding born out of the unity of mind and heart is critical to achieving success. The Atharvaveda mantras found in 6.64 bring out this issue unambiguously[5]. The mantra begins by saying, "let us all agree and be united with minds focused on one common issue". The mantra further encourages the seeker to have a resolve that is one and the same, with harmonious hearts and minds so that all may happily consent and together entertain one common purpose. In another mantra (Atharvaveda 7.12), the prayer says, "Let all the members of the cabinet of the king have same voice and thoughts so that the king is profited by the advice given by the eminent scholars in this cabinet."

The Atharvaveda-saṁhitā has some special features because of which it stands apart from the other three Vedas. It deals more with the worldly things 'here and now' than the 'hereafter'. It facilitates this process with the sacrifices which are a means to them. A major part of this Veda is concerned with diseases and their cure, rites for prolonging life, rites for fulfilling one's desires, building construction, trade and commerce, statecraft, penances, and propitiatory rites. Atharvaveda is the only source of knowledge that gives vast detail about the earth starting with the idea that the earth is not a mere physical structure. More specifically, the content in Atharvaveda can be classified under various heads based on the theme of the issues discussed in the mantras. Table 2.1 has a listing of these heads. The Saṁhitā portion also contains high philosophical ideas as elicited in the Upaniṣads. Even the literary style is more sophisticated compared to the other three Vedas. On account of these differing subject matters and language sophistication, some scholars conjecture that this work is of a later origin and perhaps included in the Vedic corpus much later.

TABLE 2.1 A Thematic Classification of Ideas Presented in the Atharvaveda

Name	*Description*
भैषज्यानि (Bhaiṣajyāni)	Mantras which refer to various medicines for different diseases
आयुष्याणि (Āyuṣyāṇi)	Mantras for long life
पौष्टिकानि (Pauṣṭikāni)	Mantras to be recited at the time of ploughing the field, on the occasion of house construction or while exercising certain commercial transactions
प्रायिश्चित्तानि (Prāyścittāni)	Mantras which do away with the bad and inauspicious matters
स्त्रीकर्माणि (Strīkarmāṇi)	Mantras are exclusively related to the life of woman folk, e.g. marriage, love, child welfare and so on
राजकर्माणि (Rājakarmāṇi)	Mantras on political notions of the Vedic times
ब्रह्मण्यानि (Brahmaṇyāni)	Mantras on philosophical speculations of the Atharvaveda

2.3 THE FOUR DIVISIONS OF EACH VEDA

Close scrutiny of the Vedic corpus reveals to us certain aspects of the knowledge presented. Some portions of the Veda are hymns sung in praise of the devatās while certain others consist of specific know-how, instructions, and dos and don'ts related to conducting of yajña and related rituals. Other components provide food for thought encouraging deep inquiry into human nature, and our relationship with the Universe and God. Clearly, the intended audience for these is different and so are the objectives. There were different groups of devatās addressed in these portions as well. Sage Vyāsa organised each Veda into distinct portions considering these issues. These divisions in each Veda were based on the material presented, the primary objective and use of the material, the target audience, and the focus. Accordingly, each Veda is further sub-divided into a two-level hierarchy as shown in Figure 2.2. At the first level, we have the Mantras and Brāhmaṇas. While the Mantras portion, also known as Saṁhitā has the hymns in praise of devatās, the Brāhmaṇas have the remaining portions of the Veda. While the Brāhmaṇas have substantive content addressing the ritualistic aspects, one can still distinctively cull out two other portions within the Brāhmaṇas, namely Āraṇyakas and Upaniṣads. We can therefore divide the Vedas into four portions: Saṁhitā, Brāhmaṇa, Āraṇyaka, and Upaniṣads.

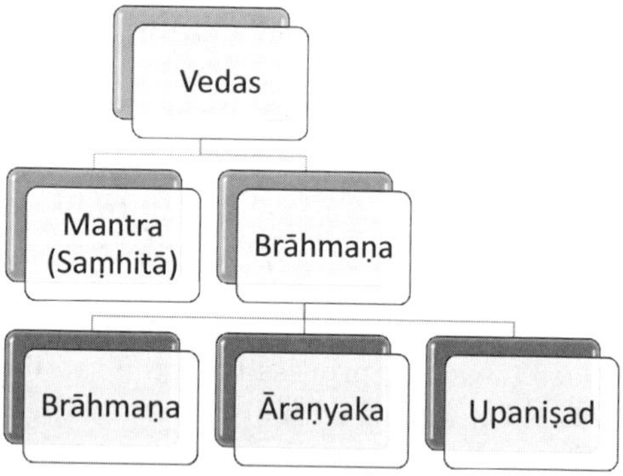

FIGURE 2.2 Classification of Each Veda into Sub-components

Table 2.2 provides a summary of the current status of the Vedic repository classified in this fashion[6]. As evident from the table, in all the four Vedas, there have been a number of śākhās (or schools of study) that were existing as per the documentation available. However, we seem to have lost most of this over time. The two portions of Yajurveda, Kṛṣṇa and Śukla have their own unique set of Mantras, Āraṇyakas, Brāhmaṇas, and Upaniṣads.

TABLE 2.2 A Summary of the Content in the Four Veda

	Ṛg Veda	*Yajur Veda*	*Sāma Veda*	*Atharva Veda*
No. Mantras (Saṃhitā)	10,552	7,154 **Kṛṣṇa:** Taittirīya, Kāṭhaka, Maitrāyaṇīya **Śukla:** Vājasaneya (Mādhyandina, Kāṇva)	1,875; only 75 unique	6,077 (last kāṇḍa is heavily borrowed from Ṛgveda)
Śākhās	21, only 5 available now	**Kṛṣṇa:** 85, only 4 available now **Śukla:** 17, only 2 available now	1000, only 3 available now	9, only 2 available now
Āraṇyakas	Aitareya, Śāṅkhāyana	**Kṛṣṇa:** Taittirīya **Śukla**: Bṛhadāraṇyaka	Talavakāra or Jaiminīya	None
Brāhmaṇas	Aitareya, Kauṣītakī	**Kṛṣṇa:** Taittirīya **Śukla:** Śatapatha	9 Brāhmaṇas (Tāṇḍya-mahā-brāhmaṇa important)	Gopatha Brāhmaṇa
Major Upaniṣads	Aitareya, Kauṣītakī	**Kṛṣṇa:** Taittirīya, Kaṭhopaniṣad **Śukla:** Bṛhadāraṇyaka, Īśāvāsya	Chāndogya, Kena	Praśna, Muṇḍaka, Māṇḍūkya

Saṃhitā

Saṁhitā constitutes the main portion of each Veda and consist of a certain number of mantras mostly presented in metrical form. For example, in the Ṛgveda-saṁhitā, there are 10,552 mantras. The ṛṣis who discovered these mantras are identified in the sūkta itself. The sūkta also indicates the devatā to whom it is addressed, and the metre used to compose these mantras. In the Ṛgveda-saṁhitā, we find a mention of over 400 ṛṣis and several categories of devatās. Four main devatās, viz., Indra, Agni, Varuṇa, and Mitra have been given the pride of place in the hymns. Nearly one-third of the mantras are addressed to Indra and a quarter to Agni. The ṛṣis are identified with a first name and a second (perhaps last) name. For example, the names of ṛṣis found include Viśvāmitra Gāthin (male ṛṣi) Dakṣiṇā Prājāpatyā (Female ṛṣi). Figure 2.3 shows the organisation of the Ṛgveda-saṁhitā. There were over 25 women ṛṣis who have composed hymns in the Ṛgveda.

- Each Veda consists of Saṁhitā, Brāhmaṇa, Āraṇyaka and Upaniṣads.
- There were 25 women rishis who have composed hymns in Ṛgveda.

Brāhmaṇas

The term Brāhmaṇa here does not relate to the modern word 'Brahmin', used to denote a caste. These are a collection of knowledge mainly confined to the issue of rituals and rites written in prose. If we go by the sheer volume, Brāhmaṇas form a large portion of the Vedic repository. Typically, one or more Brāhmaṇas are associated with every Veda and contain commentaries and explanatory notes pertaining to the mantras used. They describe different types of yajñas with all the ritualistic details. For example, the Aitareya-brāhmaṇa describes the Soma sacrifice, Agnihotra, etc. Similarly, the Kauṣītaki-brāhmaṇa provides details on food sacrifice, full-moon sacrifices, and sacrifices of the seasons. In the Śatapatha-brāhmaṇa details of Agnicayana, Aśvamedha-yajña, Upanayana, and Svādhyāya are explained.

Several aspects of the rituals are mentioned in the Brāhmaṇas portion. First, the reasons for performing rituals are mentioned. Besides, Brāhmaṇas prescribe the mantras from the Saṁhitās that need to be recited for the rituals. For example, in the Taittirīya-brāhmaṇa, we have the following instruction. One shall sprinkle water around (the food) reciting the mantra ऋतं त्वा सत्येन परिषिञ्चामि (*ṛtaṃ tvā tvartena pariṣiñcāmi*) in the evening. During the day the mantra सत्यं त्वर्तेन परिषिञ्चामि (*satyaṃ tvartena pariṣiñcāmi*) shall be recited (for the same action). The Brāhmaṇa also provides detailed instructions for preparing the altar for the yajña. An example from the Taittirīya portion illustrates this. The passage describes the process of building the main sacrificial platform (Uttara-vedī). A pit shall be dug up to the height of yajamāna's knee. It shall be filled with water up to the height of his ankle. After covering it with lotus flowers, leaves and lotus sticks the Agni has to be brought in for beginning the yajña[7]. As we can see, the content of Brāhmaṇas is technical material pertaining to the rituals. Therefore, several portions of the Brāhmaṇa literature may not appeal to a general reader.

Sāmaveda has the largest collection of Brāhmaṇas (eight). Tāṇḍya-mahā-brāhmaṇa, also known as Pañcaviṃśa-Brāhmaṇa, contains 25 chapters. The details on the conduct of Somayaga are given in this Brāhmaṇa. Another Brāhmaṇa known as Ṣaḍviṃśa-brāhmaṇa contains 26 chapters as the name suggests. This deals with several topics including the origin of Agni, provisions for śāntis for various ills affecting the nation, such as misfortunes like the untimely death of people, diseases, bad dreams, diseases affecting elephants and other animals, splitting of earth, earthquake, fire in earth, earth expelling water and inundation of earth in running water. The importance of the seven notes-based singing, the Sāma-gāna for those who cannot perform yajñas, and the description of the Sāmaveda tradition are available in the Sāma-vidhāna-brāhmaṇa. The importance of the guru-śiṣya tradition and various vidyās such as Madhu-vidyā and Śāṇḍilya-vidyā are discussed in Chāndogya-brāhmaṇa. The lineage of the ṛṣis can be found in the Vaṃśa-brāhmaṇa.

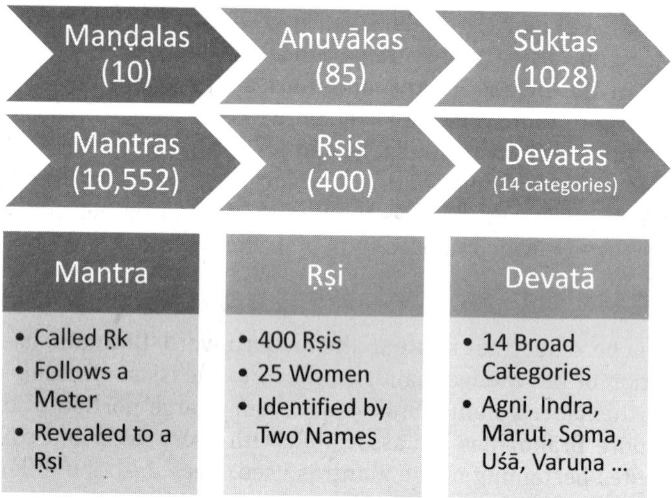

FIGURE 2.3 Organization of Ṛgveda Saṃhitā

Āraṇyakas

Āraṇyakas form the third portion of Vedas and their content appears to be similar to that of Brāhmaṇas, as both of them discuss yajña practices. However, there is an important difference that one can notice. While Brāhmaṇas approach the yajña from the perspective of the rituals to be conducted, Āraṇyakas take a philosophical approach while discussing a yajña. Their main interest is to explain the symbolism and philosophical aspects of a ritual. Not only the prescribed rituals, but the daily chores such as breathing, and eating have been given a deep meaning. It is mentioned that one must perform these activities contemplating the meaning behind these activities and such a person will be rewarded for that[8].

> ♦ Āraṇyakas inquire into the philosophical aspects of a yajña.
> ♦ Āraṇyakas are considered as bridge between Brāhmaṇas and Upaniṣads.
> ♦ Āraṇyakas represent the Upāsanā kāṇḍa of the Vedas.

Because of the philosophical nature of its content, the Āraṇyakas are supposed to be learnt from a guru in a secluded place far from the human dwellings and crowded places of living. This is suggestively indicated as a forest area (Araṇya) and hence this portion is called Āraṇyakas. Āraṇyakas are usually found at the end of Brāhmaṇas and have Upaniṣads in their concluding portion. Thus, Āraṇyakas expose the profound thought process involved in rituals helping one purify his mind paving the way for the next stage of life. Hence, Āraṇyakas are considered a bridge between Brāhmaṇas and Upaniṣads. There are currently six Āraṇyakas available, these are associated with the first three Vedas. No Āraṇyaka is associated with Atharvaveda.

Upaniṣads

By far the loftiest thoughts of the Vedic seers find their expression in the group of compositions known as Upaniṣads. These are philosophical treatises dealing with the ultimate problems of life that every one of us confronts. The word Upaniṣad means sitting near a Guru and receiving his wisdom through a teaching-learning process. The Upaniṣads have taken different approaches to impart knowledge. In the Praśna-upaniṣad, the entire teachings are organised into six questions asked one each by the six students and the Guru's reply to it. On the other hand, the Kaṭha-upaniṣad begins with a story of a young boy Naciketas and the journey of his quest to know the truth about death and immortality. Chāndogya-upaniṣad and the Bṛhadāraṇyaka-upaniṣad are large compositions bringing several messages, anecdotes, and teachings on multiple issues that we face. The profound conversation between King Janaka and Yājñavalkya appears in the Bṛhadāraṇyaka-upaniṣad, at the end of which the sage says, "Janaka, you indeed obtained fearlessness". The Indian emblem 'Satyam Eva Jayate' is a bold proclamation in the Muṇḍaka-upaniṣad.

Notwithstanding these differences in the styles of presentation, the common theme of the Upaniṣads seems to be focused on the knowledge of the Brahman. There is an in-depth study of human beings in terms of their nature and psychology. There are vivid descriptions of the relationship between the living being (jīva), the Universe that forms the context for the living being (Jagat), and the all-encompassing diving force at play (Īśvara). The Upaniṣads contain the Mahā-vākyas (meaning the ultimate pronouncements or sentences of truth).

Originally over 1,180 Upaniṣads were supposed to have been part of the Vedic corpus. Unfortunately, as in the case of other components of the Veda, several of them are lost. Currently, we have been able to locate around 200 Upaniṣads. These are found mostly in the Brāhmaṇa portion of the Vedas. Figure 2.4 lists the 108 Upaniṣads under the four Vedas. Among the available Upaniṣads, 10 are considered principal Upaniṣads as they have been commented upon and often used to support arguments in related subjects by great spiritual masters in the country. Besides, another 4–5 Upaniṣads are also popular as they too have been frequently mentioned by the spiritual masters.

Ṛg Veda – 10 Upaniṣads

Yajur Veda
- Śukla – 19 Upaniṣads
- Kṛṣṇa – 32 Upaniṣads

Sāma Veda – 16 Upaniṣads

Atharva Veda – 31 Upaniṣads

FIGURE 2.4 108 Upaniṣads in the Four Vedas

The Upaniṣads not only provide deep spiritual and philosophical insights into life. They also serve as a fountainhead of knowledge and wisdom for us to lead a successful life in this world as we pursue material benefits for a happy life. A case in point is the last anuvāka in Śikṣāvallī of Taittirīya-upaniṣad that has a typical setting of the final address by the Guru to his batch of students as they pass out of the gurukula. This can be considered equivalent to the modern-day convocation address. A reading of this portion reveals to us that there are seven waves of thought that the Guru wants to leave with the students as they step into the outside world[9]. These are summarised below:

- Advice on one's own mode of living with reference to society and oneself
- Regulating one's relationship with the previous generation and the present elders

- Relationship with oneself and one's teachers
- One's attitude towards the learned and the wise in society
- Charity and the laws of giving
- Remedy for doubts regarding one's own duty and conduct in life
- Doubts regarding one's relationship with others falsely accused in the world

Know the Charioteer of Your Life Journey

The most inert part of us is the body. There is no motive force to the body by itself. Only because of the *prāṇa* the body gets activated. This is the reason for a person who is so fond of halwa unable to open his/her eyes wide after he/she is dead when a cup of oven fresh halwa is brought in front of him/her. The *indriyas* are better than the body. This is because our *indriyas* can travel far and wide and reach places. For example, in a split moment the *indriyas* can take us to a beautiful tourist spot in the US.

The *manas* is truly a super-*indriya*, as it can do all the functions of the *indriyas* when none of them are at work. Otherwise, how can we explain watching vividly our favourite movie in dream or taking our son to a cricket match in a dream. The *buddhi* is considered superior to the *manas* as it has the capability to analyse and decide what is right and what is wrong. The *manas* can only deploy instructions to the *indriyas* and the body. It is poor in deciding what is right and what is wrong. This is the reason for the Kaṭha-upaniṣad teaching, know the buddhi to be the charioteer of your life (बुद्धिं तु सारथिं विद्धि–*buddhiṃ tu sārathiṃ viddhi).*

The soul is superior to everything as it is the very storehouse of energy (*prāṇa*) without which none of the above can perform. It is like having number of electrical gadgets, but they will work only as long as there is electricity. Once you pull the plug everything comes to a grinding halt.

2.4 VEDĀṄGAS

During the earlier times, the Vedic language was easy to recite and understand. With passing time natural changes occurred in the spoken language of the people and it slowly drifted away from the Vedic language. Therefore, these people needed support to read and comprehend the Vedic text. Because of the importance of Vedic texts, a lot of effort has gone into preserving the text in its original form. A specific body of knowledge, practices, and tools and techniques was created to preserve the Vedas and appropriately use them. Eventually, these became an integral part of the Vedic literature, hence called Vedāṅgas, literally limbs of the Vedas.

- The loftiest thoughts of the Vedic seers are found in the Upaniṣads.
- Upaniṣads pertains to the jñāna-kāṇḍa of the Vedas.
- There are 10 principal Upaniṣads.
- The Mahā-vākyas are found in the Upaniṣads.

One can identify six complementary requirements for the preservation of Vedic corpus and the practices prescribed therein and the proper use of them.

- Vedic texts being oral in nature have to be preserved in their original form. This requires listening to the sounds properly and reproducing them the same way they were heard.
- The words and sentences of the text had to be understood, without any ambiguity.
- The metres to which the mantras are set had to be systematically understood and their rules followed properly.
- There is a need for a complete guide to lead a life as prescribed in the Vedas. This requires unambiguously spelt out practices and norms for various activities to be performed.

- A properly structured set of instructions to perform the rituals prescribed in the Vedas, right from building the yajña-śālā to carrying out elaborate rituals such as the soma-yāgas need to be established.
- A method to fix the time to do all these prescribed activities is also necessary.

There are six Vedāṅgas; Śikṣā, Vyākaraṇa, Nirukta, Chandas, Kalpa, and Jyotiṣa addressing these roles. Figure 2.5 lists these roles of the Vedāṅgas. Though the seeds of all these Vedāṅgas are found in Vedas itself, people wrote elaborate texts on these topics to make them clear and updated them as and when required. Thus, we have several texts authored by various people at different points in time explaining these. While these Vedāṅga texts help us decipher the Vedic texts and know their practical applications, they have other values of practical applicability in other fields also. Works such as Pāṇini's Aṣṭādhyāyī, Piṅgala's Chandaḥ-śāstra, śulba-sūtras and many works related to jyotiṣa-śāstra are some of the examples. We will see the applications of these in other chapters.

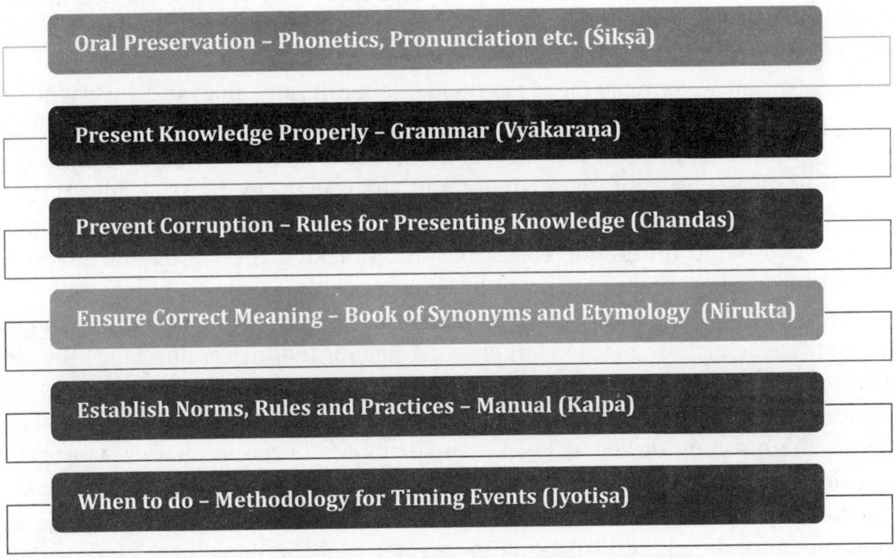

FIGURE 2.5 The Role of Vedāṅgas

2.4.1 Śikṣā

The word śikṣā means 'to acquire knowledge'. In the Vedic context the discipline that deals with pronunciation is called Śikṣā. This was the first thing students were taught in the ancient educational system; hence it is called śikṣā. This has been described in Taittirīya Upaniṣad very briefly[10]. It was further elaborated in śikṣā-śāstra. Śikṣā essentially is the science of pronunciation. As preserving the Vedic text from any form of corruption is the most important thing, śikṣā directs our attention to all the details of the process involved in pronunciation. Śikṣā-śāstra, therefore, is a systematic approach to the art and practice of phonetics. It defines the characteristics of the basic units of the sound of the language known as varṇa and explains what causes the sound pertaining to a varṇa to emanate. There are several books on śikṣā-śāstra. The most popular work today is 'Pāṇinīya-śikṣā' which is attributed to Pāṇini, the famous grammarian.

The nāda (sound) generated by the confluence of air and space (in the vocal cord) takes the form of a varṇa (the smallest component of a language) by the contact made between various parts of the tongue and the places of articulation[11]. With these fundamentals of origin of the sound from the oral cavity, it further describes the process of pronouncing the varṇas. We shall see some more details of this in Chapter 5 of the book.

2.4.2 Vyākaraṇa

Grammar is the foundational aspect of any language. Therefore, to understand the Vedic repository, rules of grammar are required. Vyākaraṇa is one of the Vedāṅgas that deals with this issue. The tradition of Vyākaraṇa dates to the Vedic period[12]. However, the credit goes to Pāṇini for presenting a structured work of the Sanskrit Vyākaraṇa. Despite several other works on Vyākaraṇa, Pāṇini's Aṣṭādhyāyī became popular because of its brevity. Another great master of Vyākaraṇa, Patañjali wrote a commentary on Aṣṭādhyāyī, known as Mahā-bhāṣya, which provided logical and philosophical support to the work and firmly established the monumental work of Pāṇini.

The term Vyākaraṇa essentially means to divide, separate, and analyse. Sanskrit grammar is unique in that it devises unambiguous and rule-based methods to construct a word. Therefore, with a knowledge of this, one can not only generate new words but also 'reverse engineer' by breaking a word into certain components and then analysing it. The advantage of this method lies in the fact that although words are infinite in the language, their components are fewer, repetitive based on unique conditions. So, breaking the words and mapping their components with meanings makes language learning easy. Vyākaraṇa is highly structured and rule-based that it has several features that parallel what has been applied in the modern-day data processing. These include methods of creating lists-based processing logic, the use of an algorithmic approach to process language, application of recursive logic to process data, etc. A unique feature of Pāṇini's work is that he gave the rules of svaras, which are very essential in determining the meaning of Vedas. Even a small change in the intonation can bring about a significant difference in the meaning at times. Out of 4000 sūtras, Pāṇini has dedicated over 400 to give the rules of svaras.

We shall see more details of Vyākaraṇa in Chapter 5 of the book.

2.4.3 Nirukta

Nirukta is etymology, which is extracting the meaning of a word using linguistic theories and considering phonetic changes. While issues with understanding the meaning of the Vedas are addressed by vyākaraṇa-śāstra, Nirukta engages in the same task with a different approach. The objective of vyākaraṇa is to ascertain a correct form of a word to express a certain meaning. On the other hand, Nirukta analyses a given word to extract its meaning. Yāska, wrote Nirukta in the 5th century BCE, which is a commentary on Nighaṇṭu, a collection of rarely used words in the Vedas. Nighaṇṭu can be equated with a thesaurus, where synonyms of various genres are collated. Some of these words are used in a totally different sense in a Vedic context. Therefore, the study of the Nirukta is indispensable for the understanding of the Veda[13]. Like other Vedāṅgas, this developed as a major discipline on its own in later times. It appears that 'Nirukta' is the most ancient work available on etymology anywhere in the world.

There are 1770 words grouped in three parts namely Naighaṇṭuka-kāṇḍa consisting of three chapters, Naigama-kāṇḍa, and Daivata-kāṇḍa each consisting of one chapter. Naighaṇṭuka consists of 69 groups of synonyms. Words with multiple meanings are organised in each of these groups. Table 2.3 has details on the overall organisation of Nighaṇṭu. The need for a work such as Nirukta arises primarily on account of different meanings associated with words in the context of Vedic texts. In the absence of this, there is a danger of associating a word with a wrong meaning and interpreting the Vedic text wrongly. Sometimes, the error can mislead the knowledge seeker and present the Vedic corpus in poor light simply on account of ignorance. (See box for an example of this).

TABLE 2.3 Organisation of Nighaṇṭu

Kāṇḍa	*Adhyāya*	*No. of Words*	*Contents*
Naighaṇṭuka	1	415	17 groups of synonyms
	2	516	22 groups of synonyms
	3	410	30 groups of synonyms
Naigama		278	Words with multiple meanings in 3 groups
Daivata		151	Names of Devatās in 3 groups

In the process of arriving at the meanings associated with the words, a four-step approach is employed:

- Observe the repeated occurrence of letters or a group of letters in different words
- Observe the repetition of the same meaning in different words
- Map repeating sounds with the repeating meanings
- Assign meaning to the component of a word

Risks in Translating Vedic Literature

The words used in Vedic literature sometimes have unique meanings, very different from the normal usage. Nirukta plays a crucial role in ensuring the correct meaning is obtained by providing a list of synonyms of words used in Vedic literature. In the absence of this knowledge, there is a risk of wrong translation as is evident from the translation of Tāṇḍya-brāhmaṇa by a Dutch scholar, W. Caland. See below the original mantra and the translation adopted.

यावद्वै सहस्रं गाव उत्तराधरा इत्याहुस्तावदस्मात् लोकात् स्वर्गो लोक इति तस्मादाहुः सहस्रयाजी वा इमान् लोकान् प्राप्नोति ||
(Tāndya-brāhmana 16.8.6)

yāvadvai sahasraṃ gāva uttarādharā ityāhus-tāvadasmāt lokāt svargo loka iti tasmād-āhuḥ sahasra-yājī vā imān lokān prāpnoti ||

The translation reads as follows:

The world of heaven is as far removed from this (earthly) world, they say, as a **thousand cows standing one above the other** (emphasis added). Therefore, they say, "He who sacrifices with a sacrifice at which a thousand dakṣiṇās are given reaches these worlds".

For any sensible-minded person, such a translation will be unacceptable. How can one stack 1000 cows one over the other and hope to assume this is the distance between the earth and heaven?

There are 21 synonyms listed in a group in Nirukta. They are 'Gauḥ, gmā, jmā, kṣmā, kṣā, kṣamā, kṣoṇiḥ, kṣitiḥ, avaniḥ, urvī, pṛthvī, mahī, ripaḥ, aditiḥ, iḻā, nirṛtiḥ, bhūḥ, bhūmiḥ, pūṣā, gātuḥ, gotrā'. In normal usage, the word 'गौः (gauḥ)' stands for a cow. Whereas in Veda it can mean many other things such as earth, a cow, a moving object, the Sun. All these meanings are extracted from a modified form of the verbal root 'गम् (gam)'. The meaning of the root 'gam' is to move. The same has been taken as the verbal root 'गो (go)'. Hence, the earth is called 'go' as the beings here move on it. The 'cow' is called 'go' since it used to go out far away from the village for grazing. The Sun is called 'go' because it appears to be moving. Now of these meanings whichever suits a context must be taken. The meaning of "go" in this mantra must relate to the earth and not to the cow.

This example demonstrates how a lack of knowledge on Nirukta can mislead Indologists in their efforts to translate the Vedic corpus into English and other foreign languages.

Source for the translation: Caland, W. (1982), *Pañcaviṃśa-brāhmaṇa,* Satguru Publications, New Delhi, p. 440.

2.4.4 Chandas

Rhythm is the key aspect of an oral tradition be it Vedic mantras or music. The saṁhitā portion of the Vedas are almost entirely in prosody. Therefore, the study of the metres to which they are set is important. Chandas is the metre of poetic composition. There are many references to these metres and their characteristics in Vedic texts themselves. The major treatise on Vedic metres currently in usage is Chandaḥ-śāstra by Piṅgala written around 300 BCE. This also lays the foundation for many other metres used in classical Sanskrit literature.

Before we get to know the details of Chandas, let us get introduced to some basic terms. A chandas (metre) can be viewed using a three-level hierarchy. See Figure 2.6 for a schematic representation of the same. The lowest unit or building block of a metre is a syllable, called akṣara. A certain number and pattern of akṣaras make up a pāda (quarter) and a certain number of padās make a metre. Let us consider the following mantra from the Ṛgveda to understand this concept: स्वादिष्ठया मदिष्ठया पवस्व सोम धारया । इन्द्राय पातवे सुतः ॥ This mantra indeed consists of three pādas, each of eight syllables. This metre is called gāyatrī metre. The details are in Table 2.4.

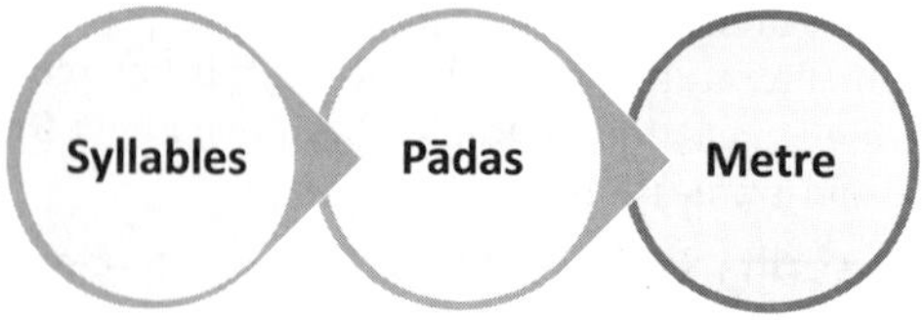

FIGURE 2.6 The Hierarchical Structure of Metre

Metres in the Vedas

There are seven main metres used in the Vedas. The majority of these metres have four quarters in them. Some metres have three. There are also some exceptions with varying numbers of pādas. The Vedic metres differ from one another on two counts; the number of pādas that constitute a metre and the number of syllables in each pāda. Table 2.5 has the details of chandas and the numbers pādas and syllables for each of them. There is much more variety of chandas with varying numbers of pādas and syllables in them. Further, by the addition or removal of syllables from an existing metre, some more variations are created, making it a large variety in reality. Chandas plays a crucial role in the preservation of the Vedic corpus and loss-less transmission down the generations. Any addition or removal of even a single syllable from a mantra will become evident at once as the rhythm of the mantra will be lost. Thus, chandas helps in preserving intact not only the Vedic texts but also any literature set in prosody. This also helps in removing the doubts in meaning and makes the recitation of mantras joyful.

TABLE 2.4 Three Level Hierarchy of a Metre – An Illustration

Mantra	स्वादिष्ठया मदिष्ठया पवस्व सोम धारया । इन्द्राय पातवे सुतः ॥ svādiṣṭhayā madiṣṭhayā pavasva soma dhārayā | indrāya pātave sutaḥ ||
Metre	Gāyatrī Metre
Pādas (3) – each of 8 syllables	Pāda 1: स्वादिष्ठया मदिष्ठया (svādiṣṭhayā madiṣṭhayā); Pāda 2: पवस्व सोम धारया (pavasva soma dhārayā); Pāda 3: इन्द्राय पातवे सुतः (indrāya pātave sutaḥ);
Syllables (24)	svā-di-ṣṭha-yā-ma-di-ṣṭha-yā-pa-va-sva-so-ma-dhā-ra-yā-in-drā-ya-pā-ta-ve-su-taḥ

We shall see more details and other applications of the chandaḥ-śāstra in Chapters 6 and 8 of the book.

TABLE 2.5 Details Pertaining to Vedic Metrers

Chandas	*Number of pādas*	*Number Syllables per pāda*
Gāyatrī	3	8 + 8 + 8 = 24
Uṣṇih	3	8 + 8 + 12 = 28
Anuṣṭup	4	8 + 8 + 8 + 8 = 32
Bṛhatī	4	8 + 8 + 12 + 8 = 36
Paṅkti	5	8 + 8 + 8 + 8 + 8 = 40
Triṣṭup	4	11 + 11 + 11 + 11 = 44
Jagatī	4	12 + 12 + 12 + 12 = 48

2.4.5 Kalpa

Kalpa focused on several operational aspects of the issues discussed in the Vedic corpus. In simple terms, Kalpa can be thought of as a guide or a user manual that provides instructions and directions to lead all aspects of life including personal, family, and social dimensions. A study of the Kalpa-sūtras sheds light on all important aspects of ancient Indian life as it covers household life, ceremonies, and law. One can see elaborate descriptions of different types of yajñas and methods of performing

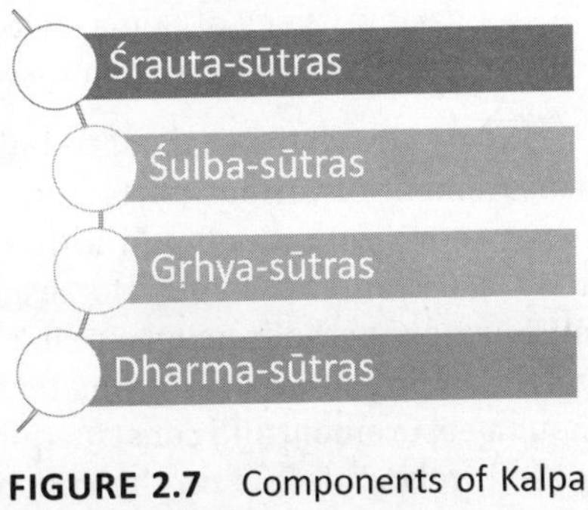

FIGURE 2.7 Components of Kalpa

these, their frequency, etc. There are vivid descriptions of how to prepare the Agni, the types of people involved, and their roles, the type of offerings to be made, and other resources required.

There are different schools of Vedic studies and each of these schools has its own set of basic instructive texts called Kalpa-sūtras. These comprise four types of texts (see Figure 2.7 for a pictorial representation of the same):

- Śrauta-sūtras – Guide for Vedic rituals
- Śulba-sūtras – Rules for measurements and construction of fire altars, sacrificial place, etc.
- Gṛhya-sūtras – Deals with house-hold ceremonies
- Dharma-sūtras – Guide for social duties

Śrauta-sūtra provides details of various rituals prescribed in the Vedas. These are to be performed at various intervals. Some of them are quite elaborate and require a large number of people to perform (such as Aśvamedha or Rājasūya-yajña). These are invariably done by kings as the outlay is huge. Some rituals are performed for many days and months. The role of Śrauta-sūtras is to provide some sort of an operations manual laying out minute details. It systematically describes all the activities in a ritual, the roles to be taken by various people, the sequence of various activities, etc. One is supposed to maintain three *Śrauta-agnis* to perform *yajñas* as per Śrauta-sūtra; Gārhapatya (maintained always), Āhavanīya and Dakṣiṇāgni. Gārhapatya is a circular one placed on the west, whereas Āhavanīya is square and is on the eastern side. On the other hand, a Semi-circular altar called Dakṣiṇāgni is in the south. These are supposed to have an equal area and are to be built in five layers using burnt bricks of prescribed numbers and sizes. It is no wonder therefore that ancient Indians made good approximations to the value of π to solve the problem of designing a circle and square of the same area. There are many different shapes and sizes of the Vedic altar specified in Kalpa. For instance, the darśa-pūrṇa-māsa altar has a unique shape. See Figure 2.8 for an illustration of these Vedic altars.

Some of the major rituals prescribed are as follows:

- Agnihotra – this is a ritual the custodian of the three fires has to perform twice every day. An offering of ghee and milk is made into the sacrificial fire.
- Darśapūrṇamāsa-iṣṭi – This is a ritual performed on the day after the full moon and no moon days. It involves several people other than the yajamāna. Rice puddings baked in the sacrificial fire are the main offerings in this ritual.
- Cāturmāsya-iṣṭi – As the name suggests this ritual is performed once in four months.
- Various types of Somayāgas – These are the long rituals that go for more than a week. A sixteen-member team is involved, and it is an elaborate ritual.

Śulba-sūtra is the manual to build a sacrificial room and altars. Śulba means a thread. The measurement of various parts of the yāga-śālā is done with the help of a thread and two nails. As we know, with a nail and a thread we can only generate a circular shape. However, Vedic altars had complex shapes (for example a flying falcon). Ancient Indians devised interesting construction methods using circular constructions. Therefore, we can infer that they had a good understanding of cyclical geometry. Expectedly, Śulba-sūtras contain minute and basic details about measurement and construction of sacrificial place: वेदी (vedī), चिति (citi), मण्डप (maṇḍapa). These involved details on the construction of squares, rectangles, circles, etc. Further, minute details ensuring accurate construction of the altar (*vedī*) including orientation, size, shape, and areas are specified. The Kalpa specifications also require that altars of very different shapes need to be of the same area.

Gṛhya-sūtra is a guide to perform domestic ceremonies. Everyday rituals such as sandhyā-vandanam, the sixteen saṃskāras, sacraments of a person to be performed right from garbhādhāna (insemination) to antyeṣṭi (last rites), the five mahāyajñas are some of the topics covered here. To perform these domestic ceremonies, the householder (*Gṛhasta*) is required to maintain the Agni (**Gṛhyāgni**) from marriage unto death for all household rites. These rites are meant to worship the Gods through a variety of yajñas to obtain material benefits, good health, and progeny. The ceremonies are also performed for the achievement of targeted desires and results and the welfare of society in general.

Dharma-sūtra is the overall guide for an individual to be part of the society in a gainful manner. Dharma-sūtra is mainly concerned with rules of conduct that an individual needs to follow so that he/she can have a harmonious living in society. The rules for the people of different occupations, duties of an individual at various levels of his life, guidance for the rulers to rule the state are some of the important topics described here. These evolved as systematized laws which were in the form of customs. It covers issues such as lawful occupations of the class of society, stages of life, dietary laws, punishments, forms of government, taxation administration, and army. In a way, these sutras provide details on public policy, governance, and administration.

FIGURE 2.8 A Typical Set of Altars for Śrauta Rites

2.4.6 Jyotiṣa

Like several other ancient civilizations, Indians were using the natural calendar which is based on keen observation of the sky with moon and stars. This knowledge of the movement of stars and planetary bodies is called jyotiṣa. In India, the lunar and the solar cycles play a very important role in the cultural and spiritual development of the Vedic people. These practices continue even today as all the festivals and practices are inextricably linked to the Moon, the

Sun, and other stars and constellations. Having studied several aspects of the Vedic life, Lagadha composed Vedāṅga-jyotiṣa[14] at a period of time when the sun was in Dhaniṣṭhā nakṣåtra during the winter solstice. Jyotiṣa occupies a special place in the Vedāṅgas, and it is compared to the jewel on the hood of the snake and the crest of the peacock[15].

Jyotiṣa-śāstra is classified into vaidika (related to Vedas) and laukika (related to the world). The Jyotiṣa which is connected to Vedas is called Vedāṅga-jyotiṣa, which was followed by the people during the Vedic period. Vedānga-jyotiṣa is classified into two recensions: Ṛgvedic Vedāṅga-jyotiṣa – 36 verses and Yajurvedic Vedāṅga-Jyotiṣa – 43 verses. The Atharvaṇa-jyotiṣa is known to be taught by Prajāpati to Kaśyapa. Jyotiṣa-śāstra is further classified into three parts and each of those further into two each[16]. Figure 2.9 schematically presents the classification of Jyotiṣa-śāstra.

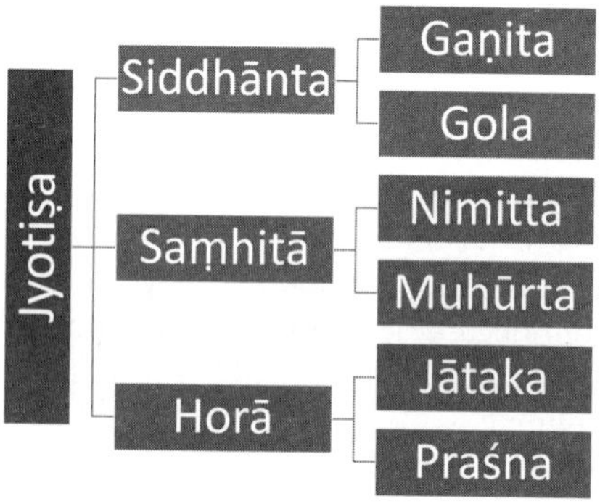

FIGURE 2.9 Classification of Topics in Jyotiṣa Śāstra

Siddhānta deals with the various measures of time; planetary theory, arithmetical computations as well as algebraical processes, location of the earth, the stars and the planets, and description and usage of instruments[17]. Saṃhitā consists of the scientific and mathematical concepts of astrology. Although listed as one of the three divisions. In reality, the issues discussed are included in all the three skandhas (Siddhānta, Saṃhitā, and Horā). It can be divided into three sections. Section 1 deals with the movement of the planets in the several Rāśis (zodiacs). These are ascertained through mathematical computation, known as tantra. In Section 2, horoscope and nimitta (Omens) issues are discussed. Section 3 provides details to precisely locate the correct position of lagna, which is the first step in Jātakarma. In addition to these issues, Saṃhitā deals with many aspects of human life. Table 2.6 lists some of the major topics discussed in the Saṃhitā.

Jyotiṣa is intricately connected to astronomy. We discuss some more details in Chapter 9 of the book.

TABLE 2.6 The Breadth of Topics Discussed in Saṃhitā

No.	*Topic Discussed*	*Details*
1	दैवज्ञ-लक्षणम् (Daivajña-lakṣaṇam)	Qualities of an Astrologer
2	ग्रहाणां चार-विचारः (Grahānāṃ cāra-vicāraḥ)	Planetary Movements
3	भूकम्पोल्कादि-लक्षण-विचारः (Bhūkampolkādi-lakṣaṇa-vicārah)	Characteristics of Earthquakes, Meteors etc.
4	वास्तु-विचारः (Vāstu-vicāraḥ)	Architecture
5	जलान्वेषण-विचारः (Jalānveṣaṇa-vicāraḥ)	Exploration of Water Springs
6	देवालय-प्रतिमादि-निर्माण-क्रमः (Devālaya-pratimādi-nirmāṇa-kramaḥ)	Temple Construction and Sculpture
7	शकुनादि-विचारः (Śakunādi-vicāraḥ)	Omens
8	शरीर-लक्षणादि-विचारः (Śarīra-lakṣaṇādi-vicāraḥ)	Body Parts – Characteristics

IKS IN ACTION 2.2

Pañca-mahā-yajña

The term Yajña typically invokes in our mind an altar in which offerings are made to a deity with a certain expectation of tangible material benefits, usually of a self-centred nature. The critical aspect of Yajña is giving (or sharing) without a sense of attachment. This puts our life into perspective and provides broad guidelines for us to imbibe the value of peaceful coexistence in our daily life. This can happen only when we are able to practice 'care and share' as a daily habit. The concept of pañca-mahā-yajña (Five Great Sacrifices) as our daily duty was instituted to practically implement this idea.

In the Indian tradition, the role of a householder was considered very important in maintaining sustenance of the society. The responsibility of a householder to provide for the sustenance of the life around him is formalised with pañca-mahā-yajña. Every householder is expected to perform this every day and if one does not undertake these yajñas he may not even qualify to be a householder.

This is a mechanism to care for and share his food with the other entities in the creation and it subtly reminds himself every day of his responsibility towards all creation. The importance of pañca-mahā-yajña is evident from the references to it found in all Indian literature: in the Vedas, the itihāsas and purāṇas, smṛtis and dharmaśāstras.

- *Bhūta-yajña* is for all created beings. By offering Bhūta-yajña, we take care of small living beings around us (such as birds, domesticated animals, worms, insects etc.). By being very conscious of the environment we can extend this idea to plants, rocks and rivers. This is the extended thinking of bhūta-yajña. This is a good approach to address the vexing ecological problems that we face today.
- By *Manuṣya-yajña* we derive the joy of helping destitute, orphans, unexpected guests, poor and the needy by offering whatever we can (in cash or kind, food) to them. This can be the bedrock of social sustainability and can positively impact the society.
- By *Pitṛ-yajña* we give away food for the sake of our ancestors and offer our respects and deep sense of gratitude to the departed souls in our family who are responsible for what we are today.
- By offering *Deva-yajña*, we express our thankfulness for what the Gods bless us (in terms of rain, and other bounties of nature) and continue to receive them in a framework of mutual dependence. The most manifested form of the Devas are the five great elements (space, air, fire, water and earth). This daily act makes us acutely aware of the need to keep our ambience in absolute rhythm with our activities and thereby minimize pollution (air, water, earth etc.) levels.
- By *Brahma-yajña* (by reciting the Vedic hymns, Upaniṣads etc. and teaching them to others) we show our enormous respect to the great seers and rishis, who gave us the wisdom that we can ever have. We also assure them by this act that this great knowledge is being handed down the generations with reverence and a sense of responsibility.

Pañca-mahā-yajña is related to the Indian understanding of human life as a gift that is sustained by all aspects of creation. Man is thus born in and lives in debt, to all creation, and it therefore becomes his duty to recognize this debt and undertake to repay it every day.

Panca-mahā-yajña, therefore, is not performed for the sake of earning merit or virtue. It is merely a matter of endeavouring to repay debts that are incurred by being born and living in the world. It is about being humanly responsible.

2.5 VEDIC LIFE: DISTINCTIVE FEATURES

So far, we have seen some of the salient features of the Vedas, the material presented in the Vedas, and the issues described in various parts of the Vedas. Based on this limited understanding, we can develop some broad ideas about certain distinctive features of the Vedic life. These can be discussed under the following heads:

Vedas Extol Living a Zestful and Exuberant Life: There is a misconception that Vedic people de-emphasized materialism and instead chose to lead a simple life with no minimum comforts. However, the mantras, the prayers, and the things that they asked for (the karma-kāṇḍa of the Vedic literature is abundant with such examples), we can infer that they aspired to live a life full of energy, enthusiasm, hopes, desire to explore and innovate to make their life better and comfortable.

Balanced Life Priorities: The Vedic thinking (discussed in great detail in the jñāna-kāṇḍa) also presents a balanced view of life, where the attainment of material riches and prosperity was considered important, but not at the cost of spiritual orientation to life. The inquisitiveness of the people and a desire to know and innovate enabled them to develop ideas, knowledge, and thought processes that addressed both material progress and spiritual progress. This resulted in the development of karma kāṇḍa and jñāna kāṇḍa aspects in the Vedic repository. As a result, the Vedic living ought to have advocated for the all-round development of an individual in terms of physical, mental, intellectual, and spiritual dimensions.

- Vedic thinking presents a balanced view of life where material prosperity and spiritual orientation are important.
- Yajña was central to day-to-day living in Vedic times.
- Vedic life recognized the overarching role of natural systems in sustainability.

Emphasis on Sustainable Living: The practicality of Vedic life proposed a model that was economically and ecologically viable, and socially sustainable. The Vedic people recognised the importance of mutual dependence and co-existence with nature and other living beings. This is well documented and articulated in the numerous hymns in *Ṛgveda* on several aspects of nature. The principle of mutual dependence pervades much more than what we normally imagine. Man, and nature have a strong relationship of mutual dependence. Living entities and non-living entities also are mutually dependent. Our ancestral wisdom and practices in everyday living seem to have understood this aspect and respected it. Numerous references convey this idea. For example, as we already saw, *in Śānti-sūkta*, the well-being of not just the living entities, but also of natural systems is sought through the prayers.

The Primacy of Agni: This is primarily because Yajña was central to day-to-day living. Every activity and celebration in life was done with Yajña and dāna (gifting). It was recognized as a way of showing reverence and gratitude to the Gods for making things happen the way they are. Agni was considered the main deity and carrier of the offering to all the other devatās. Therefore, it is not surprising that the first mantra of Ṛgveda begins with a celebration of Agni as the priest and the giver of all riches to us.[18] The notion of Yajña as conceptualized by the Vedic seers is much larger than what is ordinarily understood as an act of sacrificial offering to Agni in an altar. It was a grand principle of 'give and take' and 'live and let live' and thereby ensure social sustainability in addition to the narrow environmental sustainability that we are currently debating on.

A Life Guided by Ṛta, Satya, and Dharma: There are larger principles that shaped the paradigms of good living. Ṛta in simple terms is the cosmic order or equilibrium which ensures that the Universe functions in its natural state. The nearest English word that one can think of is rhythm. In the Rigveda, the term *Ṛta* appears as many as 390 times and has been characterised as 'the one concept' which pervades the whole of Ṛgvedic thought[19]. The actions that individuals and society take in their living are not supposed to disturb the ṛta. In Taittirīya-upaniṣad, second chapter it is mentioned that the wind blows, the sun, the fire, Indra, and the Lord of death perform their daily routines in consonance to this cosmic order[20]. While ṛta is an overarching concept, its life practice was facilitated by the concept of satya and dharma. These two operational guidelines ensure the cosmic order is not disturbed.

SUMMARY

- Vedas are the quintessential wisdom that forms the foundation for the Sanātana dharma and is considered as the ultimate reference for every aspect of living in India.
- The Ṛgveda represents the earliest sacred book of India and the biggest amongst all the four Vedas.
- The Yajurveda is more pronouncedly a ritual Veda for it is essentially used by the Adhvaryu to do all ritualistic works in a sacrifice.
- The mantras of the Sāmaveda, have musical scales, similar to the seven scales of classical music. It can be considered as the origin of Indian classical music.
- A major part of Atharvaveda is concerned with diseases and their cure, rites for prolonging life, rites for fulfilling one's desires, building construction, trade and commerce, statecraft, penances, and propitiatory rites.
- Each Veda consists of Saṁhitā, Brāhmaṇa, Āraṇyaka, and Upaniṣads.
- Mantras are also known as saṁhitās. They constitute the main portion of each Veda. In the Ṛgveda Saṃhitā, there are 10,552 mantras.
- The Brāhmaṇas contain commentaries of the mantras used and explanatory notes pertaining to the mantras. They describe different types of Yajñas with all the ritualistic details.
- Upaniṣads are philosophical treatises dealing with the ultimate problems of life that every one of us confronts.
- The common theme of the Upaniṣads focuses on the knowledge of the Brahman. The other related themes include the spiritual nature of human beings, the study of human beings in-depth in terms of their nature and psychology.
- Knowledge of the six Vedāṅgas helps one develop a correct understanding of the Vedic repository and its use.
- Śikṣā essentially is the science of pronunciation and it helps in preserving the Vedic text from any form of corruption during oral transmission.
- While the objective of Vyākaraṇa is to ascertain a correct form of a word to express a certain meaning, Nirukta analyses a given word to extract its meaning.
- A chandas (metre) can be viewed using a three-level hierarchy. The lowest building block of a metre is an akṣara. A certain number and pattern of akṣaras make up a pāda and a certain number of padās make a metre.
- Kalpa can be thought of as a guide or a user manual that provides instructions and directions to lead all aspects of life including personal, family, and social dimensions.

REVIEW QUESTIONS

1. What do you understand by the term 'Veda'? Why is the Veda called 'apauruṣeya'?
2. What are the four Vedas? What is the basis on which these have been classified?
3. Describe the organisation of a Yajña. Who is involved and what are their roles?
4. How is the knowledge further classified and arranged in each of the Vedas?

5. Prepare a one-page note outlining the salient features of Ṛgveda Saṁhitā.
6. What are some of the major areas in which Atharvaveda has contributed?
7. What are the key differences between the Saṁhitā (Mantra) portion and the Brāhmaṇa portion of a Veda?
8. What are the key contributions found in the Brāhmaṇas of Sāmaveda?
9. What is the meaning of the word 'Upaniṣad'? What is the purpose of Upaniṣads?
10. Comment on the statement, "There are several useful perspectives that one can derive from the Vedic corpus". Do you agree with this statement? Prepare a note with supporting arguments.
11. What is the relevance of Vedāṅgas in the context of the Vedic repository?
12. Comment on the statement, "Vyākaraṇa and Nirukta address the same issue of analysing words and their meanings".
13. Why is Śikśā considered an important Vedāṅga? What role does it play in the oral tradition of the Vedic preservation?
14. What are the building blocks of a chandas? Explain the main forms of chandas deployed in a Vedic repository.
15. What do you understand by the term 'śulba-sūtra'? What is the use of it?
16. Distinguish between gṛhya-sūtra and dharma-sūtra.
17. Why do the Vedic people need Jyotiṣa?
18. Briefly outline the salient features of life during Vedic times. Are there any useful ideas to imbibe in our modern-day living?

DISCOVER IKS

1. Upaniṣads comprise the highest forms of knowledge in the Vedic repository. They contain very useful ideas in one's life. What are these messages and how does it help us? Watch the video available on the following site: https://youtu.be/27INqc3-8xo. Prepare a note that covers the following aspects:
   (a) What is the positioning of the Upaniṣad in the Vedic repository? How many Upaniṣads are there?
   (b) What are the key messages found in the Upaniṣads?
   (c) What kind of transformation an individual will go through benefited from the message of Upaniṣads?
2. It has been a matter of great debate as to what is the date for the Vedic corpus. At one level, the content is valuable that any effort to fix the date accurately is less valuable. Nevertheless, there is always a curiosity to know when Vedas were recorded in history. Prof. G C Tripathi's paper on the date of Ṛgveda is available in the following video: https://youtu.be/HraoKemxmiU. Listen carefully to the talk and prepare a three-page note to answer the following questions:
   (a) What was the date assigned by the Western Indologists for the Vedic period?
   (b) Why should we consider dates earlier than what was proposed? Give some specific reasons to support the argument.
   (c) Identify four specific arguments presented in the talk that support the dating of the Ṛgveda.

SUGGESTED READINGS

Kashyap, R.L. (2016). "The Essentials of Yajurveda", *Sri Aurobindo Kapāli Sāstri Institute of Vedic Culture*, Bengaluru.

Archak, K.B. (2012). "The Āraṇyakas and Upaniṣads", Chapter IX in *Essentials of Vedic Literature*, Kaveri Books, New Delhi, pp. 302–329.

Archak, K.B. (2012). "The Brāhmaṇa Literature", Chapter VIII in *Essentials of Vedic Literature*, Kaveri Books, New Delhi, pp. 271–301.

Arkasomayaji, D. (1980). "Siddhānta Śiromaṇi of Bhākarācārya", Rashtriya Sanskrit Sansthan, New Delhi.

Bhagwat, B. (2009). "Kalpa-Vedāṅga: Origin and Development", Adarsha Sanskrit Shodha Samstha, Vaidika Samshodhana Mandala, Pune.

Holay, P.V. (1989). "Vedic Astronomy (Vedanga Jyotisha: A Prehistoric Puzzle)", Shri Baba Saheb Apte Smarak Samitee, Nagpur.

Kanshi Ram (2010). *The Ladghu-siddhānta-kaumudī of Varadaraja: A Primer on Pāṇini's Grammar*, Volumes I–III, Motilal Banarsidass Publishers (P), Delhi.

Kashyap (2013). "The Essentials of Yajurveda", *Sri Aurobindo Kapāli Sāstri Institute of Vedic Culture*, Bengaluru.

Kashyap, R.L. (2003). "Introduction to Rig Veda" in Secrets of Rig Veda: First 121 Sūktas, *Sri Aurobindo Kapāli Sāstry Institute of Vedic Culture*, pp. 1–44.

Sastri, V.S. and Bhat, M.R. (1946). "Varahamihira's Brihat Samhita", M.B.D Electric Printing Works, Bengaluru.

Raman, B.V. (1996). *Prasna Marga (Part–1)*, Motilal Banarsidass, Delhi.

Sarup, L. (2002). *The Nighaṇṭu and The Nirukta of Śrī Yāskācārya*, Motilal Banarsidass, New Delhi.

Shashi Prabha Kumar (2002). "Bhumisūkta of the Atharvaveda", in Vedic Traditions and Modern Crisis, Pushpendara Kumar (Ed.), Eastern Book Linkers, Delhi, pp. 10–22.

Swami Chinmayananda. (2013). "Taittirīya Upaniṣad: Self Beyond Sheaths", Central Chinmaya Mission Trust, Mumbai.

Swami Harshananda. "A Bird's Eye View of Vedas". Ramakrishna Math. Bengaluru. Http://rkmathbangalore.org/Books/ABirdsEyeViewOfTheVedas.pdf. Last accessed on Oct. 1, 2021.

Swami Ranganathananda (2019). "The Message of the Upaniṣads", Advaita Ashrama, Kolkata.

Vaidya, C.V. (1930). *History of Sanskrit Literature, Vol. 1: Śruti (Vedic) Period*, Arya Bhushan Press, Poona.

Vedavrata Vaishmapayana (1984). *Yajurveda – Ondu Ahyayana*, (Kannada), Samaaja Pustakalaya, Dharwad.

केदारदत्त जोशी (2005): श्रीमद्दैवज्ञवराहमिहिराचार्येण विरचितं बृहज्जातकम्, मोतीलाल बनारसीदास, दिल्ली ।

ENDNOTES

1. The word Veda can be derived from five different verbal roots. They are:
   - विद्यते = अस्ति इति वेदः (विद सत्तायाम्), vidyate = asti iti vedaḥ (vida sattāyām), that which exists for ever.
   - विदन्ति = जानन्ति धर्मादीन् अर्थान् अनेन इति वेदः (विद ज्ञाने), vidanti = jānanti dharmādīn arthān anena iti vedaḥ (vida jñāne), that through which one knows things such as dharma.
   - विन्दते = विचारयन्ति एनम् इति वेदः (विद विचारणे), vindate = vicārayanti enam iti vedah, that which people think about.
   - विन्दति = प्राप्नोति अनेन इति वेदः (विद्लृ लाभे), vindati = prāpnoti anena iti vedaḥ (vidḷ lābhe), that through which we obtain dharma etc.
   - वेदयति = ज्ञापयति इति वेदः (वेद चेतनाख्याननिवासेषु), vedayati = jñāpayati iti vedaḥ (veda cetanākhyānanivāseṣu), that which makes things like dharma known to us.
2. तस्माद्यज्ञात्सर्वहुत ऋचः सामानि जज्ञिरे । छन्दांसि जज्ञिरे तस्माद्यजुस्तस्मादजायत ।
   tasmādyajñātsarvahuta ṛcaḥ sāmāni jajñire | chandāṃsi jajñire tasmādyajustasmādajāyata | (Ṛgveda 10.90.9)
   http://www.sanskritweb.net/rigveda/rv10.pdf Last accessed on Oct. 1, 2021.
3. The explanation for the term "yajus" is as follows: यजन्ति येन (मनुष्या ईश्वरं धार्मिकान् विदुषश्च) पूजयन्ति शिल्प-विज्ञान-सङ्गतिकरणं च कुर्वन्ति तत् यजुः । yajanti yena (manuṣyā īśvaraṃ dhārmikān viduṣaśca) pūjayanti śilpa-vijñāna-saṅgatikaraṇaṃ ca kurvanti tat yajuḥ |
4. Pandit Shripad Damodar Satvekar (1958), "Sāmaveda Saṁhitā, Pardi, Gujarat.
5. Extract from Atharvaveda 6.64 on oneness and mutual understanding:

   सं जानीध्वं सं पृच्यध्वं सं वो मनांसि जानताम्।
   देवा भागं यथा पूर्वे संजानाना उपासते ॥१॥
   समानो मन्त्रः समितिः समानी समानं व्रतं सह चित्तमेषाम्।
   समानेन वो हविषा जुहोमि समानं चेतो अभिसंविशध्वम्॥२॥
   समानी व आकूतिः समाना हृदयानि वः ।
   समानमस्तु वो मनो यथा वः सुसहासति ॥३॥

   saṃ jānīdhvaṃ saṃ pṛcyadhvaṃ saṃ vo manāṃsi jānatām|
   devā bhāgaṃ yathā pūrve saṃjānānā upāsate ||1||
   samāno mantraḥ samitiḥ samānī samānaṃ vrataṃ saha cittameṣām|
   samānenа vo haviṣā juhomi samānaṃ ceto abhisaṃviśadhvam||2||
   samānī va ākūtiḥ samānā hṛdayāni vaḥ |
   samānamastu vo mano yathā vaḥ susahāsati ||3||

   For details see, पद्मभूषण डा श्रीपाद दामोदर सातबलेकर **(1985).** "अथर्ववेद का सुबोध भाष्यः द्वितीय भाग्" स्वाध्याय मण्डल
6. **Swami Harshananda (1992).** "A Bird's Eye View of Vedas", Ramakrishna Math, Bengaluru.
7. जानुदघ्नीमुत्तरवेदीं खात्वा, अपां पूरयित्वा गुल्फदघ्नम्, पुष्करपर्णैः पुष्करदण्डैः पुष्करैश्च संस्तीर्य
   jānudaghnīmuttaravedīṃ khātvā, apāṃ pūrayitvā gulphadaghnam, puṣkaraparṇaiḥ puṣkaradaṇḍaiḥ puṣkaraiśca saṃstīrya (taittirīyāraṇyakam 1/22/78,79,80); **A. Mahadeva Shastri**, *The Taittiriyaranyaka*, Vol. I, Government Branch Press, Mysuru, p. 132. प्राचीनवंशं करोति prācīnavaṃśaṃ karoti (taittirīya saṃhitā 6/1/1) **A. Mahadeva Shastri**, *The Taittiriya Samhita*, Vol. X, Government Branch Press, Mysuru, 1990, p. 1.
8. (स्वर्गं लोकं गमयति) य एवं विद्वान् अश्नाति पिबति चाऽऽशयति च पाययति च (शांखायनारण्यकम् 10.8)
   (svargaṃ lokaṃ gamayati) ya evaṃ vidvān aśnāti pibati cā"śayati ca pāyayati ca (śāṃkhāyanāraṇyakam 10.8)
   Pathak, S. (1922). *Shankhayanaranyakam*, Anandashram Press, p. 39.
9. For more details on this, see **Swami Chinmayananda (2014).** "Taittirīya Upaniṣad", Central Chinmaya Mission Trust, pp. 107–133.

10. शीक्षां व्याख्यास्यामः । वर्णः स्वरः । मात्रा बलम् । साम सन्तानः । इत्युक्तः शीक्षाध्यायः । (तैत्तिरीयोपनिषद् 1.1) śīkṣāṃ vyākhyāsyāmaḥ | varṇaḥ svaraḥ | mātrā balam | sāma santānaḥ | ityuktaḥ śīkṣādhyāyaḥ | (taittirīyopaniṣad 1.1)
For more details on this, see **Swami Chinmayananda (2014).** "Taittirīya Upaniṣad", Central Chinmaya Mission Trust.

11. आकाशवायुप्रभवः शरीरात् समुच्चरन् वक्त्रमुपैति नादः । स्थानान्तरेषु प्रविभज्यमानो वर्णत्वमागच्छति यः स शब्दः ॥ (आपिशलशिक्षा 1) ākāśavāyuprabhavaḥ śarīrāt samuccaran vaktramupaiti nādaḥ | sthānāntareṣu pravibhajyamāno varṇatvamāgacchati yaḥ sa śabdaḥ || (āpiśalaśikṣā 1) **Pandit Yudhishthir Mimamsak**, Śikṣā sūtrāṇi, Ramalal Kapur Trust, 1983, p. 1.

12. वाग्वै पराच्यव्याकृताऽवदत् ते देवा इद्न्द्रमब्रुवन्निमां नो वाचं व्याकुर्विति .... तामिन्द्रो मध्यतोऽवक्रम्य व्याकरोत् तस्मादियं व्याकृता वागुद्यते । (तैत्तिरीयसंहिता 6.4.7) vāgvai parācyavyākṛtā'vadat te devā idndramabruvannimāṃ no vācaṃ vyākurviti .... tāmindro madhyato'vakramya vyākarot tasmādiyaṃ vyākṛtā vāgudyate | (taittirīyasaṃhitā 6.4.7) **Mahadeva Shastri, A. (1897),** *The Taittirīya Saṃhitā of the Kṛshṇa Yajurveda*, Vol. X, Government Branch Press, Mysuru, 1897, p. 433.

13. Sāyaṇa, in the introduction to his commentary on the Ṛgveda (ऋग्वेदभाष्यभूमिका) says: अथ निरुक्तप्रयोजनमुच्यते । अर्थावबोधे निरपेक्षतया पदजातं यत्रोक्तं तन्निरुक्तम् atha niruktaprayojanamucyate | arthāvabodhe nirapekṣatayā padajātaṃ yatroktaṃ tanniruktam—Now the purpose of the will be *Nirukta* explained. *Nirukta* is that work which explains the origin of words with certainty for the purpose of understanding.

14. **K.V. Sarma (1985).** "Vedāṅga-Jyotiṣa of Lagadha", Indian National Science Academy, New Delhi. Ṛgveda Jyotiśa. verse 6, p. 23.

15. यथा शिखा मयूराणां नागानां मणयो यथा । तद्वद्वेदाङ्गशास्त्राणां गणितं मूर्धनि स्थितम् । (याजुषज्यौतिषम् ४) yathā śikhā mayūrāṇāṃ nāgānāṃ maṇayo yathā | tadvadvedāṅgaśāstrāṇāṃ gaṇitaṃ mūrdhani sthitam | (yājuṣajyautiṣam 4) **Sudhakara Dvivedin, M. (1908).** *Yājuṣa Jyautiṣam*, Medical Hall Press, Banares, p. 4.

16. स्कन्धत्रयात्मकं ज्योतिश्शास्त्रमेतत् षडङ्गवत् । गणितं संहिता होरा चेति स्कन्धत्रयं मतम् ॥ skandhatrayātmakaṃ jyotiśśāstrametat ṣaḍaṅgavat | gaṇitaṃ saṃhitā horā ceti skandhatrayaṃ matam || Ch 1, verse 5, p. 2. For details see, पुन्नशेरी नम्पि नीलकण्ठशर्मा, "प्रश्नमार्गः", श्रीगीर्वाणवाणीपुस्तकशाला, पाल्काट् । For English version you may refer to — https://archive.org/details/PrasnaMargaBVR/ Last accessed on Oct. 1, 2021.

17. त्रुट्यादिप्रलयान्तकालकलना मानप्रभेदः क्रमाच्चाराश्च द्युसदां द्विधा च गणितं प्रश्नास्तथा सोत्तराः ।
भूधिष्ण्यग्रहसंस्थितेश्च कथनं यन्त्रादि यत्रोच्यते सिद्धान्तः स उदाहृतोत्र गणितस्कन्धप्रबन्धे बुधैः ॥
truṭyādipralayāntakālakalanā mānaprabhedaḥ kramāccārāśca dyusadāṃ dvidhā ca gaṇitaṃ praśnāstathā sottarāḥ |
bhūdhiṣṇyagrahasaṃsthiteśca kathanaṃ yantrādi yatrocyate siddhāntaḥ sa udāhṛtotra gaṇitaskandhaprabandhe budhaiḥ ||
Arkasomayaji, D. (1980). "Siddhānta Śiromaṇi of Bhaskarācārya, Rashtriya Sanskrit Santhan, New Delhi, p. 2.

18. The first mantra of Ṛgveda: अग्निमीळे पुरोहितं यज्ञस्य देवमृत्विजम् | होतारं रत्नधातमम् || agnimīḻe purohitaṃ yajñasya devamṛtvijam | hotāraṃ ratnadhātamam || Ṛg Veda 1.001.01. http://www.sanskritweb.net/rigveda/rv01-001.pdf Last accessed on Oct. 1, 2021.

19. **Ramakrishna, G. (1965).** "Origin and Growth of the Concept of *Ṛta* in Vedic Literature". Doctoral Dissertation: University of Mysuru.

20. भीषाऽस्मा द्वातः पवते । भीषोदेति सूर्यः । भीषाऽस्मादग्निश्चेन्द्रश्च । मृत्युर्धावति पञ्चम इति । bhīṣā'smād vātaḥ pavate | bhīṣodeti sūryaḥ | bhīṣā'smādagniścendraśca | mṛtyurdhāvati pañcama iti | Taittirīya Upaniṣad, 2.7. For more details on this, see **Swami Chinmayananda (2014).** "Taittirīya Upaniṣad", Central Chinmaya Mission Trust.

CHAPTER

3 Philosophical Systems

LEARNING OUTCOMES

After finishing this chapter, you will be able to:

- Understand the broad classification of Indian philosophical systems
- Understand salient features of the philosophical systems belonging to the Vedic school
- Understand salient features of the philosophical systems belonging to the non-Vedic school

This is an idol of Pārśvanātha, the 23rd Tīrthaṅkara of Jainism found in Madhya Pradesh. According to the Jain sources Pārśvanātha lived during 872 BCE–772 BCE. He was born to King Aśvasena and Queen Vāmadevī of Vārānasi and belonged to the Ikṣvāku dynasty. He lived as a prince until the age of 30 after which he renounced the world to become an ascetic. According to the Śvētāmbara texts, Pārśvanātha's four restraints were non-violence (ahimsā), non-possession (aparigraha), non-stealing (asteya) and truth (satya).

IKS IN ACTION 3.1

Father–Son Conversation in Taittirīya-upaniṣad

Upaniṣads mainly confine the discussions to questions that are existential in nature and constantly coaxes an individual to deeply introspect about his 'true identity'. In this process, they provide several alternative means to inquire into various questions that are philosophical in nature. Origin of the Universe, the human beings and other living organisms, goals of life, the variant and invariant aspects of the reality, the planes of existence of an individual, the notion of Brahman and means of knowing the Brahman are some of the themes discussed.

Due to the peculiar nature of the subject matter, Upaniṣads employ several methods of imparting knowledge. Teaching in the Upaniṣads is often done in a conversational mode between the teacher and the student. In this format, the student is constantly persuaded to develop deep understanding through self-inquiry, rather than giving a ready answer to the question that he is facing. The conversation between the father–son duo in Taittirīya-upaniṣad is an example.

Bhṛgu desirous of knowing Brahman approaches his father Varuṇa and asks him to share with him the knowledge of Brahman. The father curtly replies, "that from which all the beings are born, that which supports their life, that into which they enter and dissolve in the end is Brahman". Further, he tells his son that it can only be known through deep introspection with a single pointed focus (tapas).

Bhṛgu starts the tapas and realises that beings are born out of food (here food does not have the literal meaning, it points to the larger system of entities manifesting in this earth, sustaining life forms), they live with the support of food, and when they die they become one with the food, therefore food must be Brahman.

However, after some more introspection, he was unsatisfied and approached his father again. The father merely repeated, "it can only be known through deep introspection with a single-pointed focus (tapas)." Bhṛgu was therefore sent back to do more tapas to know Brahman.

This process continued and Bhṛgu developed a deep understanding of the layers of the consciousness such as vital force (prāṇa), mind (manas), and knowledge (vijñāna) as Brahman respectively, only to refute it later by self-reflection. He finally realised that bliss (ānanda) is Brahman and since he did not have any more confusions and conflicts in his mind he concluded that bliss is Brahman.

This episode highlights how the Vedic seers continuously strived to know the truth with repeated questioning of their understanding and experiences. The episode also brings out the fact that philosophical questions such as understanding 'the reality' is to be deeply contemplated and experienced by the self through whatever effort it requires, and a knowledgeable person can at best be only a guide to him.

This episode brings another interesting aspect of acquiring knowledge. An elderly, wise and knowledgeable individual has an opinion about the deep philosophical issue. However, by merely accepting it one does not get the knowledge. Rather it needs efforts on the part of the seeker to self-reflect to gain the knowledge.

This importance given to one's own experience of truth has been the hallmark of Indian philosophical systems.

Let us consider a simple question, "Why should one not commit theft?" A religious preacher would point to the injunctions that prohibit such actions. On the other hand, a modern-day rationalistic professional would say, "because it is against the law of the land and would introduce costs and complexities in life". The important point is why should the God or the sovereign issue commands? If there is something fundamental to an individual that can resolve such questions that one faces in life, then it provides a sound footing to handle dilemmas and confusions in life. At a fundamental level, the goal of every individual is to be happy and progress or evolve in life. Nobody is known to have been longing for 'sadness and grief' in their life. Despite this simple truth, the notion of happiness is not yet well understood despite being the most fundamental instinct of living beings.

All the endeavours of living beings are towards avoiding what is unpleasant and attaining what is pleasant. In the search for answers to several questions in life, we inquire into certain fundamental questions about life; Who am 'I' and what is my source and destination? What are the nature of the Universe and that of God? Are there any universal laws for an individual to remain happy forever? A study of such questions of existential nature belongs to the domain of philosophy. Philosophical thinking flourished in different parts of the world. In this chapter, we shall see some details of the Indian philosophical systems.

In the process of finding answers to the questions that we raised, people have developed their holistic understanding of the world and shown the way to attain the final goal of life. This holistic view of the world is called Darśana. The word Darśana is derived from the Sanskrit root, *dṛś*, 'to see', suggesting that these philosophical systems provide a true worldview and a vision for life and help us resolve the issues that we face in our life. These different traditions of Darśanas or 'Schools of Philosophy' have enriched Indian thinking and had their influence on all aspects of life, including worship, rituals, art, literature, and medicine.

3.1 INDIAN PHILOSOPHICAL SYSTEMS – DEVELOPMENT AND UNIQUE FEATURES

Phrases such as 'happiness' require 'individuals' as the unit of analysis. Further, as one inquires deep into this question, the focus shifts to the Universe, which provides the context for the individual to make his inferences and choices in life. Inevitably, the notion of an all-pervading Universal force (known as the Divinity or God) also becomes another important dimension in this analysis. Therefore, the study of these aspects invariably happens in a context consisting of three aspects: an individual (Jīva), the Universe (Jagat), and the God (Īśvara).

The beginning of the philosophical inquiry in India can be traced to the Vedic literature. The Ṛgvedic seers enquired about the forces causing natural phenomena such as rain, day and night, growth of life and recognised these forces as Dvatās (Gods). Hence, we see several mantras in praise of these Devatās, describing their nature and deeds. The Ṛgvedic hymns delved into a rigorous inquiry trying to realise the root force behind the functioning of these Devatās. They realised the concept of 'Puruṣa', all-encompassing supreme being, and proclaimed, all this is nothing but the Puruṣa' (पुरुष एवेदं सर्वं – Puruṣa evedaṃ sarvam). According to them, knowing the Puruṣa is the path to overcome death, and there is no other way to go.

- The philosophical systems provide a true worldview and a vision for life and help us resolve the issues that we face in our life.
- The teachings of the Upaniṣads suggest that attainment of the knowledge of Brahman is the highest goal of human life.

We also see a serious inquiry about the origin of the Universe in the famous Nāsadīya-sūkta of Ṛgveda. The sūkta is inquisitive about what was there in the beginning, before this whole world was created? It begins with the speculation that there was neither existence (Sat) nor non-existence (Asat) and then proceeds to ask several questions; What was it covered with? Was it covered with water? The darkness? Who knows and who can articulate where from this creation came into existence? The one who created this, does he know this or does he not? Upaniṣads are full of such philosophical inquiries and thoughts. What was described as 'Puruṣa' in the Ṛgveda is referred to as 'Brahman' in the Upaniṣads. Brahman and Ātman are the concepts that are discussed widely in the Upaniṣads. The teachings of the Upaniṣads suggest that attainment of the knowledge of Brahman is the highest goal of human life. Brahman is said to be beyond words, cannot be expressed completely by anyone, hence Upaniṣads have adopted various ways to describe Brahman. It is said to be the ultimate reality to be known, for after having known it nothing else remains to be known. The Upaniṣads further clarify that this truth cannot be known using the sensory organs as we would in the case of any worldly entity, it called for preparing oneself to 'experience' the truth from within. The ultimate goal of the Upaniṣads is to prepare an individual for this eventual 'personal experience'. There are multiple passages and anecdotes to drive the goal of one's life, the path of liberation, and the means of achieving it.

The philosophical inquiry continued in India further giving rise to several schools, each one developing its own understanding of the world. Each of these schools presents its view with rigorous intellectual exercise and uncompromised importance on the self-experience. The ideas were codified in the basic texts of their Darśana, in the form of sūtras in most cases, and commentaries and sub-commentaries through the guru-śiṣya tradition spread the thoughts further. Before we study the specific characteristics of the schools and their differences it is important to know certain broad parameters and unique aspects of the Indian philosophical thought and their relevance to various schools of thought.

- Unlike the Western counterpart, the Indian philosophical thought is closely intertwined with religious thought.
- The ultimate goal of the human life is clearly spelt out and the path for attaining the same is also articulated in all the darśanas.

- Unlike the Western counterpart, the Indian philosophical thought is closely intertwined with religious thought. The philosophical systems provide a broad basis for addressing larger existential issues of individuals and the religious schools draw from these to suitably configuring socio-cultural practices, norms of behavior, ethical standards, and values that shape one's life. The religious dimension provides the operating principles for the mundane life based on these specific configurations.
- Two generic classes of philosophical systems could be thought about; Vedic schools of philosophy (which has six schools of thought as we will see shortly), and Non-vedic schools (Jaina philosophy, Buddhist philosophy, and the Cārvāka philosophy).
- Despite the classification based on the religious dimension, all the religious-oriented philosophical systems agree on certain common parameters. This includes the notion of accumulation of fruits of action (Karma), birth–death cycle (Saṃsāra), and the notion of free will. On account of this, they all talk about the common goal of liberation (Mukti)[1]. These aspects distinguish the Indian philosophical systems from the Western.
- The ultimate goal of the human life is clearly spelt out and the path for attaining the same is also articulated in all the darśanas. However, the darśanas have divergent opinions on the specifics of these.
- Since darśana is all about knowledge, the term knowledge (Jñāna) and other related terminologies are well defined. Furthermore, all the Indian schools of philosophy also employ epistemological tools (Pramāṇa) for the establishment of valid knowledge. However, each school differs from the other in the set of epistemological tools considered for analysis.

- Despite divergent views on some of the philosophical concepts and foundational premises between these schools, there is a healthy culture of respectful and peaceful coexistence of these schools of thought. There was no effort to demean, dismiss or downgrade one school by the other using any emotional, dogmatic, irrational, or unscientific methods. Instead, there was a healthy tradition of the followers of the schools to engage in dialogues and debates. These demanded a highly advanced intellectual exercise, be it writing a book refuting the argument of the opposing point of view or engaging in the dialogue following strict rules of debate.
- One of the Vedic schools, Nyāya provided a de facto framework for all other schools to engage in such intellectual debates. In fact, the ontologies of Sāṃkhya and Vaiśeṣika schools have been adopted by most other schools with little modifications to suit their theories.
- There are historical accounts of several such debates that took place throughout the country for several centuries, which continue even today. Thus, several schools of philosophy have flourished in India, interacted extensively with each other, and have organically co-existed for millennia.

Figure 3.1 provides a schematic of the classification of the Indian philosophical systems into Vedic and non-Vedic systems.

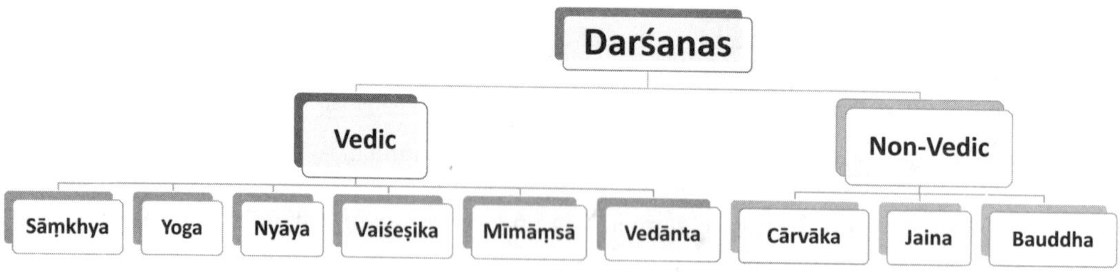

FIGURE 3.1 Indian Philosophical Systems (Darśanas)

3.2 VEDIC SCHOOLS OF PHILOSOPHY

Figure 3.2 summarises some of the salient features of vedic schools. All the schools have a common goal, i.e. to answer questions such as, "Who am I?" and "What is the process for final liberation? To know oneself and escape from the cycle of birth–death and get liberated, correct knowledge needs to be obtained. The schools differ in presenting how to obtain the right knowledge and use it as the means for the ultimate liberation of the 'self'. The context for the philosophical discussion is the three inter-related concepts of God - Universe - Individual". All schools have employed several constructs involving these three entities in discussing the path for self-evolution and liberation. The six schools differ in the approach taken to reach the goal. In the case of Sāṃkhya and Yoga systems, there is the centrality of the role of the 'matter' in this journey. A good understanding of the evolution of nature leading to the context paves the way for liberation. On the other hand, Nyāya and Vaiśeṣika systems have prominently focused on the importance of obtaining the 'right knowledge' in the journey of liberation. Therefore, these systems elaborately focused on getting the right knowledge of oneself and the other entities. On the other hand, Mīmāṃsā and Vedānta proposed that the Vedic repository provides guidance for an individual in his journey of liberation. While Mīmāṃsā stressed on the importance of the ritualistic part of the Vedic corpus (Karma-kāṇḍa), Vedānta emphasised on the knowledge leading to self-experience (Jñāna-kāṇḍa).

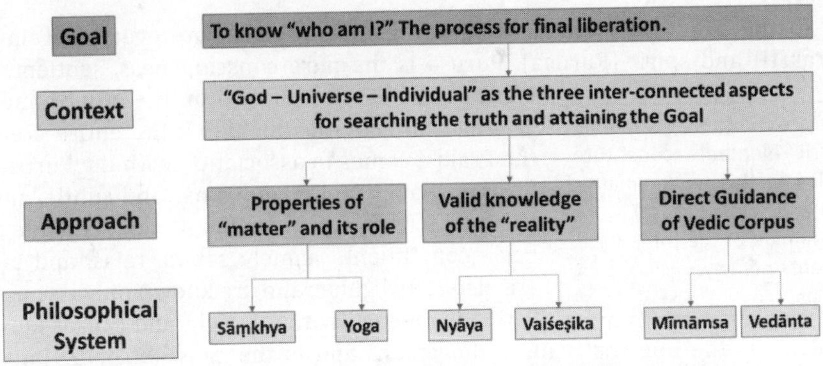

FIGURE 3.2 The Six Vedic Schools – Some Salient Aspects

3.3 SĀṂKHYA AND YOGA SCHOOLS OF PHILOSOPHY

The Sāṃkhya–Yoga philosophical system begins with the basic premise that Prakṛti, the source of the physical Universe and the 'self' are sub-ordinate to a larger force called Puruṣa and liberation of the 'self' involves getting to know the right knowledge of the Puruṣa and Prakṛti and their relative role. According to these schools of thought, Prakṛti is the first cause of the Universe (of everything except the spirit) and entirely accounts for whatever is physical, both matter and force. Prakṛti is conceived as constituted of the tri-guṇas. The evolution of Prakṛti causes the Universe, bringing the multi-various forms and entities. The preponderance of the three Guṇas and their role in establishing the link between Puruṣa and Prakṛti is another common ground for both the philosophical systems. Finally, both the schools agree that the process of final liberation involves the realisation of the true nature of Prakṛti and Puruṣa.

- Vedic schools acknowledge the authority of the Vedic text whereas Non-Vedic schools dont.
- The context for philosophical discussion is three inter-related concepts of God – Universe – Individual.

Sāṃkhya school does not acknowledge the existence of an ultimate God (Īśvara). On the other hand, Yoga acknowledges the existence of a supreme being. Yoga has an emphasis on a more structured, practical methodology for cessation of all activities of the mind. To facilitate this process, the Yoga system of philosophy provides a practical step-by-step approach for this journey. On the other hand, Sāṃkhya school emphasises more of contemplation and analysis leading to experiential knowledge.

3.3.1 Sāṃkhya-darśana

Although sage Kapila is supposed to be the author of the Sāṃkhya system, there is no available evidence or material to substantiate this. The earliest authoritative material available on Sāṃkhya is the Sāṃkhya-kārikā by Īśvarakṛṣṇa. This is a work in seventy verses and has a lucid exposition of the Sāṃkhya system. Sāṃkhya argues that the root cause of all pains and sufferings is the lack of the correct knowledge (Saṃkhyā). By a proper understanding of the ontology of Sāṃkhya-darśana, the causes of pain and the way to end it can be explained.

Puruṣa and Prakṛti

According to the Sāṃkhya system, two basic elements constitute everything in this world, matter (Prakṛti) and spirit (Puruṣa). Puruṣa is the pure consciousness, sentient, changeless, eternal, and passive. Prakṛti on the other hand is the root cause of all activities including the entire creation. When the Prakṛti comes in association with the Puruṣa it assumes diverse shapes and forms, gross and subtle, and manifests as body, senses, and the mind. Prakṛti is made of the three basic constituents namely sattva, rajas, and tamas. These are also called guṇas and are known only through inference. Sattva is the faculty that is light and causes knowledge and pleasure. Rajas is the one that causes movement and is the cause of pain. Tamas is heavy, causes ignorance and indifference. Before the manifestation of the Prakṛti its constituents, sattva, rajas, and tamas are in equilibrium.

> ♦ According to the Sāṃkhya system, two basic elements constitute everything in this world, matter (Prakṛti) and spirit (Puruṣa).
> ♦ Prakṛti is made of the three basic constituents namely sattva, rajas, and tamas.

The evolution of the Prakṛti results in creating the following elements:

- 'Mahat', which is also called 'Buddhi'
- Ego or Self-consciousness (Ahaṅkāra), which introduces the sense of 'I' and 'mine'
- Mind (Manas), the master of the organs and the conduit between the internal and external instruments
- Five sense organs (Jñānendriyas): ears, skin, eyes, tongue, and nose
- Five organs of action (Karmendriyas): the mouth (speech), hands, feet, sex organs, and anus
- Five generic classes attributable to objects perceived through sense organs (Tanmātras): sound, touch, form or colour, taste, and smell
- Five gross elements (Bhūtas): ether, air, fire, water, and earth.

Figure 3.3 illustrates the evolution of Prakṛti described above. When the Puruṣa is conditioned by the twenty-three elements it becomes embodied and perpetually transmigrates from one body to another. This is the cause of all mundane existence, and this continues so long as it does not discriminate the difference between Puruṣa and Prakṛti[2]. Once the conscious self comes out of this ignorance, through the actual knowledge, he realises that he is separate from the Prakṛti and its manifestations, he does not feel pleasure, etc. anymore. This is called liberation (kaivalya, i.e. mokṣa) in the Sāṃkhya doctrine.

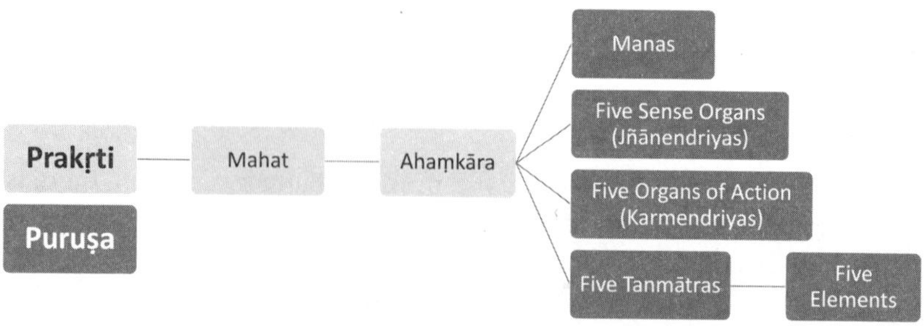

FIGURE 3.3 Prakṛti and its Evolution into Matter

3.3.2 Yoga-darśana

Yoga as a school of philosophy is said to have been founded by Patañjali through his yoga-sūtras. Some Indologists assign a date of 5th century CE for this work, while many others associate this work with Patañjali, who wrote the Mahābhāṣya for Pāṇini's grammar. In that case, it will be dated to the 2nd century BCE. Yoga serves as a methodology to the realisation of the difference of Prakṛti and Puruṣa. It elaborately establishes the necessary practices an individual needs to go through to have the realisation of this separation. The unique thing that establishes Yoga as distinct among the other darśanas is its emphasis on understanding the mind, its various states, its cognitive activities, and methods to control it. The other schools have a difference of opinion on the matters of epistemology, and the concept of mokṣa with Yoga. However, they accept methods prescribed in yoga to gain control over the mind.

- Twenty-three elements act as the seed, out of which the body (consisting of the internal (subtle) instruments and the external (gross) organs) is produced.
- According to Yoga philosophy, to attain liberation, an individual must focus on the physical, psychological, and moral states of his being.

The Yoga philosophy rests on the basic premise that if a person wants to understand his true nature, and experience bliss eventually, he must focus on the physical, psychological, and moral states of his being and make simultaneous progress on all the three. To achieve this, the basic prescription is to develop the capacity for single-pointed concentration of the mind. Therefore, Yoga-sūtras begin with the definition that 'Yoga is the cessation of mental modifications'[3]. Unless a person arrives at this stage, it will not be possible for him to understand the notion of 'existence' and the secrets of nature may not get revealed. Therefore, the operational part of Yoga provides a practical set of 'actionable' steps that an aspirant can go through sequentially to reach this state. Further Yoga system observes that a journey of constant practice with dispassion makes a person perfect[4] and he will feel within himself the universal truth with no sense of separateness. While Sāṃkhya prescribes a method of analysis and contemplation, yoga argues for mind control through sustained practices as prescribed. Yoga system provides an eight-step process to gradually attain complete cessation of the activities of mind. Figure 3.4 presents the eight steps in a pictorial fashion. The details of the eight steps are as follows:

- **Yama:** The ultimate journey to complete cessation of the activities of the mind starts with the first step which is forbearance or control over mind, body, and speech. Five activities are prescribed for practice in this stage; abstaining from harming (Ahiṃsā), speaking the truth (Satya), not stealing others' belongings (Asteya), keeping away from lust (Brahmacarya), and resisting from accumulating wealth (Aparigraha). According to Patañjali these are to be followed irrespective of time, place, and status by a sādhaka and this is called mahāvrata (greatest of all austerities).
- **Niyama:** The five kinds of forbearances, specified in the previous step relate to abstaining from negative injunctions. On the other hand, in the second stage, five kinds of observances, which are positive commands are prescribed. The five niyamas include cleanliness of body and mind (Śauca), being happy with what one possesses (Santoṣa), tolerating heat, cold and other physical difficulties and purifying the senses and the body (Tapas), the study of Vedas and other scriptures (Svādhyāya) and the meditation of Īśvara (Īśvara-praṇidhāna). These two stages are to be practiced by an individual (sādhaka) at all times.

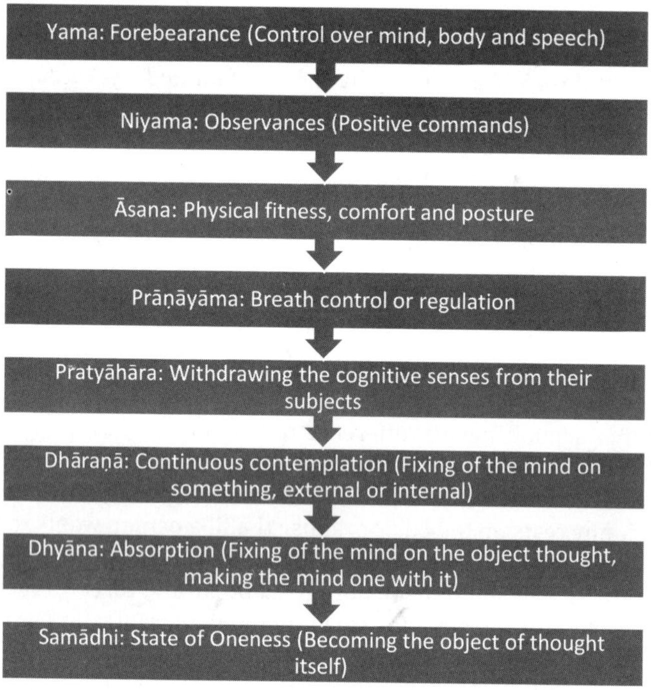

FIGURE 3.4 Aṣṭāṅga-yoga – Eight Step Process

- **Āsana:** A meditative posture where one can sit comfortably for a long time. Various modes of keeping the body in position at the time of performing Yoga are prescribed in this stage. As per Yoga-sūtra, the sitting posture must be firm and comfortable for the sādhaka[5].

  Henceforth, all the aspects are of the meditation process.

- **Prāṇāyāma:** This stage focuses on controlling the breath. Breath is directly connected to the mind, hence having it in control is the key to controlling the activities of the mind. Patañjali prescribes four types of prāṇāyāma practices meant for controlling the mind.
- **Pratyāhāra**: The next stage of practice is withdrawing the cognitive senses from their objects, bringing them to a state as if they were not in contact with their objects.

  The first five stages are external in terms of effort (bahiraṅga-prayoga) and help the practitioner purify the thoughts and regulate or moderate the inner self by avoiding various distractions. Once this stage is reached, it will be possible to focus on internal efforts (antaraṅga-prayoga). The last three stages of the 8-step process provide details on this.

- **Dhāraṇā:** Focusing the mind on an object, not letting it perceive other objects by constant practice is the next stage of this process.
- **Dhyāna:** Meditating continuously on an object without break is called dhyāna. It is achieved by fixing the mind on something, external (such as a picture, OM sign, a deity)

or internal (using the tip of the nose, and the space between the two eyebrows as a reference to focus inside or visualising OM in the mind).

- **Samādhi:** The final step in this process is being completely absorbed into the object of meditation, known as samādhi. Once a person reaches here the activities of his mind completely cease, leading to the realisation that he, the conscious 'self' is separate from Prakṛti and thus becomes liberated, free from all pains.

Cessation of all mental activities of the mind, the final goal of Yoga has a very useful bye product. On account of the journey in the path of mind control, Yoga can help an individual address the vexing problem of stress-induced lifestyle leading to health and wellness challenges that modern society is facing. We have briefly touched upon this aspect in the chapter on health, wellness, and psychology. The current popularity of Yoga globally stems from this aspect of Yoga practices. However, it must be understood that Yoga is a way of life, with a grand purpose of realising the true nature of oneself and liberate one from the limitations of the mundane world arising out of the duality of pain and pleasure. We should never lose sight of this ultimate objective of Yoga as a darśana.

3.4 NYĀYA AND VAIŚEṢIKA SCHOOLS OF PHILOSOPHY

The Nyāya–Vaiśeṣika school begins the analysis with the world that an individual experiences. By experience, it means all varieties of valid knowledge, whether perceptual or non-perceptual. Therefore, it starts with the assumption that whatever is obtained by uncontradicted experience must necessarily be real. The Nyāya–Vaiśeṣika school proceeds with an analysis of the experience to understand reality, also known as *knowable*. Both Nyāya and Vaiśeṣika set out in their journey of the 'knowable' and define various categories to describe the same. In this process, two possibilities emerge to conduct the study in greater detail. The first is about the 'ways of knowing the reality' and the second is about the 'objects in the reality that is knowable'. The Vaiśeṣika school provided a greater emphasis on the latter, mainly studying the reality itself in its various aspects. On the other hand, the Nyāya school proposed a methodology for an investigation into the problem of knowledge in its relation to reality. Nyāya and Vaiśeṣika schools place a greater emphasis on obtaining the 'right' knowledge for liberation.

3.4.1 Nyāya-darśana

The original theory of Nyāya school is found in the Nyāya-sūtra, a set of aphorisms developed by Gautama. The aphorisms are organised into five chapters, each having two sections. It was followed by many other seminal works. By the end of the 11th century CE, Gaṅgeśopādhyāya took different positions on some of the theories of the school and established a new school called Navya-nyāya, meaning the new Nyāya School. The navya-nyāya school has contributed extensively to discussing details of inference and verbal cognition. These discussions and the set of technical terminologies developed by the navya-nyāya school have influenced the other schools so much that in the later times all other schools laid significant emphasis on these topics and adopted the language to discuss the issues in their respective schools as well.

- Yoga provides a structured and practical eight-step process to gradually reach a stage of complete cessation of the activities of Citta.
- Yoga can help an individual address the vexing problem of stress-induced lifestyle leading to health and wellness challenges that modern society is facing.

The Nyāya philosophy starts with the proposition that one attains liberation only when he/she acquires the knowledge of the truth. The knowledge of the truth drives away miseries and an individual escapes the cycle of birth–death leading to final liberation. The Nyāya system, therefore, placed enormous emphasis on the means of obtaining 'right knowledge'. Therefore, the unique contribution of Nyāya school is its detailed inquiry of knowledge (Pramā) and valid cognition and its means (Pramāṇa). The elucidation of the correct way of thinking and arriving at the right conclusions, the art of debating, well laid out rules for a debate to arrive at the most reasonable conclusion are the main contributions of the Nyāya school. On account of this, Nyāya is commonly understood as 'argumentation'. Nyāya concepts and the art of debating to establish true knowledge became a useful tool for all philosophical systems. It has over time assumed the de facto methodology to establish valid knowledge. The details of these have been explained in Chapter 7 (see Section 7.4 for details).

The goal of a self is not to gain pleasure but to be liberated from all kinds of pains once for all. This is liberation according to the Nyāya school. Īśvara or the God is the creator of this universe. An individual self must try to gain the correct knowledge of the self, i.e., he/she is not the body, or the mind or the senses, which often people mistake 'the self' out of delusion. Then he ceases to have attachment for the fruits of his/her actions, as all actions an individual does are aimed at gaining worldly pleasure. When he/she gives up the desire for the results of his/her actions, he/she no more accumulates the effects of his actions, good or bad, which are the cause of the birth and death cycle. When an individual has finished experiencing the effects of his/her past actions there is no reason for his/her birth and he/she will be free from the birth–death cycle. Ultimately, he/she will be free from pain.

3.4.2 Vaiśeṣika-darśana

Vaiśeṣika was propounded by Kaṇāda and his work was organised into aphorisms in ten chapters, each consisting of two sections. The exact date of the work is hard to establish, however, it is believed that it is earlier than Nyāya-sūtras. A 5th century CE commentary on Vaiśeṣika-sūtras was authored by Praśastapāda. A lucid exposition of the Vaiśeṣika-sūtras is available in the commentaries Kiraṇāvalī (Udayana, 984 CE), Kandalī (Śrīdhara) and Upaskāra (Śaṃkara Miśra, 1650 CE).

- Nyāya deals with 'ways of knowing the reality' and Vaiśeṣika with 'objects in the reality that is knowable'.
- Nyāya and Vaiśeṣika schools place a greater emphasis on obtaining the 'right' knowledge for liberation.

The word Vaiśeṣika is derived from the word 'Viśeṣa', meaning difference or unique attributes in a thing. According to this school, diversity, not unity is fundamental to the Universe. Although the multitude of forms and shapes exist they can be reduced to certain types. It is by virtue of this that 'knowables' are divisible into three classes of dravya, guṇa, and karma, but also into sub-classes such as cows, redness, or moving. While there is some sense of 'alikeness' in the manner described above, it must be remembered that if there are two things that resemble each other in every aspect, there must be something distinctive since there are 'two' of them. This is the basic concept of Viśeṣa. The Vaiśeṣika school proceeds along this line to systematically present the 'knowables' that form all the 'real' entities in the universe using certain categories.

Sāmānya is a very important aspect in the scheme of classification. When we classify many things into one category, consider the similarities. For example, we designate a particular set of animals 'cow' because they bear some common features in them. That common quality, which we shall say 'cowness', is the reason for all such animals being referred to as a cow. It is an inherent property that all these animals carry by their nature. This is called sāmānya or jāti. Because of this sāmānya, we can group, categorise or generalise things.

Viśeṣa essentially becomes important as Vaiśeṣika school considers all entities to be different from each other. In this approach, the difference is sought by an examination of the constituents of an object of knowledge. Proceeding in this manner, differentiating by examining the constituents finely, the Vaiśeṣika school eventually proposes smallest particles, ultimate atomic material (Paramāṇus), which have no more constituents and are therefore not further divisible. Hence Vaiśeṣika school accepts a property called 'viśeṣa' in Paramāṇus. Using this elemental matter, the difference of the constituted bodies can be explained.

- The word Vaiśeṣika is derived from the word 'Viśeṣa', meaning difference or unique attributes in a thing.
- The Vaiśeṣika school presents the 'knowable' that form all the 'real' entities in the Universe using certain categories.

According to Kaṇāda, six sub-categories constitute existence, and knowledge of them is considered the essence of the supreme good[6]. The final liberation or salvation comes as a result of real knowledge produced by proper understanding of the six categories listed out in the sūtra. The desire to end the misery of the birth–death cycle leads one to acquire the knowledge of the categories from a master. This knowledge removes ignorance once and for all and the individual is free from love, hate, etc. As he does not accumulate further any merit or demerit on account of this clarified knowledge of the reality, in a certain finite cycle of birth–death, he would have exhausted all the exiting merits and demerits, leading to final liberation.

3.5 PŪRVA-MĪMĀṂSĀ AND VEDĀNTA SCHOOLS OF PHILOSOPHY

In Chapter 2, we discussed the details about the Vedic corpus, where we remarked that the Brāhmaṇas, the portion of the Vedic corpus has details on the ritualistic aspects. This portion is typically referred to as karma-kāṇḍa of the Vedic corpus. On the other hand, the Upaniṣads lay greater emphasis on the knowledge of the Brahman. This portion is known as Jñāna-kāṇḍa. The Pūrva-mīmāṃsā school has established its tenets based on the karma-kāṇḍa and the Vedānta school has established its tenets based on the Jñāna-kāṇḍa. The Vedānta school is also known as Uttara-mīmāṃsā on account of its reliance on the latter portion of the Vedic corpus.

Both the schools share common beliefs in several of the philosophical principles. This includes the notion of ātman, the existence of karma, rebirth, and long and seemingly endless cycles of birth–death. Therefore, the common goal of these two systems is to liberate the ātman from the clutches of birth–death. However, the major difference lies in the path to liberation. In the case of the Pūrva-mīmāṃsa school, it is believed that ultimate liberation is obtained only by engaging in various rituals that purify the karma and extinguishing them eventually. In the case of the Vedānta school, it is argued that total detachment from worldly activities is the only way to exhaust all the karma. With a purified mind one will then experience the ultimate knowledge 'within' to liberate.

3.5.1 Pūrva-mīmāṃsā-darśana

The word 'Mīmāṃsā' conveys different meanings: reflection, consideration, profound thought, investigation, examination, and discussion. In the context of the Pūrva-mīmāṃsā school of philosophy, Mīmāṃsā means 'reflection' or 'critical investigation' and is primarily based on a tradition of deep contemplation on the meanings of Vedic texts which it relies on as the authority for its principles. Between the Saṃhitā and Brāhmaṇas, the Pūrva-mīmāṃsā school places greater emphasis on the Brāhmaṇas and draws from it substantially as it is the part of Vedic corpus that has elaborate procedural details on the Vedic rituals.

- The Pūrva-Mīmāṃsā school has established its tenets based on the karma-kāṇḍa and the Vedānta school has established its tenets based on the Jñāna-kāṇḍa.
- The Pūrva-Mīmāṃsā text provides rules for the interpretation of the Vedas and provides philosophical justifications for the observance of Vedic rituals.

The Pūrva-mīmāṃsā school was established by Jaimini, who is said to be the student of Veda Vyāsa. Jaimini presented his aphorisms numbering over 2500 in twelve chapters, which are further divided into sixty sections. The text provides rules for the interpretation of the Vedas and also provides philosophical justifications for the observance of Vedic rituals, by offering meaning and significance of Vedic rituals to attain Mokṣa. Over the centuries many commentaries were written on this text, most important being the Śābara-bhāṣya written by Śabara-svāmin, the only extant commentary on all the 12 chapters of the Mīmāṃsā-sūtras. The major commentaries were written on the text as well as the *Śābara-bhāṣya* by Kumārila Bhaṭṭa and Prabhākara Miśra. These texts have collectively put together robust rules of language analysis which enables one to not only examine injunctive propositions in any scripture but also examine the alternate related or reverse propositions for better understanding.

The main aim of the school is to ascertain the meanings of the Saṃhitā and Brāhmaṇa portions of Veda which lay importance on the karma, performing rituals, and thereby attaining dharma, a quality that is acquired by an ātman which prompts him to the respective results such as svarga (heaven). Eventually, with purified actions, such a person attains liberation (mokṣa). If one does not resort to dharmic actions, then he is likely to continue in the cycle of birth and death (Figure 3.5). Hence it is also called 'dharma-mīmāṃsā'. Along with ascertaining the intended meaning of the Vedic texts in this context, the school provides a philosophical explanation as to how a ritual performed results in the desired outcome. Thus, it holds the status of a darśana or school of philosophy.

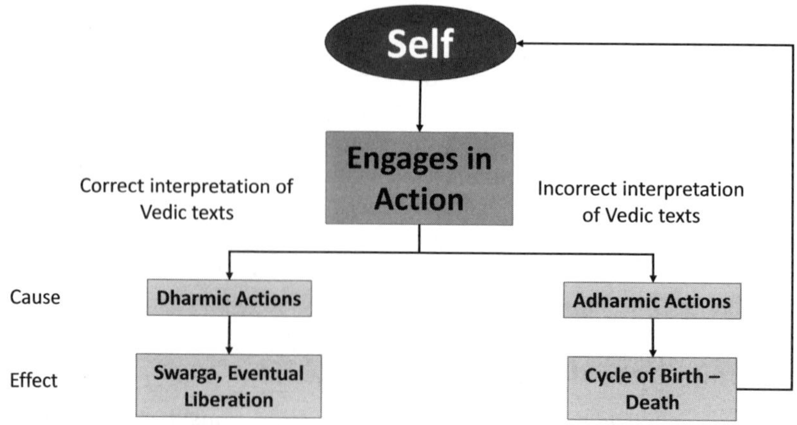

FIGURE 3.5 Pūrva-Mīmāṃsā – The Path to Liberation

The ultimate goal of human beings is to seek joy and happiness in this life and the next. The Pūrva-mīmāṃsā school of philosophy argued that this is possible only when one engages in actions that are considered as dharmic. Such actions are prescribed by the Vedic texts, and it is important to properly interpret and understand the Vedic sentences, words, and meaning. Therefore, it is not surprising that the Pūrva-mīmāṃsā school was centrally concerned with the philosophy of language and correct interpretation. On account of a special emphasis on the correct interpretation of the Vedic text concerning various injunctions and prohibitions, Pūrva-mīmāṃsā developed into a good source for hermeneutics. The laws of interpretation formulated by Jaimini and his successors are quite general and applicable to literary works outside the Vedas too. Pūrva-mīmāṃsā principles are widely utilised for arriving at a right interpretation of all old texts, particularly legal treatises, and the legal fraternity could greatly benefit from the knowledge of Pūrva-mīmāṃsā.

3.5.2 Vedānta (Uttara-mīmāṃsā-darśana)

All Vedānta schools of philosophy derive a considerable part of their material from the Upaniṣads. The Vedānta schools of philosophy rely on three major texts, known as 'Prasthāna-traya', for establishing their tenets. This includes the Brahma-sūtras, a collection of about 550 aphorisms written by Bādarāyaṇa (Vyāsa), the Bhagavadgītā, and the Upaniṣads. The proponents of the Vedānta schools have written detailed commentaries on the three major texts and through that have established the unique position that they have taken in articulating their version of Vedānta philosophy. In general, the Vedānta schools look upon Brahman as the absolute and are predominantly philosophical in their approach. In this section, we shall briefly see the salient aspects of Advaita, Viśiṣṭādvaita, and Dvaita schools of Vedānta.

- Pūrva-mīmāṃsā texts have put together robust rules of language analysis which enables one to examine injunctive propositions in any scripture.
- Pūrva-mīmāṃsā principles are widely utilised for arriving at a right interpretation of all old texts, particularly legal treatises.
- Liberation in Advaita vedānta is not reaching something new but recognising what is one's inherent nature.

Advaita-vedānta

Śaṇkara (688–720 CE) propounded a monistic philosophy, known as Advaita, with the conception of the attribute-less God (Nirguṇa-brahman) as the ultimate reality. The Upaniṣads describe Brahman as one without form, name, and attributes, known as Nirguṇa-brahman. Śaṇkara built further on Gauḍapāda's foundational work through his kārikā (set of verses) for Māṇḍukya-upaniṣad and gave more strength and structure and formalised the Advaita-vedānta. The other main aspects of his philosophy include the doctrine of Māyā, the identity of the Jīva with the Brahman and the conception of mokṣa as the merger of Jīva in Brahman. Śaṇkara's commentary on the Brahma-sūtra is remarkable for its charming style and highly logical and consistent arguments. Śaṇkara has made immense contributions to Indian philosophical thought through numerous commentaries and independent works which run to several thousand pages. He also made robust institutional arrangements in different parts of India to preserve and promote Advaita-vedānta. The conceptualisation of Nirguṇa-brahman of Śaṇkara at the outset will resemble closely the śūnya-vāda (emptiness or nothingness) of Buddhists. However, through elaborate commentaries on the three major texts, Śaṇkara has established the uniqueness of his proposition.

Advaita-vedānta postulates oneness of Jīva and Īśvara and in its scheme of things, the Īśvara and the Jīva deploy similar mechanisms to create the world. Īśvara represents the macrocosm with Māyā as the mechanism to create the physical universe. In the same manner, the Jīva uses avidyā (ignorance) as the mechanism to create its own world constituting a parallel microcosm. Viewed in this manner, the Māyā is the cosmic illusion for the Īśvara to create the world out of himself and avidyā the source of illusion for the Jīva. The rejection of the world as something illusory by the Advaita school does not advocate cessation of all worldly activities that we undertake under the garb of it being 'unreal'. Instead, it brings out different stages of knowledge an individual experiences. So long as the identity of oneself with the Brahman is not realised, the empirical world, the activities, and the knowledge about these are true. This is similar to the conditional knowledge of the dream until one wakes up. Therefore, in the Advaita school, two types of knowledge are proposed corresponding to two realms of reality that we have. One is a transactional reality of the changing world and the associated empirical knowledge that we have which helps us to engage in day-to-day chores of life. On the other hand, once the identity with the Brahman is realised, it produces the knowledge that reality is changeless and that oneness of everything is eternal. And this knowledge of oneself results in realisation of bliss. This is considered to be the 'true or ultimate' knowledge.

Śaṇkara proposed a two-stage approach to realise 'true' knowledge. By merely engaging in the world of activities, one does not obtain the 'true' knowledge. In the first stage, we need to engage with the world and perform the required activities to purify the mind. This is the karma yoga that Bhagavadgīta advocates. Once the mind is purified, the second stage is to engage in deep contemplation and self-reflection on one's real nature which will reveal the 'true' knowledge[7]. Liberation in Advaita-vedānta is not reaching something new but recognising what is one's inherent nature. The two-stage approach indeed leads an individual towards this goal.

Viśiṣṭādvaita-vedānta

Rāmānuja (1017–1137 CE) proposed the philosophy of Viśiṣṭādvaita. Viśiṣṭādvaita combines the bhakti (devotion) to a personal God, who has name, form, and shape and who saves his devotees from all miseries of the world and the impersonal God as espoused in the Upaniṣads into a single school of thought. Brahman or the ultimate reality is also referred to as Viṣṇu, etymologically meaning that which pervades everything. The attribute-less Brahman held by the Advaita school is rejected as metaphysical abstraction and Brahman is conceived, by Viśiṣṭādvaitin as God with attributes like possessing a bodily form, with infinite good qualities and glories.

The ultimate is one, according to Viśiṣṭādvaita, but is not the attributeless. Matter, Jīva and Īśvara are three entities recognised in the Viśiṣṭādvaita school. The matter and the Jīva are considered as the body of Īśvara and are sustained by Him and existing entirely for Him. In other words, the three are inseparable unity, the first two being subjected to the restraint of the third in all its forms. This is the core of the conceptualisation in Viśiṣṭādvaita. An example of a blue lotus illustrates this concept. The blueness (a quality) is quite distinct from the lotus (substance). However, blueness depends on the substance for its very being and is not considered external to the lotus. According to Rāmānuja, all things are eventually forms of Īśvara and all names are his only. Every word is a symbol of Īśvara and points to Him only.

According to Viśiṣṭādvaita school, the cycle of birth and death and the associated sorrows are due to the forgetfulness of an individual of the relation between them and Nārāyaṇa. One attains freedom by gaining knowledge of the nature of self and attaining the feet of the Lord in his abode, Vaikuṇṭha. However, in order to gain this knowledge, each Jīva has to put forth the effort to attain liberation. The nature of the effort to be invested requires a continuous and unwavering meditation with love on the Supreme Being. This is referred to as Bhakti in the Viśiṣṭādvaita school. Bhakti is generated with total observance of religious duties as prescribed in the scriptures. The concept of total surrender to the Lord (Prapatti) is also considered as the direct means to liberation. In fact, according to Viśiṣṭādvaita, both bhakti and prapatti are two sides of the same coin and hence they function as the direct means to attain the feet of the Lord.

- Viśiṣṭādvaita combines the bhakti to the personal God, and the impersonal God as espoused in the Upaniṣads into a single school of thought.
- The quintessential aspects of Dvaita Vedānta are that Viṣṇu is the supreme God, the world is real and there is a difference between God and the jīvas.

Dvaita-vedānta

Madhvācārya (1238–1317 CE) is the founder of the Dvaita-vedānta school. Madhvācārya established Udupi as the center of the Dvaita-vedānta. Like the other schools of Vedānta, Madhva derives his philosophical tenets from prasthānatraya, the purāṇas and Mahābhārata. The quintessential aspects of Dvaita-vedānta are that Viṣṇu (Hari) is the supreme God, the world is real and there is a difference between God and the jīvas. All jīvas are dependent upon Viṣṇu, and liberation consists in the enjoyment of bliss that is inherent in oneself. Finally, pure devotion is the means of attaining it.

The Dvaita school clearly admits two independent and mutually irreducible substances that make up the Universe: the Jīva and the Īśvara. However, of the two, Īśvara is independent whereas the jīvas are dependent on Īśvara. The matter making up the physical universe was considered real, unlike illusory as in the case of Advaita. The Dvaita school also refuted the idea of Viśiṣṭādvaita that the matter and the jīvas are different yet form a part of Īśvara. For Dvaita, there are clear differences among them, despite being dependent on Īśvara. Difference (Bheda) is the very essence of Dvaita philosophy. Madhva advocated five-fold differences: between Jīva and Īśvara, among jīvas, Jīva and matter, Īśvara and matter and one material thing and another.

Madhva acknowledged the reality of human misery and bondage. According to Dvaita-vedānta, it is the Īśvara who causes the individual to be unaware of the relationship of the ātman with God. Hence, he alone eventually brings liberation through his grace. In this process, a self-effort on the part of the Jīva is an essential component. According to dvaita school, the jīvas have an innate nature (svarūpa) that never changes. This svarūpa should not be confused with the habits of a person at a superficial level. This acts at a deep level and thus differentiates one jīva from another. The purpose of the creation is to allow this and provide a conducive environment to the jīva to manifest to fullest of his nature. Depending on the composition of sattva, rajas and tamas guṇas in the svarūpa, the jīvas perform karmas accordingly and attain mokṣa, or niraya (hell), or be bound in the birth–death cycle for ever. The doctrine of jīvas that are liberated and those eternally damned has a parallel in Jaina's religious thought also.

In Dvaita-vedānta, liberation is achieved through the knowledge of the greatness of Īśvara. Similar to Viśiṣṭādvaita philosophy, Īśvara in Dvaita-vedānta is a personal God with attributes of name and form, who can be reached through devotion (bhakti). Through bhakti combined with meditation, one can dispose oneself to the experience and grace of Īśvara. The Jīva, on his part, must prove himself worthy of it by good works (karma), acquisition of right knowledge (jñāna-yoga), and single-minded devotion (bhakti-yoga).

Table 3.1 provides a comparative picture of the three schools of Vedānta, summarising the salient aspects of the schools on several elements of philosophical thinking.

TABLE 3.1 A Comparison of the Salient Features of the Three Schools of Vedānta

No.	*Criterion*	*Advaita*	*Viśiṣṭādvaita*	*Dvaita*
1	Basic reference for establishing the tenets	Upaniṣads, Bhagavadgītā, Brahmasūtra	Upaniṣads, Bhagavadgītā, Brahmasūtra	Upaniṣads, Bhagavadgītā, Brahmasūtra
2	Concept of Īśvara	One, attribute-less (Nirguṇa-brahman)	One, personal God (Saguṇa-brahman) – Viṣṇu or Nārāyana, Independent, Ultimate	One, personal God (Saguṇa-brahman) – Viṣṇu or Hari, Independent, Ultimate
3	Concept of Universe (Jagat)	Not real, mere illusory experience, made of Prakṛti and Guṇas	Real, made of Prakṛti and Guṇas	Real, made of Prakṛti and Guṇas
4	Jīva–Jagat–Īśvara relation	All are one and the same – Brahman	All are part and parcel of Īśvara, Jīva and Jagat depend on Iśvara	All are uniquely different, Jīva and Jagat depend on Īśvara
5	Valid means of knowledge (Pramāṇas)	Perception, inference, comparison, verbal testimony, presumption, non-apprehension	Perception, inference, verbal testimony	Perception, inference, verbal testimony
6	Liberation	Experience oneness with Brahman	Attaining the feet of the Lord	Knowledge of the greatness of Īśvara
7	Path to liberation	Jñāna-yoga, Karma-yoga as a pre-requisite	Bhakti and Prapatti, Bhakti-yoga, Karma-yoga and Jñāna-yoga as pre-requisites	Bhakti-yoga, Grace of Īśvara

3.6 NON-VEDIC PHILOSOPHICAL SYSTEMS

There are other philosophical systems that lie outside of the realm of the Vedic corpus. These philosophical systems did not consider the Vedas as an authoritative text and are called Non-Vedic philosophical systems. These schools do not also accept the entity Īśvara. Jaina, Bauddha, and Cārvāka schools are prominent among them.

3.6.1 Jaina School of Philosophy

The word 'Jaina' is derived from the Sanskrit root 'ji', to conquer, essentially indicating someone who has successfully subdued his passions and obtained mastery. The Jaina school considers twenty-four Tīrthaṇkaras, starting from Vṛṣabhadeva to Mahāvīra as prophets and masters of the philosophy. Tīrthaṅkaras appear periodically in the world to educate and lead people to cross over the ocean of rebirth. This is similar to the notion of avatāra-puruṣas, who by their conduct and teaching help the human beings cross the ocean of samsāra (endless birth–death cycle). Although in contemporary terms Mahāvīra is well-known among the twenty-four, he is regarded as the last of the twenty-four Tīrthaṅkaras. According to the Jain tradition, Mahāvīra lived during the 6th century BCE. There is a vast literature in which the doctrines of the school are recorded. The details of the Jaina school of philosophy can be found in the canonical texts of Jainism, which are largely based on the teachings of the Tīrthaṅkaras. Jain philosophy refuses to acknowledge the authority of the Vedas and the notion of a supreme God, however, several concepts in the Jaina school are in line with the Vedic schools of philosophy.

During the early part of the common era (during 4–5th century CE), two sects of Jains, Śvetāmbaras (white-clad ascetics) and Digambaras (sky-clad ascetics) emerged. There are some differences between the two sects in certain aspects such as rituals, ascetic practices, and monastic organisation. Despite this, on matters of philosophical principles and concepts, they remain similar. Both the sects accept the authority of the Tattvārthasūtra, composed by Umāsvāti during 2nd–3rd century CE. The Tattvārthasūtra has been commented upon by both Śvetāmbara and Digambara scholars over the centuries and is, therefore, an important Jain text.

According to Jain ontology, the fundamental categories of being are a soul (Jīva), a matter of which the substances in the world are formed (Pudgala), space (Ākāśa), time (Kāla), the principle of motion (Dharma), and the principle of rest (Adharma). Jīvas are infinite and so are the material particles. These particles also possess innumerable qualities and jīvas with their limited ability cannot describe them completely. Hence our knowledge of any substance is not absolute but relative. Jaina school proposes a methodology to address this issue and argues that capturing reality perfectly with the language is not possible.

This is analogous to six blind men trying to describe how an elephant looks like. Each one of them will describe an elephant in a manner that is both right and not right. It is right in a limited sense and not right if we take it as the ultimate description of the elephant. However, the description of the reality can be sufficiently enhanced through appropriate qualification of the claim made. This approach is known as 'syād-vāda', meaning conditional predication. 'Syāt' in Sanskrit essentially means, 'maybe'. In this context, it would mean, 'in a certain sense of the term' or 'from a certain point of view'. Using this concept, the Jaina school lists seven possibilities for the truth values. With syādvāda, Jain philosophers are able to analyse claims made by various systems of thought and show them to be relative assertions of the truth as understood by the Jain tradition.

- The concept of rebirth and other world is completely dismissed in Cārvāka philosophy.
- The details of the Jain School of Philosophy can be found in the canonical texts of Jainism, which are largely based on the teachings of the Tīrthaṅkaras.

In the Jaina school, the cycle of birth–death is attributed to tiny particles of matter (Pudgala) that have embedded themselves into the Jīva. This is called karma and in the Jain philosophy, spiritual growth is to overcome this karma. An analogy of a wet cloth explains how karma affects the Jīva. Just as a wet cloth becomes sticky when worn, the kārmic matter gets attached to the Jīva. The passions that we get attracted to are compared to the water in a wet cloth. A wet cloth attracts dust, in the same manner, the Jīva attracts karma. According to the Jain philosophy, the passions are evoked by experiences, which arise due to the kārmic particles that have previously bonded with the Jīva. Just as the seeds ripen eventually and bear fruit, the karma is supposed to have an impact on the jīva in terms of some experiences. These experiences could be pleasant, painful, or neutral, and evoke corresponding passions of attraction, aversion, or indifference. The passions, in turn, attract more kārmic particles or seeds, and the entire process repeats itself.

The Jīva is stuck in bondage with the matter because of his karma and passions. Hence freeing the self from the matter is the way to liberation. The association of the matter with Jīva is due to the ignorance about himself and the world. The real knowledge which can destroy the ignorance is not easily obtained by the Jīva, for that he has to listen to the teachings of the great masters, the Tīrthaṅkaras, who are liberated from the bondage. From a practical point of view, the goal is to purify the Jīva of kārmic matter, in a way by cleaning the karma so that the Jīva can radiate in its inherent blissful nature. To achieve this goal, Jain philosophy considers ascetic practices as essential. Since karma is considered as a physical substance that has bonded with the Jīva, Jain philosophy puts special emphasis on ascetic practices in terms of what one must and must not do, as a means to 'clean up' the karma. Three gems are prescribed: right faith, right knowledge, and right conduct[8]. Right faith is given utmost importance as any activity undertaken with false convictions loses much of its value. Right knowledge pertains to a good understanding of the Jain philosophy. Right conduct is also placed huge importance in the Jain philosophy. The individual must control the passion with the right conducts, of which ahiṃsā is the most important one.

Once a person begins to diligently practice the ethical restraints and prescribed ascetic disciplines, the karmas slowly drop away and the pure knowledge, which is the inherent nature of the Jīva begins to radiate. By these practices, the passions can be calmed and through a two-way process of cleaning existing kārmic matter and preventing further accumulation of karma, the Jīva can attain the final goal.

3.6.2 Bauddha School of Philosophy

The Bauddha (or Buddhist) school of philosophy is largely based on the teachings of Gautama Buddha. Buddha was born as Siddhārtha during the 4th–5th Century BCE. Although Buddhism originated in ancient India, it later spread to several parts of Asia. There are two forms of Buddhism: the northern form and the southern form. The form of Buddhism prevailing in Nepal, Tibet, China, Korea, Vietnam, Taiwan, Singapore, and Japan is called northern Buddhism (also synonymous with Mahāyāna) while the form prevailing in Sri Lanka and other parts of Southeast Asia including Cambodia, Laos, Myanmar, and Thailand is called Southern Buddhism (also known as Theravāda). The earliest form is the northern version, and it includes several sub-traditions such as Zen, Nichiren, and Shingon. Kaniśka is supposed to have convened a great council of the Northern Buddhists in the 1st century CE. Lalitavistara is an important work composed by the Northern Buddhists sometime during the 2nd–4th century CE. Tibetan Buddhism drifted away from the primitive Buddhism in India and is supposed to have adopted forms and ceremonies, which were unknown to Gautama and his followers.

According to early texts, Gautama was moved by the suffering of life and death. Further, on account of rebirth, this suffering is experienced in an endless cycle of birth–death. His enlightenment showed him the path for liberation from this suffering forever, by reaching a state of Nirvāṇa. Gautama's teachings were initially oral and in the later period, they developed into a complete philosophical system with several treatises written by the followers of the school.

Buddha's philosophy focuses on the means of ending the suffering of the individuals. It is based on four noble truths (catvāri ārya-satyāni). Figure 3.6 graphically illustrates this. These are elaborated as follows:

1. *There is suffering:* According to Gautama, "Birth is suffering, decay is suffering, illness is suffering, death is suffering. The presence of objects we hate is suffering, not being

able to obtain what we desire is suffering." The Buddhist philosophy argues that the human being is a compound of five aggregates and clinging to them leads to suffering. The five aggregates include the following:

(a) The form made of four elements (earth, water, fire, and air), five sense organs, five attributes of matter (smell, form, sound, taste, and touch), two distinctions of sex (male, female), three essential conditions (thought, vitality, and space) and two means of communication (gesture and speech)

(b) Consciousness

(c) Feeling: Sensations of pleasure and pain

(d) Formation

(e) Perception and Potentialities which lead to good or bad results

These five aggregates include all physical and mental elements and powers of man and are impermanent in nature. Consciousness arises from other aggregates and mental factors from the contact of consciousness and other aggregates.

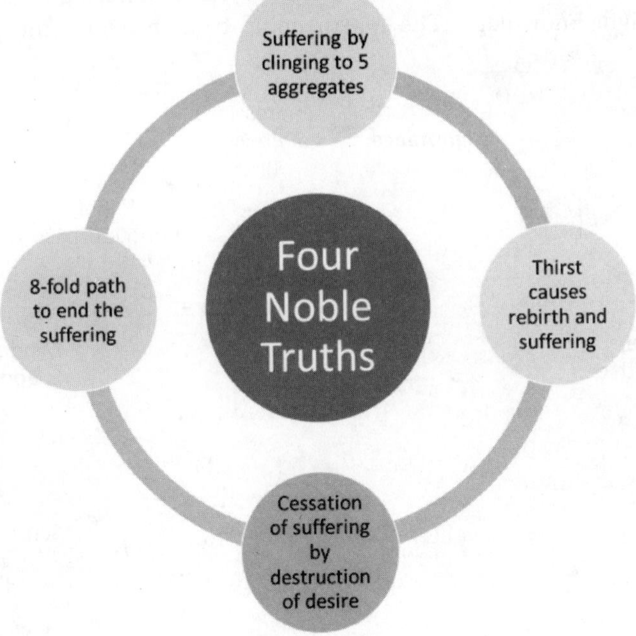

FIGURE 3.6 The Four Noble Truths of Buddhism

2. *There is the cause of suffering:* In Gautama's words, "Thirst leads to rebirth accompanied by pleasure and lust, thirst for pleasure, thirst for existence, thirst for prosperity".
3. *The cessation of suffering:* The cessation of suffering will be possible with the complete cessation of thirst, which amounts to the absence of passion and complete destruction of desire.

4. *There is a path to end the suffering*: Buddhist philosophy prescribes a holy eight-fold path that enables one to lead a holy moral life and that will lead one to the final goal of liberation. The eightfold path includes right views, right resolve, right speech, right conduct, right livelihood, right effort, right mindfulness, and right concentration.

The ultimate goal in Buddhist philosophy is to reach Nirvāṇa. Nirvāṇa is not a state reached after death, but something that is attainable in this very life. It is the sinless calm state of mind attained due to freedom from desires and passions, a state of perfect peace, goodness, and wisdom. Once a person reaches Nirvāṇa, the cycle of birth and death ends, which is the final goal leading to liberation.

As we have seen above, the central issue in Buddhist philosophy is to strive for the cessation of suffering. The Buddhist philosophy systematically argues how sufferings happen using a cause – effect cycle. Figure 3.7 illustrates this cycle leading to suffering. As seen in the figure, the root cause of the suffering is ignorance. Due to ignorance, the impressions of the previous birth lead to initial consciousness. The body and the mind and the sense organs evolve out of this consciousness. Once the sense organs are in contact with the senses and gather the experiences of life, the thirst for enjoyment drives the process leading to rebirth and suffering. Therefore, the only way to break this cycle of suffering is to remove ignorance by acquiring the right knowledge. The fourth noble truth provides the path for removing ignorance.

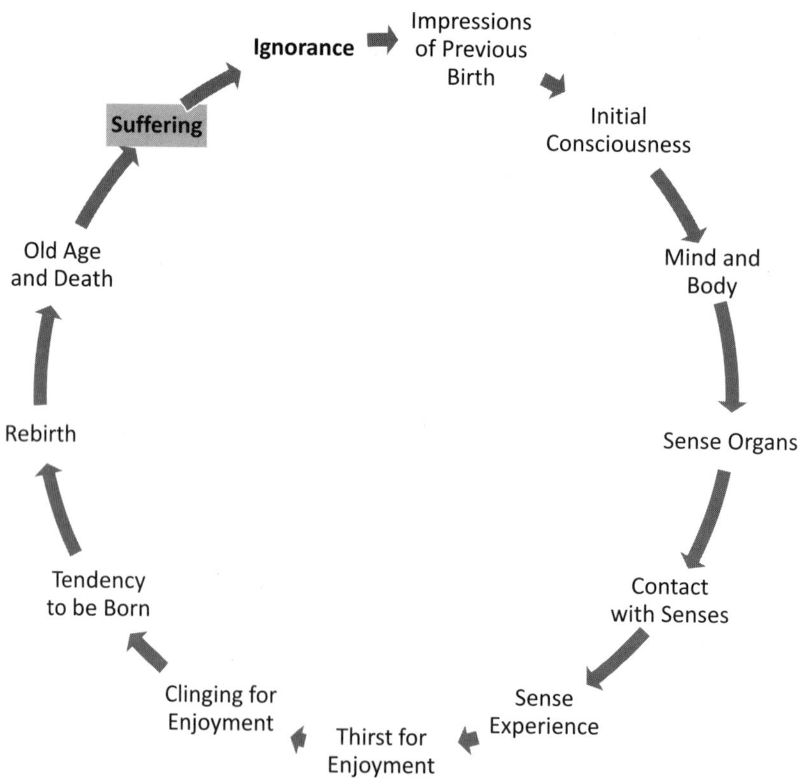

FIGURE 3.7 The Cycle of Suffering

3.6.3 Cārvāka School

Cārvāka school of thought closely maps to the trait of materialism, which emphasises a life of enjoyment based on certain principles and assumptions about life. The word Cārvāka literally means 'sweet-tongued' (cāru–vāka), in a way signifying that the ideas appear attractive at the outset. This is because the system only advocates two of the four puruṣārthas, pleasure and wealth as objectives of living. This school is also called Lokāyata. The Cārvāka school considers matter as the ultimate reality and rejects the idea that there is a divine or a transcendental power behind the matter, called Prakṛti conceptualised by the Sāṃkhya–Yoga school.

Philosophical systems in India had systematic methods for the presentation of the key concepts in the system. This began with a book (of aphorisms), followed by a growing literature of a few commentaries and sub-commentaries. In the case of the Cārvāka system, we do not seem to have such extensive literature on the school. No text of the Cārvāka school is available to us today which discusses its tenets in totality. The absence of canonical texts and a lineage of followers who were able to establish the tenets of the system by constantly engaging in intellectual debates to establish their tenets were perhaps responsible for its decay. Current discussions on this school of thought are based on the scanty fragments available for some analysis[9]. The available material is from texts such as Sarva-darśana-saṅgraha where, during the discussion of these schools, some verses are quoted. Mostly we get to know about this system through refutations from its opponents.

- ♦ No text of the Cārvāka school is available to us today which discusses its tenets in totality.
- ♦ According to Cārvāka school whatever is directly perceivable can only be accepted as valid means of knowledge.

One of the major differences of the Cārvāka school with that of the Vedic schools of philosophy pertains to what is the accepted means of valid knowledge. Unlike all other schools, Cārvāka school considers only direct perception through senses as pramāṇa[10]. Essentially this implies that whatever is directly perceivable can only be accepted as valid means of knowledge. This has significant implications for metaphysics. On account of this, Cārvāka school considers matter as the only reality using which the world is made of. Furthermore, the world is constituted of only four basic categories, namely, earth, water, fire, and air, which are all physical and directly perceivable. Ether or space is not accepted as the fifth element because it is not perceivable. Other entities such as the sky, ātman, mind, īśvara, dharma, reincarnation, svarga, and mokṣa that the other schools have accepted are rejected in the Cārvāka system. In essence, all transcendental entities are dismissed using the argument that only direct perception provides valid knowledge.

There is nothing called ātman other than the body. Cārvāka considers the four basic elements of the world as the basic constituents of the body too. When the individual constituents exist in a disjointed state, they are bereft of life and consciousness. However, when these come together the body is formed, and by a peculiar combination of these constituents, the life-breath and consciousness appear in the body. The concept of rebirth and other world is completely dismissed in this system. The arguments primarily stem from the limitation of using only direct perception as the means of valid knowledge. There are no means available for determining the existence of the 'other world'. Moreover, there is no 'other world' because of the absence of any 'otherworldly' being. Since the existence of consciousness in the other world cannot be substantially established through direct perception, which is the only means available for valid knowledge, these ideas are dismissed.

Once a person dies, there is no afterlife. Pleasure and pain are felt in no other place than the body, hence there is no need to accept something called the ātman which is not confirmed by the pramāṇa. Since there is no ātman and there is no rebirth the highest goal of human life is to have pleasure in one's lifetime. Once the body is confined to the flames and burnt to ashes, how can it ever return, therefore enjoyment 'here and now' is the goal of living in this school of thought[11].

The salient features of the Cārvāka school of philosophy are summarised in Figure 3.8.

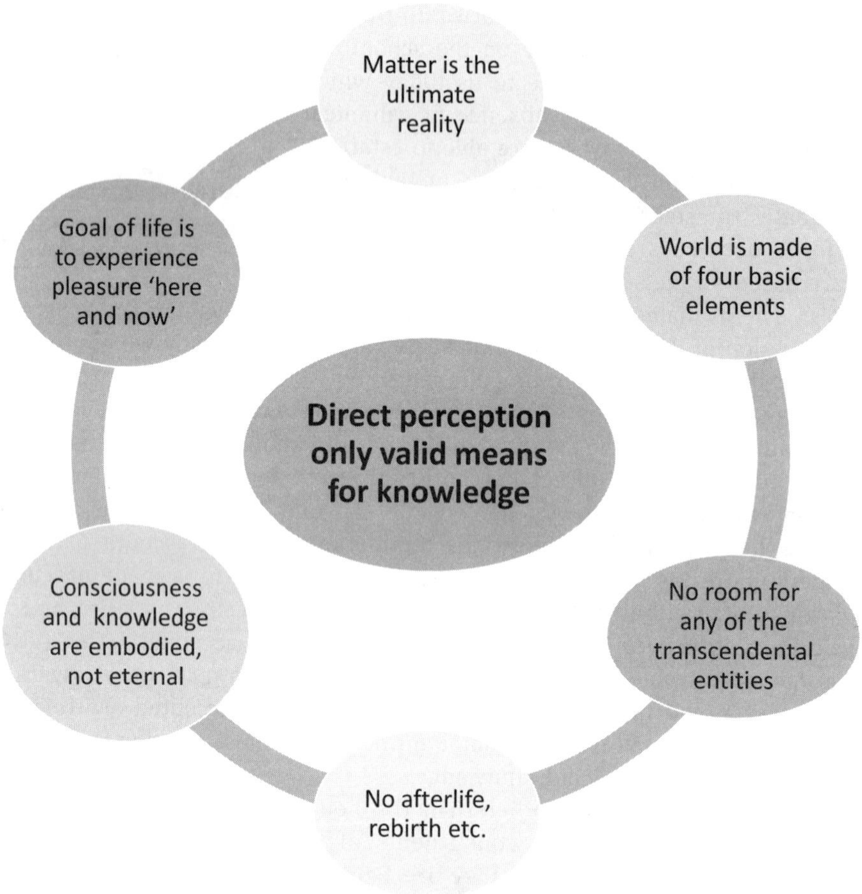

FIGURE 3.8 Salient Features of the Cārvāka School

SUMMARY

- The word darśana is derived from the Sanskrit root, dṛś, 'to see', suggesting that these philosophical systems provide a true worldview and a vision for life and help us resolve the issues that we face in our life.
- The philosophical inquiry in India gave rise to several schools, each developing its own understanding of the world. Each of these schools presents its view with rigorous intellectual exercise and uncompromised importance on the self-experience.
- While the Hindu schools of philosophy rely on the Vedic text as the ultimate authority and means for valid knowledge, the Jaina philosophy relies on their canonical texts.
- The schools differ in presenting how to get the right knowledge and use it as the means for the ultimate liberation of the soul. The context for the philosophical discussion is the three inter-related concepts of God–Universe–Individual.
- According to the Sāṃkhya system, two basic elements constitute everything in this world, matter (Prakṛti) and spirit (Puruṣa). Puruṣa is the pure consciousness, sentient, changeless, eternal, and passive. Prakṛti on the other hand is the root cause of all activities including the entire creation.
- The unique thing that establishes Yoga as distinct among the other darśanas is that its emphasis on understanding the mind, its various states, its cognitive activities, and methods to control it. Yoga provides a structured and practical eight-step process to gradually reach a stage of complete cessation of the activities of Citta.
- Both Nyāya and Vaiśeṣika set out in their journey of the 'knowable' and define various categories to describe the same. The first is about the 'ways of knowing the reality' and the second is about the 'objects in the reality that is knowable'.
- The Nyāya system, placed enormous emphasis on the means of obtaining 'right knowledge'. Therefore, the unique contribution of Nyāya school is its detailed inquiry of knowledge (Pramā) and valid cognition and its means (Pramāṇa).
- The main aim of Pūrva-mīmāṃsā school is to ascertain the meanings of the Saṃhitā and Brāhmaṇa portions of Veda which lay importance on the karma, performing rituals, and thereby attaining dharma as a means for liberation.
- The Vedānta schools of philosophy rely on three major texts: Brahma-Sūtras, Bhagavadgītā, and the Upaniṣads, known as prasthānatraya, for establishing their tenets.
- Śaṇkara propounded a monistic philosophy, known as Advaita, with the conception of the attribute-less God (Nirguṇa-brahman) as the ultimate reality.
- Matter, Jīva and Īśvara are three entities recognised in the Viśiṣṭādvaita school. Though all the three are ultimate, the first two have an absolute dependent relationship with Īśvara.
- The Dvaita school clearly admits three independent and mutually irreducible substances that make up the Universe: the Jīva, the Jagat and the Īśvara.
- Several concepts of the Jain school of thought is in line with the Vedic schools of philosophy. With syād-vāda, Jain philosophers are able to analyze claims made by various systems of thought and show them to be relative assertions of the truth as understood by the Jain tradition.
- Buddha's philosophy focuses on the means of ending the suffering of the individuals. It is based on four noble truths. The ultimate goal in Buddhist philosophy is to reach Nirvāṇa. Nirvāṇa is not a state reached after death, but something that is attainable in this very life.
- Cārvāka school of thought closely maps to the trait of materialism, which emphasises a life of enjoyment based on certain principles and assumptions about life. The concept of rebirth and other world is completely dismissed in this system.

REVIEW QUESTIONS

1. Outline the salient features of Indian philosophical systems.
2. What do you understand by the terms 'Vedic and Non-Vedic' schools of philosophy?
3. What are some of the common features of the Vedic schools of philosophy? What are their main differences?
4. Prepare a one-page note each, enumerating the commonalities and differences between the following schools of philosophy:
   (a) Sāṃkhya and Yoga
   (b) Nyāya and Vaiśeṣika
   (c) Pūrva-mīmāṃsā and Vedānta
5. Briefly describes the salient aspects of the following schools of philosophy:
   (a) Sāṃkhya
   (b) Yoga
   (c) Nyāyā
   (d) Vaiśeṣika
   (e) Pūrva-mīmāṃsā
6. What do you understand by the term 'Puruṣa and Prakṛti'?
7. Explain the term 'Aṣṭāṅga-yoga'.
8. What are the key differences and commonalities among the three schools of Vedānta?
9. What is the role of karma-yoga, bhakti-yoga, and jñāna-yoga according to the three Vedānta schools?
10. Comment on the statement, "Cārvāka school of philosophy has taken a different approach compared to other philosophical systems".
11. Prepare a one-page note outlining the salient features of the Jaina school of philosophy.
12. Compare and contrast the Jaina school of philosophy with that of the Buddhist school.
13. What are the key recommendations of the Buddhist school of philosophy to attain liberation?

DISCOVER IKS

1. All Indian schools of philosophy share certain common features. They also differ on certain aspects and these features primarily drive the basic tenets of a particular school of thought. The video in the link below is a talk by Swami Tadatmananda on the relative differences between Buddhism and the Advaita-vedānta school of philosophy: https://www.youtube.com/watch?v=Wq2eukYfRoA. After watching the video, prepare a two-page note to answer the following questions:
   (a) What are some of the common ideas between Advaita and Buddhism?
   (b) What are the key differences between the two schools?
2. One of the distinguishing aspects of Jainism is their special attention to ahiṃsā. It is their core belief that we need to make several lifestyle changes to truthfully practice ahiṃsā. There are other such principles such as non-possession (aparigraha). These practices are to be diligently followed by the followers of Jain irrespective of whether the person is a householder or a monk. The video traces the life of Jain monks and provides some clues as to how they practice these virtues in their daily life: https://www.youtube.com/watch?v=jqExyhLTFaA
   After watching the video, develop a three-page note to answer the following questions:
   (a) What do you understand by the 'low consumption' principle? What are its implications for modern living, both to an individual and the society?

(b) Does the ascetic life of a Jain monk promote environment-friendly practices? How?

(c) How do the Jain monks contribute to the wellbeing of the members of the society?

SUGGESTED READINGS

Balasubramanian, R. (2000). "Introduction". In Chattopadhyana (Ed.). History of Science, Philosophy and Culture in Indian Civilisation. Volume II, Part 2: Advaita Vedanta. Delhi: Centre for Studies in Civilisations.

Bhaduri, S. (1975). "Studies in Nyāya - Vaiśeṣika Metaphysics", Bhandarkar Oriental Research Institute, Pune.

Bhattacharya, R. (2002). "Cārvāka Fragments: A new collection", *Journal of Indian Philosophy*, **30**, pp. 597–640.

Bose, R.C. (1884). *The Hindu Philosophy: Orthodox Systems*, Funk & Wagnalls, London.

Chattopadhyaya, J-C. (1912). *The Hindu Realism*, The Indian Press, Allahabad.

Dasgupta, S. (1975). *A History of Indian Philosophy - Volume I*, Motilal Banarsidass, New Delhi.

Hiriyanna, M. (1994). *Outlines of Indian Philosophy*, Motilal Banarsidass, New Delhi.

Jain, P.S. (2000). "Collected Papers on Jaina Studies", Motilal Banarsidass, Delhi.

Jain, P.S. (2001). "Collected Papers on Buddhist Studies", Motilal Banarsidass, Delhi

Madhavacarya, S. (1978). *Sarvadarsana-samgraha*, V.S. Abhyankar (Ed.), 3rd ed., Bhandarkar Oriental Research Institute.

Puligandla, R. (2005). *Fundamentals of Indian Philosophy*, DK Print World, New Delhi, pp. 119–208.

Radhakrishnan, S. (1958). *Indian Philosophy - Volume 2*, The Macmillan Company, New York.

Raghavendrachar, H.N. and Wadia, A.R. (1941). "The Dvaita Philosophy and Its Place in the Vedanta", The University of Mysuru.

Sethia, T. (2004). *Ahiṃsā, Anekānta and Jainism*, Motilal Banarsidass, New Delhi.

Sharma, C. A. (1964). *Critical Survey of Indian Philosophy*. Motilal Banarasidass Publication.

Sinha, N. (1990). *The Nyāya Sūtras of Gotama*, Motilal Banarsidass Publishers, Delhi.

Sinha, N. (2008). *The Vaiśeṣika Sūtras of Kanāda*, Cosmos Publications. New Delhi.

Srinivasa-Chari, S.M. (2004). *Fundamentals of Visistadvaita Vedanta: A Study Based on Vedanta Desika's Tattvamuktakalapa*, Motilal Banarasidass, Delhi.

ENDNOTES

1. For some more details on this, see, **Dasgupta, S. (1975).** *A History of Indian Philosophy - Volume I*, Motilal Banarsidass, Delhi, pp. 71–75.
2. This idea is brought out concisely in the Gītā in Chapter 13, verse 21: पुरुषः प्रकृतिस्थो हि भुङ्क्ते प्रकृतिजान्गुणान् । कारणं गुणसङ्गोऽस्य सदसद्योनिजन्मसु ॥१३–२१॥ puruṣaḥ prakṛtistho hi bhuṅkte prakṛtijānguṇān | kāraṇaṃ guṇasaṅgo'sya sadasadyonijanmasu ||13–21|| See for details, **Swami Chinmayananda (2002).** "The Holy Geeta", Central Chinmaya Mission Trust, Mumbai, pp. 880–882.
3. For details see, **Arjunwadkar, K.S. (2006).** "Yogasūtras of Patañjali", Bhandarkar Oriental Research Institute, Pune, p. ४. योगश्चित्तवृत्तिनिरोधः ॥ yogaścittavṛttinirodhaḥ || Yoga Sūtra 1.2.

4. *Ibid.* p. १८. अभ्यासवैराग्याभ्यां तन्निरोधः ॥ abhyāsavairāgyābhyāṃ tannirodhaḥ || Yoga Sūtra 1.12.

5. *Ibid.* p ११२. स्थिरसुखमासनम् ॥ sthirasukhamāsanam || Yoga Sūtra 2.46. The same idea is conveyed in verse 11 in Chapter 6 of Bhagavadgīta, "शुचौ देशे प्रतिष्ठाप्य स्थिरमासनमात्मनः । नात्युच्छ्रितं नातिनीचं चैलाजिनकुशोत्तरम् ॥६–११॥ śucau deśe pratiṣṭhāpya sthiramāsanamātmanaḥ | nātyucchritaṃ nātinīcaṃ cailājinakuśottaram ||6–11|| See for details, **Swami Chinmayananda (2002).** "The Holy Geeta", Central Chinmaya Mission Trust, Mumbai, pp. 390–392.

6. This is formally stated in the Vaiśeṣika Sūtra 1.1.4 as, धर्मविशेषप्रसूताद् द्रव्य-गुण-कर्म-सामान्य-विशेष-समवायानां पदार्थानां साधर्म्य-वैधर्म्याभ्यां तत्त्वज्ञानात् निःश्रेयसम् । dharmaviśeṣaprasūtād dravya-guṇa-karma-sāmānya-viśeṣa-samavāyānāṃ padārthānāṃ sādharmya-vaidharmyābhyāṃ tattvajñānāt niḥśreyasam | See for details, **Sinha, N. (2008).** *The Vaisesika Sutras of Kanada*, Cosmo Publications, New Delhi, pp. 8–9.

7. See for example in Vivekacūḍāmaṇi the verse: चित्तस्य शुद्धये कर्म न तु वस्तूपलब्धये । वस्तुसिद्धिर्विचारेण न किञ्चित् कर्मकोटिभिः ॥ cittasya śuddhaye karma na tu vastūpalabdhaye | vastusiddhirvicāreṇa na kiñcit karmakoṭibhiḥ || 11 See for details, **Swami Chinmyananada (2003).** "Vivekachoodamani", Central Chinmaya Mission Trust, Mumbai, pp. 21–22.

8. सम्यग्दर्शन-ज्ञान-चारित्राणि मोक्षमार्गः । samyagdarśana-jñāna-cāritrāṇi mokṣamārgaḥ | Umāsvāti, as quoted in **Hiriyanna, M. (1994).** *Outlines of Indian Philosophy*, Motilal Banarsidass, New Delhi, p. 166.

9. To know more about this limitation and the current attempts to cull out and make sense of the existing material on Cārvāka school, see, **Bhattacharya, R. (2002).** "Cārvāka Fragments: A New Collection", *Journal of Indian Philosophy*, **30**, pp. 597–640.

10. *Ibid.* प्रत्यक्षमेकमेव प्रमाणम् ॥ pratyakṣamekameva pramāṇam ||

11. यावज्जीवं सुखं जीवेत् नास्ति मृत्योरगोचरः । भस्मीभूतस्य देहस्य पुनरागमनं कुतः ॥ yāvajjīvaṃ sukhaṃ jīvet nāsti mṛtyoragocaraḥ | bhasmībhūtasya dehasya punarāgamanaṃ kutaḥ || **Uma Shankar Sharma (2008)**: "Sarva-Darśana-Saṅgraha", Chowkhambha Vidyabhawan, Varanasi. Chapter 1, Verse 1, p. 3.

CHAPTER

4 Wisdom through the Ages

LEARNING OUTCOMES

After finishing this chapter, you will be able to:

- Develop a basic understanding of Purāṇic repository and the issues addressed
- Understand the broad contents and the importance of the two epics (Rāmāyaṇa and Mahābhārata) and their contrasting approach
- Familiarise with the issue addressed in Nīti-śāstras in general and get to know some popular nīti texts
- Appreciate the complementary role played by Subhāṣitas in sharing some useful nuggets for life

Pañcatantra tales have captured the imagination of a vast section of the population. Temple architecture depicts stories from Pañcatantra such as this one found in the North face of Temple 2, Nalanda (7th century CE). In this tale, the turtle is escaping from hunters (not shown) thanks to two geese, who bear it on a stick that it is grasping with its jaws. Unfortunately, it opened its mouth to boast of the escapade, which caused him to fall to its death (food for the hungry boy and girl below).

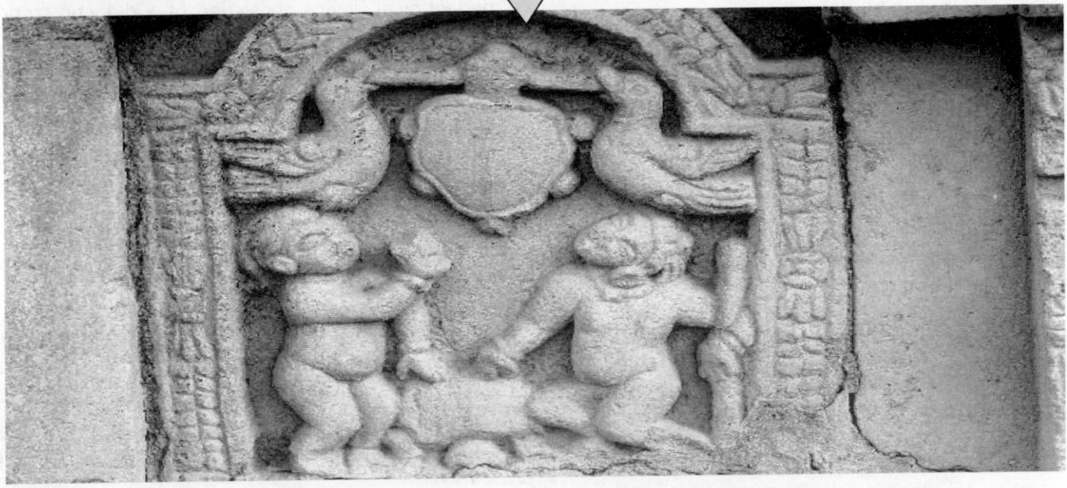

Source: Own work, G41rn8, https://commons.wikimedia.org/wiki/File:Nalanda_Temple_2_ei7-42.jpg

IKS IN ACTION 4.1

Pañcatantra – A Treatise on Statecraft through Stories

Stories provide a powerful medium through which one can impart education on a wide range of topics. In modern parlance, stories are used as the basis for discussing several real-world problems. In business schools, case study methodology employs stories to analyze complex problems and various decision options. A shining example of the use of stories for education is Pañcatantra, a collection of stories involving the animal kingdom as the main characters. The text was composed by Viṣṇu Śarman.

The genesis of this text provides interesting insights into the power of storytelling in bringing the accumulated wisdom to use. Once a king named Amaraśakti had three sons who were ill-minded and were disinterested in learning. The king engaged a teacher, Viṣṇu Śarman to make his sons educated and eligible in governance at the earliest.

As the sons of kings were not interested in studying the śāstras Viṣṇu Śarman discovered a unique way to teach through stories. It had five Tantras (books), which is a set of stories on Governance, life, ethics, and morals, popularly known as Pañcatantra. It is also considered an important text on Nīti-śāstra.

Pañcatantra is considered as a Nīti-śāstra' as it deals with wise conduct in life. It consists of several stories nested within stories. Most of the characters are animals, who talk to each other and conduct several chores of life. Human characteristics, behaviour patterns, and even ethical values are ascribed to these animals. The wisdom in the five books deals with the following:

- *Book 1:* The Loss of Friends (*Mitra-bheda*) consists of a series of fables describing the conspiracies and causes that lead to close friends breaking up.
- *Book 2:* Winning of Friends (*Mitra-lābha*) emphasises the importance of friendships, teamwork, and alliances.
- *Book 3:* Crow and the Owl (*Kākolūkīyam*) contains eighteen fables discussing certain aspects of war and peace. It suggests that a battle of wits is a more potent force than a battle of swords.
- *Book 4:* Loss of Gains (*Labdha-praṇāśam*) presents some examples and consequences thereby highlighting actions to avoid.
- *Book 5:* An ill-considered action (*Aparīkṣita-kārakam*) presents twelve fables about hasty actions and jumping to conclusions without establishing facts and proper due diligence. Unlike the previous four books, the characters in book five are human beings.

The first book is the longest covering nearly half of the content and the last two books together constitute about 7% of the total.

The charm of Pañcatantra lies in its ability to frame complex matters of statecraft into stories involving nīti and daṇḍa from the Arthaśāstra and presenting them in easily relatable fables. For example, in Book 3 concerning the war of crows and owls, six acts of royal policy, i.e. Ṣāḍguṇya (peace, war, change of base, entrenchment, alliance, and duplicity) discussed in Arthaśāstra are quoted.

Much of the wisdom of ancient Indian statecraft which we are aware of today is attributable to Arthaśāstra. Pañcatantra, seems to have derived its inspirations and main messages from it. It is an intelligent rendering of Arthaśāstra using a pedagogical device consisting of a didactic story-based tool.

The captivating style of storytelling adopted by Pañcatantra has attracted people far and near. It appears that there is a version of Pañcatantra in every major language of India, and there are several versions of the text in more than 50 languages around the world.

Human beings face several challenges and moral and ethical dilemmas in life and constantly look for guidance to resolve them. These issues are eternal, not problems that only the current day society faces. Therefore, in our search of the answers to these, it is desirable to look back and understand how the ancestors addressed them. Every society develops a body of knowledge over time shaped by the accumulated experiences of the people, the inferences made out of situations faced and experiments conducted. These experiences and insights are shared and stored for posterity through literary works. These showcase the cultural practices, history, science and technology, social customs, assumptions about life, and know-how prevailing at the time the literary work was created. It also describes how some of the challenges faced in personal and public life were addressed. It also provides useful guidelines to society on several issues faced in life. Therefore, the accumulated body of literature serves as the repository of wisdom through the ages. With a long civilizational history, India has a rich body of knowledge and experience that has developed within the society from time to time, captured through a variety of literary works.

In this chapter, we shall have a glimpse of the ancestral wisdom recorded in various literary works. Purāṇas and Itihāsas mainly belong to this category. The origin of the purāṇas and the itihāsas is not yet accurately established, but it dates back to antiquity. However, in the recorded history, several literary works have been authored in Sanskrit and other regional languages.

4.1 PURĀṆAS – AN ENCYCLOPAEDIC WORK

In the history of Indian literature, Purāṇas occupy a unique position. Purāṇas are literary sources of ancient and medieval Indian history and culture. The larger purāṇic repository provides valuable insights into the history of philosophy and religion and is a storehouse that provides deep insights into all aspects and phases of the society, culture during the ancient and medieval times in India. Purāṇas may be described as a popular encyclopaedia of ancient and medieval religious, philosophical, historical, personal, social, and political aspects of the society.

On account of the diversity of people and issues discussed in the purāṇic repository, purāṇas have significantly contributed to maintaining cultural pluralism in India. They played a role in influencing classical art forms such as dance and music and promoted social practices such as the celebration of various festivals. The spread of the cultural heritage among the masses has been, to a large extent, possible only because of the purāṇas. Moreover, they have played a pioneering role in motivating and practice of ethical and moral values in the life of the people of Indian society using stories as a powerful medium to communicate the ideals. In this manner, Purāṇas have provided means by which socio-cultural values can blend with religious values and dharmic living.

- Purāṇas are one of the most important literary sources of ancient and medieval Indian history and culture.
- Purāṇas are companion texts to help understand and interpret the Vedas.

The literary beauty and story-based narration of ideas through a living tradition called 'Hari Katha' have perennially attracted the masses towards the purāṇas. On account of this, purāṇas enjoy a unique position in the sacred literature of ancient India. They closely align with the epics in form and substance. Purāṇas enable us to know the true import of the ethos, philosophy, and religion of the Vedas and are companion texts to help understand and interpret the Vedas. Therefore, knowledge of purāṇas is considered very important as evident from the following observations in some of the purāṇas[1]:

- Skanda-purāṇa – Śruti (Vedas), Śmṛti (Rāmāyaṇa, Mahābhārata, Manu-smṛti, etc.) and Purāṇas are the three eyes of a scholar. One who sees the world with his three eyes is indeed a part of the supreme God.
- Viṣṇu-purāṇa – One who is a scholar of all the four Vedas, with its limbs and Upaniṣads in it and has no knowledge of purāṇas cannot be ever considered as a complete scholar.

As in the case of the knowledge repository in ancient India, purāṇas have been handed down the ages through oral transmission. Therefore, it is difficult to establish the period to the authorship of the purāṇas as there is no explicit mention. The current version available to us may be the work of several authors over the centuries.

However, using the information available in the purāṇas, Indology researchers have made some guesstimates. Accordingly, some of the extant Mahā-purāṇas, e.g. Vāyu, Brahmāṇḍa, Viṣṇu, Matsya, and Mārkaṇḍeya are assignable to a period of 300 CE – 660 CE, while some others such as Agni, and Garuda are assignable to a period of 600 CE – 900 CE and some others are assigned to a period later than 900 CE[2]. On the other hand, in the Indian tradition, the antiquity of purāṇas is established using the available material in the purāṇas itself. According to Matsya Purāṇa, Brahmā remembered the purāṇas even before the Vedas[3]. We find mention of the word Purāṇa in the Vedas and Upaniṣads. For instance, in Chāndogyopaniṣad it is mentioned that Purāṇas can be considered as the fifth Veda[4].

- The study of purāṇas and the upa-purāṇas are critical to developing an understanding of the social and religious institutions in India.
- Purānas deal with the evolution of the Universe, the recreation of the Universe, genealogies of Gods and seers, and the history of royal families.

4.1.1 Mahā-purāṇas, Upa-purāṇas and Sthala-purāṇas

The purāṇic repository can be broadly divided into three: Mahāpurāṇas (Major Purāṇas), Upa-purāṇas (Minor Purāṇas), and Sthala-purāṇas (Regional Purāṇas).

- ***Mahā-purāṇas***, the largest ever composed literature in the history of mankind, is very extensive. There are 18 mahā-purāṇas[5] and they are estimated to contain among them about 400,000 verses. The text in the purāṇas is in the form of several stories, anecdotes, clarificatory concepts, and rules and observances in a narrative style, often in the form of a reply to some issues raised during a dialogue between the narrator (usually a sage) and the audience (usually a group of sages, king, etc.). It enumerates stories of different forms of God, various kings, ṛṣis, etc. Usually, each purāṇa focuses on one form of the trinity (Brahma, Viṣṇu, Śiva) and celebrates the chosen trinity by narrating events centred around that trinity. For instance, Bhāgavata-purāṇa and Viṣṇu-purāṇa hold Viṣṇu as a supreme being, and Liṅga-purāṇa holds Śiva, etc. Some major purāṇas have discussed in detail a focused theme. For example, the Skanda, Padma and Bhaviṣya purāṇas deal mainly with pilgrimage to holy places (Tīrta-māhātmyas). The purāṇas cover a wide range of issues including the creation process and the geographical details, various rituals and their importance, certain scientific aspects, etc.
- ***Upa-purāṇas*** are similar to the mahā-purāṇas but are smaller in size. The number of Upa-purāṇas cannot be established with certainty, though about 100 of them can be enumerated from different sources. However, it is generally believed that there are eighteen of them which are prominent[6]. The list of upa-purāṇas differ in the verses found in various purāṇas[7]. So, it is difficult to list the names of eighteen upa-purāṇas exactly.
- ***Sthala-purāṇas*** are a class of sacred works which provide a connection between the land and the Divine in many forms. Each sthala-purāṇa is focused on a specific geographical place (usually a famous temple in a city) and weaves information centred

around the presiding deity. Various local cultural practices, customs, festivals, and other social practices form part of the narration. There are many Sthala-purāṇas in different languages of India. However, most of them are in Sanskrit.

Figure 4.1 graphically illustrates the purāṇic repository and its key features.

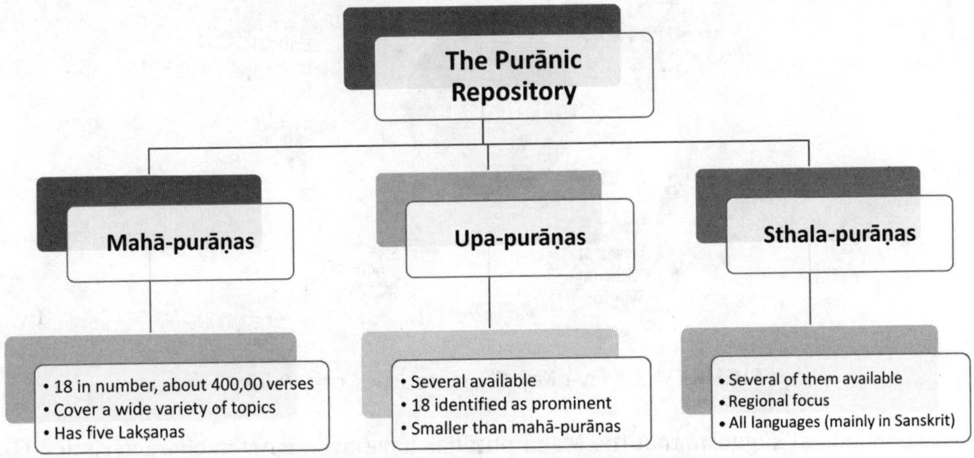

FIGURE 4.1 The Purāṇic Repository

4.1.2 Contents of the Purāṇas

There are five key characteristics of purāṇas, known as pañca-lakṣaṇas[8]. Figure 4.2 graphically presents these five key characteristics, and the details are as follows:

- **Sarga** (Creation of the Universe). Bhāgavata describes the process as, "from the agitation of the original modes within the unmanifest material nature, the mahat-tattva arises. From the mahat-tattva comes the element false ego, which divides itself into three aspects. This threefold false ego further manifests as the subtle forms of perception, the senses, and the gross sense objects. The generation of all these is called creation"[9].
- **Prati-sarga** (Destruction and re-creation). It means the creation after creation, i.e., the continued creation of primitive matter. Just as a seed produces additional seeds, activities that promote material desires in the performer produce moving and non-moving life forms[10].
- **Vaṁśa** (Dynasties of kings). Bhāgavata-purāṇa defines Vamśa as the present, past and future lineage of kings created by the Brahman[11].
- **Manvantara** (Times of different Manus). It is a time span that plays an important role in the understanding of human history. Purāṇas give the detailed explanation of the various activities that happened and different people (king, ṛṣis) that existed during the time of each Manu[12].
- **Vaṃśānucarita** (Stories of the dynasties). It also includes the stories of the ṛṣis and kings born in those dynasties[13].

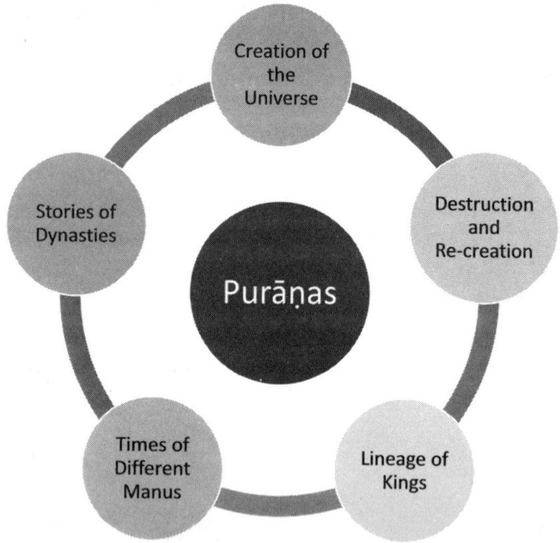

FIGURE 4.2 Five Key Characteristics of a Purāṇa

An expanded list suggests that the Mahā-purāṇas have to satisfy ten characteristics, (Daśa-lakṣaṇas), which additionally includes means of livelihood (vṛtti), protection of the universe (rakśā), final emancipation (mukti), jīva, unmanifest (hetu), and Brahman (apāśraya). Besides the ten characteristics, the Purāṇas deal with the glorification of Brahmā, Viṣṇu, Sūrya and Rudra, and the four puruṣārthas (dharma - righteous conduct and living, artha - economics and polity, kāma - desire, and mokṣa - emancipation). The Puränas also contain, besides, a good deal on supplementary topics of religious instruction, rituals, dāna, vratas, bhakti, yoga, and also medicine, music, grammar, poetics, metrics, dramaturgy, astronomy, and astrology, architecture, sculpture, iconography, polity, raja-dharma, etc. Table 4.1 provides a compact summary of the issues discussed in the 18 mahā-purāṇas other than those covered under the purāṇa-lakṣaṇas described above.

4.1.3 Issues of Interest in the Purāṇas

Food and Medicine

Purāṇas contain several discussions related to the plant kingdom and its role in addressing food, medicine, and other requirements of society[14]. The Brahmavaivarta-purāṇa (4.126: 53–54) mentions a balanced diet consisting of staple food, vegetables, and fruits that people used to take. Based on the information available in the Agni, Matsya, and Brahmavaivarta purāṇas we can infer that rice (dhānya), wheat (godhuma), barley (yava), pulses (śimbi or śamīdhānya), and sesame (tila) are some of the common plants used for edible purposes. Different varieties of dhānya are mentioned in the purāṇas in connection with religious and dietic purposes. As per Agni-purāṇa, paddy was cultivated on a large scale and people performed

- Purāṇas contain several discussions related to the plant kingdom and its role in addressing food, medicine, and other requirements of society.
- Several purāṇas have discussions on astronomy, which consists of ideas in the pre-telescope age.

religious functions for enhancement of production of rice. The Agni-purāṇa contains a description of different varieties of dhānya or paddy, such as deva-dhānya, śyāmakā, and nīvāra. The Matsya-purāṇa mentions eighteen varieties of rice and during the religious performance called aśvaratha-dāna, the observer had to fill in four jars with eighteen types of rice. In the Agni-purāṇa, wheat and barley are recommended as food items for the observers of religious vows, particularly yava is eaten by the performers of prājāpatya-vrata. In the Agni-purāṇa, there are mentions of tax on profits out of pulse production. It is stated that one-eighth of the produced pulses are paid to the king as tax. This indicates to us that pulses were being cultivated on large scale.

TABLE 4.1 Topics Discussed in the Eighteen Mahā-purāṇas (other than those related to purāṇa-lakṣaṇas)

Sl. No.	*Purāṇa (Alphabetical)*	*No. of Verses*	*Topics Discussed*
1.	Agni	16,000	Rules for worship of various deities, installations of images in temples, astrology, architecture, sculptures, medicine, toxicology, principles of dramaturgy, human physiology, figures of speech
2.	Bhāgavata	18,000	Most authoritative text dealing with Śrī Kṛṣṇa's life, cosmic form of God, Creation of the world, Uddhava-gītā, list of kings who ruled after Kṛṣṇa, a graphic description of Kaliyuga
3.	Bhaviṣya	14,500	Sixteen Saṃskāras, Rules concerning the studies of Vedas, Varṇāśrama Dharma, Vratas, Types of Dāna, Genealogy of the king who will come in future and the way they will rule the country
4.	Brahma	10,000	What is available seems to be more a compilation of chapters from Mahābhārata, Viṣṇu, Mārkaṇḍeya and Vāyu purāṇas
5.	Brahmāṇḍa	12,000	The subjects of creation, Geography of earth and Bhāratavarṣa, 14 worlds, types of hells, Balineese translation of this purana has been found in the Bali island
6.	Brahmavaivarta	18,000	Āyurveda, Sandhyā ritual, Śāligrāma worship, Kaliyuga, Greatness of Tulasi leaves, Code of conduct for married women and widows, Greatness of Bhārata-deśa, building construction
7.	Garuḍa	18,000	Cosmography, astronomy, astrology, omens, portents, medicine and knowledge of precious stones, journey of Jīva after death, rites to be performed during and after death, Torments of hell, encounter with pretas, Yamaloka
8.	Kūrma	18,000	Duties of four Varṇas and Āśramas, evolution of Prakṛti, Description of physical features of the world and of Jambūdvīpa, Division of Vedas, after death rites, Prāyaścittas, Pralaya
9.	Liṅga	11,000	Music and its propagation, Dānas and their fruits

(*Contd.*)

TABLE 4.1 Topics Discussed in the Eighteen Mahā-purāṇas (other than those related to purāṇa-lakṣaṇas) (*Contd.*)

Sl. No.	*Purāṇa (Alphabetical)*	*No. of Verses*	*Topics Discussed*
10.	Mārkaṇḍeya	9,000	Descriptions of hell, Karma and its fructification, Some aspects of Yoga, Geographical description of Bhārata and Śrāddhas
11.	Matsya	14,000	Vratas, Dānas and their fruits, Greatness of holy places like Prayāga, Vārāṇasī and river Narmadā, Duties of Kings, Omens, Iconography, Building construction
12.	Nāradīya	25,000	Vratas, Duties of the people of various Varṇas and Āśramas, Six Vedāṅgas, Ritualistic workshop, Fasting on Ekādaśī days
13.	Padma	55,000	Description of the various worlds of goblins, gandharvas, heaven and so on, Varṇāśrama dharmas, Vratas and religious vows, Various types of Śrāddhas
14.	Skanda	81,000	Detail descriptions of several places of pilgrimage such as Kāśī, Purī and Ujjainī, Advaita-vedānta, Various aspects of Lord Śiva, and methods of meditations, Satyanārāyaṇa-vrata
15.	Vāmana	10,000	Certain vratas, Doctrine of karma
16.	Varāha	24,000	Dharma śāstra, Pilgrimage, Dāna, images of deities and their worship, Aśouca, Śrāddha, Theory of karma, Hells, Cosmology, Sins
17.	Vāyu	24,000	Geographical descriptions of places, Importance and greatness of Gaya, Famous places of pilgrimage, Details of Śrāddha ceremony, Science of music
18.	Viṣṇu	23,000	Four Varṇas and Āśramas, The impact of Kali, Philosophical ideas of the Vedas, Description of Kaliyuga, Bhakti as main Sādhanā

Source: Compiled from **Swami Harshananda (2000).** "The Purāṇas", Sri Ramakṛishna Math, Bengaluru.

The use of plants, herbs, and shrubs is evident from the descriptions in the Agni-purāṇa. Thirty-six plants such as harītakī, nimba, akṣa, and bald were used in combinations for the treatment of different kinds of diseases. Similarly, in the Matsya-purāṇa nearly seventy-five plants have been identified to have great medicinal value (mahauṣadha). According to the Agni-purāṇa, the physicians used to treat mental disorders with certain plants and herbs such as hiṅgu, suvarcalā, and vyoṣa. In the Agni-purāṇa, mention is also made of food and medicine of the elephants. It is observed that plants such as yava, vrīhi, ṣaṣṭika, śāli, godhūma, and ikṣu are recommended diet for the elephants.

Foetal Development: Stages of Growth

The development of a human embryo from the time of fertilisation until the child is born has been well understood in modern medicine. The current state of knowledge informs us of certain key stages of this growth. We see a parallel in Bhāgavata-purāṇa on this subject. In Bhāgavata-purāṇa, the third book discusses the creation process in some detail. In the 31st chapter of Book 3, the first eight verses describe the various stages of the growth of an embryo into a new-born child[15]. The stages described in Bhāgavata-purāṇa can be illustrated using a simple graphic. Figure 4.3 depicts the various stages mentioned in the purāṇa.

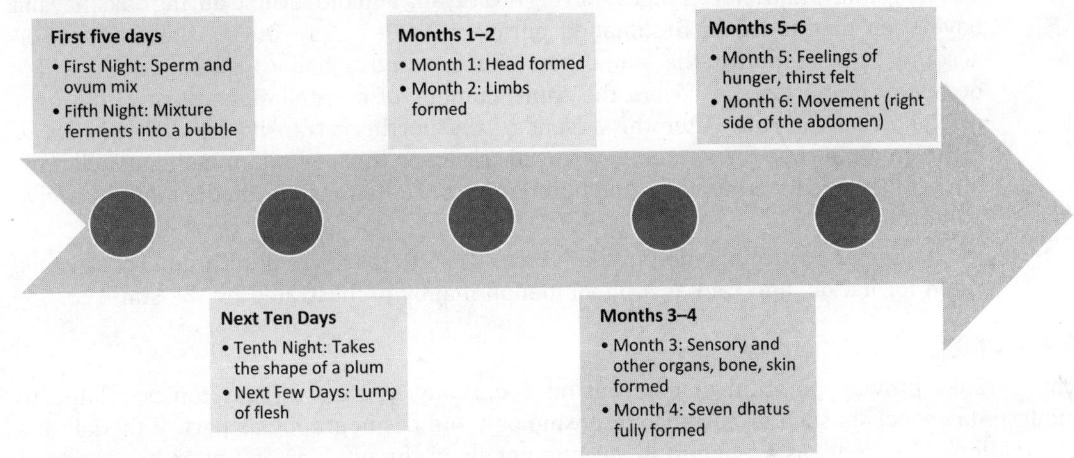

FIGURE 4.3 Stages of Foetal Development as per Bhāgavata-purāṇa

As seen from the figure, in the first ten days after fertilisation, a lump of flesh is formed. In two months, the head and the limbs are formed. During third and the fourth month the sensory organs and the seven basic materials that make up the human body are formed. In the sixth month the foetus begins showing movements. Similar descriptions are found in Agni-purāṇa and they are mostly aligned to the discussion in the āyurvedic texts. Moreover, these descriptions are also aligned with the current understanding of the medical field. However, with the available gadgets and improvements in scanning and other technology we are able to obtain a much deeper undersatnding of the stages of embryonic growth today.

Astronomy

Several purāṇas have discussions on astronomy, which consists of ideas in the pre-telescope age. The issues discussed invariably cover the movement of the Sun and the Moon, day and night, months, and seasons, which are basic ideas in astronomy. There are also discussions related to the stellar region, size of the planets, and other stars. These discussions are found in Brahmāṇḍa-purāṇa (Chapters 21–24), Matsya-purāṇa (Chapters 124–128), Viṣṇu-purāṇa (Chapters 2.08–2.12) and Vāyu-purāṇa (Chapters 1.50–1.53). It may, however, be noted that some of the ideas especially related to the distances, positions, and dimensions of planets are inaccurate and outdated due to the advance of modern astronomy. Let us see a glimpse of some of the details available in Brahmāṇḍa-purāṇa[16]:

- All the basic details of astronomy are specified in Chapter 21. This includes the notion of a solar month and solar year, ṛtus, dakṣiṇāyana and uttarāyaṇa, names of twelve solar months and their mapping to ṛtus, equinoxes and the need for intercalary months in a lunisolar calendar.

- It is mentioned (21.49–21.50), "the two quarters of the East and West are remembered through sunrise and sunset. While it blazes in front, (the heat is felt) behind as well as at the sides. Where the sun appears rising, it is remembered as sunrise to the people there-of, where the sun vanishes, it is called sunset to the people there".
- The longer days and shorter nights during uttarāyaṇa and the reverse during dakṣiṇāyana have been mentioned in Brahmāṇḍa purāna (21.89–21.93), During dakṣiṇāyana, on account of his rapidity, the sun covers thirteen and a half constellations in a day of twelve Muhūrtas. He covers the same number of constellations during the night of eighteen Muhūrtas. After the advent of the northern transit, the day consists of eighteen Muhūrtas, the sun of slow speed traverses thirteen and a half constellations during the day. He covers the same number of constellations during the night of twelve Muhūrtas.
- In Chapter 22 of Brahmāṇḍa-purāṇa (verses 23–60), there are descriptions of seasons, cloud formation, and rain as a phenomenon mainly orchestrated by the sun.

Geography

The purāṇas provide plentiful of material on geography as they discuss topics related to modern-day concepts such as cosmogony, cosmology, and cosmography as part of the lakṣaṇa of 'creation'[17]. These ideas are found in varying details in the purāṇas and often are repeated in all the purāṇas with similar details. There are descriptions of the origin of the Universe and the Earth, oceans, and continents, mountain systems of the world regions, and geography. There are also specific references to the geography of Bhārata. The facts contained require careful analysis and understanding, as at first reading it is like fantasy. However, certain assumptions that the purāṇas have made in their analysis of geography and people concur with current-day thinking and therefore it merits serious study of the matter.

Padma-purāṇa: Composition of the Foetus at the Time of Child Birth

In the Padma-purāṇa, Book 2, Chapter 66, there is a conversation between Yayāti and Mātali (the charioteer of Indra). In response to a question that Yayāti asked about the birth and death of human beings, there is an elaborate description of the birth and death process. In verses 22–38, there is a description of the formation of the embryo and the subsequent stages of foetal development similar to what we find in Bhāgavata-purāṇa.

Further, there are vivid descriptions of the composition of the foetus at the time of its ejection out of the uterus (childbirth) in verses 49–66. Some salient aspects are as follows:

- Bones in the skeleton – three hundred and a hundred more (i.e. four hundred).
- Muscles – five hundred. It is all around covered with small soft hair numbering three crores and half. The body is full of crores of these gross and subtle, visible and invisible, fleshy tubular organs from within.
- The teeth – thirty-two, the nails – twenty.
- The quantity of various substances: bile (in the body) – one kuḍava (160 gms); phlegm – half an āḍhaka (1.28 kg), marrow – five palas (200 gm) and the buttocks are half of it, lump of flesh – five palas; fat – ten palas; thick blood – three palas; the quantity of marrow is four times that of the blood; semen – half a kuḍava.

Source: पद्मपुराणम्/खण्डः २ (भूमिखण्डः)/अध्यायः ०६६ - विकिस्रोतः (wikisource.org); https://www.wisdomlib.org/hinduism/book/the-padma-purana/d/doc364273.html.

The purāṇas, for example, state that the common origin of all races of mankind is traceable to one centre on this planet. This is in accordance with the modern monogenist theory which holds that all varieties of humankind are of single zoological species. The purāṇas also present a picture in which the seven different human groups radiated from the common centre and occupied seven different regions of the world. In each of these regions, human society and civilization developed independently. This view also aligns with the current understanding that the world's primary races belong to seven different climatic regions. The purāṇas state that one of the seven regions is 'Jambū-dvīpa', where the Indian sub-continent (Bhārata-varṣa) is situated.

4.2 ITIHĀSA AS A SOURCE OF WISDOM

The etymology of the word 'Itihāsa' provides us some clues about the nature of the work and its contents. It consists of three components, Iti + ha + āsa (verily did exist thus). For example, 'Rāma lived thus', 'Rāvaṇa lived thus', 'This happened thus', etc. The literature which explains such examples is considered as Itihāsa. This gives us a sense that the word Itihāsa points to 'history'. However, other definitions convey a wider meaning:

- The text which teaches the four goals of life Dharma, Artha, Kāma, and Mokṣa, and which deals with the stories of historical happenings is called Itihāsa[18].
- The text explains the story of the past[19].
- According to Kauṭilya – history (*Purāṇa, Itivṛtta*), tales (*Ākhyāyikā*), illustrative stories (*Udāharaṇa*), *Dharmaśāstra*, and *Arthaśāstra* are (known by the name) *Itihāsa*[20].

Therefore, Itihāsa is not merely a collection of stories related to some past events but an attempt to see the events through the lens of the four puruṣārthas. Though several literary works are considered as Itihāsa, only Rāmāyaṇa and Mahābhārata, also known as the great epics, have earned the fame of two major Itihāsas. At a cursory glance, the separation as Purāṇas and Itihāsas may appear artificial as both focus on narrating stories related to past events. However, there are crucial differences between the two:

- Itihāsas are narrations of the story in which the narrator has been part of the story. Thus, Mahābhārata authored by Vyāsa is a story in which he was also one of the actors, living at the time. So is the case of Vālmīki in Rāmayāṇa.
- In this sense, Itihāsas are not stories about a distant past, as in the case of purāṇas. Therefore, there is a certain element of historicity to the Itihāsas, and these have been dated. On the other hand, establishing the period for the events narrated in purāṇas is not possible as they belong to a very distant past.
- In the case of purāṇas, there are five lakṣaṇas, as we have already seen. This involves discussion of multiple lineages of royal dynasties and ṛṣis. In contrast, in the two Itihāsas, there is a clear focus respectively on a single dynasty. In the case of Rāmāyaṇa, the focus is on the 'solar dynasty' (Sūryavaṃśa) and in the Mahābhārata, the focus is on the 'lunar dynasty' (Candravaṃśa). While the main flow of the story revolves around these respective dynasties, there are narrations of some purāṇic stories in Itihāsas, more as a side story to illustrate some point.

- By design, itihāsas are not meant to be encyclopaedic in nature, as in the case of purāṇas. Instead, the focus is to narrate the events using the lens of the four puruṣārthas, as mentioned already.

4.2.1 Uniqueness of the Two Epics

The two Itihāsas, Rāmāyaṇa and Mahābhārata occupy a very unique position in the repository of the Indian knowledge system. According to Swāmī Vivekānanda, "The Rāmāyaṇa and the Mahābhārata are the two encyclopaedias of the ancient Āryan life and wisdom, portraying an ideal civilisation, which humanity has yet to aspire after[21]". It is a common belief in Indian society that there is nothing that is not covered in these two Itihāsas. Several poetries, texts on governance, and Nīti śāstra have been inspired by these two epics.

> - Itihāsa is stories related to some past events viewed through the lens of the four puruṣārthas.
> - Rāmāyaṇa and Mahābhārata are the two encyclopaedias of the ancient Aryan life and wisdom, portraying an ideal civilisation.

These two great epics have deeply influenced Indian society for time immemorial as is evident from several references to the heroes of the two epics in local matters throughout the length and breadth of the Indian sub-continent. In several sthala-purāṇas and local traditions we find examples such as, 'This rock contains the footprints of Rāma, the mountain got split into two unable to bear the weight of the Hanumān, the colour of this river is like this because mother Sītā had washed her clothes here, this is the cave where the Pāṇḍavas spent their Vanavāsa' etc. Another indication of the great influence of the two epics on society is evident from the fact that classical Indian literary works (mahā-kāvyas) selected the themes from the two epics. For example, Kālidāsa's Raghuvamśa, Kumāradāsa's Jānakī-haraṇa, and Bhaṭṭi's Bhaṭṭi-kāvya (Rāvaṇa-vadha) were based on Rāmāyaṇa. Similarly, Kālidāsa's Abhijñāna-śākuntala, Bhāravi's Kirāṭārjunīya, Māgha's Śiśupāla-vadha and Śrīharṣa's Naiṣadhīya-carita were based on Mahābhārata. Similarly, Indian drama, dance, movies, and other performing arts have been greatly influenced by the two epics.

The two epics have taken different approaches to the issues of life and the propagation of dharma. Interestingly, both address the core issue of dharma and one's duty towards personal, family, and societal obligations. However, by virtue of the differences in their approach, they arrive at different answers to the questions related to the core issue. On account of this, they together serve as a complete handbook for sensible and successful living. Rāmāyāṇa's approach is simple and straightforward. Despite personal sufferings, the duty is clear, there is little scope for ethical and moral dilemmas. Presenting an ideal picture of life, characters, relationships, and values to be pursued, it seeks to elevate an individual to a higher plane. This approach may be called a 'normative approach' to life (essentially suggesting what ought to be done in life).

On the other hand, Mahābhārata has taken what could be referred to as a 'descriptive approach'(essentially illustrating what the reality of life is). In this approach, by presenting the actual happenings, and the moral and ethical dilemmas that different actors faced, the epic showcases the decisions made by the actors and their consequences in the long run. Rather than presenting the ideals of life, it shows life as it unfolds with all its attendant challenges. Instead of informing us what our duty is it merely presents using several anecdotes and nested stories of how various people discharged their duties and what the fallout was. The richness and the value of Mahābhārata lie in the fact that it is full of unsolved ethical riddles and dilemmas, which are faced by ordinary human beings, wherein the characters struggle to find what is right, make mistakes and suffer the full consequences.

It is remarkable to notice that the two epics are not a mere piece of poetic or literary work but a living entity deeply engaging the lives of the Indian people, not merely of the intelligentsia, but also all sections of the society.

4.3 RĀMĀYAṆA – KEY ISSUES AND MESSAGES

Rāmāyaṇa consists of seven books which are called as Kāṇḍas, organized into 645 sargas (chapters), and 23,672 verses[22]. Rāmāyaṇa is a vast history of Rāma but also consists of many other related instances. In India, we have more than 40 versions of Rāmāyaṇa in different languages like the Rāmacaritmānas in Avadhi, Kamba-Rāmāyaṇa in Tamil, Mādhava-Kaṇḍali in Assamese, Goṇa Budha's Raṅganātha-Rāmayaṇa in Telugu, Jagamohan-Rāmāyaṇa in Oriya, Narahari-Rāmāyaṇa in Kannada, etc. The story of Rāma is very well known in most parts of the world. It has been directly written in various languages in different countries. The Vālmīki-Rāmāyaṇa is translated into many languages. There are reportedly around three hundred versions of Rāmāyaṇa.

- Rāmāyaṇa takes a 'normative' approach while Mahābhārata takes a 'descriptive' approach to present events and decisions.
- Indian drama, dance, and other performing arts have been greatly influenced by the two epics.

The Rāmāyaṇa has played a significant role in cultural transmission between Indian and other Asian countries. It has travelled from India to other parts of Asia such as China, Turkestan, Burma, Thailand, Java, Cambodia, and Vietnam during the first millennium in the common era. It is interesting to note that almost all countries of Southeast Asia have the Rāma story, albeit in varied forms and content. From the thirteenth century onwards, several Thai kings assumed the title Rāma, and the tradition continues. According to Rāmāyaṇa, Rāma is not merely a human being but an embodiment and a living entity of the concept of Dharma. Through several characters, Rāmāyaṇa provides a rich opportunity to understand important aspects of peaceful co-existence and successful living.

Bāla-kāṇḍa deals with the birth and childhood of Rāma and his brothers. Viśvāmitra takes along with him Rāma and Lakṣmaṇa through the forest and eventually to Mithilā leading to the marriage of Sītā to Rāma. One of the key messages that we can glean from this section is the importance of knowing the reality of life in the country by the king-to-be. Rāma's journey with Viśvāmitra through the forest and villages provided an opportunity for him to develop this awareness. This reminds us of an important principle stated in modern management parlance. In organisations, great leadership requires a good knowledge of the issues and realities facing the organisation. This is obtained only when the leader can connect to the people and situations in the organisation physically and mentally.

Ayodhyā-kāṇḍa describes the fast turn of events in which an attempt to coronate Rāma as the next heir apparent to the throne by Daśaratha was aborted and instead Rāma was sent to the forest for 14 years. The desire of Kaikeyī, the mother of Bharata, to install him as the prince went futile and Daśaratha died of shock and grief. This part of the epic provides an important message to the young that life can be a roller coaster ride. There is a need to understand that events in everyone's life can turn adversarial at any time. Rāma demonstrates that to be successful in life one needs to develop mental equanimity to handle such unforeseen events in one's life. The conversation between Bharata and Rāma in the final portions of this section contains a wealth of information on public policy and administration and larger issues of life. We shall see some aspects of this in Chapter 14 of the book.

Araṇya-kāṇḍa marks the large-scale elimination of rākṣasas in the forest in response to the request from the sages and seers, who were constantly harassed by these people. It also portrays the forceful and deceitful kidnapping of Sītā by Rāvaṇa. The epic reminds us that goodness needs to be protected from evil forces, lest it becomes unsustainable. The conversation between Mārīca and Rāvaṇa brings out an important message that while removing a poisonous snake that entered into a freshwater lake, the only source of livelihood for a village, 1000's of good fishes will die for no fault of theirs. It reminds us of the fact that an evil action by an individual will eventually trigger a bigger reaction from the system and in the process of eliminating the evil, several good aspects may also get destroyed.

Kiṣkindhā-kāṇḍa describes the efforts on the part of Rāma and Lakṣmaṇa to find Sitā, who was abducted by Rāvaṇa. The grief that they underwent was unparalleled as they were not even aware of what happened to Sītā and who took her away. The meeting of Sugrīva through the good offices of Hanumān instilled a ray of hope in their search. The epic brings out the importance of friendship, especially during periods of great distress. It also reminds us of the importance of wise counselling to wade through the challenges that we face in our life.

Sundara-kāṇḍa describes the heroic efforts of Hanumān to cross the ocean to locate Sītā and repose confidence in her. It also describes the horrifying state of affairs for Sītā and the grit and determination with which she faced the situation. The epic beautifully demonstrates the strength of the character of Sītā, with which she discovered unlimited courage within and faced the adversity with grace and purpose, which can inspire the womenfolk. Other messages include statecraft, dealing with ambassadors, and the presence of mind and tact required for assessing the strength of an opponent.

- Rāmāyaṇa encourages one to contemplate the difference between 'goodness' and 'greatness' and the need to have both to succeed in life.
- Mahābhārata discusses a variety of topics including law, philosophy, religion, and custom.

Yuddha-kāṇḍa describes the royal battle that erupted between Sugrīva's army and Rāvaṇa's army, eventually culminating in large-scale elimination of rākṣasas and their king Rāvaṇa as prophesied by Mārīca. Before the war, Rāvaṇa bluntly dismissed several attempts by his near and dear to instil a sense in him by pointing to the dangers of the imminent war and the need to broker peace with Rāma by giving up Sītā. Kumbhakarṇa while advising Rāvaṇa says, "One who takes the timely advice concerning Dharma, Artha and Kāma, about the pros and cons of the deeds from his scholarly councilors never gets into trouble[23]". Rāvaṇa was very powerful but he flouted the advice given by his great councilors and well-wishers like Vibhīṣaṇa, Mārīca, Mālyavān, Maṇḍodarī, and Kumbhakarṇa. This led to his downfall and death. The epic forcefully brings out the benefits of wise counseling and the dangers of hasty decisions arising out of an inebriated mind polluted with attachment to unreasonable and unethical desires.

Uttara-kāṇḍa has a collection of related stories not necessarily in chronological order as in the case of the previous books. It discusses certain events in Rāvaṇa's life not mentioned in the other books, the birth of Lava and Kuśa and the departure of Rāma and Sītā from this world.

The epic throughout depicts Rāma and Rāvaṇa in equal measure when it comes to 'greatness'. While describing the final encounter between Rāma and Rāvaṇa, Vālmīki remarks that it is impossible to find a simile to describe this, "Just as the vastness of the sky has no simile and the majesty of the ocean also does not have a simile, in the same manner, there is no analogy to the war between Rāma and Rāvaṇa"[24]. Unfortunately, in terms of 'goodness' Rāvaṇa was no match to Rāma. The epic encourages one to contemplate the difference between 'goodness' and 'greatness' and the need to have both to succeed in life. Greatness without 'goodness' is a definite recipe for eventual disaster as demonstrated by Rāvaṇa. On the other hand, 'goodness' without 'greatness' may not enable one to cover much ground in his life.

Rāma lived as a human being and faced the day-to-day situations and conflicts of life from an ordinary human plane. This makes Rāmāyaṇa special and realistic. The Rāmāyaṇa describes situations where there are moral dilemmas and conflicts between two rights. Rāma was a perfect role model, who lived through the good and enjoyable aspects of life yet in a detached fashion. He was detached from power, wealth, greed, fame, and desire. He was kind and even minded to even those who opposed him. His ability to face difficult situations in life and to gracefully handle success and tragedy is an excellent example to follow.

The Vālmīki Rāmāyaṇa beautifully depicts the dharma of personal life, family life, and social order. The human characters in Rāma, Lakṣmaṇa, Bharata, Kausalyā, and Sītā bring out the principles of ideal living. It emphasises the ideal relationships between father and sons, between brothers, and between husband and wife. The concept of dharma as the way to a meaningful life could be a powerful theme for positive mental health. Many conflicts in life can be resolved using this perspective. These anecdotes, the conflicts, and the situations are the real nuggets of wisdom for the people.

4.4 MAHĀBHĀRATA – A SOURCEBOOK FOR WORLDLY WISDOM

Mahābhārata is one of the greatest poems in the world composed by Kṛṣṇa-Dvaipāyana (Veda Vyāsa). According to the available information, the original and first version of Mahābhārata which was named Jaya consisted of only about 8,000 verses and then the second version had 24,000 verses which was named Bhārata[25]. It was later extended to up to 1,00,000 verses and was named Mahābhārata. There are multiple versions of the epic available. Therefore, a long-term project under the auspices of BORI, Pune started in 1919 under V.S. Sukthankar to prepare a critical edition of Mahābhārata. After collating information from 1,259 manuscripts and several years of work, a critical edition was finally brought out in 1966. The critical edition consists of more than 89,000 verses organised under 18 Parvas.

Multiple editions and redaction of the original text of Mahābhārata may appear to be a cause of concern. However, the leading incidents and characters of the epic have not been affected and modified by these changes. Mahābhārata depicts the trials and tribulations that various kings of the lunar dynasty went through. To a large extent, the story focuses on the dispute between two families, the Kauravas and the Pāṇḍavas. The story unfolds steadily in the text but is interspersed throughout with a variety of topics including law, philosophy, religion, custom, and to some extent, geography and cosmography combined with several episodes and legends adding to the richness of the message. Table 4.2 concisely presents the broad organisation of the text and the major topics covered in the text. The issues discussed in the text can be broadly summarized under five major heads:

- Mahābhārata presents the reality of life and allows the reader to pick up their lessons for living.
- Nītis are the guidance to take a person to the path of Dharma by providing the right perspectives about life and life goals.

TABLE 4.2 Overall Structure and Organisation of Topics in Mahābhārata

Parva	*Chapters*	*Topics*
Ādi	225	Stories of the ancestors of Kuaravas and Pāṇḍavas; birth and education of Kauravas and Pāṇḍavas; enmity in their childhood; Draupadī's marriage to Pāṇḍavas; Arjuna's marriage to Subhadrā
Sabhā	72	Performance of Rājasūya yajña by Yudhiṣṭhira, the game of dice maneuvered by Duryodhana, and its tragic consequences
Āraṇyaka	299	Pāṇḍavas in exile, several stories from the past, famous conversation known as Yakṣapraśna
Virāṭa	67	The stay of the Pāṇḍavas incognito in the kingdom of Virāṭa, Virāṭa's princess Uttarā gets married to Abhimanyu
Udyoga	197	Unsuccessful peace parleys and preparation for the war, the famous discourse of sage Sanatsujāta to Dhṛtarāṣṭra
Bhīṣma	117	Śrimad *Bhagavad Gītā,* detailed descriptions of the first ten days of the war, Bhīṣma mortally wounded by Arjuna
Droṇa	173	The heroic exploits of Droṇa and his death through a planned strategy; Abhimanyu's tragic death
Karṇa	69	Death of Duśśāsana, the second of the Kaurava brothers, fall of Karṇa at the hands of Arjuna
Śalya	64	Final encounter between Bhīma and Duryodhana on the last day of the war
Sauptika	18	The gruesome massacre of the Pāṇḍava army in the night during sleep by Aśvatthāman
Strī	27	Lamentations of the widows of the dead warriors
Śānti	353	Wonderful discourses on all aspects of *dharma* by the patriarch Bhīṣma at the request of Yudhiṣṭhira, Bhīṣma's demise, Yudhiṣṭhira's coronation
Anuśāsana	154	
Āśvamedhika	96	Departure of Śrī Kṛṣṇa to Dvārakā, Aśvamedha Yajña by Yudhiṣṭhira
Āśramavāsika	47	Departure of Dhṛtarāṣṭra to the forest along with Gāndhārī, and Kuntī; their subsequent death in a forest fire
Mausala	9	Mutual destruction of the Yādava heroes and also the death of Śrī Kṛṣṇa
Mahāprasthānika	3	Final journey of the Pāṇḍavas, their death on the way, Yudhiṣṭhira alone reaching heaven
Svargārohaṇa	5	
Harivamśa	118	Supplement of Mahābhārata
Total	**2,113**	

- **Birth, Childhood, and Rivalry:** The first two books (Ādi Parva, Sabhā Parva) of Mahābhārata describe in detail the ancestral details of the Kauravas and the Pāṇḍavas, their birth, the early childhood and the growing rivalry between the cousins of the two families. The growth both in terms of the popularity of Yudhiṣṭhira and his brothers, their material richness, and the respect that they commanded was seeding jealousy in Duryodhana and his brothers and it reached a flashpoint in winning a dubious game of dice and ill-treating Draupadī. Ironically, several elders, well-wishers, and wise people were party to this treacherous act, directly or otherwise.
- **Banishing of Pāṇḍavas:** The next two books (Āraṇyaka-parva, Virāṭa-parva) of Mahābhārata vividly describe a variety of events that unfolded when the Pāṇḍavas were asked to go to the forest as a compromise to restore their kingdom if they successfully complete the tenure. The possibility of unfair treatment on their return ought to have loomed large in their minds, based on their past experience of dealing with the Kauravas. Therefore, Pāṇḍavas make use of this time to gather enough strength and support, and blessings from the Gods to face any eventual war.
- **Failed peace parleys leading to war:** The next seven books starting from Udyoga-parva describe the failed efforts from both the Pāṇḍava side and the Kaurava side to avert the war. It is followed by the description of the great war that erupted which resulted in total decimation of the Kaurava side, killing of Duryodhana and all his Kaurava brothers and several prominent people on both sides of the army and ends with the lamentations of the widows and the ladies.
- **Post-war conversation between Yudhiṣṭhira and Bhīṣma:** The next two books (Śānti-parva and Anuśāsana-parva) deals with the detailed conversation between Yudhiṣṭhira, who is going to be coronated as the king and Bhīṣma, who was waiting to give up his mortal body. This section has a wealth of information on several aspects of life, public policy and administration, spiritual and materialistic aspects of life, dharma, and the other puruṣārthas.
- **The exit of the characters:** In the last five books starting from Āśvamedhika-parva, the epic discusses the role of Yudhiṣṭhira as a king. This is followed by the exit of all the remaining main characters, including Pāṇḍavas, Draupadī, Kṛṣṇa, Dhṛtarāṣṭra, and his cohort Gāndhārī, and Kuntī.

At a cursory level, Mahābhārata comes across as merely a historical account of two warring factions of the Pāṇḍavas and Kauravas. However, a deeper understanding of the epic will reveal to us several hidden treasures of wisdom. Rather than presenting an ideal, 'textbook' like character, the epic presents the reality of life and allows the reader to pick up their lessons for living. The epic presents almost all the characters in different shades of grey bringing out the reality of decision making and the dilemmas that one faces in life. Even Kṛṣṇa, who is considered omnipotent and worshipped as a god is portrayed objectively. There is no attempt to conceal the faults of the characters, be it the venerable characters such as Bhīṣma and Droṇa or others such as the Pāṇḍavas. For example, as a part of the dying declaration, Duryodhana puts to shame some acts of the Pāṇḍavaṣ and points to the instances wherein they had cheated.

The topics in Mahābhārata has such a wide canvas that it covers several aspects of human life. Therefore, it is observed, "whatever knowledge related to Dharma, Artha, Kāma, and

Mokṣa found in the Mahābhārata may be found elsewhere; but what is not here may not be found elsewhere"[26]. Unlike the Rāmāyaṇa, the stories in Mahābhārata are several, rich in their content and cover a wide range of issues of interest to the society. Anuśāsana-parva provides several interesting insights on governance and administration. In the conversation with Bhīṣma, Yudhiṣṭhira brings forth the moral conflicts and ethical dilemmas that he faces in administration and seeks advice from Bhīṣma. In this Parva, several principles related to Dharma and Artha could be found, which will find relevance in modern-day economics. It also deals with the issue of charity and provides guidelines for individual duty and code of conduct.

The Conversation between Yudhiṣṭhira and a Yakṣa

The episode involving the conversation between Yudhiṣṭhira and a Yakṣa found at the end of Āraṇyaka-parva is an example of using interesting episodes as a medium to share wisdom of life. In an episode in which the Pāṇḍavas visit a mysterious lake to quench their thirst, Yudhiṣṭhira was the last to visit the lake, the other four were found lying dead near the lake. Yudhiṣṭhira encounters a Yakṣa and there is an interesting conversation between the two.

During the conversation more than 126 questions were asked by Yakṣa, one of which was, "What is the most surprising aspect of life?" He replied, "Everyday several living beings die and yet those who are still living desire a state of permanence in the world". This encourages one to reflect the necessity of developing a certain level of detachment thereby keeping aloof from overindulgence in materialistic pursuits. The epic has several such life-altering lessons. Table 4.3 has a list of some of the questions and the answers.

TABLE 4.3 Some Words of Wisdom from the Yakṣa-praśna Episode

Sl. No.	*Questions*	*Answers*
1	What remains the most beneficial when it is falling?	Rain
2	What remains the most beneficial even when it is thrown away?	Seed
3	What is faster than the wind?	Mind
4	What is more numerous than grass?	Thoughts
5	Who is the householder's friend?	Wife
6	Who is the friend of a dying person?	Charity
7	What is the greatest gain?	Health
8	What is the greatest happiness?	Contentment
9	By giving up what can one become lovable?	Pride
10	By giving up what will one not feel sad?	Anger
11	By giving up what can one become wealthy?	Desire
12	By giving up what can one become happy?	Greed

Dharma is arguably a very intricate subject, not lending itself to a simple bulleted set of ideas. Communicating this idea is not an easy task. On the other hand, staying away from the subject citing reasons of complexity in articulating the idea is not desirable. The value of several anecdotes and nested stories in Mahābhārata lies in tiding over this difficulty. For example, during the conversation with Yudhiṣṭhira, Bhīṣma brings the point that dharma varies in accordance with the demands of the situation: there is one dharma for moments of strength and another in times of distress. Even during a famine, a king ought to maintain his exchequer and his army. Since wealth is critical to meet these requirements, a king may use appropriate means to obtain wealth. In such cases, he will not incur sin. Figure 4.4 provides a summary of some salient aspects of the two epics as evident from the above discussions.

Rāmāyaṇa	Mahābhārata
• 645 chapters, 23,672 verses • Normative approach – Ideal 'textbook' like characters of life • All countries in SE Asia have the Rāma story • Descriptions of dharma of personal and social life and social order • Notion of 'Goodness' vs 'Greatness'	• 1,00,000 verses, over 2,000 chapters • Descriptive approach – Characters present the 'reality' of life • Rich collection of deeply nested stories • Many niggets of wisdom – Yakṣa Praśna, Vidura Nīti, Sanat-sujātīya, Bhagavad Gītā

FIGURE 4.4 A Summary of Some Salient Aspects of the Two Epics

4.5 NĪTI-ŚĀSTRAS – COLLECTION OF SNIPPETS OF WISDOM

Nītis are the guidance to take a person to the path of Dharma by providing the right perspectives about life and life goals. The role of Nīti-śāstras lies in providing us helpful directions to resolve such conflicts by providing snippets of wisdom culled out or distilled from past experiences. It utilises verses set in poetic metres to concisely communicate the message.

In the ancient Indian tradition, our ancestors have identified four principal categories of goals in human life, i.e., Dharma, Artha, Kāma, and Mokṣa. Dharma is the enabling framework for leading a life that is self-fulfilling, sustainable, and allows us to peacefully co-exist with other human beings and other entities in the Universe. It is the nearest explanation for modern-day sustainability and provides us a broad set of dos and don'ts to achieve our life goals. It is the basic edifice for us to build our mansion of life. Artha provides the material wherewithal to fulfil our life goals, needs, and wants. It addresses the larger issues of health, well-being, and wealth required to satisfy our basic needs and other wants not in conflict with dharma. Kāma on the other hand, is the set of desires, spanning across biological, physiological, and intellectual dimensions that one wants to attain in this lifetime. Mokṣa is the liberation of an individual from the clutches and limitations of life and a means to realize one's true potential and nature.

Nīti texts take into consideration the very basic rule that all puruṣārthas are generally guaranteed to the individuals by the Universe. It is when he realises all the four types of goals his life is comprehensively successful. Nīti acknowledges this fact and helps people in attaining the same. However, Nīti-śāstras predominantly focus on the first three goals. It is assumed that once an individual perfects the art of managing the first three goals in a non-conflicting manner, his transition to the fourth will happen seamlessly and naturally. In practical life, human consequences are so varied that it is not so easy to pursue one goal at a time. There arise conflicts between various goals and one finds it difficult to resolve decision-making dilemmas in such situations. To steer individuals forward in these tough situations, Nīti-śāstra provides helpful guidelines that help in the trajectory of an individual's life such that such conflicts are reduced to the minimum.

Though there is no specific classification of nīti texts, from the available works we can broadly classify the repository into two: Nitya or Sāmānya-nīti and Rāja-nīti.

- The unit of analysis in the case of Sāmānya-nīti is the individual. These texts focus on the good aspects of living and provide helpful insights on the code of conduct, differences between a wise person and others, good and evil, and success and failure in life pursuits. It also emphasises the need for harmonious living in society. For example, the Nīti-śataka, which is an example of a text dealing with sāmānya-nīti is about various aspects of conduct in this world, such as the nobleness in adhering to a promise, the value of learning, self-respect, valour, moral courage, and large-mindedness. Besides, it inculcates the importance of industry, the traits of a true friend, and a contrast of the conduct of a virtuous and noble-hearted man with that of a selfish and little-minded individual.
- Rāja-nīti, on the other hand, focuses on the political and administrative tasks and provides ideas on several dos and don'ts for a king. Arthaśāstra of Kauṭilya is an important text in this category. Other works such as Pañcatantra, Nīti-sāra of Kāmandaka are inspired by Kauṭilya's work. The Śānti-parva of Mahābhārata is a nīti text as it has considerable material on both Rāja-nīti and Sāmanya-nīti. Similarly, the kaccit sarga in Ayodhyā-kāṇda in Rāmayāṇa has a good discussion on Rāja-nīti. We discuss Rāja-nīti aspects in detail using Arthaśāstra in Chapter 14 of the book.

Strict separation of a text into one of the two may not be possible always. For example, Vidura-nīti and Pañcatantra have topics relevant to both. However, wiṭhin any text, we can delineate these two themes unambiguously. Table 4.4 has a brief list of some popular nīti texts for Sāmānya-nīti. The Rāja-nīti texts will be covered in Chapter 14 of the book.

TABLE 4.4 Some Selected Nīti Texts

Sl. No.	*Name of the Work*	*Brief Description of the Work*
1	Bārhaspatyam	Must be considered only as redactions of the śāstra by later scholars since both these belong to oral tradition. Many of the chapters and contents are extinct.
2	Śukra-nīti	
3	Vidura-nīti (Part of Mahābhārata)	Consists of 700 verses – response to Dhṛtarāṣṭra's questions – discusses elements of good living, personal values, governance and administration, etc.
4	Pañcatantra (Viṣṇu Śarman)	Uses a series of fables to impart principles of ethical and moral living as well as elements of Rāja-Nīti.
5	Hitopadeśa (Nārāyaṇa Paṇḍita)	Inspired by Pañcatantra, follows a similar style of using fables to impart principles of ethical and moral living and elements of Rāja-nīti.
6	Nīti-śataka (Bartṛhari)	A collection of 100 verses to impart various aspects of Sāmānya-nīti.

Virtues of a Wise-person: Insights from Vidura-Nīti

Vidura-nīti is a collection of 700 verses in Mahābhārata, that describes the conversation between Dhṛtarāṣṭra and his brother Vidura, just before the war is to begin. In fact, Dhṛtarāṣṭra knew for certain that the war is imminent as multiple efforts to broker peace between the cousins failed. He became restless and suffered from sleeplessness. When he mentioned his difficult situation to Vidura, and sought his wise counselling. Vidura replied, "A thief, a lustful person, one who has lost all his wealth, one who failed to achieve success, weak and one who has been attacked by a strong person suffer from sleeplessness". He hoped that the king does not belong to any of these six categories, nor is he suffering from coveting other's wealth. Over the next 680 verses he discussed several aspects of good living and good governance.

During the course of the long conversation, Vidura enumerated several qualities of a wise person. Chapter 1 of Vidura-nīti (verses 20–34) has these details. Some of them are listed below:

- One who knows the self and follows the Dharma;
- One who is not drawn by the anger, exultation, pride, shame, stupefaction and vanity;
- One who practices the good and denies the bad;
- One whose work or secrecy is not known by others till it is accomplished;
- One whose acts are not disturbed by cold, heat, fear, lust, prosperity or adversity;
- Never disregards any task; never strives for unattainable; never grieves for the lost and never loses the senses during hard times;
- Neither revels at honours nor grieves at slights, remains cool and unagitated like a lake in the course of Ganga;
- Strives, having commenced anything till it is completed, who never wastes his time, and who has his soul under control;
- Exerts to the best of his might, and disregards nothing as insignificant.

For details, see, **Rustagi, U. (2009).** *"Viduranītiḥ",* J.P. Publishing House, Delhi. ISBN: 81-86702-47-4. Ch. 1, Verses 20–34, pp. 14–18.

4.6 SUBHĀṢITAS – A COLLECTION OF INSIGHTS FROM VARIOUS TEXTS

The concise poetry which communicates the valuable message for the society with some interesting examples is called Subhāṣita. These are nothing but a systematic collection of insightful observations, examples, and principles of life from a wide range of existing repositories of literary works. The motivation for putting together an interesting collection of ideas from various texts comes from certain traditions of the great Indian poets. The poets had two objectives in mind while composing kāvyas; one is to narrate the chosen story, the other is to contextually bring worldly wisdom using the opportunity of narrating certain incidents in the story. Consider the following examples:

EXAMPLE 1: In the first chapter of Sundara-kāṇḍa, Vālmīki vividly narrates the journey Hanumān takes in crossing over the seas to reach Laṇkā in search of Sītā. As soon as Hanumān reaches the other side of the shore Vālmīki makes an interesting observation about the qualities required to achieve success in one's pursuits:

> *O Lord of vānara he in whom these four qualities, fortitude, vision, intelligence, and dexterity exist, as in you, will achieve his mission and not get lost.*[27]

EXAMPLE 2: In Kirātārjunīya, a mahākāvya by Kavi Bhāravi, one finds several valuable messages for us to imbibe. Consider the following messages:

A blade of grass bends due to a lack of strength. It flies away due to a lack of weight. A person without honor is just like grass – without inner strength or weight.[28]

If a person does not deal with a crook in a crooked manner, he is bound to face defeat. They must be paid back in their coin. Just like how an arrow most certainly hits (and hurts) that part of the body (however small) that is not covered by the armor, such deceitful people search and hurt only those people who are innocent and incapable of deceit themselves.[29]

When Bhīma was impatient and angry and was keen to take revenge on Kauravas, Yudhiṣṭhira had to calm him and at this point, the following message appears:

One should not act in haste. Action without thinking leads to big problems. Wealth chooses that person by itself who acts after thinking properly.[30]

EXAMPLE 3: In the Udyoga-parva of Mahābhārata we find the following interesting message:

There are five kinds of people who one must listen to and respect – knowledgeable people, pious people, close friends, well-educated people, and the elderly. Such a person will never tread along the wrong path due to ignorance.[31]

It follows from the above examples, that it may be worthwhile to cull out such insightful messages from the existing repository and present such a collection in a single work. Subhāṣitas are such classified messages, usually organized under specific themes. For example, all ideas on friendship from various texts could be arranged into one section, the value and importance of dharma into another, and so on. The themes and messages found in subhāṣitas are so varied that it forms nearly an encyclopaedic material when it comes to inculcating values, ideas, and principles for sensible living and addressing several issues that we may face in our lives. The genre is also varied ranging from prayers to the almighty, imparting serious values through curt messages, fun-filled verses, pun, and riddles. We shall see some examples.

Verses Teaching Morality, Philosophical Truths

- Subhāṣitas are systematic collection of insightful observations, examples, and principles of life from existing repositories of literary works.
- The themes and messages in subhāṣitas form nearly an encyclopedic material inculcating values, ideas, and principles for sensible living.

Scriptures and disciplines are many, obstacles are many during learning those and the lifespan is very less. So, one must try to learn the quintessence of the scriptures like the way the swan takes (separates) milk from the water.[32]

Sorrow is involved in both earning the wealth as well as protecting it. There is a pain in gain and pain in loss. Alas! Wealth is always the resort of grief.[33]

The flowers in the palms make both the palms fragrant. Similarly, the affection of people with a kind heart is equal on both sides. They never differentiate one from another and respect and love all equally.[34]

Verses for Pure Entertainment

O king of doctors, the brother of Yamarāja (lord of death) salutations to you. Yama only takes life, but you take both lives as well as the money.[35]

Brain Teasers, Puzzles Based on a Pun

This is a poem in which there is a dialogue between Gopī and Kṛṣṇa where Gopī is enquiring something and Kṛṣṇa answers that. Whatever Kṛṣṇa answers have two meanings and Gopī takes the second meaning and plays with Kṛṣṇa: [36]

(*Gopī*) : *Who is tapping the door with the finger?*
(*Kṛṣṇa*) : *O! crooked one! (I am) Mādhava. (also means spring season).*
(*Gopī*) : *Is it the spring season?*
(*Kṛṣṇa*) : *No, (I am) Cakrapāṇi the one with the Discuss in hand (also means a potter).*
(*Gopī*) : *Is it a potter?*
(*Kṛṣṇa*) : *No, I am the one bearing the world.*
(*Gopī*) : *Is it the serpent Śeṣa (who is said to bear the world on his head), the king of the snakes?*
(*Kṛṣṇa*) : *Not me, (I am) the one who controlled the frightful snake (Kāliya).*
(*Gopī*) : *Are you Garuda, the king of birds? (and enemy of snakes).*
(*Kṛṣṇa*) : *No, (I am) Hari (also means a monkey).*
(*Gopī*) : *Are you the chief of the monkeys?*

In this manner, Lord Padmanābha who was not smart and was unable to respond to the Gopī may protect me.

Though the messages contained in the subhāṣitas are relevant to all the ages in the Indian tradition these are specially taught to the children so that they can imbibe these ideas while they are young and can make use of this wisdom all through their life. As a part of subhāṣitas, they are taught the value of Dharma, knowledge, truthfulness, friendship, love towards nation, etc.

Selected Subhāṣitas on Friendship from Nīti-śataka

There are several subhāṣitas that extol the virtues of friendship, the value, and true attributes of good friendship. We shall see some of them from Nīti-śataka here:

आरम्भगुर्वी क्षयिणी क्रमेण लघ्वी पुरा वृद्धिमुपैति पश्चात् ।
दिनस्य पूर्वार्द्धपरार्द्धभिन्ना छायेव मैत्री खलसज्जनानाम् ॥ 50
Ārambha-gurvī kṣayiṇī krameṇa laghvī purā vṛddhim-upaiti paścāt |
dinasya pūrvārddha-parārddha-bhinnā chāyeva maitrī khala-sajjanānām || 50

In the morning the shadows are long but get shortened as the day advances. In the afternoon shadows are short but they get longer and longer as the Sun gradually sets. The friendship with wicked persons and the noble ones decreases and increases accordingly.

संतप्तायसि संस्थितस्य पयसो नामापि न श्रूयते मुक्ताकारतया तदेव नलिनीपत्रस्थितं दृश्यते |
अन्तः सागरशुक्तिमध्यपतितं तन्मौक्तिकं जायते प्रायेणाधममध्यमोत्तमजुषामेवंविधा वृत्तयः || 59
saṃtaptāyasi saṃsthitasya payaso nāmāpi na śrūyate muktākāratayā tadeva nalinī-patra-sthitaṃ dṛśyate |
antaḥ sāgara-śukti-madhya-patitaṃ tan-mauktikaṃ jāyate prāyeṇādhama-madhyamottama-juṣāmevaṃ-vidhā vṛttayaḥ || 59

A drop of water instantly gets burnt up when it falls on a hot piece of iron. But when it falls on a lotus leaf it shines like a pearl. When it falls into an oyster shell in the ocean on the day of Svāti nakṣatra, it becomes a real pearl. The quality of contact determines the result of such a company.

दुर्जनः परिहर्तव्यो विद्ययालंकृतोऽपि सन् | मणिना भूषितः सर्पः किमसौ न भयंकरः || 43
durjanaḥ parihartavyo vidyayālaṃkṛto'pi san | maṇinā bhūṣitaḥ sarpaḥ kimasau na bhayaṃkaraḥ ||43

A wicked person should be avoided even if he is learned. Is not a Cobra dangerous even if it is wearing a jewel on its head?

SUMMARY

- With a long civilisational history, India has a rich body of knowledge and experience that has developed within the society from time to time, captured through a variety of literary works.
- The purāṇic repository provides valuable insights into the history of philosophy and religion and is a storehouse that provides deep insights into all aspects of the society, culture, and living during the ancient and medieval times in India.
- The purāṇic repository can be broadly divided into three; Mahā-purāṇas (Major Purāṇas), Upa-purāṇas (Minor Purāṇas), and Sthala-purāṇas.
- There are 18 mahā-purāṇas and they are estimated to contain among them about 4,00,000 verses. The number of Upa-purāṇas cannot be established with certainty, though about 100 of them can be enumerated from different sources. There are many Sthala-purāṇās, in different languages of India, most of them in Sanskrit.
- The purāṇas contain several discussions related to the plant kingdom and its role in addressing food, medicine, and other requirements of society. Since one of the themes of the purāṇas is to describe the creation process, there are discussions on the stages of childbirth from the embryo.
- Several purāṇas have discussions on astronomy, which consists of ideas in the pre-telescope age. The purāṇas have material on geography as they discuss topics related to modern-day concepts such as cosmogony, cosmology, and cosmography.
- Itihāsa is not merely a collection of stories related to some past events but an attempt to see the events through the lens of the four puruṣārthas.
- The two epics, Rāmāyaṇa and Mahābhārata address the core issue of dharma and one's duty towards personal, family, and societal obligations, but have taken different approaches.
- The Vālmīki Rāmāyaṇa beautifully depicts the dharma of personal life, family life, and social order. The anecdotes, the conflicts, and the situations discussed in the epic are the real nuggets of wisdom for the people.
- Mahābhārata depicts the trials and tribulations that various kings of the lunar dynasty went through. The story is interspersed throughout with a variety of topics including law, philosophy, religion, custom, and to some extent, geography and cosmography.
- Nītis are the guidance to take a person to the path of Dharma by providing the right perspectives about life and life goals. They do not take the approach of storytelling but utilizes verses set in poetic metres to concisely communicate the message.
- Sāmānya-nīti focuses on the good aspects of living and provides helpful insights on the code of conduct, differences between a wise person and others, good and evil, and success and failure in life pursuits.
- Rāja-nīti focuses on the political and administrative tasks and provides ideas on several do's and don'ts for a king.
- The concise poetry which communicates the valuable message for the society with some interesting examples is called Subhāṣita. These are nothing but a systematic collection of insightful observations, examples, and principles of life from a wide range of existing repositories of literary works.
- The messages contained in the subhāṣitas are relevant to all the ages in the Indian tradition; these are specially taught to the children so that they can imbibe these ideas while they are young and can make use of this wisdom all through their life.

REVIEW QUESTIONS

1. Briefly enumerate the composition of the purāṇic repository. What is the use of the purāṇic knowledge?
2. What are the key characteristics of purāṇas? How does it help in assessing the contents of a purāṇa?
3. Prepare a one-page note on the issues of contemporary interest discussed in purāṇas.
4. What are some of the observations found in the purāṇas on foetal development?
5. What do you understand by the term 'Itihāsa'? What issues do they address?
6. What are the main differences between a purāṇa and an itihāsa?
7. Comment on the statement, "Rāmāyaṇa and Mahābhārata have taken contrasting approaches to bring the key message of life".
8. Summarize the key messages found in each kāṇḍa of Rāmāyaṇa. What is the relevance of these messages to contemporary society?
9. Briefly enumerate the key contributions of Mahābhārata to contemporary society.
10. What do you mean by nīti-śāstra? What are the key issues discussed in such works?
11. How is sāmānya-nīti different from rāja-nīti? What are some of the contributions of these texts? Provide one or two examples.
12. Comment on the statement, "Subhāṣitas, Nīti-śāstras, and Itihāsas address the same set of issues".

DISCOVER IKS

1. Rāmāyaṇa is a historical story that has closely weaved into the cultural ethos and daily lives of the Indian society for time immemorial. Despite this, using the current lens to analyse such historical works, we face certain questions. Watch this video of Sadguru Jaggi Vasudev, in which some of these issues come up as part of the discussion: https://www.youtube.com/watch?v=PoSlhO-hYC8. After watching the video, prepare a two pages note to answer the following questions.
    - Is it fair to consider Rāmāyaṇa a mythological story or a historical piece? Why or why not?
    - Are there interesting take-aways from Rāmāyaṇa, even though the epic is more than 6000 years old?
    - Despite certain difficult events described in the Rāmāyaṇa, why is the character Rāma worshipped even today?

SUGGESTED READINGS

Acharya, N.R. (2011). *Subhāṣita-ratna-bhāṇḍāgāra*, Chaukhamba Sanskrit Series Office, Varanasi. ISBN: 81-7080-003-X.

Ali, S.M. (1966). *The Geography of the Puranas*, People's Publishing House, New Delhi.

Arya, R.P. (2013). *Rāmāyaṇa of Vālmīki*, Parimal Publications, Delhi.

Charan, C. (2019). *Wisdom from the Ramayana: On Life and Relationships*, Prakash Books, India, New Delhi, ISBN: 978 93 8814 482 7.

Debroy, B. (2015). *The Mahabharata*, Penguin Books, India Haryana, ISBN: 9780142425144.

Dutt, M.N. (2018). *Mahābhārata*, Ten Volumes, Parimal Publications, Delhi, ISBN: 978-81-7110-197-9.

Goswami, A.P. and Seth, N. (2019). *Song of the Free: Teachings of the 24 Gurus of Dattatreya*, Bloomsbury India, New Delhi, ISBN: 978-9-3884-1472-2.

Kale, M.R. (2015). *The Hitopadeśa of Nārāyaṇa*, Motilal Banarsidass Publishers, New Delhi, ISBN: 978-81-208-0602-3.

Kale, M.R. (2008). *Pañcatantra of Viṣṇu Śarmā*, Motilal Banarsidass Publisher, Delhi. ISBN: 978-81-208-0219-3

Kangle, R.P. (2019). *Kautilīya Arthaśāstra*, Motilal Banarsidass Publishers, New Delhi, ISBN: 978-81-208-0039-7.

Lal, K. (2005). *The Śukranīti*, J.P. Publishing House, Delhi, ISBN: 81-86702-19-9.

प्रो० गङ्गाधर पण्डा (2006): संस्कृत वाङ्मय का बृहद् इतिहास, त्रयोदश-खण्ड पुराण, उत्तरप्रदेश संस्कृत संस्थान, लखनऊ.

Pt. Thaneshchandra Upreti (2003). *Viṣṇumahāpurāṇam of Maharṣi Vedavyāsa*, Parimal publications, Delhi, 1/2/23 pp. 13.

Pūjyaśrī Candraśekharendra Sarasvatī Svāmī (2018). *Hindu Dharma: The Universal Way of Dharma*, Bharatiya Vidya Bhavan, Mumbai.

Pusalker, A.D. (1955). *Studies in the Epics and Purāṇas*, Bharatiya Vidya Bhavan, Mumbai.

Rajagopalachari, C. (2018). *Ramayana*, Bharatiya Vidya Bhavan, Mumbai.

Rajagopalachari, C. (2019). *Mahabharata*, Bharatiya Vidya Bhavan, Mumbai.

Rustagi, U. (2009). *Viduranītiḥ*, J.P. Publishing House, Delhi, ISBN: 81-86702-47-4.

Srinivasa Sastri, V.S. (2013). *Lectures on the Ramayana*, The Samskrita Academy, Madras, Chennai–4.

Srinivasan, A.V. (2016). *Yakṣa Prashna: A Fable from the Mahabharata*, Vision Books, New Delhi. ISBN: 81-7094-967-X.

Swami Harshananda (2005). *The Rāmāyaṇa of Vālmīki*, RamaKṛṣṇa Math, Bengaluru, ISBN: 81-7907-051-4.

Swami Harshananda (2000). *Mahābhārata: A Study*, RamaKṛṣṇa Math, Bengaluru.

Swami Suparnananda (2016). *The Cultural Heritage of India (Vol. II)*, The RamaKṛṣṇa Mission Institute of Culture, Gol Park, Kolkata, ISBN: 81-85843-03-1.

Tagare, G.V. (2000). *The Brahmānda Purāna: Part I*, Chapters 21–24, Motilal Banarsidass, Delhi, pp. 198–247.

ENDNOTES

1. See for example the following verses from various purāṇas: (i) श्रुतिस्मृतिपुराणानि विदुषां लोचनत्रयम् । यस्त्रिभिर्नयनैः पश्येत् सोंऽशो माहेश्वरो मतः ॥ śrutismṛtipurāṇāni viduṣāṃ locanatrayam | yastribhirnayanaiḥ paśyet so'ṃśo māheśvaro mataḥ || (skandapurāṇam 4.1.27. **Kshemaraj Krishnadas (2006)**. *The Skandamahāpurāṇam* (श्रीस्कन्दमहापुराणम्), Nag Publishers, Jawaharanagar, Delhi–110007, 4.1.27. (ii) यो विद्याच्चतुरो वेदान् साङ्गोपनिषदो द्विजः । न चेत् पुराणं संविद्यान्नैव स स्याद्विचक्षणः ॥ (विष्णुपुराणम्) yo vidyāccaturo vedān sāṅgopaniṣado dvijaḥ | na cet purāṇaṃ saṃvidyānnaiva sa syādvicakṣaṇaḥ || **Hari Narayan Apte (1905).** "Vāuyupurāṇam", Anandashram Mudranalaya, Pune, Ch. 1, Verse 200, p. 8.
2. For details see, **Vinit, N. (2007).** "Nilamatapurana: A Study of Temples and Tirthas", Proceedings of the Indian History Congress, Vol. 68, Part One (2007), pp. 149–152.
3. प्रो० गङ्गाधर पण्डा **(2006).** संस्कृत वाङ्मय का बृहद् इतिहास, त्रयोदश-खण्ड पुराण, उत्तरप्रदेश संस्कृत संस्थान, लखनऊ । पुराणं सर्वशास्त्राणां प्रथमं ब्रह्मणा स्मृतम् । अनन्तरं च वक्त्रेभ्यो वेदास्तस्य विनिर्गताः ॥ (पद्मपुराणम् सृष्टिखण्डः – अ - 104) purāṇaṃ sarvaśāstrāṇāṃ prathamaṃ brahmaṇā smṛtam | anantaraṃ ca vaktrebhyo vedāstasya vinirgatāḥ || (padmapurāṇam sṛṣṭikhaṇḍaḥ – a - 104), p. 1.

4. *Ibid.* स होवाच ऋग्वेदं भगवोऽध्येमि यजुर्वेदं सामवेदमथर्वाणं चतुर्थमितिहासं पुराणं पञ्चमं वेदानां वेदम् । (छान्दोग्योपनिषद् 7.1.1), sa hovāca ṛgvedaṃ bhagavo'dhyemi yajurvedaṃ sāmavedamatharvāṇaṃ caturthamitihāsaṃ purāṇaṃ pañcamaṃ vedānāṃ vedam | (Chāndogyopaniṣad 7.1.1) p. 1.

5. There is an easy way to remember all the eighteen mahāpurāṇas. The verse goes thus: मद्वयं भद्वयं चैव ब्रत्रयं वचतुष्टयम् । अनापलिङ्गकूस्कानि पुराणानि प्रचक्षते ॥ madvayaṃ bhadvayaṃ caiva bratrayaṃ vacatuṣṭayam | anāpaliṅgakūskāni purāṇāni pracakṣate || This verse can be referred in श्रीमद्देवीभागवतपुराणम्, गीताप्रेस गोरखपुर । Skandha 1, Ch. 3. Verse 2.

   https://archive.org/details/DeviBhagavataWithHindiTranslationVolume1GitaPress/page/n1/mode/2up. Last accessed on Oct. 1, 2021.

6. प्रो० गङ्गाधर पण्डा **(2006).** "संस्कृत वाङ्मय का बृहद् इतिहास, त्रयोदश-खण्ड पुराण", उत्तरप्रदेश संस्कृत संस्थान, लखनऊ। अष्टादशपुराणानामेवमेवं विदुर्बुधाः । एवञ्चोपपुराणानाम् अष्टादश प्रकीर्तिताः ॥ (ब्रह्मवैवर्तपुराणम् श्रीकृष्णजन्मखण्डः 131/22) aṣṭādaśapurāṇānāmevamevaṃ vidurbudhāḥ | evañcopapurāṇānām aṣṭādaśa prakīrtitāḥ || (brahmavaivartapurāṇam śrīkṛṣṇajanmakhaṇḍaḥ 131/22)

7. *Ibid.* pp. 23–25.

8. *Ibid*, p. 6. The following verse summarises the five key characteristics of a purāṇa: सर्गश्च प्रतिसर्गश्च वंशो मन्वन्तराणि च । वंशानुचरितं चैव पुराणं पञ्चलक्षणम् ॥ वायुपुराणम् ४/१० sargaśca pratisargaśca vaṃśo manvantarāṇi ca | vaṃśānucaritaṃ caiva purāṇaṃ pañcalakṣaṇam || राधावल्लभ त्रिपाठी (२०१२), "श्रीवायुमहापुराणम्". राष्ट्रीय संस्कृत संस्थान, नई दिल्ली. 4/10.

9. प्रो० गङ्गाधर पण्डा **(2006).** "संस्कृत वाङ्मय का बृहद् इतिहास, त्रयोदश-खण्ड पुराण", उत्तरप्रदेश संस्कृत संस्थान, लखनऊ ।. अव्याकृतगुणक्षोभान्महतस्त्रिवृतोऽहमः । भूतमात्रेन्द्रियार्थानां सम्भवः सर्ग उच्यते ॥ avyākṛtaguṇakṣobhā nmahatastrivṛto'hamaḥ | bhūtamātrendriyārthānāṃ sambhavaḥ sarga ucyate || Bhāgavata purāṇa 12/7/11, p. 7.

10. **Pt. Thaneshchandra Upreti (2003).** *Viṣṇumahāpurāṇam of Maharṣi Vedavyāsa*, Parimal Publications, Delhi, 1/2/23 p. 13.

11. प्रो० गङ्गाधर पण्डा **(2006)**, "संस्कृत वाङ्मय का बृहद् इतिहास, त्रयोदश-खण्ड पुराण", उत्तरप्रदेश संस्कृत संस्थान, लखनऊ ।. राज्ञां ब्रह्मप्रसूतानां वंशस्त्रैकालिकोऽन्वयः । rājñāṃ brahmaprasūtānāṃ vaṃśastraikāliko'nvayaḥ | p. 8. Bhāgavata purāṇa, 12/7/16.

12. **Pt. Thaneshchandra Upreti (2003).** *Viṣṇumahāpurāṇam of Maharṣi Vedavyāsa*, Parimal Publications, Delhi, Introduction, pp. 65–66.

13. *Ibid.* राज्ञां ब्रह्मप्रसूतानां वंशस्त्रैकालिकोऽन्वयः । वंशानुचरितं तेषां वृत्तं वंशधराश्च ते ॥ rājñāṃ brahmaprasūtānāṃ vaṃśastraikāliko'nvayaḥ | vaṃśānucaritaṃ teṣāṃ vṛttaṃ vaṃśadharāśca te || Bhāgavata purāṇa 12/7/16.

14. For more details on the food and medicinal aspects found in the purāṇas see, **Bhattacharjee, S. (2010).** "Studies on flora in the puranas with special reference to Matsya Purana, Agni Purana, and Brahmavaivarta Purana", Unpublished Ph.D thesis, Department of Sanskrit, Gauhati University.

15. Śrimad Bhāgavata Mahāpurāṇa (Part–I) Gita press, Gorakhpur. (Rendered into English by – **C.L. Goswami, M.A. Shastri).** The relevant verses from Bhāgavata-purāṇa are reproduced here:
    श्रीभगवानुवाच - कर्मणा दैवनेत्रेण जन्तुर्देहोपपत्तये । स्त्रियाः प्रविष्ट उदरं पुंसो रेतःकणाश्रयः ॥ 3.31.1
    कललं त्वेकरात्रेण पञ्चरात्रेण बुद्बुदम् । दशाहेन तु कर्कन्धूः पेश्यण्डं वा ततः परम् ॥ 3.31.2
    मासेन तु शिरो द्वाभ्यां बाह्वङ्घ्र्याद्यङ्गविग्रहः । नखलोमास्थिचर्माणि लिङ्गच्छिद्रोद्भवस्त्रिभिः ॥ 3.31.3
    चतुर्भिर्धातवः सप्त पञ्चभिः क्षुत्तृडुद्भवः । षड्भिर्जरायुणा वीतः कुक्षौ भ्राम्यति दक्षिणे ॥ 3.31.4
    मातुर्जग्धान्नपानाद्यैरेधद्धातुरसम्मते । शेते विण्मूत्रयोर्गर्ते स जन्तुर्जन्तुसम्भवे ॥ 3.31.5
    कृमिभिः क्षतसर्वाङ्गः सौकुमार्यात्प्रतिक्षणम् । मूर्च्छामाप्नोत्युरुक्लेशस्तत्रत्यैः क्षुधितैर्मुहुः ॥ 3.31.6
    कटुतीक्ष्णोष्णलवणरूक्षाम्लादिभिरुल्बणैः । मातृभुक्तैरुपस्पृष्टः सर्वाङ्गोत्थितवेदनः ॥ 3.31.7
    उल्बेन संवृतस्तस्मिन्नन्त्रैश्च बहिरावृतः । आस्ते कृत्वा शिरः कुक्षौ भुग्नपृष्ठशिरोधरः ॥ 3.31.8
    śrībhagavānuvāca –
    karmaṇā daivanetreṇa janturdehopapattaye | striyā: praviṣṭa udaraṃ puṃso reta:kaṇāśraya: || 3.31.1

kalalaṃ tvekarātreṇa pañcarātreṇa budbudam | daśāhena tu karkandhū: peśyaṇḍaṃ vā tata: param || 3.31.2
māsena tu śiro dvābhyāṃ bāhvaṅghryādyaṅgavigraha: | nakhalomāsthicarmāṇi liṅgacchidrodbhavastribhi: || 3.31.3
caturbhirdhātava: sapta pañcabhi: kṣuttṛḍudbhava: | ṣaḍbhirjarāyuṇā vīta: kukṣau bhrāmyati dakṣiṇe || 3.31.4
māturjagdhānnapānādyairedhaddhāturasammate | śete viṇmūtrayorgarte sa janturjantusambhave || 3.31.5
kṛmibhi: kṣatasarvāṅga: saukumāryātpratikṣaṇam | mūrcchāmāpnotyurukleśastatratyai: kṣudhitairmuhu: || 3.31.6
kaṭutīkṣṇoṣṇalavaṇarūkṣāmlādibhirulbhaiḥ | mātṛbhuktairupaspṛṣṭa: sarvāṅgotthitavedana: || 3.31.7
ulbena saṃvṛtastasminnantraiśca bahirāvṛta: | āste kṛtvā śira: kukṣau bhugnapṛṣṭhaśirodhara: || 3.31.8

16. For a full reading of the astronomical details available in Brahmāṇḍa purāṇa, see **Tagare. G.V. (2000).** *The Brahmāṇḍa Purāṇa: Part I*, Chapters 21–24, Motilal Banarsidass, Delhi, pp. 198–247.
17. For a comprehensive coverage of the geography in purāṇas refer, **Ali, S.M. (1966).** *The Geography of the Puranas*, People's Publishing House, New Delhi.
18. **Kṛṣṇadas, K. (1998).** *The Viṣṇudharmottarapurāṇam*, Nag Publishers, Delhi. धर्मार्थकाममोक्षाणामुपदेशसमन्वितम् । पूर्ववृत्तकथायुक्तम् इतिहासं प्रचक्षते ॥ dharmārthakāmamokṣāṇāmupadeśa samanvitam | pūrvavṛttakathāyuktam itihāsaṃ pracakṣate || 3/15/1-2, p. 312.
19. **Pt. Haragovinda Śāstrī (2012).** "Nāmaliṅgānuśāsana or Amarakośa of Amarasiṃha", Chaukhambha Sanskrit Sansthan, Varanasi, ISBN: 81-86937-29-3. इतिहासः पुरावृत्तम् । 1/6/4.
20. **Kangle, R.P. (2019).** *Kauṭilīya Arthaśāstra (Part II)*, Motilal Banarsidass Publishers, New Delhi. ISBN: 978-81-208-0039-7. पुराणमितिवृत्तमाख्यायिकोदाहरणं धर्मशास्त्रमर्थशास्त्रं चेति इतिहासः । purāṇamitivṛttamākhyāyikodāharaṇaṃ dharmaśāstramarthaśāstraṃ ceti itihāsaḥ | 1/5/14, p. 7.
21. **Swami Suparnananda (2016).** "The Cultural Heritage of India (Vol. II)", The Ramakrishna Mission Institute of Culture, Gol Park, Kolkata, Chapter 2, p. 14.
22. The Rāmāyaṇa edition of Gītā Press Gorakhpur is based on the critical edition of Rāmāyaṇa brought out by the Oriental Research Institute. MS University, Baroda. The number of sargas and verses is based on this.
23. **Ravi Prakash Arya (2013).** *Rāmāyaṇa of Vālmīki*, Volume–III, Parimal Publications, Delhi. काले धर्मार्थकामान् यः संमन्त्र्य सचिवैः सह। निषेवेतात्मवान् लोके न स व्यसनमाप्नुयात्। हितानुबन्धमालोच्य कार्याकार्यमिहात्मनः । राजा सहार्थतत्त्वज्ञैः सचिवैः स हि जीवति ॥ kāle dharmārthakāmān yaḥ saṃmantrya sacivaiḥ saha | niṣevetātmavān loke na sa vyasanamāpnuyāt | hitānubandhamālocya kāryākāryamihātmanaḥ | rājā sahārthatattvajñaiḥ sacivaiḥ sa hi jīvati || Yuddhakāṇḍa, Ch. 63, Verses 12–13, p. 170.
24. See the 107th sarga in yuddhakāṇdam for details:
गन्धर्वाप्सरसां संघा दृष्ट्वा युद्धमनूपमम् | गगनं गगनाकारं सागरः सागरोपमः || 6.107.52
रामरावणयोर्युद्धं रामरावणयोरिव | एवं ब्रुवन्तो ददृशुस्तद्युद्धं रामरावणम् || 6.107.53
gandharvāpsarasāṃ saṃghā dṛṣṭvā yuddhamanūpamam | gaganaṃ gaganākāraṃ sāgaraḥ sāgaropamaḥ || 6.107.52
rāmarāvaṇayoryuddhaṃ rāmarāvaṇayoriva | evaṃ bruvanto dadṛśustadyuddhaṃ rāmarāvaṇam || 6.107.53
25. चतुर्विंशतिसाहस्रीं चक्रे भारतसंहिताम् । उपाख्यानैर्विना तावद् भारतं प्रोच्यते बुधैः ॥ caturviṃśatisāhasrīṃ cakre bhāratasaṃhitām | upākhyānairvinā tāvad bhārataṃ procyate budhaiḥ || Mahābhārata, Ādi Parva, 1/102. See for detail, **M.N. Dutt (2018).** *Mahābhārata*, Volume–I, Parimal Publications, Delhi.
26. *Ibid*. p. 173. धर्मे चार्थे च कामे च मोक्षे च भरतर्षभ । यदिहास्ति तदन्यत्र यन्नेहास्ति न तत्क्वचित् ॥ dharme cārthe ca kāme ca mokṣe ca bharatarṣabha | yadihāsti tadanyatra yannehāsti na tatkvacit ||

27. यस्य त्वेतानि चत्वारि वानरेन्द्र यथा तव । धृतिर्दृष्टिर्मतिर्दाक्ष्यं सः कर्मसु न सीदति || yasya tvetāni catvāri vānarendra yathā tava | dhṛtirdṛṣṭirmatirdākṣyaṃ saḥ karmasu na sīdati || Vālmīki Rāmāyaṇam 5.1.198

28. शक्तिवैकल्यनम्रस्य निःसारत्वाल्लघीयसः | जन्मिनो मानहीनस्य तृणस्य च समा गतिः || śaktivaikalyanamrasya niḥsāratvāllaghīyasaḥ | janmino mānahīnasya tṛṇasya ca samā gatiḥ || 1.106. See for details, **Kale, M.R. (2008).** *Pañcatantra of Viṣnuśarman*, Motilal Banarasidass Publishers, Delhi. p. 17.

29. व्रजन्ति ते मूढधियः पराभवं भवन्ति मायाविषु येन मायिनः | प्रविश्य हि घ्नन्ति शठास्तथाविधान् असंवृताङ्गान् निशिता इवेषवः || vrajanti te mūḍhadhiyaḥ parābhavaṃ bhavanti māyāviṣu yena māyinaḥ | praviśya hi ghnanti śaṭhāstathāvidhān asaṃvṛtāṅgān niśitā iveṣavaḥ|| 1.30. See for details, **Kale, M.R. (2008).** *The Kirātārjunīyam of Bhāravi,* Canto I–III, Motilal Banarsidass Publishers, Delhi, Ch. 1, Verse 30.

30. *Ibid.* सहसा विदधीत न क्रियाम् अविवेकः परमापदां पदम् । वृणुते हि विमृश्यकारिणं गुणलुब्धाः स्वयमेव सम्पदः ॥ sahasā vidadhīta na kriyām avivekaḥ paramāpadāṃ padam | vṛṇute hi vimṛśyakāriṇaṃ guṇalubdhāḥ svayameva sampadaḥ || 2.30

31. प्रज्ञावृद्धं धर्मवृद्धं स्वबन्धुं विद्यावृद्धं वयसा चापि वृद्धं | कार्याकार्ये पूजयित्वा प्रसाद्य यः संपृच्छेन्न स मुह्येत् कदाचित् || prajāvṛddhaṃ dharmavṛddhaṃ svabandhuṃ vidyāvṛddhaṃ vayasā cāpi vṛddhaṃ | kāryākārye pūjayitvā prasādya yaḥ saṃpṛcchenna sa muhyet kadācit || Viduranīti 8.23 https://archive.org/details/HindiBookVidurNeetiCompleteByGitaPress/page/n137/mode/2up. Last accessed on Oct. 1, 2021.

32. अनेकानि च शास्त्राणि स्वल्पायुर्विघ्नकोटयः । तस्मात्सारं विजानीयात् क्षीरं हंस इवाम्भसि ॥ anekāni ca śāstrāṇi svalpāyurvighnakoṭayaḥ | tasmātsāraṃ vijānīyāt kṣīraṃ haṃsa ivāmbhasi || Garuḍapurāṇam 2.49.84. Source: https://sa.wikisource.org/wiki/गरुडपुराणम्/ प्रेतकाण्डः_(धर्मकाण्डः)/ Last accessed on Oct. 1, 2021.

33. अर्थानाम् अर्जने दुःखम् अर्जितानां च रक्षणे । आये दुःखं व्यये दुःखं धिगर्थाः कष्टसंश्रयाः ॥ arthānām arjane duḥkham arjitānāṃ ca rakṣaṇe | āye duḥkhaṃ vyaye duḥkhaṃ dhigarthāḥ kaṣṭasaṃśrayāḥ || 1.164. See for details, **Kale, M.R. (2008).** *Pañcatantra of Viṣṇuśarman*, Motilal Banarsidass Publishers, Delhi, p. 29.

34. अञ्जलिस्थानि पुष्पाणि वासयन्ति करद्वयम् । अहो सुमनसां प्रीतिर्वामदक्षिणयोः समम् ॥ añjalisthāni puṣpāṇi vāsayanti karadvayam | aho sumanasāṃ prītirvāmadakṣiṇayoḥ samam || 1.03. For details see, **Acharya, N.R. (2011)**. "Subhāṣita-ratna-bhāṇḍāgāra", Chaokhamba Sanskrit Series Office, Varanasi. Ch. 1, सज्जनप्रशंसा, Verse 3, p. 45.

35. वैद्यराज नमस्तुभ्यं यमराजसहोदर । यमस्तु हरति प्राणान् वैद्यः प्राणान् धनानि च ॥ vaidyarāja namastubhyaṃ yamarājasahodara | yamastu harati prāṇān vaidyaḥ prāṇān dhanāni ca || Sambhashana Sandesha, Samskrita Bharati, Bengaluru. May 1999, p. 21.

36. अङ्गुल्या कः कवाटे प्रहरति कुटिलो माधवः किं वसन्तः नो चक्री किं कुलालः न हि धरणीधरः किं फणीन्द्रो द्विजिह्वः । मुग्धे घोराहिमाथी किं खगपतिः नो हरिः किं कपीन्द्रः इत्थं लक्ष्म्या कृतः असौ हतवचनः पातु लक्ष्मीधवो वः ॥ aṅgulyā kaḥ kavāṭe praharati kuṭilo mādhavaḥ kiṃ vasantaḥ no cakrī kiṃ kulālaḥ na hi dharaṇīdharaḥ kiṃ phaṇīndro dvijihvaḥ |
mugdhe ghorāhimāthī kiṃ khagapatiḥ no hariḥ kiṃ kapīndraḥ itthaṃ lakṣmyā kṛtaḥ asau hatavacanaḥ pātu lakṣmīdhavo vaḥ || Ukti Pratyukti Stotra of Sri Vadirajayati. Verse 1. https://www.transliteral.org/pages/z140123204237/view. Last accessed on Oct. 1, 2021.

Part 2

FOUNDATIONAL CONCEPTS FOR SCIENCE AND TECHNOLOGY

CHAPTER

5 Linguistics

LEARNING OUTCOMES

After finishing this chapter, you will be able to:

- Know the phonetical foundations in Sanskrit Language
- Understand the basic structure of Sanskrit Language Grammar
- Get introduced to some computational aspects in Aṣṭādhyāyī
- Appreciate the importance of verbs in Sanskrit Language
- Appreciate the potential Sanskrit offers in Natural Language Processing

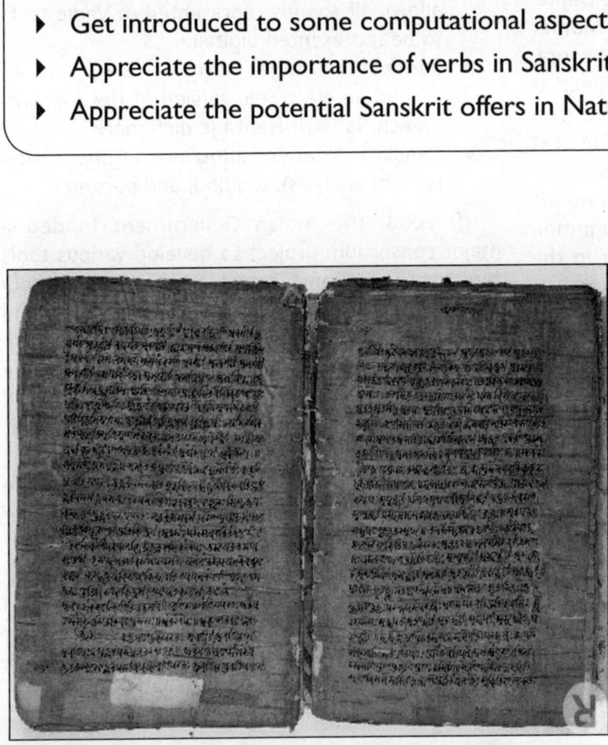

Pāṇini's work on Sanskrit grammar has been acclaimed to be one of the best in the field of linguistics and phonetics. It has attracted wide attention. The accompanying figure is a Birch bark manuscript from Kashmir of the Rupavatara, a grammatical textbook based on the Sanskrit grammar of Pāṇini. It was composed by Dharmakīrti, a Buddhist monk from Ceylon. The manuscript was transcribed in 1663.

Source: https://commons.wikimedia.org/wiki/File:Birch_bark_MS_from_Kashmir_of_the_Rupavatra_Wellcome_L0032691.jpg

IKS IN ACTION 5.1

An Ecosystem for Sanskrit Language Processing

Sanskrit is one of three ancient languages (Sanskrit, Greek, and Latin). While ancient Greek and Latin have attained a state of being classical, the Sanskrit language continues to be in use, albeit in specific domains by researchers. However, in recent years, there is a growing recognition of the importance of Sanskrit and several initiatives have been taken globally to integrate technology aids with the Sanskrit language. One of them is building an ecosystem of language processing tools for Sanskrit.

The efforts began in the late 1980s with several projects digitizing Sanskrit texts. Over a decade a central registry of digitized Indic texts was created in some Universities in Germany. Another progress in this direction was the creation of digital dictionaries of Sanskrit. The Cologne Digital Sanskrit Lexicon project digitized the Monier Williams' Sanskrit-English Dictionary. Furthermore, the Digital Dictionaries of South Asia project at the University of Chicago digitized Apte's and MacDonell's Sanskrit-English dictionaries. The next step in this direction was the processing of Sanskrit text using machine coding. Academy of Sanskrit Research, Melkote developed a verbal cognition generator for Bhandarkar's Sanskrit primer in the early 1990s.

By the turn of the century, the Ministry of Communication and Information Technology, Government of India set up the Technology Development for Indian Languages program (TDIL). Further to this, there has been a spurt in the efforts evident from multiple national and international collaborative efforts in this direction.

The Akshar Bharati group developed 'Anusāraka', a language accessor that employed techniques based on Pāṇini's Aṣṭādhyāyī. Beginning in 2002, a suite of Sanskrit computational tools was developed through several research projects. These activities were carried out in several Universities including the Jawaharlal Nehru University (JNU), Delhi, University of Hyderabad, IIT Kanpur, IIT Bombay, and Rashtriya Sanskrit Vidyapeeth, Tirupati. Globally other Sanskrit linguistic researchers were collaborating on these.

Some of the fallout of these are developing several tools and enabling infrastructure for Sanskrit, including the following:

- Web-based Sanskrit reader program at Brown University called Kramapāṭha
- A digital edition of Whitney's roots that served as the source of verbal stems for their inflectional generation software
- The International Digital Sanskrit Library Integration project in the Classics Department at Brown University
- Sanskrit Library Phonetic encoding that allows all sounds represented in Vedic texts to be represented digitally
- Sanskrit Heritage platform (SH), centred around an electronic version of the Sanskrit-French Sanskrit Heritage dictionary
- Sanskrit Reader, allowing segmentation (sandhi analysis), tagging, and parsing

In 2008, the Indian Government funded a major consortium project to develop various tools for analysis of Sanskrit text and a Sanskrit-Hindi machine translation system. Sanskrit scholars and computational linguists collaborated to develop several tools for enhancing the Sanskrit language processing ecosystem. This includes Sandhi and Samāsa analyser, morphological analyser, generator, sandhi joiner and splitter, and full-fledged parser.

These tools are now being deployed in some machine translation systems for testing and further refinements.

Source: Adopted from, **Goyal, P., Huet, G., Kulkarni, A., Scharf, P., and Bunker, R. (2012).** "A Distributed Platform for Sanskrit Processing", *Proceedings of COLING 2012: Technical Papers*, pp. 1011–1028, Mumbai, December 2012.

Language has been the most effective tool for our communication since time immemorial. Though some basic communications can be handled with gestures, language becomes inevitable for communicating elaborate and complex ideas. Scientific discovery, the advancement of knowledge, and collaborative working require a common method of communication. Language plays this role in a civilized society. As technology and computing prowess improves, we are able to develop better applications using Artificial Intelligence (AI) techniques. This requires us to develop efficient Natural Language Processing (NLP) capabilities. A systematic study of languages and their capabilities for the evolving requirements has become important for the science and technology community. In this chapter we shall see the developments in the field of language in ancient India.

5.1 COMPONENTS OF A LANGUAGE

How many of us can read and comprehend Shakespeare's works or native Indian literature such as Rāmāyaṇa? Even if we know the language in which these texts are written, it is difficult because the way the language is used undergoes changes. Language is a tool used by everyone in a community and it is very difficult to maintain it unchanged. We also notice that there are differences in the same language spoken by people from different regions. For these reasons, the literature in a language becomes incomprehensible for the people in the future though they may be speaking the same language. Studying the structure of the language helps us not lose the underlying principles that govern a language and ensures that the received wisdom from the ancestors is not lost. It also maintains continuity in language processing.

- One of the Vedāṅgas known as Vyākarana focuses on linguistics and phonetics aspects of Sanskrit language.
- Aṣṭādhyāyī is considered a fine creation of human intelligence and the best available descriptive model of a language.

Communication is key to trade, science and technology, and societal progress. It hinges on our ability to effectively process language as it is central to all human transactions and pursuits. Language processing has two dimensions: receptive and productive. The *receptive* part of a language deals with the ability of an individual to receive language inputs from multiple sources and process them to decipher the intended message and comprehend them. On the other hand, the *productive* part of the language is to transmit back to others for their consumption. The focus in the former is on listening and reading, whereas it is on speaking and writing in the latter. Viewed from another perspective, sound (listening and speaking) and script (reading and writing) are the essential elements of a language. Therefore, language processing can be represented in a 2×2 framework as shown in Figure 5.1. Linguistics addresses all these aspects of a language. Phonetics will cater to the receptive and productive aspects of the sound and a syntactic structure will cater to the scriptural aspects of a language.

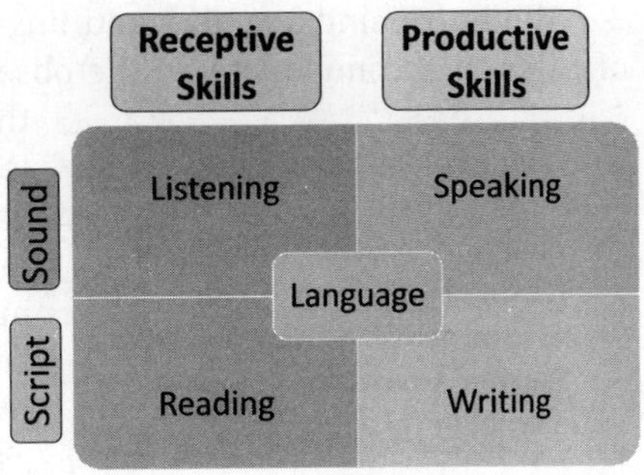

FIGURE 5.1 Components of a Language

Linguistics is a branch of language research that provides a scientific study of a language. It is a systematic study of language to understand speech sounds, grammatical structures, and meaning. It helps us analyse the language form and meaning and identify systematic methods integral to the language to derive the word forms and their meaning using structured rules and syntax. The earliest approach to a systematic treatment of linguistics is attributed to the Indian grammarian, Pāṇini who lived in the 6th Century BCE. His work on Sanskrit grammar known as *Aṣṭādhyāyī* is a magnum opus on linguistics. In this chapter, we shall see more details of this.

5.2 PĀṆINI'S WORK ON SANSKRIT GRAMMAR

In the Indian context, since the preservation of the Vedas and their meanings was given utmost importance, linguistics was an important discipline to be studied by everyone. Therefore, three out of the six Vedāṅgas, Śikṣā, Nirukta and Vyākaraṇa focused on phonetical and syntactical aspects of language. We have briefly discussed these in Chapter 2 of the book. On account of this, people who are trained in Sanskrit can comprehend ancient works such as the epics, Purāṇas, and the Vedic corpus in a lossless fashion. Several Indian languages have borrowed the syntactical and phonetic structure, language organization, and vocabulary from Pāṇini's work on Sanskrit grammar. Therefore, it is important to know this in some detail. Pāṇini's work on Sanskrit grammar, known as Aṣṭādhyāyī is a process of refinement and syntactical structuring, because of which the language is called Saṃskṛta (the refined one) and is considered a fine creation of human intelligence.

The Aṣṭādhyāyī is the culmination of a long grammatical tradition. Pāṇini brought his originality into action and proposed a structure to the grammar. It is the best available descriptive model of a language. The greatness of this work is evident from the fact that it has overshadowed all previous attempts on Sanskrit grammar. Moreover, several commentaries written on it is another proof of its prominence in linguistics. In the 4th century BCE, Kātyāyana composed a commentary (vārtika) on the Pāṇinian work, which served to provide further explanation of the work and clarified certain aspects of Aṣṭādhyāyī. Another great work on Aṣṭādhyāyī is the Mahābhāṣya, a commentary on the Aṣṭādhyāyī, authored by Patañjali in the 2nd century BCE. These three are considered the three great sages (tri-muni) of Sanskrit grammar.

While framing a syntax and linguistic framework, the challenge is to find an efficient set of rules to accommodate all the observed patterns and variations in the use of a language by the society. If there are robust rules that govern a language leaving several exceptions, it is not useful. At the other extreme, to accommodate all the possible variations, the number of rules framed may become very large, again making it less useful. The greatness of the Sanskrit grammar laid out by Pāṇini lies in its astuteness in establishing a framework for linguistics that nicely balanced both these aspects. Pāṇini did not write a set of rules and insisted that everyone must follow these rules to use the language. This is possible when we develop a computer language such as C or Python, for example. Pāṇini composed 3,983 rules (known as Sūtras) to mostly accommodate the patterns and variations in the Sanskrit language and arranged them in eight chapters (therefore, the name Aṣṭādhyāyī). Each chapter is further divided into 4 quarters, thereby making it 32 quarters in all.

> ♦ Pāṇini composed 3,983 rules to accommodate all the patterns and variations in Sanskrit language.
> ♦ The basic approach of Pāṇini and its distinguishing features make Sanskrit a powerful language and eternal in its appeal.

The basic approach of Pāṇini and its distinguishing features make Sanskrit a powerful language and eternal in its appeal. These are summarized below:

- The entire vocabulary of the Sanskrit language could be created using the 3,983 rules. Few exceptions need special handling. The rules are aphorisms (known as sūtras), which are easy to commit to memory. What it implies is that if someone is familiar with these rules and how it needs to be applied, then it amounts to gaining unambiguous mastery of Sanskrit language. The educational systems in India until the introduction of the Macaulayan system of education in the 19th century CE ensured this mastery for the students.

- Language processing and word generation are strictly rule-based and derivative in nature. Proper application of the rules must result in a valid word (a form of the language). There is no need to make any additional assumptions to derive any word.
- The derivation of words using the rules could be done using a step-by-step process. What it implies is that technically speaking, starting from a base (a verbal root or a nominal root), the logic can search if any of the 3,983 rules could be applied to the current transformation of the root. If the rule can modify, the current structure it will perform the operation. The procedure stops when none of the rules could be further applied. The result is the final word. This rule-based recursive structure to language is highly amenable to computer-based processing. The entire process is logical, unambiguous, and rational.
- The entire scheme for word generation follows a highly modular approach. Two basic components form a word. Each word is formed out of a base (verbal or nominal root). To this, one or more suffixes are added to generate the word. On account of this method of word generation, the Sanskrit language inherently has a high degree of 'patterns' of words and word structures that makes it very efficient to develop vocabulary.
- Since the entire language is rule-based, the vocabulary is not fixed or static for the language. The rules can be used for generating new words, as long as the rules are not violated. Therefore, the language is dynamic, can construct new words as demand arises, and can maintain its relevance. This has implications for lexicographic studies in the language.
- Aṣṭādhyāyī deploys several interesting data structures and computational elements that make it unique among linguistic studies.

These distinguishing aspects of Pāṇini's grammar are discussed in detail in Sections 5.5, 5.6 and 5.8 of this chapter. Since language has the phonetic (sound) and Syntactical (Structural) dimensions, we shall see Panini's approach to these in the Sanskrit language.

5.3 PHONETICS IN SANSKRIT

Phonetics is the study of sounds in a language, particularly the production of sound in a language and how it communicates the language corresponding to the scripts of the language. It also addresses the issue of how the sound is perceived in the language. Phonetics in the Sanskrit language has been addressed in some detail, since this is vital, because the ancient Indian knowledge tradition is oral. The entire transmission of the Vedas from time immemorial has survived several thousand years on account of scientific methods of oral rendering. This has been possible on account of a well-developed science of phonetics. This also ensured an impeccable textual transmission superior to the classical texts of other cultures. This is perhaps the reason for UNESCO recognizing Vedas in the form of oral knowledge as a heritage for preservation.

The science of the study of sound, known as Śikṣā, forms one of the six of the Vedāṅgas. In Chapter 2, we have a brief introduction to Śikṣā. In the earliest Indian traditions, Prātiśākhyas address the issue of how sounds are produced. Ṛgveda-prātiśākhya and the Taittirīya-prātiśākhya are the earliest works on the subject. According to these texts, the sound primarily arises at the junction of the throat and the chest due to the movement of the breath in the body and manifests in the oral cavity at different locations resulting in various sound patterns[1]. Pāṇini has discussed this aspect in the Aṣṭādhyāyī by way of some sūtras specifying phonetic rules as well as a separate addendum, known as Pāṇinīya-śikṣā. According to Pāṇini, the origin of the sounds specific to the letters (varṇas) in Sanskrit is related to many dimensions. First among them is the origin of the sound in the oral cavity. Six locations have been identified and the sounds either originate due to these locations or a combination of these. Using a set of sūtras he establishes this idea[2]. Figure 5.2 illustrates this pictorially. In addition to what is shown in Figure 5.2, combinative locations generate other sounds as listed below:

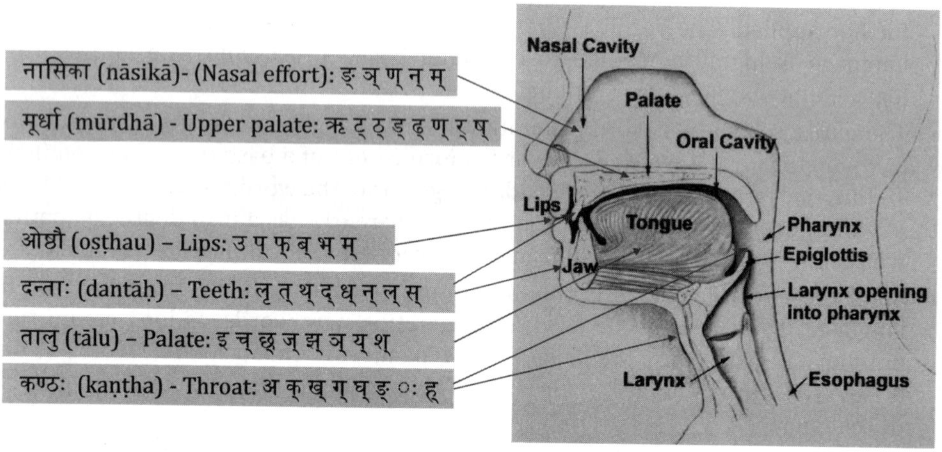

Source: https://commons.wikimedia.org/wiki/File:Illu01_head_neck.jpg

FIGURE 5.2 Origin of Sound from the Oral Cavity – An Illustration

According to Pāṇini, the vowels have a temporal factor in the production of the sound[3]. Three variations have been specified; short (hrasva), long (dīrgha), and prolate or elongated (pluta). If we take the duration of the utterance of a short vowel as one unit, the long will be two units and the prolate will be three units. One can represent them using the letters as अ, आ, अ३. The use of the prolated version is not unusual in practice, during times of doubt, caution, while calling from far and exclamation we tend to utter the vowel for a longer duration than usual. Another variation concerning the vowels is the nature of effort employed in using the nasal cavity in addition to the oral cavity. Aṣṭādhyāyī establishes that in our effort to utter a vowel, it is possible to use either the oral cavity alone or the nasal cavity along with the oral cavity. When a vowel is uttered by the joint effort of the oral and the nasal cavities, it is referred to as nasal (anunāsikā)[4]. By contrast, a vowel uttered only using the oral cavity may be referred to as non-nasal (nir-anunāsikā). The third aspect of the vowel sound is the

- The entire transmission of the Vedas from time immemorial has been possible on account of a well-developed science of phonetics.
- Vowels have a temporal factor in the production of the sound.
- Three variations have been specified for the pitch of the vowel sound.

pitch used to utter the vowel. Three variations have been specified for the pitch of the sound[5]. A vowel uttered with high pitch is called उदात्त (udātta), with low pitch अनुदात्त (anudātta) and that with a combination of high and low pitches known as स्वरित (svarita).

Based on the above classifications of different types of efforts with which a vowel sound can be produced, the vowels can have 18 different possibilities (temporal factor (3) × pitch factor (3) × nasal factor (2). In some cases, the variations are only twelve, as there are no short (ए, ऐ, ओ, औ) or long (ऌ) form available. The Sanskrit grammar makes further fine distinctions using a few factors on how the variations in the observed sound with respect to the letters manifest. Five categories have been identified for the nature of the internal effort made in producing the sound and three categories for the external effort. For example, the strength of the breath that we emit when we utter a consonant produce variations. It can be mild breath (अल्पप्राण - alpa-prāṇa) such as क, ग, प, ब, and strong breath (महाप्राण - mahā-prāṇa) such as ख, घ, फ, भ. The three letters य्, व्, ल् are both nasal and non-nasal.

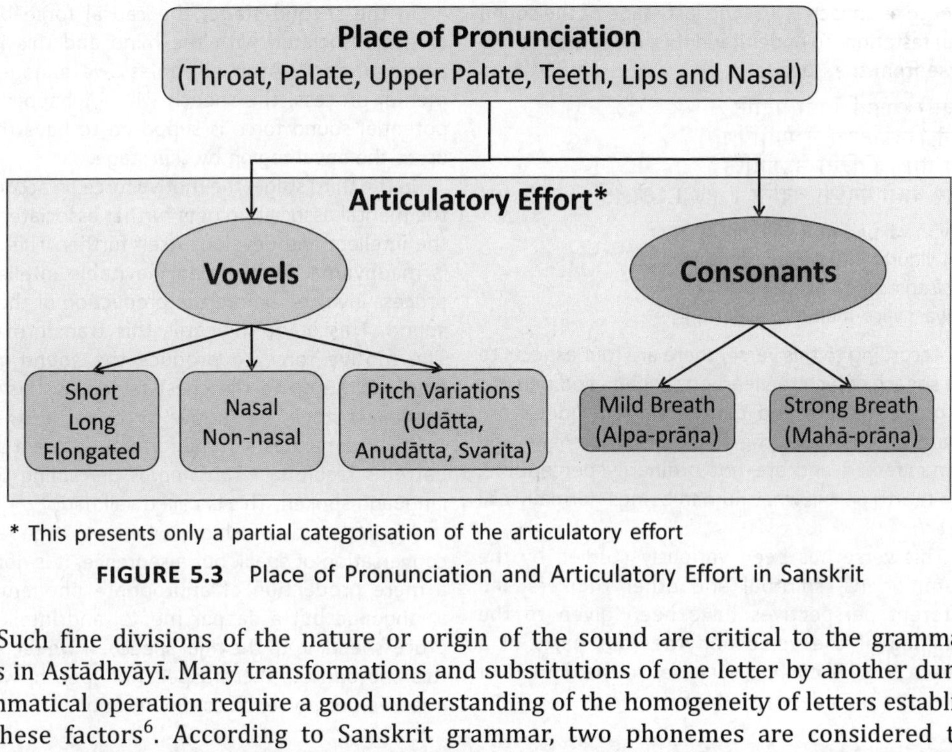

FIGURE 5.3 Place of Pronunciation and Articulatory Effort in Sanskrit

Such fine divisions of the nature or origin of the sound are critical to the grammatical rules in Aṣṭādhyāyī. Many transformations and substitutions of one letter by another during a grammatical operation require a good understanding of the homogeneity of letters established by these factors[6]. According to Sanskrit grammar, two phonemes are considered to be homogenous if they are produced with the same articulatory effort and at the same place of pronunciation. Figure 5.3 captures the essence of these discussions and presents the phonetic concepts in the Sanskrit language.

These aspects of Sanskrit grammar demonstrate the use of linguistic and phonetic concepts. It has been possible to impart a good phonetical training to language aspirants and closely monitor and rectify the phonetical inaccuracies using these principles. The accrued benefits include arresting deterioration of pronunciation over time and across cultures and ensuring that the oral tradition preserves and transmits the language components intact.

IKS IN ACTION 5.2

Four Stages of Speech

Phonetics is about understanding the origin of sounds produced in a linguistic context. Modern studies on phonetics try to analyse how the sound originates from the oral cavity and produces the desired effect with respect to the letter pronounced in a language. As we have already discussed, in the Indian phonetical tradition, the origin of the sound related to various letters of the Sanskrit language is traced to specific parts of the oral cavity.

However, if we look at other sources in the Indian knowledge system, we notice that what we refer to as speech is just the last stage of the sound manifestation. To understand this, let us look at one verse from the Ṛgveda:

चत्वारि वाक्परिमिता पदानि
तानि विदुर्ब्राह्मणा ये मनीषिणः ।
गुहा त्रीणि निहिता नेङ्गयन्ति
तुरीयं वाचो मनुष्या वदन्ति ॥ RV 1.164.45

catvāri vākparimitā padāni
tāni vidurbrāhmaṇā ye manīṣiṇaḥ |
guhā trīṇi nihitā neṅgayanti
turīyaṃ vāco manuṣyā vadanti ||

According to this verse, there are four aspects to the speech, which the learned scholars and persons of great wisdom and contemplation indeed are aware of. Of these (the first) three are deep and unmanifested and are not ordinarily perceptible. The fourth part is what human beings normally call *speech*.

This verse has been variously studied by the grammarians, spiritual and other thinkers and different perspectives have been given to the meaning of the verse and the four parts of a speech. We will confine ourselves to the domain of phonetics to understand the four parts.

According to Indian traditions, the speech indeed goes through four stages before it finally comes out in the form of sound. These stages are Parā, Paśyantī, Madhyamā and Vaikharī. The first stage is Parā, where the sound is resident as a completely unmanifested, potential form, capable of transforming into other forms whenever required. It is the potential force that exists in all of us as a causal force behind the entire gamut of speech that we produce. It resides as a storehouse of potential sound energy in us.

In the second stage, the causal force begins to get associated with the mind and this is the paśyantī stage. After all, unless we engage in a mental process, the speech will not happen. The potential sound force is supposed to have moved up to the naval region by this stage.

In the third stage, the motive force on account of the mental association gets further associated with the intellect and develops itself further. This stage is madhyamā. There is an inevitable intellectual process involved before the production of the final sound. This stage indicates this transformation. The motive force to produce the sound would have moved up to the chest region by this time. In the last stage, the motive force reaches the oral cavity and manifests in the form of different sound patterns resulting in phonemes pertaining to the language spoken. This is called vaikharī.

Viewed in this fashion, when we engage in a conversation or speak out a sentence, it is not only a mere production of appropriate phonemes of a language but a deeper mental and intellectual process leading to the final speech. If we recognise this and address these aspects of speech, we can improve our power of communication.

5.4 PATTERNS IN SANSKRIT VOCABULARY

The ultimate building block of any language is the word. The words are combined in several ways to communicate ideas and transact knowledge. Understanding the mechanics of word formation and generation is at the core of language mastery. Sanskrit grammar addresses this aspect, in a unique way. Sanskrit vocabulary has set patterns that are unambiguously generated by clear syntactical rules. Before we see the word generation logic, let us understand the patterns with an example.

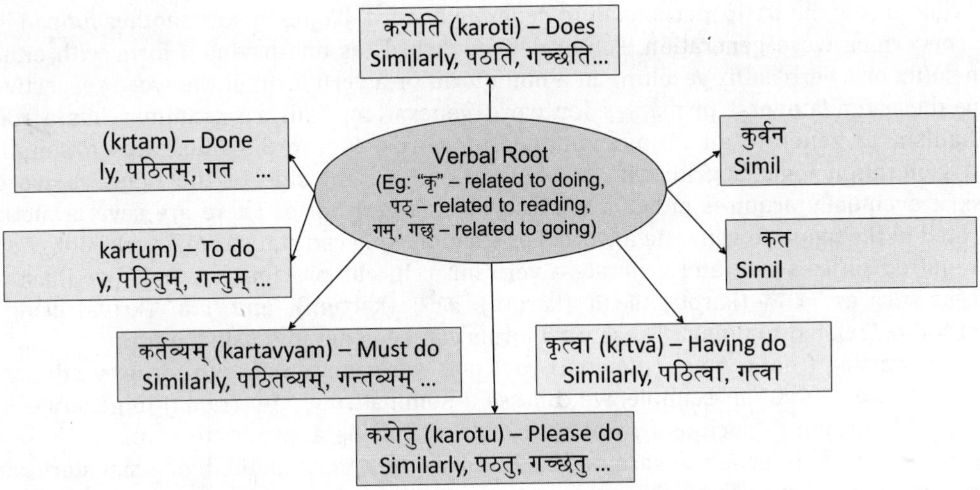

FIGURE 5.4 Word Patterns in Sanskrit Language – An example

We shall start with a verbal root, say "kṛ" (कृ) meaning 'in the context of or related to doing'. Using the grammatical rules we can generate alternative forms of this verb. Figure 5.4 illustrates this example for eight forms: does, doing, doer, having done, please do, must do, to do, and done. For the other verbs, we can obtain equivalent words that will be similar to this structure. Figure 5.4 indicates this with two more examples for each form. We see a set pattern of word formation in these examples that is easy to understand and quick to build a vocabulary of the language. The question of interest to us is, "How is it possible to generate such patterns"?

The answer to this lies in the basic mechanism used to generate a word. The words in Sanskrit can be divided into two categories: Noun forms (known as Subanta) and verb forms (known as Tiṅanta). Irrespective of whether a word has a noun form or a verb form[7], it can be expressed in a simple equation as follows:

Word = Base + Suffix

After adding a suffix to the base, relevant grammatical rules are invoked to generate the final word. If required additional suffixes can also be added to generate the word. This standardized structure lends itself to a natural process of patterns of vocabulary that we saw in Figure 5.4. To return to the example, as a simple illustration, the patterns happen in the following fashion:

$$\left\{\begin{array}{l}\text{कृ (kṛ)}\\ \text{पठ् (paṭh)}\\ \text{गछ्छ (gach)}\\ \ldots\end{array}\right\} + \left[\text{तु (tu)}\right] = \left\{\begin{array}{l}\text{करोतु (karotu)}\\ \text{पठतु (paṭhatu)}\\ \text{गच्छतु (gacchatu)}\\ \ldots\end{array}\right\}$$

Base + Suffix = Word

While this is the basic mechanism to generate a word, Pāṇini makes another fundamental rule concerning word generation. Pāṇini defined "word" as one having a form with either a noun suffix or a verb suffix resulting in a noun form or a verb form of the word respectively[8]. Using these fundamental principles for word generation, Sanskrit grammar has a robust mechanism to generate an infinite number of words. Figure 5.5 illustrates a simplified word generation logic in Sanskrit. As shown on the right side of the figure, a word in Sanskrit eventually acquires either a noun form or a verb form. There are several methods indicated in the figure to generate a word. For example, one can start from a verb root and add the required suffixes to it and generate a verb form. In our previous example, we illustrated patterns such as करोति (karoti), करोतु (karotu), कर्तुम् (kartum), and कृत्वा (kṛtvā) using the verb root kṛ, "related to doing". These verb forms can be generated using appropriate suffixes. Similarly, starting from a noun (nominal) root one can generate a noun form by adding the required suffixes to it. For example, we can use a nominal root 'राम' (rāma) to generate noun forms in the meanings such as by Rāma, Rāma's, etc. using appropriate suffixes. These two are direct methods to generate verb and noun forms. In a third method, one can start with a verbal root and add a suffix to generate a noun root. Once the noun root is available, one can generate a noun form in the manner described earlier. In our earlier example on patterns of words, we used the root kṛ, 'related to doing' and illustrated kartā, 'doer' as one of the patterns. In the reverse pass, one can start from a noun form and add a suffix to get a verbal root. With the verbal root, one can subsequently generate a verb form. This way it is possible to perform a crossover operation from noun to verb and vice versa. In addition to what is illustrated here, there are many more ways of generating words. One method is to generate compound nouns using several nouns, known as samāsa. We shall see the computational logic of samāsa in another section. Table 5.1 provides additional explanations and some illustrative examples.

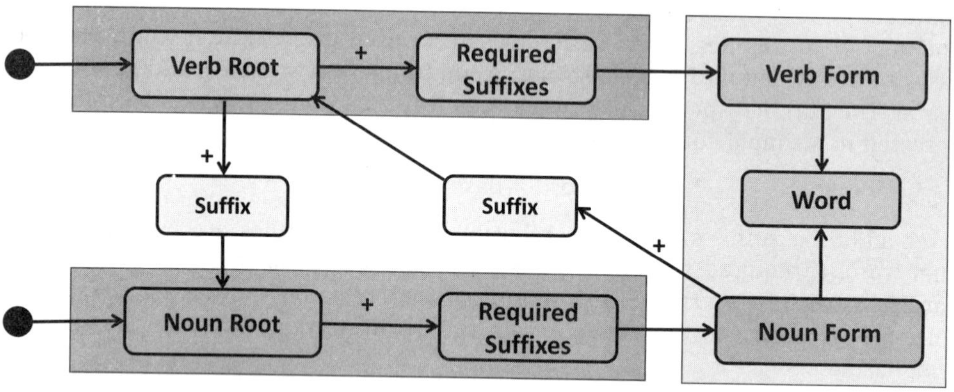

FIGURE 5.5 The Word Generation Scheme in Sanskrit Grammar

5.5 COMPUTATIONAL CONCEPTS IN AṢṬĀDHYĀYĪ

One of the distinguishing aspects of the Pāṇinian approach to linguistics and Sanskrit grammar is the use of several features, which in modern parlance map to certain computational concepts. These aspects make the Sanskrit grammar highly structured, easily amenable for machine coding, and enhancing its attractiveness for current-day applications such as Natural Language

TABLE 5.1 The Word Generation Scheme in Sanskrit Grammar – Examples

Base to Generate a Word	*Role of the Suffix (pratyaya)*	*Examples*	*Remarks*
Nominal Root	For generating singular, dual and plural of seven cases of nouns the relevant suffixes are applied.	*For the nominal root* 'राम' (Rāma), we can generate: रामः – रामौ – रामाः *(Rāmaḥ – Ramau – Rāmāḥ)* रामेण – रामाभ्यां – रामैः *(Rāmeṇa – Rāmābhyām – Rāmaiḥ)*, etc.	Additional suffixes for generating feminine forms can be added.
Verb Root	For generating singular, dual and plural of 1st person, 2nd person and 3rd person of verbs the relevant suffixes are applied.	*For verbal root* 'पठ्' (paṭh) the present tense forms can be generated: पठति – पठतः – पठन्ति (paṭhati – paṭhataḥ – paṭhanti) पठसि – पठथः – पठथ (paṭhasi – paṭhathaḥ – paṭhatha) पठामि – पठावः – पठामः (paṭhāmi – paṭhāvaḥ – paṭhāmaḥ)	Relevant suffixes for generating 10 verb forms (6 tenses and 4 moods) can be added.

Processing. A computer language is a formal language with its own vocabulary and syntax for representing knowledge precisely. Instructions are given in the form of algorithms. Pāṇini's grammar shares many of these characteristics. He has developed a meta-language for the knowledge processing system created by him. The following are some of the important common aspects seen in a computer language and Pāṇini's rules.

- Sanskrit grammar has a robust mechanism to generate an infinite number of words.
- Sanskrit grammar of Pāṇini rests on a fundamental set of sūtras known as Māheśvara sūtras.

- Vocabulary exclusively meant for his work
- Abbreviated forms (mnemonics) for brevity and better retention of ideas
- Exclusive syntax for Aṣṭādhyāyī
- An algorithmic approach to word generation
- Recursive logic

Māheśvara-sūtras

The entire Sanskrit grammar of Pāṇini rests on a fundamental set of sūtras known as Māheśvara sūtras. These sūtras, 14 in number, present the basic letters of Sanskrit uniquely as shown below:

1. अइउण् ।	2. ऋऌक् ।	3. एओङ् ।	4. ऐऔच् ।
5. हयवरट् ।	6. लण् ।	7. ञमङणनम् ।	8. झभञ् ।
9. घढधष् ।	10. जबगडदश् ।	11. खफछठथचटतव् ।	12. कपय् ।
13. शषसर् ।	14. हल् ।		
1. a-i-uṇ.	2. ṛ-ḷk.	3. e-oṅ.	4. ai-auc.
5. ha-ya-va-raṭ.	6. la-ṇ.	7. ña-ma-ṅa-ṇa-nam.	8. jha-bhañ.
9. gha-ḍha-dhaṣ.	10. ja-ba-ga-ḍa-daś.	11. kha-pha-cha-ṭha-tha-ca-ṭa-tav.	12. ka-pay.
13. śa-ṣa-sar.	14. hal.		

A closer look at the sūtras reveals several intriguing aspects:

1. The first four sūtras cover the alphabets in the normal order.
2. Sūtras 5 to 14 present the consonants in a somewhat obscure order than what we are normally used to. For example, the "ka" varga letters are not in the same sūtra. They are distributed across sūtras 7, 9, 10, 11, and 12. The other groups' elements are also scattered across sūtras.
3. Each sūtra ends with a termination which is a consonant. For example, the first sūtra which presents the three vowels "अ – इ – उ" ends with a consonant "ण्". Similarly, the 13th sūtra presents three consonants, "श – ष – स" and ends with a consonant "र्". These consonants are to be considered merely as a termination tag. While at the outset it does not appear to yield any meaning or serve any purpose, this arrangement allows for extensive use of mnemonics and present various rules concisely.

Each sūtra listed above can be viewed as an ordered set of letters. For example, set 1 = {अ, इ, उ}, after removing the consonant tag at the end. Set 10 = {ज, ब, ग, ड, द}. Pāṇini uses several mnemonics by combining these sets and naming them with the beginning of the letter in the first set and the terminating tag in the last set. Here are a few examples:

अच् (ac) = {All vowels}; हल् (hal) = {All consonants}[9]; खय् (khay) = { ख, फ, छ, ठ, थ, च, ट, त, क, प}– this represents column 1–2 of consonants. जश् (jaś) = {ज, ब, ग, ड, द} – column 3 of the varga consonants and अट् (aṭ) = {अ, आ, ... र}.

While it is a matter of research to understand why the letters are listed in a particular fashion, it is noteworthy that the sūtras in the Aṣṭādhyāyī use the Māheśvara-sūtras to concisely specify various rules to be deployed.

Use of Mnemonics

In the previous discussion, we saw how the letters were jointly represented using mnemonics obtained out of combining sets of letters specified through the Māheśvara-sūtras. This feature is extensively used by Pāṇini to describe various rules in Aṣṭādhyāyī. In Section 5.4, we discussed two sets of suffixes for generating noun forms and verb forms. These two are concisely represented by two mnemonics, sup and tiṅ respectively. The 'sup' has a set of 21 suffixes to be used for generating the singular, dual, and plural forms of the seven cases of the nouns[10]. Similarly, the 'tiṅ' is a mnemonic that represents a set of 18 suffixes to generate the verb forms[11].

We shall take an example to understand the power and conciseness of mnemonics used in Sanskrit grammar. Consider the following sūtra in Aṣṭādhyāyī: इको यणचि (iko yaṇ-aci)[12]. This sūtra simply instructs that if 'ik' is followed by 'ac', then replace 'ik' by 'yan'. Let us understand this in some detail. There are three mnemonics in this sūtra: इक् (ik), यण् (yaṇ), and अच् (ac). Using Māheśvara-sūtras, the set represented by the three mnemonics can be enumerated as follows:

इक् (ik) – **A** = {इ, उ, ऋ, ऌ}; Let a member of this set be denoted as 'a'
अच् (ac) – **B** = {set of vowels – अ, इ, उ, ऋ, ऌ, ए, ओ, ऐ, औ}; Let a member of this set be denoted by 'b'
यण् (yaṇ) – **C** = {य्, व्, र्, ल्}; Let a member of this set be denoted by 'c'

This sūtra provides a rule for the juxta-positioning of two letters. Let the first letter be denoted by 'a' and the letter that follows it be denoted by 'b'. The rule specifies applicable transformation for this operation. Essentially, this sūtra informs that when we attempt juxtaposition of 'a' and 'b', then 'a' is replaced by 'c', Since **A** and **C** have the same set of entries, the replacement is always the corresponding member in **C**.

EXAMPLE: Consider the juxta-positioning of 'प्रति' and 'एकम्'. Using the above rule, the transformation will happen as follows:

प्रति = (प्रत् + इ); एकम् = (ए + कम्).
As per our discussion above in this example a = इ, b = ए, and c = य्.
Therefore, the combined word is = प्रत् + य् + ए + कम् = प्रत्येकम्

The beauty of this cryptic sūtra lies in the fact that such detailed instructions have been codified in a very simple way using mnemonics. This makes it easy to remember and apply the rules once we are familiar with this. Pāṇini has utilised several mnemonics to codify detailed and apparently complex instructions using such an approach throughout Aṣṭādhyāyī.

Rule-based Grammatical Operations

Pāṇini's system of applying grammatical conditions to derive words works exactly like a rule-based engine. There is a situation presented and in the process of addressing the situation, the sūtras bring the necessary and sufficient conditions to address the requirement. The sūtras astonishingly resemble 'if this condition is satisfied, then perform this operation'. Viewed in this manner, Sanskrit grammar is entirely derivational. What it means is that every conceivable word in the Sanskrit language can be derived using a set of rules in a strict algorithmic fashion. Figure 5.6 illustrates the Pāṇinian algorithm for handling grammatical operations. As seen from the figure, the process starts with reading the relevant inputs for the grammatical operation. For instance, if the idea is to generate the seven cases of the word 'Rāma', then the procedure starts with reading 'Rāma' as the base on which relevant grammatical operations need to be applied. The procedure scans the 3,983 sūtras to find out if any sūtra will be relevant to accomplish the task. If a sūtra makes itself available, then the relevant grammatical operations arising out of the sūtra will be performed. At the end of this operation, the input base might have undergone some transformation. Therefore, the procedure will return to the header and scan again for a possible sūtra that may perform a further transformation. The logic repeats in this fashion, taking one sūtra at a time for performing the grammatical operation. As the algorithm proceeds, every one of the 3,983 sūtras presents themselves for possible application. As long as any of the sūtras are applicable, the algorithm will proceed one step. When no more sūtras are applicable, the algorithm stops, and the final word is generated. This algorithm merely illustrates the concept, however, actual implementation of this algorithm will be more efficient. See the box on 'Rule-based derivation of a word' for an illustration of this[3]. One can make the following observations based on the illustration in the box:

- Sūtras from different locations of Aṣṭādhyāyī under various headings come to operate where their conditions are satisfied.
- The final form remaining after all the operations are carried out is the word and it becomes eligible for use.

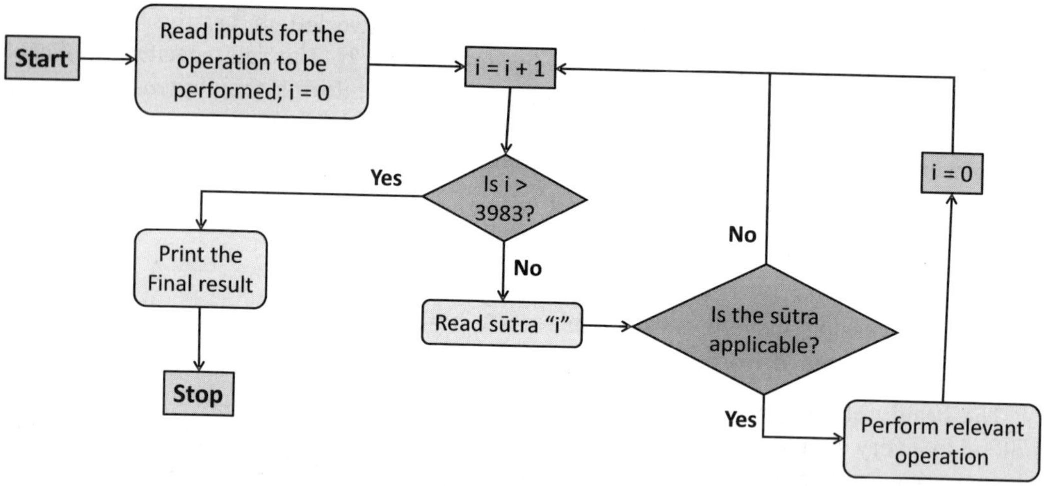

FIGURE 5.6 The Pāṇinian Algorithm for Grammatical Operations

Algorithm for Compound Words

Aṣṭādhyāyī employs recursive logic to process grammatical applications. One such example is the logic of samāsa, a method of creating compound words from a group of nouns. We shall see the samāsa principle and the application of recursive logic adopted for samāsa. In Sanskrit, like in other languages, it is possible to combine one or more noun forms into a single noun. This process of creating a single noun from several nouns is called samāsa and the resultant word is a compound word.

Let us take an example to understand this. Let us consider two words, 'शास्त्रे' (śāstre) and 'निपुणः' (nipuṇaḥ) and see how to get a compound word out of this. Let us designate the first word as pūrva-pada (PP) and the second word as uttrara-pada (UP). In our chosen example, PP = 'śāstre' and UP = 'nipuṇaḥ'. If we may recall, the noun forms in Sanskrit are generated from a noun root using some suffixes. Therefore, the suffixes of the two words can be dismantled to obtain the respective noun roots. Let us replace the PP and UP with their respective noun roots. In our example, the revised PP and UP are: PP = शास्त्र (śāstra) and UP = निपुण (nipuṇa). The compound word can be obtained by merely combining the PP and UP. In our example, the combined word now becomes 'शास्त्रनिपुण' (śāstranipuṇa).

This combined new form is indeed a new noun root. In this simple example, the combination of the two words is straightforward. No issues are arising out of the juxta-positioning (known as sandhi in Sanskrit) of the last letter of the first word with the first letter of the second word. If required relevant grammatical rules can be applied to address these. Out of this noun root, the usual noun forms can be further generated using the suffixes as we have discussed in Section 5.4. For example, the first case, singular masculine, and feminine forms can be derived using the grammatical rules and suffixes as शास्त्रनिपुणः (śāstranipuṇaḥ) and शास्त्रनिपुणा (śāstranipuṇā) respectively. Pāṇini has set up a procedure by which any number of noun words can be combined using the above logic in a recurring fashion. Let us, therefore, generalise the logic for combining '*n*' words into a single compound word:

Rule-based Derivation of a Word – An Example

Let us take a noun root = 'राम' (Rāma) and see how to derive the singular – third case form रामेण (rāmeṇa) (i.e. 'by Rama').

Step 1: noun root = 'राम'. At this stage, rule 4.1.2 of Aṣṭādhyāyī becomes applicable (this rule is meant to supply the appropriate suffix). As per the rule, add the suffix 'टा' to the stem 'राम' to generate the required word.

Step 2: राम + टा = ? At this stage, rule 7.1.12 of Aṣṭādhyāyī becomes applicable (this rule provides certain conditions for replacing 'टा' with any other applicable suffix). As per the rule, the suffix 'टा' will be replaced with 'इन' to generate the required word.

Step 3: राम + इन = ? At this stage rule 6.1.87 of Aṣṭādhyāyī becomes applicable (this rule provides certain conditions for combining two juxtaposed letters). As per the rule, when a vowel 'अ' is followed by another vowel 'इ', both will be replaced by a new vowel 'ए' as per guṇa sandhi. Since there is a 'अ' at the end of the word 'राम' followed by 'इ', this will apply.

Step 4: रामे + न = ? At this stage, rule 8.4.2 of Aṣṭādhyāyī becomes applicable (this rule provides certain conditions for replacing 'न्' with 'ण्').

Step 5: रामे + ण = ? At this stage, none of the 3,983 rules become applicable. Therefore, the algorithm will stop. The final word therefore is 'रामेण'.

- Pāṇini's system of applying grammatical conditions to derive words works exactly like a rule-based engine.
- Aṣṭādhyāyī employs recursive logic to process certain grammatical applications.

Suppose we want '*n*' nouns to be combined into a single compound word. Since the noun roots can be extracted from each of these nouns, we shall use $\boldsymbol{W} = \{w_1, w_2, w_3, \ldots, w_n\}$ to denote the set of '*n*' noun roots that are to be combined into a single compound word. Let $PP = w_1$. The recursive logic for generating the noun root for compound word can be stated using the following relationship:

$S_n = S_{n-1} + w_n$, where S_n denotes the noun root for the compound word at the *n*th stage of recursion.

A procedure for obtaining the noun root of the compound word based on the above recursive logic can be as follows:

Step 1: *Read inputs:* n, $\boldsymbol{W} = \{w_1, w_2, w_3, \ldots w_n\}$.
$PP = w_1$
$S_1 = PP$
For $i = 2$ *to* n
Do
Step 2: $UP = w_i$
$S_i = S_{i-1} + UP$ *(using relevant grammatical rules to combine)*
Step 3: $PP = S_i$
$i = i + 1$
Go to Step 2
End Do

Final $S_n = w_1 + w_2 + w_3 + \cdots + w_n$
Noun root for the compound word = S_n.

Following illustrative example will clarify this operation.

Let us form a compound word using a set of nouns to express an idea. A lamp (दीपः - dīpaḥ) placed (स्थितस्य - sthithasya) in the belly (उदरे - udare) of a pot (घटस्य - ghaṭasya), that is pierced with several holes (नानाछिद्रस्य - nānāchidrasya). The noun forms that are to be combined is the set {नानाछिद्रस्य, घटस्य, उदरे, स्थितस्य, दीपः}. The noun roots of these words are {नानाछिद्र, घट, उदर, स्थित, दीप} ({nānāchidra, ghaṭa, udara, sthita, dīpa}). Using the above recursive logic, we have the following:

n = 5 and **W** = {नानाछिद्र, घट, उदर, स्थित, दीप}.
PP = नानाछिद्र; S_1 = नानाछिद्र

Iterating the logic 7 times, we obtain the values for S as follows:

i = 2; S_2 = नानाछिद्रघट
i = 3; S_3 = नानाछिद्रघटोदर (change of letter on account of sandhi rule)
i = 4; S_4 = नानाछिद्रघटोदरस्थित
i = 5; S_5 = नानाछिद्रघटोदरस्थितदीप

The logic ends.

Final noun root for the compound word = नानाछिद्रघटोदरस्थितदीप (nānāchidraghaṭodarasth itadīpa), which refers to a lamp placed in the belly of a pot with multiple holes.

5.6 LOGIC FOR SENTENCE CONSTRUCTION

Words are to be grammatically correct in terms of construction. However, a mere assemblage of grammatically correct words together to form a sentence may not in itself guarantee that the intended message is communicated. Only when an action (expressed through a verb) is associated with a group of words in a sentence either explicitly or implicitly, will the sentence bear some meaning. Let us see an example. Suppose someone asks, "What did you have for breakfast?" and the listener replies 'dosa'. Although the listener used a single word 'dosa', it makes eminent sense, because other useful and contextual words were inferred. The word 'dosa' was unambiguously connected with the action of 'eating' to complete the sensemaking process. On the other hand, if we merely say "dosa" without a context, then the listener will be left with a question, "dosa, what?" His ability to make sense will be possible only when he hears a verb along with other necessary words such as, 'I ate dosa, I cooked dosa, I liked dosa, etc.'

- Kāraka is a concept that helps to link the words in a sentence to the kriyā (action).
- Kriyā and kāraka are the essential elements of any sentence. All other elements that are found in a sentence are woven around these two.

Therefore, for a sentence to be complete there must be a verb, implicit or explicit, denoting an action. On the other hand, a verb alone cannot make a meaningful sentence. If somebody says, 'comes', the listener will experience ambiguity. If the verb is associated with a noun such as, 'Ram comes' the listener is satisfied, as it makes enough sense for the listener. Hence it follows that a sentence must have a verb (denoting an action) and other words denoting participant and other attributes connected to the action. The other words in the sentence need to be logically connected with one another and with the verb (the action indicated).

The manner the Sanskrit language has addressed this differs from a language such as English. The Sanskrit language uses a concept called kāraka to provide in-built mechanisms for constructing unambiguous and grammatically correct sentences. A participant involved in the action in some manner is called कारक (kāraka). Kāraka helps to link the words in a sentence to the action (kriyā). Viewed in this manner, kriyā and kāraka are the essential elements of any sentence. All elements found in a sentence are woven around these two[14]. The importance of this is best understood by taking an example from the English language and comparing it with the Sanskrit equivalent:

English: The fat boy eats the tasty food with the hand.

Sanskrit: स्थूलः बालकः स्वादु भोजनं हस्तेन खादति ।
(sthūlaḥ bālakaḥ svādu bhojanaṃ hastena khādati).

There are four components in the sentence: 'fat boy', 'tasty food', 'eats', and 'with the hand'. Let us reorder the words in the sentences in some arbitrary fashion. Table 5.2 gives a sample of five alternatives for grouping the words in the sentence. The sentences in English have not only lost their intended meaning but also communicate illogical and inappropriate ideas. On the other hand, the sentences in Sanskrit have preserved the original meaning and intent. Notwithstanding where we place the words in the Sanskrit sentence the meaning does not get distorted on account of the two features that we notice in the Sanskrit language:

TABLE 5.2 Issues in Sentence Formation – An Illustration

Sl. No.	*Sentence in English*	*Sentence in Sanskrit*
1	The **fat boy** eats the *tasty food* with the hand	**स्थूलः बालकः** स्वादु भोजनं हस्तेन खादति । **sthūlaḥ bālakaḥ** *svādu bhojanaṃ* hastena khādati
2	The **fat** hand eats the *tasty food* with the **boy**	**स्थूलः** हस्तेन खादति स्वादु भोजनं **बालकः** । **sthūlaḥ** hastena khādati *svādu bhojanaṃ* **bālakaḥ**
3	The **fat** *food* eats the tasty hand with the **boy**	**स्थूलः** भोजनं खादति स्वादु हस्तेन **बालकः** । **sthūlaḥ** *bhojanaṃ* khādati *svādu* hastena **bālakaḥ**
4	The *food tasty* eats the **fat** hand with the **boy**	स्वादु भोजनं खादति **स्थूलः** हस्तेन **बालकः** । *svādu bhojanaṃ* khādati **sthūlaḥ** hastena **bālakaḥ**
5	The *tasty* **boy** eats the **fat** *food* with the hand	स्वादु **बालकः** खादति **स्थूलः** भोजनं हस्तेन । *svādu* **bālakaḥ** khādati **sthūlaḥ** *bhojanaṃ* hastena

(a) The case variations are in-built with the noun form making it always a single word. For example, hastena is a single word meaning 'by hand'. Therefore, it does not matter where we place this word. This is on account of the basic methodology of word generation as explained in Section 5.4. The case variations are presented as suffixes and are assimilated into the noun root eventually.

(b) The stipulated number, gender, and case agreements connect the related words even if they are placed in different parts of a sentence. For example, sthūlaḥ and bālakaḥ are 'first case singular' forms and they will always be linked to each other, notwithstanding where we place these words in the sentence.

In order to understand this, let us consider a sentence, "The technician removes the machine from the office in the morning with a truck". This sentence can be written in Sanskrit as, "यन्त्रकारकः प्रातःकाले यन्त्रं वाहनेन कार्यालयात् अपाकरोति" । (yantrakārakaḥ prātaḥ kāle yantraṃ vāhanena kāryālayāt apākaroti). Using this example, let us try to understand the kārakas and their relationship to the kriya. The kriyā in this sentence is 'removing'. Therefore, every kāraka will create a direct or indirect link to the kriyā. The six kārakas are related to vibhaktis (cases) in Sanskrit. The six kārakas (and the corresponding cases) are as follows:

- कर्तृ (kartṛ – doer)–One in whom the cause of action is resident. In the above example, the technician is where the cause of the action is resident. This is the first case.
- कर्म (karma–the locus of the result of an action)–In the example, the removing action is directed at the machine and the effect of removing will be on the machine. This represents the second case.
- करण (karaṇa – instrument)–That which aids in the attainment of the action. In our example, the act of removing was facilitated by a truck. This is the third case.
- संप्रदान (sampradāna – receiver)–That with which the karma is desired to get associated. We have not included this in our example.
- अपादान (apādāna – reference point in separation)–That which is stable in separation. The office was the point of separation for the machine, as it is being removed away from the office. This is the fifth case.
- अधिकरण (adhikaraṇa – the locus of the action)–That which provides the substratum, context, or reference to the action performed through kartṛ or karma. In our example, the action is performed in the morning, which provides the context. This represents the seventh case.

Figure 5.7 pictorially illustrates the kārakas using the above example. The sixth case can attach itself to any of the six kārakas through some associative properties. In the above example, if we want to say, "the machine was removed using friend's vehicle", then the sixth case of 'friend' is attached to 'vehicle'. Similarly, if we want to say, "the machine was removed from Ram's office", then the sixth case of 'Ram' is attached to 'office' and so on. So, each word in a sentence presents a meaning, and all these meanings are related to action directly or indirectly.

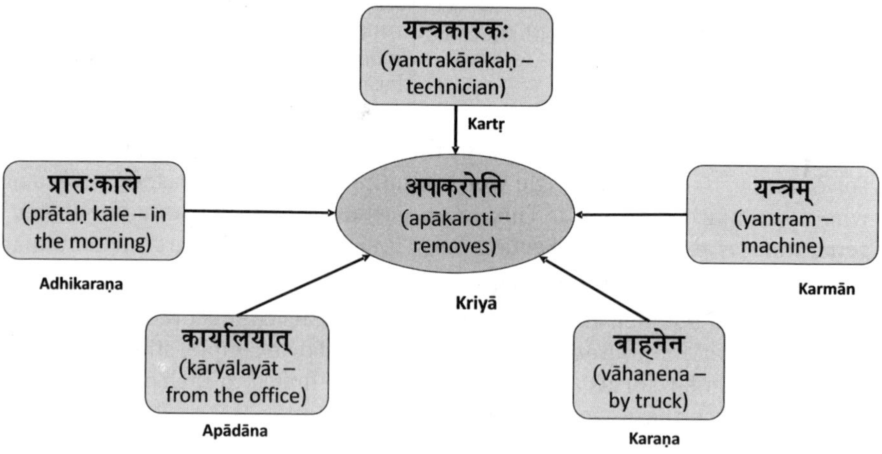

FIGURE 5.7 An Illustration of Kārakas

In the Sanskrit language, we have singular, dual, and plural forms for both the verbs and nouns. Moreover, the nouns have gender variations (masculine, feminine, and neuter) and the verbs have person variations (I person, II person, and III person). The grammatical rules require that the adjectives must agree in gender, case, and number. On account of the kāraka concepts and these linguistic arrangements, the sentence formation has a robust structure and leaves very little scope for ambiguity in the Sanskrit language.

5.7 IMPORTANCE OF VERBS

Language is required only when we are in action. If no one is engaged in action, then there is no need for language. Therefore, several aspects of linguistics revolve around the action. The discussion in the previous section establishes the importance of a verb (kriyā-pada) in Sanskrit language. The importance of the verbs is further amplified by the fact that several of the noun roots are also derived from the verb roots only.

Since most words (both the verb forms and noun forms) originate from the verb roots (dhātus) there is another important dimension to this. In Sanskrit, we often find several synonyms for a word. Each synonym for a word is often derived from a dhātu. With a good knowledge of the dhātus, we will be able to make sense of the synonyms, and the unique meaning the word derives from a particular synonym. This significantly increases the power of the language as one can choose an appropriate synonym depending on the context in which the word is intended to be used. Let us consider an example to understand this idea. Let us consider the English word 'fire'. A common word in Sanskrit that may come to one's mind is 'Agni'. Amarakoṣa, a book of synonyms in Sanskrit provides 34 synonyms for the word Agni[15]. Let us take four synonyms from this list and understand the meaning of the synonym:

1. वह्निः (vahniḥ) – this is derived from the dhātu, 'वह् (vah)', meaning 'to carry'. Therefore, when we need to communicate 'fire' in a sense of a carrier (as in the case of āhuti being offered to the Gods in a sacrifice) this term may be appropriate.
2. पावकः (pāvakaḥ) is derived from the dhātu 'पू (pū)' meaning 'to purify'. When we need to discuss the role of fire as a purifying agent, then the choice of this synonym becomes most appropriate.
3. शुष्मा (śuṣmā) is derived from the dhātu 'शुष् (śuṣ)' meaning 'to dry or shrink'. Fire plays the role of drying air, water, etc. or shrink organic objects by evaporating the water by virtue of the heat it possesses. In such contexts, the use of this synonym may be appropriate.
4. दहनः (dahanaḥ) is derived from the dhatu 'दह् (dah)' meaning 'convert or burnt to ashes'. Where we need to communicate the power of fire to burn off objects, this synonym communicates the idea appropriately.

These examples demonstrate to us that the knowledge of the dhātus from which a noun has been derived is important to communicate an idea most appropriately.

Prefixes for Verb Forms

Most of the word generation as we have seen so far works on adding suffixes to the base and applying relevant grammatical rules to arrive at the final word. However, prefixes (upa-sargas), are appended to the verb forms in order to create additional words. There are 22 prefixes, and one or more of these could be prefixed to a verb form. By adding the prefixes, it is possible to express the meaning in many ways[16]. Let us take an example of a verb root, 'कृ' (kṛ)[17], which can be used to generate many verb forms using prefixes. Figure 5.8 is a sample illustration of the use of prefixes. We observe the following from Figure 5.8:

- Prefixes can be added to several verb forms. adhikaroti, upakaroti and apakaroti represent three different verb forms of the dhātu.
- More than one prefix can be added to the verb form. For example, pratyupa-karoti has two upa-sargas prefixed to the verb form 'karoti'. These are 'prati' and 'upa'.

> ♦ Prefixes, known as upa-sargas, are appended to the verb forms in order to create additional words.

In general, when we add a prefix to the verb form the meaning gets modified in many ways:

- The original meaning is further strengthened or emphasized by the addition of a prefix. For example, स्मरति (smarati) means remembers, and with the addition of the prefix सम् (sam), संस्मरति (saṃsmarati) means remembers well.
- Prefixes can bring an expanded improvisation of the original intent. The verb form 'karoti' means 'does an action'. The prefix 'upa' means near and 'apa' means away. By adding this to the verb form 'karoti', we can qualify the nature of action. Upa-karoti indicates an action that took one closer to the result (conducive action, help, etc.), and apa-karoti indicates an action that took one away from the desired effect.
- Certain prefixes can directly bring the opposite meaning to the original verb form. For example, the prefix 'prati' means the opposite (in a directional sense also). Prati-karoti means doing an action in return for the action done.
- More than one prefix can be used. For example, Pratyupa-karoti deploys two prefixes, 'prati' and 'upa', which means doing a conducive action in return (such as returning the help earlier received from someone).

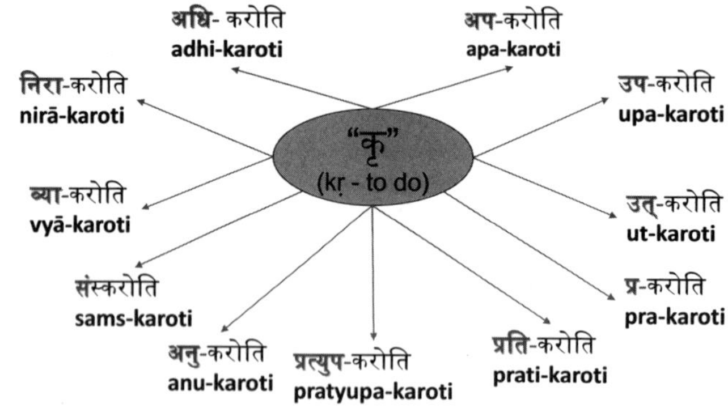

FIGURE 5.8 An Illustration of the Use of Suffixes

5.8 ROLE OF SANSKRIT IN NATURAL LANGUAGE PROCESSING

Natural Language Processing (NLP) is a branch of linguistics mainly concerned with the processing of natural language data using computers and programming techniques. NLP uses ideas from linguistics, computer science, and artificial intelligence to make natural languages understandable to computers so that computers can carry out actions based on the instructions given in a natural language. NLP can be viewed as two sub-sets of ideas; Natural Language Generation (NLG) is the process of generating text from a meaning representation using a logic implemented through computer programs. On the contrary, Natural Language Understanding (NLU) is the process of distilling the meaning from natural language data using computer programs. Both are complementary and together form the basic building block for NLP. NLG is comparatively easier to handle than NLU, especially when there are rule-based constructs, lexicon, and syntax available for word generation. NLU, on the other hand, involves handling ambiguities.

The underlying logic for NLP is to develop a robust methodology for information processing, coding the information, disintegrating them into components, identifying appropriate meanings for the components, and ensuring that the meanings are related unambiguously, and the meaning logically flows through the sentence. This is easier said than done and there are multiple challenges that one will face in the process. Let us take an example of a sentence in Sanskrit to understand some of these issues. In the sentence, 'बालः वृक्षस्य फलं 'खादति' (bālaḥ vṛkṣasya phalaṃ khādati) meaning, 'the boy eats the fruit of the tree', there are some components and information encoded. One of the achievements of the early linguists was that they understood this process of information encoding and they assigned those meanings to the components. As we have discussed in Sections 5.5 and 5.6, the rule-based derivational aspect of the language and the principles of kāraka and vibhakti will enable us to distill the components and the encoded information in the sentence as given below:

बाल	सुँ	वृक्ष	स्य	फल	अम्	खाद्	ति
bāla	Sŭ	Vṛkṣa	Sya	phala	am	khād	Ti
boy	–	Tree	of	fruit	karma	To eat	Kartṛ
Noun base	1st case suffix	Noun base	6th case suffix	Noun base	2nd case suffix	Verb base	Present tense suffix

Once it is understood, the next step is to see how these meanings are related and then how the information flows through the sentence. In the above example, these are merely pieces of information in the sentence. They don't readily convey the information the speaker wants to convey. It is important to know how they are related. The structure of the sentence conveys that also. The concepts encoded in the components of these words have expectancy for other meanings by nature, and hence they find their connection with other meanings. This expectancy is called ākāṅkṣā. Indian linguists had understood that a kriyā (action) always has expectancy for kārakas (participants). So, whenever there is a 'kriyā' indicated by a dhātu, it is looking for a kāraka to relate. When a kāraka is found marked by the second case, since it also has expectancy, they both find the relation. Hence we come to know that it is the fruit that the boy is eating not the tree because the tree is not related to eating directly. Such groups of semantically connected concepts help in understanding the actual meaning of a sentence. This is the way the information in a sentence is processed in a natural language. This understanding is very essential for the processing of information by computer.

It is difficult for a computer to comprehend natural languages like humans as there are many ambiguities in natural languages. Humans can easily resolve these ambiguities as they are aware of the context whereas a computer-based program cannot. This inability to comprehend natural language results in a huge limitation in achieving the goal of NLP based activities such as machine translation. When we use words having multiple meanings there is ambiguity about the meaning. Humans can remove the ambiguity by the context. 'Payas' is the common word for both water and milk in Sanskrit. When we hear the sentence 'नद्यां पयः प्रवहति' (nadyāṃ payaḥ pravahati), meaning 'payas' is flowing in the river, we immediately resolve the ambiguity because we know for the fact that it is water that flows in a river, not milk. Indian linguists have described fourteen such determiners to fix the meaning of a word in case of multiple meanings. These are useful features for word sense disambiguation routines used in NLP.

> - The principles of kāraka and vibhkti will enable us to distill the components and the encoded information in a sentence.
> - Indian linguists have described fourteen determiners to fix the meaning of a word in case of multiple meanings.

Studies show that certain inherent advantages in the structure of the Sanskrit language and the grammar makes it an attractive candidate for NLP and artificial intelligence-related work[18]. The Sanskrit language has certain inherent features such as being derivational and inflectional. With the availability of a full-fledged generative grammar for Sanskrit in the form of Aṣṭādhyāyī, it lends itself as a potential candidate for use in NLP. Despite these inherent advantages, there is still a lot of work to accomplish to use Sanskrit in NLP. In recent years several researchers across the world have worked in the area of Sanskrit computational linguistics and have developed several components that can aid NLP using Sanskrit. This includes a variety of word generators (including compound words), morphological analysers, parsers, and creating lexical resources.

SUMMARY

- As technology and computing prowess improves, we are able to develop better applications using Artificial Intelligence (AI) techniques. This requires us to develop efficient Natural Language Processing (NLP) capabilities.
- Aṣṭādhyāyī is considered a fine creation of human intelligence and is the best available descriptive model of a language.
- Pāṇini developed 3,983 rules (known as Sūtras) to accommodate most of the patterns and variations in the Sanskrit language and arranged them in eight chapters.
- According to earliest Indian texts, the sound primarily arises at the junction of the throat and the chest due to the movement of the breath in the body and manifests in the oral cavity at different locations resulting in various sound patterns.
- According to Sanskrit grammar, two phonemes are considered to be homogenous if they are produced with the same articulatory effort and at the same place of pronunciation.
- The entire scheme for word generation follows a highly modular approach. Two basic components form a word: base (prakṛti) and suffix (pratyaya).
- Each word is formed out of a base (verbal or nominal root). To this, one or more suffixes are added to generate the word.
- One of the distinguishing aspects of the Pāṇinian approach to linguistics and Sanskrit grammar is the use of several features, which in modern parlance map to certain computational concepts.
- Pāṇini's system of applying grammatical conditions to derive words works exactly like a rule-based engine. Every conceivable word

in the Sanskrit language can be derived using a set of rules in a strict algorithmic fashion.

- Aṣṭādhyāyī employs recursive logic to process grammatical applications. One such example is the logic of samāsa, a method of creating compound words from a group of nouns.
- On account of the kāraka concepts and linguistic arrangements, the sentence formation has a robust structure and leaves very little scope for ambiguity in the Sanskrit language.
- Most words (both the verb forms and noun forms) originate from the dhātus. In Sanskrit, we often find several synonyms for a word. Each synonym for a word is often derived from a dhātu.
- Prefixes, known as upa-sargas, are appended to the verb forms in order to create additional words.
- There are 22 upa-sargas and these could be prefixed to a verb form. By adding the upa-sargas, it is possible to express the meaning in many ways.
- Studies show that certain inherent advantages in the structure of the Sanskrit language and the grammar makes it an attractive candidate for NLP and artificial intelligence-related work.

REVIEW QUESTIONS

1. Briefly sketch the organization of Aṣṭādhyāyī. How many rules are there in Aṣṭādhyāyī?
2. What are the unique aspects of the Sanskrit grammar propounded by Pāṇini?
3. How is the issue of phonetics addressed in the Sanskrit language? What are the phonetical aspects pertaining to vowels specified in Sanskrit grammar?
4. What do you understand by the terms 'Prakṛti' and 'Pratyaya'? What is the relevance of these in Sanskrit grammar?
5. Write short notes on the following:
   (a) The Sanskrit language is derivational in nature
   (b) Sanskrit grammar is rule-driven
   (c) Sanskrit grammar is modular in structure
6. What are the ways by which one can generate noun forms in Sanskrit? Similarly, explain how verb forms can be generated.
7. What do you understand by the terms 'subanta' and 'tiṅanta'?
8. What do you understand by the term 'mnemonics'? How is it used in Aṣṭādhyāyī?
9. Briefly explain how compound words are generated in the Sanskrit language.
10. What is the relevance of the 'kāraka' concept? Using an example identify the kārakas in your example. Can you also relate it to the vibhaktis?
11. Write a one-page note explaining why knowledge of dhātus in Sanskrit is very valuable.
12. What happens when an upa-sarga is prefixed to a verb form? Give some examples in support of your argument.
13. Comment on the statement, "The Sanskrit language has potential for use in NLP and AI applications".

14. Fill in the table with suitable examples:

Sl. No.	*Criteria*	*Examples*
1	Sound originating in the throat	
2	Sound originating in the lips	
3	Sound originating in the teeth	
4	Sound originating in the lips and teeth	
5	Anunāsikā	
6	Mahā-prāṇa	
7	Pluta	

DISCOVER IKS

1. There is widespread confusion among the people about the relevance of the Sanskrit language. There is a feeling that an ancient and classical language may not be relevant in the modern-day jet setting style of living driven by advances in Science and Technology. Watch this video of Dr. Sampadananda Mishra: https://www.youtube.com/watch?v=AmUJLx47ZDg. After watching the video prepare a two-page note to answer the following questions:

(a) Is Sanskrit a sectarian, old language difficult to learn? Why or why not?

(b) Can Sanskrit generate new vocabulary required today (say a word for Camera or Microphone)? How does it enable this process?

(c) What are some of the unique aspects of Sanskrit phonetics?

2. Watch this video on the topic, *The beauty of Sanskrit* by Abhinav Seetharaman. https://youtu.be/cubtefu3IjsTBA

(a) What are some of the commonalities and differences between Sanskrit, Ancient Greek, and Latin?

(b) Does learning Sanskrit provide any advantages linguistically?

(c) Should Sanskrit remain static in terms of its use and application? What are the ways by which Sanskrit is being promoted globally?

SUGGESTED READINGS

Bharati, A., Chaitanya, V. and Sanghal, R. (2004). *Natural Language Processing: A Paninian Perspective*, Prentice-Hall of India, New Delhi.

Dalai, B.K. (2007). *Studies in Indian Linguistics*, Bharatiya Kala Prakashan, Delhi.

हरेकान्त मिश्रः (2011). "बृहद्धातुकुसुमाकरः" चैखाम्बा संस्कृत प्रतिष्ठान्, दिल्लि.

Huet, G., Kulkarni, A. and Scharf. P. (2009). *Sanskrit Computational Linguistics.* Springer.

Kak, S, Bhate, S. (1993). "Pāṇini's Grammar and Computer Science", Annals of Bhandarkar Oriental Research Institute, Pune.

Kanshi Ram (2010). *The Ladghusiddhāntakaumudī of Varadaraja: A Primer on Pāṇini's Grammar*, Volumes I–III, Motilal Banarsidass Publishers, Delhi.

महाबलेश्वरभट्टः, जी. (1999). "कारकम्", संस्कृतभारती, बेङ्गलूरु.

महाबलेश्वरभट्टः, जी. (1994). "समासः", संस्कृतभारती, बेङ्गलूरु.

रमाशंकर मिश्र (2010). "अष्टाध्यायीसूत्रपाठः" मोतिलाल बनारसीदास, वाराणसी.

Sampad and Vijay (2011). "The Wonder that is Sanskrit", Sri Aurobindo Society, Puducherry.

श्रीधरानन्दशास्त्री (2010). "श्रीवरदराजविरचिता लघुसिद्धान्तकौमुदी", मोतीलाल बनारसीदास, दिल्लि.

ENDNOTES

1. For details see, **Cardona, G. (2014).** "Some contributions of ancient Indian thinkers to Linguistics", Chapter 1 in Sanskrit and Development of World Thought, V Kutumba Sastry (Ed.), D K Printworld, pp. 1–22, New Delhi.
2. For details on all sūtras referred in this chapter, see, **Kanshi Ram (2010).** *The Laghusiddhāntakaumudī of Varadarāja: A primer on Pāṇini's Grammar*, Volumes I–III, Motilal Banarsidass Publishers, Delhi. The relevant sūtra in Aṣṭādhyāyī is तुल्यास्यप्रयत्नं सवर्णम् । tulyāsyaprayatnaṃ savarṇam | 1.1.9. As a further explanation to this sūtra, the following sūtras can be found in laghu-siddhānta-kaumudi of Varadarāja: अकुहविसर्जनीयानां कण्ठः, इचुयशानां तालु, ऋटुरषाणां मूर्धा, ऌतुलसानां दन्ताः, उपूपध्मानीयानां ओष्ठौ ञमङणनानां नासिका च, एदैतोः कण्ठतालु, ओदौतोः कण्ठोष्ठम्, वकारस्य दन्तोष्ठम्, जिह्वामूलीयस्य जिह्वामूलम्, नासिकाऽनुस्वारस्य, इति स्थानानि। akuhavisarjanīyānām kaṇṭhaḥ, icuyaśānāṃ tālu, ṛṭuraṣāṇāṃ mūrdhā, ḷtulasānāṃ dantāḥ, upūpadhmānīyānāṃ oṣṭhau ñamaṅaṇanānāṃ nāsikā ca, edaitoḥ kaṇṭhatālu, odautoḥ kaṇṭhoṣṭham, vakārasya dantoṣṭham, jihvāmūlīyasya jihvāmūlam, nāsikā'nusvārasya, iti sthānāni|
3. The relevant sūtra that communicates this idea is ऊकलोऽज्झस्वदीर्घप्लुतः । ūkalo'jjhrsvadīrghaplutaḥ | 1.2.27.
4. The relevant sūtra is मुखनासिकावचनोऽनुनासिकः । mukhanāsikāvacano'nunāsikaḥ | 1.1.8.
5. The sūtras are उच्चैरुदात्तः । uccairudāttaḥ | (1.2.29), नीचैरनुदात्तः । nīcairanudāttaḥ | (1.2.30), and समाहारः स्वरितः । samāhāraḥ svaritaḥ | (1.2.31).
6. This is established by the sūtra, तुल्यास्यप्रयत्नं सवर्णम् । tulyāsyaprayatnaṃ savarṇam | 1.1.9.
7. In this chapter, we use the term 'noun form' to indicate a wide variety of nouns and their variations that could be generated using the 'sup' and other noun generating suffixes. Similarly, using 'tiṅ' and a host of their suffixes several verbs, participles and other transformations of the verb root is possible. All these are collectively referred as 'verb form'.
8. This is established through the sūtra, 'सुप्-तिङ्-अन्तं पदम्' (sup-tiṅ-antam padam) 1.4.14. By this it is implied that whatever be the word generated in Sanskrit, it has to be 'sup' ending or 'tiṅ' ending. 'sup' is a set of pratyayas to generate a noun form (known as subanta). On the other hand, 'tiṅ' is the set of pratyayas to generate a verb form (known as tiṅanta).
9. There is a rule in Aṣṭādyāyī that prevents us from taking the last sūtra as the substitute for हल्.
10. This is given in sutra 4.1.2, स्वौजसमौट्छष्टाभ्याम्भिस्ङेभ्याम्भ्यस्ङसिभ्याम्भ्यस्-ङसोसाम्ङ्योस्सुप् । svaujasamau ṭchaṣṭābhyāmbhisṅebhyāmbhyasṅasibhyāmbhyas-ṅasosāmṅyossup | For Aṣṭādhyāyī sūtras you can also refer, डा. रमाशङ्कर मिश्र (२०१०). "अष्टाध्यायीसूत्रपाठः", मोतिलाल बनारसीदास, दिल्ली.
11. This is as per the sutra 3.4.78, तिप्तस्झिसिप्थस्थमिब्वस्मस्-तातांझथासाथांध्वमिड्वहिमहिङ्. । tiptasjhisipthasthamibvasmas-tātāṃjhathāsāthāṃdhvamiḍvahimahiṅ. |
12. This sūtra is the 77th in the first quarter of Chapter 6 of the Aṣṭādhyāyī (6.1.77).
13. The rules used for the derivation of the word given in the box are as follows: स्वौजसमौट्छष्टाभ्याम्भिस्ङेभ्याम्भ्यस्ङसिभ्याम्भ्यस्ङसोसाम्ङ्योस्सुप् । svaujasmauṭchaṣṭābhyāṃbhisṅebhyāṃbhyasṅasibhyāṃbhyasṅsos āmṅyossup | 4.1.2, टाङसिङसामिनात्स्याः । ṭāṅasiṅasāminātsyāḥ | 7.1.12, अदेङ् गुणः । adeṅ guṇaḥ | 1.1.2 and अट्कुप्वाङ्नुम्व्यवायेऽपि । aṭkupvāṅnumvyavāye'pi | 8.4.2.

14. 'Ākhyātaṃ sāvyayakārakaviśeṣaṇaṃ vākyam' (Kātyāyana Vārtika 2.1.1), A tiṅānta verb along with indiclinables, kārakas and other modifiers is a vākya. There are many theories proposed in sentence by Bhartṛhari in his philosophical treatise on vyākaraṇa 'Vākyapadīya'.

15. For a good discussion of this idea, please see, **Sampad and Vijay (2011).** "The wonder that is Sanskrit", Sri Aurobindo Society, Puducherry, pp. 21–28.

16. उपसर्गेण धात्वर्थो बलादन्यत्र नीयते । प्रहार-आहार-संहार-विहार-परिहार-वत् ॥ upasargeṇa dhātvartho balādanyatra nīyate | prahāra-āhāra-saṃhāra-vihāra-parihāra-vat || https://sa.wikipedia.org/wiki/उपसर्गाः Last accessed on Oct. 1, 2021.

17. The exact representation of the dhātu is "डुकृञ् (dukṛñ). It has three portions in it; 'du', 'kṛ' and 'ñ'. The first and the third letters are not part of the dhātu. They are tags used by Pāṇini to facilitate the process of invoking relevant grammatical rules while the dhātu is being processed for word generation. Essentially the dhatu is 'kṛ' which means 'in the sense of doing'.

18. See for example, **Briggs, R. (1985).** "Knowledge Representation in Sanskrit and Artificial Intelligence", The AI Magazine, **6**(1), pp. 33–39, where he argues the potential of Sanskrit language for such applications.

CHAPTER

6 Number System and Units of Measurement

LEARNING OUTCOMES

After finishing this chapter, you will be able to:

- Understand the key features of Indian Numeral System
- Appreciate the key role it has played in advancement of Science and Technology
- Understand how measures for time, length and weight were in use in Ancient India
- Decipher the unique methods deployed to represent numbers
- Identify the roots of modern-day binary system

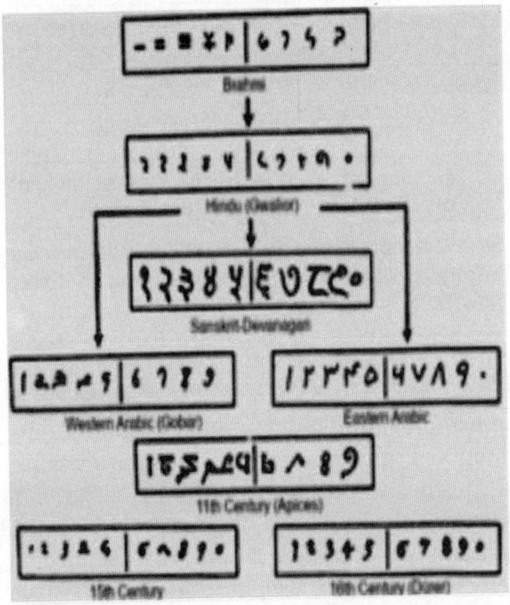

The greatest contribution of India to the development of Science and Technology lies in establishing a robust and a mature number system much before the dawn of the Common Era (CE). This was inherited up by the Arabic world during 8th Century CE and eventually passed on to the West by 11th Century CE.

IKS IN ACTION 6.1

Gautama-Buddha on Number System

An episode in the life of Buddha narrated in Lalitavistara provides an amazing account of the prevalence of a robust number system 2500 years ago. When Siddhārtha, son of king Śuddhodana was to marry Yaśodharā, he was subjected to a svayaṃvara. He was competing with 500 others and the contest included, among other things, testing of knowledge of arithmetic.

A great mathematician of the time, by name Arjuna asked Siddhārtha to express numbers beyond koṭi (10^7) in multiples of 100. In his reply to this question Siddhārtha went on up to 10^{53}, spelling out unique names to each number he uttered. For example, 10^{53} was 'tallakṣaṇa'. Starting from koti (10^7), the series up to tallakṣaṇa (10^{53}) constituted 23 levels.

The astonishing part of this conversation was when Siddhārtha mentioned that this number tallakṣaṇa forms part of one counting system. He further went on to show that eight more such counting systems are also available to enumerate higher numbers. In each system, there are 23 terms starting from the previous system. The method to get these numbers can be generalized using a formula. At the nth counting system, the last number in the series can be generated using the formula:

$$(10^{(7+46*n)})*(10^2)^{23}$$

When $n = 0$, the number is 10^{53} (tallakṣaṇa)

$n = 1$, the number is 10^{99} (dhvajāgravatī)

$n = 2$, the number is 10^{145} and so on.

Since Siddhārtha mentioned of eight more counting systems, at $n = 8$, the number is 10^{421}. Such large numbers are unheard of even today.

The existence of unique names to numbers of such a high value as early as the time of Gautama-Buddha speaks of the mathematical prowess of ancient Indians.

Intrigued by his knowledge of large numbers, Arjuna asked Siddhārtha about his knowledge of smaller numbers than unity. The reply given by Siddhārtha can be aptly summarized using the table below:

Unit of measure	*Relationship*
1 Yojana	4 Krośa from Magadha
1 Krośa	1,000 arcs
1 Arc	4 Cubits
1 Cubit	2 Spans
1 Span	12 Phalanges of fingers
1 Phalanx of a finger	7 Grains of Barley
1 Grain of Barley	7 Mustard seeds
1 Mustard seed	7 Poppy seeds
1 Poppy seed	7 Specks of dust stirred by a cow
1 Speck of dust stirred by a cow	7 Specks of dust stirred by a ram
1 Speck of dust stirred by a ram	7 Specks of dust stirred by a hare
1 Speck of dust stirred by a hare	7 Specks of dust stirred by the wind
1 Speck of dust stirred by the wind	7 Tiny Specks of dust
1 Tiny speck of dust	7 Particles of dust of first atoms (Paramāṇu)

Source: Adopted from **Ifrah, G. (2005).** *The Universal History of Numbers II*, pp. 136–139, Penguin Books, New Delhi.

The bedrock of modern scientific discoveries is the use of well-defined number systems, units of measurement, and computational mechanisms. In an era of computer-based data processing, the binary system is an important aspect of this. Another important aspect of civilizational growth is vibrant international trade. To transact through the exchange of goods and services, we need standard means of measurement, estimation, and communication. This requires a well-defined number system and units of measurement for length, weight, time, etc. We have taken all these for granted today. Little did we know or think about where all these originated and how it systematically developed through the ages. We shall dwell on these issues in this chapter.

6.1 NUMBER SYSTEM IN INDIA – HISTORICAL EVIDENCE

The contribution of ancient Indians to the development of mathematical concepts is well known and acknowledged. Many Scientists and writers from the West have time and again acknowledged this aspect, as evident from the following:

- *Based on extensive research, Ifrah presented evidence from Europe and the Arabic world to show that modern number systems indeed originated in India. Using several references during 810 CE to 1814 CE, he presented several observations of the past writers on Indian Mathematics*[1].
- *Laplace remarked, "The ingenious method of expressing every possible number using a set of ten symbols (each symbol having a place value and an absolute value) emerged in India. Its simplicity lies in the way it facilitated calculation and placed arithmetic foremost amongst useful inventions"*[2].
- *In his book on India, composed around 1030 CE, Al-Biruni wrote, "Whilst we use letters for calculation according to their numerical value, the Indians do not use letters at all for arithmetic. And just as the shape of the letters that they use for writing is different in different regions of their country, so the numerical symbols vary"*[3].

A number of archaeological excavations supplement the linguistic evidence provided by the early writers about the maturity and supremacy of the Indian mathematical foundations. The street widths in the Indus-Sarasvati Civilization were highly standardized. For instance, Kalibangam, a city in the Indus-Sarasvati Civilization (in Rajasthan, India) had street widths of 1.8 m, 3.6 m, 5.4 m, and 7.2 m[4]. These were built to the standard dimensions of 1, 2, 3, and 4 Dhanus (this unit is discussed in the section on measurements) respectively. Such widths are found at other sites also. The excavations at Harappa, Mohenjo Daro, Dholavira, and Lothal show that several constructions were done using fired bricks of standard dimensions and fine geometrical and material quality. The dimensions of the bricks appeared to be standard with length × width × depth in the ratio 4:2:1. Weights and linear measures follow a similar pattern wherever they have been found. This similarity across sites spanning several hundred kilometres points to the use of a standard unit of measurement at that time, signalling mature mathematical thinking.

> - Several archaeological excavations provide evidence about maturity and supremacy of Indian mathematical foundations.
> - An inscription on a temple wall in Gwalior dating back to the ninth century CE is considered the oldest recorded example of a zero.

In the Arthaśāstra, there is a mention of two types of Dhanus as units for measuring lengths and distances: Dhanus = 96 Aṅgulas; Gārhapatya-dhanus = 108 Angulas used for measurement of roads and distances. The oldest dated Indian document containing a number written in the place-value format that we use today is a legal document dated 594 CE from the Bharukachcha (or Broach) region in Gujarat. In an inscription at Gwalior dated 'Saṃvat 933' in the Vikrama calendar (876 CE) the numbers 50 and 270 were recorded with a small circle appearing at the appropriate positional place for zero. An ancient Indian scroll known as the Bhakshali manuscript was discovered in a field in 1881 giving evidence of the earliest recorded use of zero. Carbon dating of this has revealed that it was probably written in the third or fourth century CE. An inscription found on a temple wall in Gwalior dating back to the ninth century CE is another recorded example of a zero.

6.2 SALIENT FEATURES OF THE INDIAN NUMERAL SYSTEM

The Indian numeral system has a long history. The origin and the evolution of the numbers could be traced from the time of the Vedic period. Being an oral tradition, unique and unambiguous names were to be attributed to the numbers. Therefore, as early as the Ṛgveda time, the use of numbers and unique names prevailed. The Sanskrit language has unique names for numbers starting from one and going up to very large numbers. The first nine digits have unique names (ekam, dve, trīṇi, catvāri, pañca, ṣaṭ, sapta, aṣṭa, nava). There are unique names for numbers from 10 to 100 in steps of ten (*daśa, viṃśati, triṃśat, catvāriṃśat, pañcāśat, ṣaṣṭi, saptati, aśīti, navati, śata*). Beyond 100, there has been the use of names for numbers, which extend up to very large numbers. There are many references in the literature pointing to the use of such number names.

Spread of Indian Mathematical Concepts Westwards

The catalogue of the Sui dynasty (610 CE), pointed to several Chinese translations of Indian works on astronomy and mathematics, which are no longer extant. Records from the seventh century (of the Tang dynasty), suggest that Indian astronomers were employed at the Astronomical Board of Chang-Nan to teach the principles of Indian astronomy and calendar. An Indian named Gautama-Siddhārtha was reported to have constructed a calendar along the lines indicated in the Indian *Siddhāntas*.

Indian numerals reached the court of second Abbasid Caliph al-Mansur (753–774 CE) from Sindh. Moreover, the Indian decimal place-value system reached at least a century earlier as evident from a passage attributed to Nestorian Bishop Severus Sebokht (662 CE). In a book written during 820 CE by al-Khwarizmi, the operations of addition, subtraction, multiplication, division and the extraction of square roots according to the Indian system was explained. His work was translated into Latin by 12th Century CE and there were several commentaries in Latin on this book.

In the transmission of Indian numerals to Europe, via the Arab world, Spain played an important role as it was under Muslim rule for many years. Documents from Spain and coins from Sicily provide evidence of the spread. Fibonacci (1170–1250 CE) learnt the Indian numerals during his travels in North Africa, Egypt, Syria and Sicily and wrote a book bringing these aspects into focus. The spread westwards continued slowly, displacing Roman numerals. It was simply a matter of time before the new numerals were put to use by the bankers, traders and merchants for their daily calculations.

Source: Adopted from **Joseph (2009).** "A Passage to Infinity: Medieval Indian Mathematics from Kerala and Its impact", Sage, Chapter 8.

One can appreciate the importance of the Indian numeral system in scientific development if the following contributions are understood:

- A legacy of using large numbers with unique number names for these large numbers
- Developing a robust place value system for the numerals
- The concept of zero and its use beyond being a placeholder
- A decimal system that opened vast possibilities for arithmetic operations

6.2.1 The Concept of Zero and Its Importance

Today, zero, both as a symbol (or numeral) and a concept meaning the absence of any quantity allows us to perform calculus, solve complicated equations, and to have invented computer operations using binary digits. Among the significant contributions of the ancient Indians is the concept of zero. The available evidence shows that the concept of zero was established during the period 500–300 BCE and fully developed in India by 600 CE. Ancient Indians were able to use a decimal system that allowed them to develop a method for handling large numbers. In this process, the use of number zero became inevitable as a placeholder. The number name to indicate zero is Śūnya. Piṅgala a second century BCE Indian philosopher authored Chandaḥśāstra, which dealt with the metres used in Sanskrit poetry in which the word śūnya was used, which obtained the mathematical connotation of 0. It later became its proper name as a number.

- The concept of zero was established during the period 500–300 BCE.
- Ancient Indians used a decimal system that allowed them to develop a method for handling large numbers.

Brahmagupta developed a symbol for zero in 628 CE. With this invention, zero could be used as an independent numeral for computational purposes. The real power of zero was evident when the use of zero was beyond a mere placeholder. Ancient Indians used zero in computations ahead of others and thereby elevated zero from a placeholder to a numeral. This is considered one of the greatest breakthroughs in the history of mathematics. Bhāskara II in his Bīja-gaṇita introduced the properties of zero when mathematical operators such as addition and subtraction are operated on it. He stated that its nature does not change when numbers are added to or subtracted from zero.

6.2.2 Large Numbers and Their Representation

In the Chāndogya Upaniṣad, in a conversation between Sage Nārada and Sanatkumāra, it was mentioned, "There is no joy in the finite. The Infinite alone is joy. That which indeed is the Infinite is immortal. On the other hand, that which is finite is mortal"[5]. The Ṛgveda has several names of numbers scattered through its ten mandalas. Many of them are names of compound numbers, i.e., those which are neither atomic nor a power of 10. In the Taittirīya-saṃhitā, book 7, Chapter 2, there is a reference to numbers up to 10^{13}. In a passage in the Taittirīya-upaniṣad[6], there is a reference to the quantum of happiness of an evolved individual who has transcended the realm of desires. In the description, the numbers are progressively raised by multiples of 100 up to 10 levels, suggesting a large number of 10^{20}. There is a similar passage in Bṛhadāraṇyaka-upaniṣad[7] on the same theme, which amounts to a number 10^{14}. The opening story of the chapter (IKS in Action 6.1) demonstrates the knowledge of very large numbers to Ancient Indian at the time of Gautama-Buddha. One of the reasons for the use of large numbers could be the interest in Astronomy. Astronomical calculations invariably require large numbers. The Vedic corpus, the Itihāsas, and the Purāṇas have several references to astronomical data. Table 6.1 lists some references to large numbers in ancient Indian texts.

- Brahmagupta developed a symbol for zero in 628 CE.
- There are references to large numbers in canonical works in Jainism.

TABLE 6.1 Reference to Large Numbers in Ancient Texts

Sl. No.	*Text*	*Context*	*Number Quoted*	*Remarks*
1	Lalitavistara-sūtra	Siddhārtha during Svayaṃvara	10^{421}	*Source:* Ifrah (2004)
2	Kāccāyana's Pāli Grammar		10^{140}	*Source:* Dutta and Singh (1962)
3	Rāmāyaṇa (Yuddha-kāṇḍa)	Size of Rāma's army	10^{62}	Referred to as Mahaugha
4	Jain canonical works	Śīrṣaprahelikā - An estimate of time	$(8{,}400{,}000)^{28}$	*Source:* Dutta and Singh (1962)
5	Anuyogadvāra-sūtra	Jain canonical text	10^{28}	*Source:* Divakaran (2018)
6	Taittirīya-upaniṣad - Brahmānandavallī	An inquiry into bliss	10^{21}	Deduced from the description
7	Līlāvatī	Description of place value number system	10^{17}	Referred to as Parārdha
8	Taittirīya-saṃhitā - Book 7, Chapter 2	Count of oblations to Prajāpati	10^{13}	*Source:* Divakaran (2018)
9	Anuyogadvāra-sūtra - Jain canonical work	Number of human beings in the world (100 BCE)	2^{96}	*Source:* Dutta and Singh (1962)

Ancient Indians developed a systematic approach to the number-names that they used to describe large numbers. Three categories of naming conventions were employed to develop the number names in Sanskrit:

1. All numbers in the unit digit from 0 to 9 had unique names (śūnya, ekam, dve,... nava)
2. All numbers in the range of 11 to 99 had an additive principle for naming. For example, 45 was 5 + 40 (pañca-catvāriṃśat), 18 was 8 + 10 (aṣṭā-dasa) and so on. Optionally, the subtracting principle was employed (for example, 29 is 30 - 1: ekona-triṃśat).
3. All numbers of higher powers of 10 starting from 10^2 were named using a multiplicative principle using the unit digits as factors for multiplication. For example, 8,000 was 8 × 1000 (aṣṭa-sahasram) and 70,000 was 7 × 10,000 (sapta-ayuta).

In this manner, it was possible to express any number unambiguously once the number-names for the successive powers of ten were established.

6.2.3 Place Value of Numerals

The use of large numbers requires a sophisticated number system. Otherwise, the journey towards handling large numbers and performing arithmetic operations will end soon. Ancient Indians' gift to mankind is a mature and complete place value-based numeral system. In the Agni purāṇa, it is mentioned, "... in case of multiples from the units place, the value of each place is ten times the value of the preceding place...". Similarly, in the Vāyu purāṇa it is mentioned, "... from one place to the next in the succession, the places are in multiples of ten. The eighteenth place is called parārdha..."[8]. In the Śārīraka-bhāṣya of Śaṅkarācārya a passage exists, which says, "... an individual by name Devadatta may be called differently as a father, son, son-in-law, brother, grandson, child, youth, etc., just as, although the stroke is the same, yet by a change of place it acquires values, one, ten hundred, thousand, etc. ..."[9]. There is a similar passage in a commentary on Patañjali's *Yoga-sūtra*[10]. This shows the commonplace knowledge of place value system during the 8th century CE. In his work Gaṇita-sāra-saṅgraha (850 CE), Mahāvīrācārya gives the result of an operation as एकादिषडन्तानि क्रमेण हीनानि (ekādi-ṣaḍ-antāni krameṇa hīnāni), this means beginning with one and going up to 6 and then decreasing by one. The number depicted is '12345654321' which is the square of the number '111111111'.

Place value of numerals is a concept in which a numeral or a symbol used has a unique meaning and value. With this arrangement, it will be possible to represent any conceivable number systematically. A number system with a complete definition and place value is at the foundation of mathematics. It enables easy arithmetic computations, facilitating scientific analysis. Further, it promotes unambiguous and effective communication. To understand this concept let us see an illustration.

ILLUSTRATION: Roman numerals system is based on addition (and sometimes subtraction) of seven different values; I (1), V (5), X (10), L (50), C (100), D (500), M (1000). All numbers (corresponding symbols) to be subtracted are to be written to the left of the symbol from which we subtract and additions to the right. For example, 397 is represented using notations for 100 + 100 + 100 + (100 – 10) + 5 + 1 + 1 = CCCXCVII. Let us consider three numbers 397, 928, and 107. These are represented in the Roman systems as follows: CCCXCVII, CMXXVIII, CVII. An examination of these two numbers reveals the serious limitations of the Roman system of numerals.

- The three Roman numbers have a varying number of digits (7, 7, and 3 respectively) to represent the respective three-digit number. This makes comprehension of the numbers extremely cumbersome.
- There will be a finite limit to the numbers that one can represent using Roman numerals. To represent a number 432,000 using the Roman system we need to repeat M 432 times! This is an impractical idea.
- When two three-digit numbers have an unequal number of letters and non-standard methods to represent them (as in the case of 397 and 107) it is not clear how to perform even simple mathematical operations such as addition or subtraction. Even if the two numbers have the same number of letters (as in the case of 397 and 928) it is still not possible to use any simple mathematical operations such as addition on subtraction on them to get a result. Therefore, it is not suitable for scientific purposes which often pose complex arithmetical operations.

> - Several inscriptions belonging to 595 CE to 975 CE contain numerals written in the decimal place-value notation.
> - The rest of the world has adopted decimal place-value notation after 11th century CE.

The greatest value of using zero as a placeholder is the fact that with numbers from 0 to 9 any numerical quantity can be represented with just 10 symbols. This idea has been fully embraced by the later period mathematicians and scientists and are the de facto world standard now.

6.2.4 Decimal System

The universal way of dealing with numbers today, is through the use of 10 as the base. Another aspect of the Indian system of numerals is the use of a decimal system. As Bhāskarācārya stated in his opening verses of Līlāvatī (see box on this topic for more details), his ancestors had already developed a place-value system in multiples of ten. This is indeed the decimal system. The resulting decimal number system originated in India much before the 12th–11th century BCE. Dutta and Singh in their book provide a list of 33 inscriptions and grant plates that contain numerals written in the decimal place-value notation[11]. These inscriptions range from 595 CE to 975 CE.

Bhāskarācārya's Līlāvatī on Decimal System and Place value

Bhāskarācārya, the celebrated mathematician of the 12 century CE wrote Siddhānta-śiromaṇi at the age of 36. Līlāvatī is the first of the four parts of this work that mainly deals with arithmetic, algebra and geometry. Since this is a treatise on mathematics it is natural that it begins with some definitions of number systems in vogue and units of measurement.

The 12th verse in the chapter has a description of numbers from 1 to 10^{17} in multiples of ten. Each of these is identified with a unique name. What is also interesting is that Bhāskarācārya observed in the verse that this place value system was already in use by his ancestors. The numbers mentioned in the verse are given in Table 6.2 below:

TABLE 6.2 Decimal System of Numbers Mentioned in Līlāvatī

एक (Eka) - 1 (10^0)	प्रयुत (Prayuta) - 10^6 (Million)	महापद्म (Mahāpadma) - 10^{12} (Trillion)
दश (Daśa) - 10^1	कोटि (Koṭi) - 10^7	शङ्ख (Śaṅkha) - 10^{13}
शत (Śata) – 10^2	अर्बुद (Arbuda) - 10^8	जलधि (Jaladhi) - 10^{14}
सहस्र (Sahasra) - 10^3	अब्ज (Abja) - 10^9 (Billion)	अन्त्य (Antya) - 10^{15} (Zillion)
अयुत (Ayuta) - 10^4	खर्व (Kharva) - 10^{10}	मध्य (Madhya) -10^{16}
लक्ष (Lakṣa) - 10^5	निखर्व (Nikharva) - 10^{11}	परार्ध (Parārdha) - 10^{17}

6.3 UNIQUE APPROACHES TO REPRESENT NUMBERS

Ancient Indian mathematics seamlessly integrated mathematics with literature and poetry. When dealing with mathematical concepts and numbers, poetry was freely used to communicate the same. To do this, the numbers should have equivalent representations in terms of phrases and or alphabets. We find well-developed systems to address this requirement and discuss two of them here.

- Bhūta-Saṃkhyā system is a system of expressing numbers by means of words representing certain entities.
- Kaṭapayādi system employs a technique to convert the numerals to alphabets using certain rules.

6.3.1 Bhūta-saṃkhyā System

Bhūta-saṃkhyā system is a structured approach to express numbers 0 to 9 using words representing certain entities. As the name indicates, there are two parts to the compound word;

bhūta and saṃkhyā. The former denotes some entity and the later a number. The entities are things, beings, or concepts that are commonly known to the people. For example, if we use the word 'eye' we can spontaneously associate this with the number 'two' since we know that a pair of eyes is commonplace knowledge. Similarly, if we use the word 'guṇa', we know that it represents the number 'three' since tri-guṇa is also a widely understood idea.

To broaden the scope of using this system, synonyms are also used. For example, 'chandra (moon)' is used for the number 'one'. The other synonyms of the moon such as 'śaśin', 'vidhu', 'soma' and 'indu' are also used. This allows the user to pick an appropriate word pertaining to the occasion. If the number has to be expressed in poetic verse, then the constraints on the number of letters to be used may require the user to choose one among a set of available synonyms. While it provides alternative words and greater flexibility to use the bhūta-saṃkhyā system, it nevertheless requires that the user is aware of the synonyms of the words so that the number is correctly deciphered. In this sense, there are pros and cons to this system. There are also instances of using words for fractions also; For example, kalā denotes $\frac{1}{16}$, kuṣṭha denotes $\frac{1}{12}$ and śapha denotes $\frac{1}{4}$.

No strict rules govern what qualifies to be used from the 'bhūta' list to represent the 'saṃkhyā'. The following are some of the categories from which we frequently find words being used:

(a) Word name for the number itself – śūnya, eka, dvi, tri ... nava
(b) Physical entities such as earth, moon, stars, mountain, fire, sky, direction
(c) Examples from the animal kingdom such as elephant, horse, snake
(d) Parts of the body – eyes, limbs, seven dhātus, etc.
(e) Names of Gods – Śiva, Viṣṇu, Indra, Manu, Agni, etc.
(f) Other concepts – such as seasons, month, days, five bhūtas

As evident from the above, it is an open-ended list, and it is up to the user to choose an appropriate word. It is also contingent on the part of the reader to be aware of the terms and their synonyms. The advantage of the system is the ability of the user to beautifully merge aesthetics and literature into mathematical ideas that he/she wants to communicate through some verses. Table 6.3 has a representative set of word numerals for the bhūta-saṃkhyā system.

In ancient Indian Mathematics, the numbers are mentioned in the order of units, tens, hundreds, etc. (left to right). Let us deduce the numbers for a few examples using the bhūta-saṃkhyā system given in Table 6.3:

1. वेदवेदाङ्कचन्द्राः (veda-vedāṅka-candrāḥ)
   Let us split the words and associate the numbers to them. Veda (4); Veda (4); Aṅka (9); Candra (1). Therefore, the number is 1944.
2. खाद्रिरामाग्नयः (khādri-rāmāgnayaḥ)
   Kha (0); Adri (7); Rāma (3); Agni (3). Therefore, the number is 3370.
3. भुजङ्गनन्दद्विनगाङ्गबाणषट्कृतेन्दवः(bhujaṅga-nanda-dvi-naga-aṅga-bāṇa-ṣaṭ-kṛta-indavaḥ)
   bhujaṅga (8), nanda (9), dvi (2), naga (synonym of adri) (7), aṅga (6), bāṇa (5), ṣaṭ (6), kṛta (4), indu (1). Therefore, the number is 146,567,298.

TABLE 6.3 Word Numerals for Bhūta-saṃkhyā System

Number	*Represented by (partial list only)**
0	śūnya, anata, pūrṇa, kha
1	ādi, candra, prithivī, eka
2	aśvin, pairs of limbs, ayana, dvandva, dvi
3	rāma, guṇa, loka, kāla, agni, trinetra
4	veda, śruti, yuga, āśrama, varṇa, samudra, kṛta
5	bhūta, śāstra, bāṇa, pāṇḍava, indriya
6	aṅga, ṛtu, darśana, ṣaṇmukha, ṣaṭ
7	ṛṣi, adri, svara, dhātu, chandas
8	vasu, bhujaṅga, siddhi, dik, kuñjara, nāga
9	gṛha, aṅka, nanda
10	dik, aṅgulī, avatāra, rāvaṇaśiras
11	rudra
12	āditya, rāśi
13	viśva, kāma
14	manu
15	tithi, dina
* For all the items listed, any word from the synonyms may be used	

6.3.2 Kaṭapayādi System

Another system to convert the numerals to words is to associate a number to one or more alphabets. Once this association is established, the numbers can be replaced with a corresponding alphabet. Using the alphabet in the place of numbers, one can construct words, which by deciphering one alphabet at a time will reveal the number expressed in the word. The advantage of such a system lies in representing long (or large) numbers using a nice word, which can be easily remembered. This is particularly very valuable in an oral tradition, where large numbers may have to be remembered and communicated without loss of digits. This provided a very efficient method of presenting results of complex calculations using number symbols, and then codifying and committing to memory effortlessly.

Kaṭapayādi system employs certain rules to convert the numerals to alphabets:

(a) The vowels when standing alone indicate the number zero. In all other cases when they are conjoined with consonants, they have no role in this system other than facilitating the pronunciation of the consonants when used in a word.

(b) Each consonant is uniquely associated with a number from 0 to 9. For example, the consonants ka-ṭa-pa-ya was used to denote number 1. Therefore, the system was referred to as the Kaṭapayādi system. Similarly, kha-ṭha-pha-ra denotes number 2, and so on. Table 6.4 has the details of the assignment of consonants to the numbers.

(c) As is evident from the table, more than one consonant may be associated with each of the numerals.

(d) When more than one consonant is used in conjunction, only the terminal consonant preceding a vowel is to be considered for the identification of the corresponding numeral.

(e) If a consonant is not conjoined with a vowel and stands alone, then it will not be considered for identification of the numeral.

(f) Similar to the bhūta-saṃkhyā system, the numbers are to be read as units – tens – hundreds ... etc.

TABLE 6.4 Letter Numerals of the Kaṭapayādi System

1	2	3	4	5	6	7	8	9	0
क (ka)	ख (kha)	ग (ga)	घ (gha)	ङ (ṅa)	च (ca)	छ (cha)	ज (ja)	झ (jha)	ञ (ña)
ट (ṭa)	ठ (ṭha)	ड (ḍa)	ढ (ḍha)	ण (ṇa)	त (ta)	थ (tha)	द (da)	ध (dha)	न (na)
प (pa)	फ (pha)	ब (ba)	भ (bha)	म (ma)					
य (ya)	र (ra)	ल (la)	व (va)	श (śa)	ष (ṣa)	स (sa)	ह (ha)		

The vowels and consonants (whether conjunct or otherwise) may be used to generate familiar and easy to remember words to represent numbers. The procedure to formulate and identify the number is based on the above rules. The origin of this system is traced back to the 5th century CE. Before the advent of the Kaṭapayādi system, the bhūta-saṃkhyā system was popular. Although this system is much simpler and easy to use, it was not in general use since there were four variations of the system. Table 6.4 has the letter numerals pertaining to the kaṭapayādi system.

Mādhavācārya's Approximation to π

Ancient Indian mathematicians have provided multiple approximations to π. Many of these mathematical details, numbers and computations required are invariably provided in the form of some verses using word numeral systems such as bhūta-saṃkhyā system. One such approximation is due to the 11th Century CE mathematician Mādhavācārya. The verse below contains the relevant details. Let us use the bhūta-saṃkhyā system to deciphers the details.

विबुधनेत्रगजाहिहुताशनत्रिगुणवेदभवारणबाहवः ।
नवनिखर्वमिते वृत्तिविस्तरे परिधिमानमिदं जगदुर्बुधाः ॥

vibudha-netra-gaja-ahi-hutāśana-triguṇa-veda-bha-vāraṇa-bāhavaḥ |
navanikharva-mite vṛttivistare paridhimānam-idaṃ jagadurbudhāḥ ||

In the first line of the verse, numbers are provided using the bhūta-saṁkhya system. Let us first extract the numbers. *vibudha – Devas* (33); *netra – Eyes* (2); *gaja – Elephant* (8); *ahi – Snake* (8); *hutāśana – Agni* (3); *tri* – (3), *guṇa* – (3); *veda* – (4) *bha – stars* (27); *vāraṇa – Elephant* (8); *bāhu – Hands* (2); Therefore, the number mentioned is 2,827,433,388,233. In the second line of the verse, there is again a mention of a number; *nava* – (9); *nikharva* – 10^{11}. This number is 9×10^{11}. The balance part of the śloka mentions that this is the ratio of the circumference (*paridhimānam)* to the diameter (*vṛttivistare*) of the circle. Taking this ratio will yield us the value of π.

The value of $\pi = \dfrac{2827433388233}{9\times10^{11}} = 3.14159265359222$

Source: Based on https://www.youtube.com/watch?v=Ke9yxVsUPvo&list=PLbMVogVj5nJThf31TNSQzuN7zqxe7HdRN&index=6.

Let us see some examples of this system:

1. भवति (bhavati)

   Splitting them into separate letters bha-va-ti. After ignoring all the vowels and reading from the table the corresponding numbers, we get 4 – 4 – 6. Therefore, the number is 644.

2. शक्त्यालोके (śaktyāloke – śa-ktyā-lo-ke)

   In this case, the second group has conjoined consonant 'k' followed by 't' and 'ya'. We ignore all the preceding ones and take only 'ya' for consideration. The resultant numbers for the string of these alphabets are '5 – 1 – 3 – 1'. Therefore, the number is 1315.

3. सर्वार्थशीलस्थिरः (sarvārthaśīlasthiraḥ – sa-rvā-rtha-śī-la-sthi-rah)

   As per the rules and the table, we have (7 – 4 – 7 – 5 – 3 – 7 – 2). Therefore, the number is 2,735,747.

4. आयुरारोग्यसौख्यम् (āyurārogyasaukhyam – ā-yu-rā-ro-gya-sau-khya-m)

   In this case, there is a standalone vowel (ā), which will indicate 0. There is a standalone consonant at the end (m), which will have to be ignored. The digits are 0 – 1 – 2 – 2 – 1 – 7 – 1. Therefore, the number is 1,712,210.

Knowledge of these methods of representation numbers widely prevalent during ancient times is very important to make sense of the verses that form part of the ancient Indian mathematical treatises. As an example, look at the following portion of a verse:

पञ्चभ्यः खलु शून्येभ्यः परं द्वे सप्त चाम्बरम् । एकम् त्रीणि च रूपं च ...

pañcabhyaḥ khalu śūnyebhyaḥ paraṃ dve sapta cāmbaram | ekam trīṇi ca rūpaṃ ca ...

A person conversant with the bhūta-saṃkhyā system will be able to readily recognise the number being described in the verse as five 'zeros', followed by 2, 7, and 0 and further by 1, 3, and 1. Therefore, the number specified here is 13,107,200,000.

6.4 MEASUREMENTS FOR TIME, DISTANCE, AND WEIGHT

Existence of a robust measurement system aids trade, commerce and scientific thinking. In the ancient Indian literature three fundamental physical measures for quantifying length, time, and weight are found. There are several sources in the literature specifying units of measurement for these three physical quantities. For instance, in Līlāvatī, Bhāskarācārya begins by defining certain measures for length, volume, and mass prevailing at that time[12]. Arthaśāstra has a detailed account of weights and measures prevailing during the Mauryan time. Chapters 19 and 20 of book two of Arthaśāstra provide the units of measurement for space, time, and weight and specifies methods of checks and balances to be put in place to ensure these are adhered to by the trade[13]. Āyurvedic texts provide several measures for weight as these are used to measure ingredients while preparing a mixture of different substances.

Notion of Paramāṇu

Three generic measures pertaining to length, weight, and time could be used to measure physical entities. Time is the elapsed period for the light to pass through the physical entity, the length is the measure of the size of the entity and weight is the mass of the entity. At the smallest level, there is a fundamental measure called *paramāṇu*. It is not possible to directly relate this term to the modern-day sub-atomic particle. However, it is sufficiently clear that this is a very small particle. The time taken for light to pass through a paramāṇu constitutes the smallest measure of time. Similarly, the weight and length of the paramāṇu will respectively constitute the smallest measures for weight and length. The measures for paramāṇu are as follows:

Paramāṇu (Length): 2.88×10^{-7} mm
Paramāṇu (Weight): 5.79×10^{-5} g
Paramāṇu (Time): 1.31×10^{-5} seconds

Measures for Length

Several measures have been in vogue to measure the length in ancient times in India. Arthaśāstra has elaborate measurements for length suiting different requirements. In the archaeological site at Lothal in Gujarat, an Ivory Scale was discovered. The total length of this scale is 46 mm. The smallest unit on this scale was found to be about 1.70 mm (close to one-tenth of an aṅgula of 1.6764 cm). In 1930–31 Ernest Mackay discovered a broken piece of shell bearing 8 divisions of 6.705 6 mm (0.264 inches) each, with a dot and circle five graduations apart at Mohenjo-Daro[14]. The interval of the 5 divisions, works out to 33.528 mm (1.32 inches). This measure is known as the Indus Inch. Two engravings (12th Century CE) on a wall of the temple at Thirupukkuzhi (12th Century CE) near Kanchipuram, show two scales[15] one measuring 7.24 metres in length, with markings dividing the scale into 4 equal parts, and the second one measuring 5.69 metres in length and markings dividing the scale into 4 equal parts. It was found that each division of the first scale is equal to a Dhanus of 108 Angulas of 16.764 mm each. Similarly, the second scale is precisely equal to π times Dhanus (i.e., equal to the circumference of a circle with one Dhanus as its Diameter).

The smallest measure of length is paramāṇu, as already mentioned. Based on the different conventions of measurement of length, a consolidated table of measures for length adopted during ancient times is presented in Table 6.5. As seen from the table, the range of numbers varies from 0.28 nanometres to 14.48 kilometres.

Measures for Time

Time is a very important aspect of ancient Indians for several reasons. First, the ancient Indians were interested in studying the origin of the Universe, which is one of the five defining characteristics of a Purāṇa. The notion of time becomes very relevant in this context and measures of time become inevitable. Further, studies pertaining to Jyotiṣa (Astrology and astronomy) also requires measures pertaining to time. Table 6.6 provides a comprehensive listing of time available in book three of Bhāgavata-purāṇa[16], which discusses the origin of the Universe. As is evident from the table the smallest measure of time is of the order of 10^{-5} seconds and the largest number 10^{11} human years.

TABLE 6.5 Ancient Indian Measures for Length

Unit	*Multiplier of Preceding Unit**	*No. of Paramāṇus*	*Length (in mm)*
Paramāṇu-rajas	1	1	2.8778×10^{-7}
Reṇu	7	7	2.0145×10^{-6}
Truṭi	7	49	1.4101×10^{-5}
Vātāyana-rajas	7	343	9.8709×10^{-5}
Śaśa-rajas	7	2,401	6.9096×10^{-4}
Edaka-rajas	7	16,807	4.8367×10^{-3}
Go-rajas	7	1,17,649	0.033857202
Likṣā-rajas	7	8,23,543	0.237000411
Sarṣapa	7	57,64,801	1.65900288
Yava	7	4,03,53,607	11.61302016
Aṅguli-parva	7	28,24,75,249	81.29114114
Adopted from **Ifrah, G. (2004).** *The Universal History of Numbers II*, Penguin Books, pp. 138–139.			

Unit	*Multiplier of Preceding Unit**	*No. of Aṅgulas*	*Length (in metre)*
Aṅgula	1	1	0.016764
Dhanurmuṣṭi	8	8	0.134112
Prājāpatya-hasta	3	24	0.402336
Dhanus	4	96	1.609344
Gārhapatya-dhanus	1.125	108	1.810512
Goruta	2000	216,000	3621.024
Yojana	4	864,000	14484.096
* The length of an Angula is 16.764 mm as per Indus inch.			

Measures for Weight

Excavations have established that balances have been in use from the Harappan period. The balances were made of copper or bronze pans and suspended in a metallic beam using cotton thread. There is a continuity to these practices as evident from details in the Arthaśāstra. Arthaśāstra has a vivid description of various measures for weight. Some lower denominations of weight were perhaps used to measure precious metals such as gold and silver and gems. In addition to weights, there are guidelines for the development of 16 types of balances to be used. The administration is also required to ensure verification every three months that the balances are calibrated. Literary

- A comprehensive listing of time is available in book three of Bhāgavata-purāṇa.
- Arthaśāstra has elaborate measurements for length suiting different requirements.

evidence in the Vedic corpus also points to a system of weights and balances in vogue during the Vedic times[17]. Figure 6.1 is an illustration of a balance during Mauryan time.

TABLE 6.6 Ancient Indian Measures for Time

Unit	*Multiplier of Preceding Unit**	*No. of Paramāṇus*	*Time (Seconds)*
Paramāṇu	1	1	1.3133×10^{-5}
Aṇu	2	2	2.6266×10^{-5}
Trasareṇu	3	6	7.8797×10^{-5}
Truṭi	3	18	2.3639×10^{-4}
Vedha	100	1,800	2.3639×10^{-2}
Lava	3	5,400	7.0917×10^{-2}
Nimeṣa	3	16,200	0.212750617
Kṣaṇa	3	48,600	0.638251852
Kāṣṭhā	5	2,43,000	3.191259259
Laghu	15	36,45,000	47.86888889
Nāḍikā	15	5,46,75,000	718.0333333
Muhūrta	2	10,93,50,000	1436.066667
Prahara	7.5	82,01,25,000	10770.51

Unit	*Multiplier of Preceding Unit**	*No. of Human Years*	
Māsa	1	0.08333	
Ṛtu	2	0.16667	
Ayana	3	0.50000	
Human Year	2	1	
Human Life Span	100	100	
Celestial Life Span	360	36,000	
Mahā-yuga	12000	43,20,00,000	
Kalpa	1000	4,32,00,00,00,000	

* For example, 100 Truṭis make a Vedha and 3 Ṛtus make an Ayana

A detailed description of weights is an essential requirement for Āyurveda. Without proper knowledge of weights and measures, it will be difficult to mix the ingredients correctly while formulations are prepared. Therefore, vivid descriptions are found in Āyurvedic texts on various measures for weight. These have been suitably adopted in later times. Table 6.7 shows a list of measures of weights prescribed for āyurvedic practices based on Śārṅgadhara-saṃhitā[18].

FIGURE 6.1 An Illustration of a Balance (Mauryan Time)

Source: **Sharma, V.L. and Bhardwaj, H.C. (1989).** "Weighing Devices in Ancient India", *Indian Journal of History of Science*, **24**(4), pp. 329–336.

TABLE 6.7 Ancient Indian Measures for Weight

Unit	*Multiplier of Preceding Unit**	*No. of Paramāṇus*	*Weight (Grams)*
Paramāṇu	1	1	5.787×10^{-5}
Vaṃśi	30	30	1.736×10^{-3}
Sarṣapa	9	270	1.563×10^{-2}
Yava	8	2,160	0.125
Guñja	4	8,640	0.5
Māṣaka	6	51,840	3
Karṣa	4	2,07,360	12
Pala	4	8,29,440	48
Tulā	100	8,29,44,000	4800
Bhāra	20	1,65,88,80,000	96000

* For example, 100 Palas make a Tulā and 4 Yavas make a Guñja

IKS IN ACTION 6.2

Measurement of Time—An Illustration from Purāṇa

In the Indian tradition the purāṇas are supposed to discuss certain mandatory themes. These form the five characteristics (lakṣaṇas) of a Purāṇa. One of them is to elaborately describe the origin of the Universe. In book three, Chapter 11 of Bhāgavata-purāṇa there is an elaborate discussion of time. These descriptions form part of the discussions on the origin of the Universe.

It is always the practice to first define a unit and create additional units by establishing its linkage with it. For example, in terms of the modern parlance we define a metre. Further we define a centimetre to be 1/100th of a metre, a kilometre to be 1,000 metres and so on. In Chapter 11 of Bhāgavata-purāṇa we see a similar approach to define time. In Table 6.5 we saw various measures of time and their inter-linkages.

We shall see how the first unit of measure for time has been defined. The definition is by way of the following verse:

द्वादशार्धपलोन्मानं चतुर्भिश्चतुरङ्गुलैः ।
स्वर्णमाषैः कृतच्छिद्रं यावत् प्रस्थजलप्लुतम् ॥
dvādaśārdha palonmānam
caturbhiś-caturangulaiḥ |
svarṇamāshaiḥ kṛtac-chidram
yāvat prastha-jala-plutam ||

This verse in a way sets up an experiment described as follows. Take a copper pot weighing six palas (1 pala = 48 g), which can hold water of one prastha (1 prastha = 640 g; in the case of water it is 640 ml). The vessel shall be bored at the bottom with a golden needle weighing four māṣas (1 māṣa = 1 g) and of length four aṅgulas. Leave the pot in water and start a stopwatch. Wait until the vessel is filled fully with water and it just submerges in the water. Stop the watch and record the time. This elapsed time is nāḍikā.

Source: Based on **Goswami, C.L. and Shastri, M.A. (2010).** *Śrimad Bhāgavata Mahāpurāṇa—Part-I*, Gita Press, Gorakhpur, pp. 178–182.

6.5 PIṄGALA AND THE BINARY SYSTEM

Literature can be in the form of prose or poetry. In a prose format, there are no strict rules of the format. Sentences are written with variable word lengths. Each word can also have variable syllables. However, when the literature is written using poetry several rules apply. For example, the length of each line will be specified in terms of the number of syllables. The combination of syllables in terms of vowels and consonants (long and short forms) to be used is also specified. Piṅgala, who lived during 200–300 BCE developed Chandaḥ-śāstra, which dealt with the rules governing prosody. While Chandaḥ-śāstra has much to discuss from a literature perspective, let us focus our attention on some mathematical concepts in Piṅgala's work. These include a formal definition of the word 'śūnya' to denote zero, fundamentals of the binary number system, and some interesting concepts in combinatorial mathematics. Let us discuss the binary system here.

The basic building block of poetry is a syllable. A syllable is a vowel or a vowel with one or more consonants preceding it. There are two types of syllables defined by Piṅgala in the Chandaḥ-śāstra.

(a) Laghu (Short Syllable) – Any syllable with a short vowel
(b) Guru (Long Syllable) can be of four varieties:
   1. Any syllable with a long vowel.
   2. Any short syllable followed by conjunction of consonants.
   3. Any short syllable followed by 'ṃ' known as anusvāra or visarga denoted by ':'.
   4. The last syllable in the quarter of a meter (optionally).

Let us take a verse to understand this classification.

यदा यदा हि धर्मस्य ग्लानिर्भवति भारत ।
अभ्युत्थानमधर्मस्य तदात्मानं सृजाम्यहम् ॥
yadā yadā hi dharmasya glānirbhavati bhārata |
abhyutthānamadharmasya tadātmānaṃ sṛjāmyaham ||

Using the above rules for laghu and guru, Table 6.8 has identified the syllables and the associated classification. Laghu is labelled as 'L' and guru as 'G'.

TABLE 6.8 Identification of the Laghu and Guru as per Rules of Chandaḥ Śāstra

ya	dā	ya	dā	hi	dha	rma	sya	glā	ni	rbha	va	ti	bhā	ra	ta
L	G	L	G	L	G	G	G	G	G	L	L	L	G	L	G

a	bhyu	tthā	na	ma	dha	rma	sya	ta	dā	tmā	naṃ	sṛ	jā	mya	ham
G	G	G	L	L	G	G	G	L	G	G	G	L	G	L	G

Let us denote laghu by the number '1' and guru by the number '0'. This will convert the above table into a binary word of length 16. In the course of the development of an elaborate set of rules for chandas used in the Vedic and the other contemporary poetry, Piṅgala first defined groups of three using the laghu and the guru as the basic building block for the rules for various meters. Since laghu and guru form part of a binary system, it amounts to finding out a unique set of combinations of a binary word of length three. There are eight unique binary words one can obtain and therefore, Piṅgala defined 8 categories of these building blocks, called 'gaṇas'. Table 6.9 has the names of the gaṇas and the unique combination of laghus and gurus that make up a gaṇa.

TABLE 6.9 Eight Ganas Defined by Pingala

Sl. No.	*Gana Name*	*Binary Word**
1	'Ya' Gana	100
2	'Ma' Gana	000
3	'Ta' Gana	001
4	'Ra' Gana	010
5	'Ja' Gana	101
6	'Bha' Gana	011
7	'Na' Gana	111
8	'Sa' Gana	110

* 1 = Laghu; 0 = Guru

There is an easy way to remember mnemonic for the composition of the eight gaṇas given by the following short phrase:

यमाता-राज-भान-सलगम्
yamātā-rāja-bhāna-salagam

Analysing this phrase using the rules of laghu and guru and converting it into a binary word, we get the result as presented in Table 6.10.

TABLE 6.10 Laghus and Gurus in the Phrase

ya	mā	tā	rā	ja	bhā	na	sa	la	gam
L	G	G	G	L	G	L	L	L	G
1	0	0	0	1	0	1	1	1	0

If we examine the binary word representation of the mnemonic, we notice that the last two digits, '10' is repeated. If we remove this repetition, the remaining eight digits of the binary word form a cycle for a 3-digit binary number, as indicated in Figure 6.2. Using this binary cycle, we can generate all the 8 unique combinations of the binary word of length three that constituted the eight gaṇas. For example, from the point indicated as 'begin' in Figure 6.2, we can progressively derive all the 8 gaṇas and their binary composition taking three consecutive binary numbers at a time. '1 – 0 – 0' forms the 'ya' gaṇa, '0 – 0 – 0' forms the 'ma' gaṇa and so on. Very recently, this method of string generation was discovered using combinatorial mathematics known as the De Bruijn sequence. It is a **sequence** of order '*n*' on an alphabet A of size '*k*', in which every possible length '*n*' string on A occurs exactly once as a substring. Piṅgala demonstrated this for $k = 2$ (laghu, guru) and $n = 3$ (gaṇa of size 3).

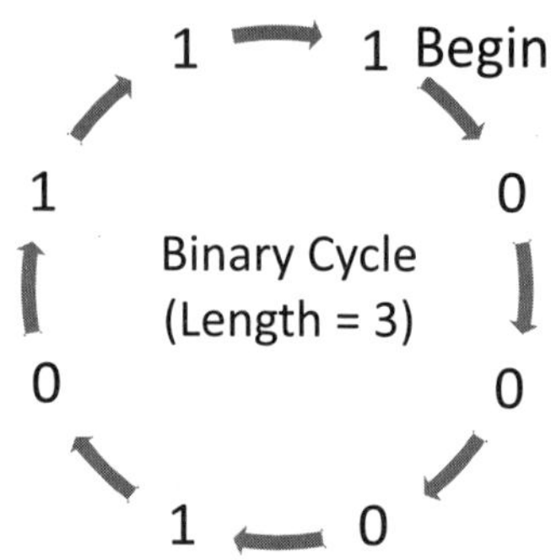

FIGURE 6.2 Binary Cycle of Length 3

SUMMARY

- The contribution of ancient Indians to the development of mathematical concepts laying the foundation for great scientific discoveries both within India and the rest of the world is significant.
- A number of archaeological excavations and linguistic evidence provided by the early writers indicate the maturity and supremacy of the Indian mathematical foundations.
- The Indian numeral system has a long history as is evident from the historical artifacts and archaeological evidence. The origin and the evolution of the numbers could be traced from the time of the Vedic period.
- The concept of zero was established in India during the period 500–300 BCE and was fully developed by 600 CE.
- Ancient Indians developed a systematic approach to the number-names that they developed to describe large numbers.
- The method of using a limited set of numerals but assigning them unique names depending on which place they occupy in the number system is an age-old practice in India.
- The decimal number system originated in India much before its use in the West.
- Three generic measures are used in the ancient Indian system pertaining to length, weight, and time.
- Several measures have been in vogue to measure the length in ancient times in India. Arthaśāstra has elaborate measurements for length suiting different requirements.
- Time is an important aspect to ancient Indians as they were interested in studying the origin of the Universe. Furthermore, studies pertaining to Jyotiṣa (Astrology and Astronomy) also requires measures pertaining to time.
- A detailed description of weights is an essential requirement for Āyurveda to mix the ingredients correctly while formulations are prepared.
- When dealing with mathematical concepts and numbers, poetry is freely used to communicate the same. Unique methods to represent numbers were developed for this purpose.
- Chandaḥ-śāstra introduces some mathematical concepts including, a formal definition of the word 'śūnya' to denote zero, fundamentals of the binary number system, and some interesting concepts in combinatorial mathematics.

REVIEW QUESTIONS

1. List down the key inferences that one can make about the Indian Mathematical tradition based on historical evidence.
2. When was the concept of zero discovered in India? What is its special significance?
3. What are the advantages of the Indian numeral system over the Roman numerals? Explain with the help of a few examples.
4. What are the key characteristics of a place-value system of numerals? How is it advantageous?
5. What do you understand by the term 'decimal system'? Where did it originate and what are its advantages?
6. Fill up the table below with correct answers.

1 Nāḍikā =	Nimeṣa
1 Dhanus =	Dhanumuṣṭi
10 Laghus =	Lava
5 Bhāras =	Karṣa
1 Goruta =	Dhanus
20 Maśakas =	Sarṣapa
1 Muhūrta	kāṣṭhā
1 Aṅguli-parva	Likṣā-rajas

7. Indian mathematics is seamlessly linked with Literature and Poetry. Comment on this statement. Provide some examples to support your arguments.
8. What are the uses of a word or letter in numeral system?
9. What are the differences between the kaṭapayādi system and bhūta-saṃkhyā system?
10. Why is a bhūta-saṃkhyā system called so?
11. Identify the numbers represented by the bhūta-saṃkhyā system in the following cases:
    (a) rāma-chandra-guṇa-nanda-ṛtu-padāḥ
    (b) guṇa-adri-tri-veda-nayana-bhujaṅgāḥ
    (c) pūrṇa-kha-śāstra-gruha-nanda-riṣi-vedāḥ
12. Use Table 6.3 and rewrite the numbers below using the bhūta-saṃkhyā system:
    (a) 179,532,695
    (b) 345,447
    (c) 670,087,123
13. For the answers that you obtained for Question 11, express them using the kaṭapayādi system.
14. Rewrite the numbers given in Question 12 using the katapayādi system.
15. Identify the numbers represented by the kaṭapayādi-system in the following cases:
    (a) आनन्दामृतम् (ānandāmṛtam)
    (b) रक्षाबन्धनोत्सवः (rakṣābandhanotsavaḥ)
    (c) कोरोना राजनाशकः (koronā rājanāśkaḥ)
16. What is a binary cycle? Can you develop a binary cycle of length four and identify all possible binary words of length four?

DISCOVER IKS

1. India's contribution to mathematics has been a subject that has attracted considerable attention in recent times. Several attempts have been made to inquire into this issue. Watch this video produced by BBC, titled, *History of Indian Mathematics, Part-1,* by pursuing the link: https://youtu.be/pElvQdcaGXE. After watching the video carefully, prepare a write-up to answer the following questions:
    (a) What are the main contributions of ancient Indians to the number system?
    (b) What is the importance of the number '0'? How does it contribute to science?
    (c) What are the contributions of Brahmagupta in use of numbers in mathematics?
2. It is often mentioned that India's contribution to the world of mathematics is the number zero. This leaves an impression there is not much other than this contribution. On the contrary, there are significant contributions by ancient Indians in a number of areas of mathematics that has led to the development of mathematical thought and applications in a significant way in the West. Watch the talk by Prof. C.K. Raju on the topic, "Not just the Zero – India's gifts of mathematics and science to the world", by pursuing the link: https://youtu.be/22uZ3D5AgaE. After watching the video, prepare a three-page note that covers answers to the following questions:
    (a) Was India's contribution to Mathematics merely the number zero or something more? Enumerate the other salient contributions and their significance.
    (b) What are the salient aspects of Mādhava's sine table (kaṭapayādi)? What is the importance of sine table?
    (c) What is the value of traditional knowledge in mathematics? Do we need it?

SUGGESTED READINGS

Bag, A.K. (1979). *Mathematics in Ancient and Medieval India*, Chaukhamba Orientalia, New Delhi.

Danino, M. (2010). *The Lost River – On the Trail of Saraswati*, Penguin Books, New Delhi.

Datta, B and Singh, A.N. (1962). *History of Hindu Mathematics: Parts I and II*, Asia Publishing House, Mumbai.

Divakaran, P.P. (2018). *The Mathematics of India: Concepts, Methods, Connections*, Springer (Hindustan Book Agency), New Delhi.

Goswami, C.L. and Shastri, M.A. (2010). *Śrimad Bhāgavata Mahāpurāṇa, Part–I*, Gita Press, Gorakhpur, pp. 178–182.

Heitzman, J. and Rajagopal, S. (2004). "Urban Geography and Land Measurement in the Twelfth Century: The Case of Kanchipuram". *The Indian Economic and Social History Review*, **41**(3), pp. 237–268.

Irfah, G. (2004). *The Universal History of Numbers*, Penguin Books, New Delhi.

Joseph, G.G. (2009). "A passage to Infinity: Medieval Indian Mathematics from Kerala and Its Impact". Sage, New Delhi.

Joseph, G.G. (2016). *Indian Mathematics Engaging with the World from Ancient to Modern Times*, World Scientific, London.

Lal, B.B. (1997). *The Earliest Civilization of South Asia*, Aryan Books International, New Delhi.

Patwardhan, K.S., Naimpally, S.A. and Singh, S.L. (2001). *Līlāvatī of Bhāskarācārya*, Motilal Banarsidass, New Delhi.

Plofker, K. (1963). *Mathematics in India*, Princeton University Press, New Jersey, USA.

Prabhakar Rao, G. (2008). *Bhaiṣajya Kalpanā Vijñānam*, Chaukhamba Publications, New Delhi, pp. 107–118.

Raju, C.K. (2007). "Cultural Foundations of Indian Mathematics", Centre for Studies in Civilization, New Delhi.

Ramasubramanaian, K. (2015). "Gaṇitānada": Selected works of Radha Charan Gupta on History of Mathematics, *Indian Society for History of Mathematics*, New Delhi.

Sarasvati Amma, T.A. (1999). *Geometry in Ancient and Medieval India*, Motilal Banarsidass, 2nd ed., New Delhi.

Shamasastry, R. (1929). *Kauṭilya's Arthaśāstra*, Wesleyan Mission Press, 3rd ed., pp. 113–121.

Sharma, V.L. and Bhardwaj, H.C. (1989). "Weighing Devices in Ancient India", *Indian Journal of History of Science*, **24**(4), pp. 336–329.

Srinivasiengar, C.N. (1988). *The History of Ancient Indian Mathematics*. The World Press, Kolkata.

Swami Chinmayananda (2014). "Taittirīya Upaniṣad", Central Chinmaya Mission Trust, pp. 222–240.

ENDNOTES

1. **Irfah, G. (2004).** *The Universal History of Numbers*, Penguin Books, pp. 11–19.
2. *Ibid.*
3. **Joseph, G.G. (2009).** "A Passage to Infinity: Medieval Indian Mathematics from Kerala and Its Impact", Sage, Chapter 8.
4. **Lal, B.B. (1997).** *The Earliest Civilization of South Asia*, Aryan Books International, New Delhi.
5. ... नाल्पे सुखमस्ति भूमैव सुखं ... nālpe sukhamasti bhūmaiva sukhaṃ 7.23; ... यो वै भूमा तदमृतं अथ यदल्पं तन्मर्त्यम् ... yo vai bhūmā tadamṛtaṃ atha yadalpaṃ tanmartyam 7.24. For more details see, **Swami Gambhirananda (2003).** "Chāndogya Upaniṣad", Advaita Ashrama, pp. 557–558.

6. For details refer, **Swami Chinmayananda (2014).** "Taittirīya Upaniṣad", Central Chinmaya Mission Trust, pp. 222–240.

7. See for details, **Swami Madhavananda (1977).** "The Bṛhadāraṇyaka Upaniṣad", Advaita Ashrama, pp. 476–480.

8. For details see, **Datta, B. and Singh, A.N. (1962).** *History of Hindu Mathematics, Parts I and II*, Asia Publishing House.

9. यथा एकोऽपि सन् देवदत्तः लोके स्वरूपं सम्बन्धिरूपं च अपेक्ष्य अनेकशब्दप्रत्ययभाग्भवति – मनुष्यः, ब्राह्मणः, श्रोत्रियः, वदान्यः, बालः, युवा, स्थविरः, पिता, पुत्रः, पौत्रः, भ्राता, जामाता इति । यथा च एकापि सती रेखा स्थानान्यत्वेन निविशमाना एक-दश-शत-सहस्रादि-शब्दप्रत्ययभेदम् अनुभवति तथा सम्बन्धिनोरेव ... yathā eko'pi san devadattaḥ loke svarūpaṃ sambandhirūpaṃ ca apekṣya anekaśabdapratyayabhāgbhavati – manuṣyaḥ, brāhmaṇaḥ, śrotriyaḥ, vadānyaḥ, bālaḥ, yuvā, sthaviraḥ, pitā, putraḥ, pautraḥ, bhrātā, jāmātā iti| yathā ca ekāpi satī rekhā sthānānyatvena niviśamānā eka-daśa-śata-sahasrādi-śabdapratyayabhedam anubhavati tathā sambandhinoreva ...

10. An early recognition of the meaning of the place-value notation is found in the following simile from *Vyāsabhāṣya* (a commentary) on the *Yogasūtra* (III. 13), dating around the beginning of the Christian Era.

    Yathaika rekha satasthan satam dasasthane dasaika caikasthane tatha caikatve 'pi stri mata cocyate duhita ca svasa cati/ स्वामी विज्ञानाश्रम (1932), "पातञ्जलयोगदर्शनम्", दी फाईन आर्ट प्रिंटिग प्रेस, अजमेर। विभूतिपादः 13, p. 299.

11. For a detailed account of this see, **Datta, B. and Singh, A.N. (1962).** *History of Hindu Mathematics: Parts I and II*, Asia Publishing House.

12. See for instance, **Patwardhan, K.S., Naimpally, S.A. and Singh, S.L. (2001).** *Līlāvatī of Bhāskarācārya*, Motilal Banarsidass, New Delhi.

13. See for more details, **Shamasastry, R. (1929).** *Kauṭilya's Arthaśāstra*, Wesleyan Mission Press, 3rd ed., pp. 113–121.

14. **Danino, M. (2010).** *The Lost River – On the Trail of Saraswati*, Penguin Books.

15. **James Heitzman and Rajagopal, S. (2004).** "Urban Geography and Land Measurement in the Twelfth Century: The Case of Kanchipuram", *The Indian Economic and Social History Review*, Sage.

16. For more details see, **Goswami, C.L. and Shastri, M.A. (2010).** *Śrīmad Bhāgavata Mahāpurāṇa – Part–I*, Gita Press, Gorakhpur, pp. 178–182.

17. For details see, **Sharma, V.L. and Bhardwaj, H.C. (1989).** "Weighing Devices in Ancient India", *Indian Journal of History of Science*, **24**(4), pp. 329–336.

18. For more details, see, "**Prabhakar Rao, G. (2008).** *Bhaiṣajya Kalpanā Vijñānam*, Chaukhamba Publications, New Delhi, pp. 107–118.

CHAPTER

7 Knowledge: Framework and Classification

LEARNING OUTCOMES

After finishing this chapter, you will be able to:

- Understand the framework for establishing the right knowledge
- Appreciate the Vaiśeṣikan approach to describe the physical reality through various components and classifications
- Get introduced to the Nyāya framework
- Familiarise with Nyāya approach for resolving ambiguity through logic and argumentation

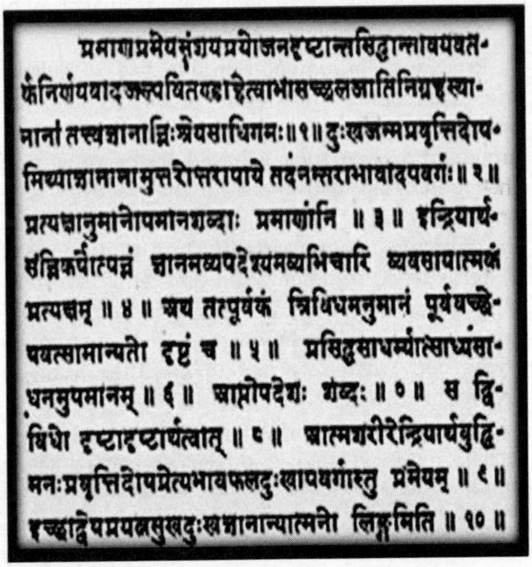
प्रमाणप्रमेयसंशयप्रयोजनदृष्टान्तसिद्धान्तावयवत-
र्कनिर्णयवादजल्पवितण्डाहेत्वाभासच्छलजातिनिग्रहस्था-
नानां तत्त्वज्ञानान्निःश्रेयसाधिगमः ॥ १ ॥ दुःखजन्मप्रवृत्तिदोष-
मिथ्याज्ञानानामुत्तरोत्तरापाये तदनन्तरापायादपवर्गः ॥ २ ॥
प्रत्यक्षानुमानोपमानशब्दाः प्रमाणानि ॥ ३ ॥ इन्द्रियार्थ-
सन्निकर्षोत्पन्नं ज्ञानमव्यपदेश्यमव्यभिचारि व्यवसायात्मकं
प्रत्यक्षम् ॥ ४ ॥ अथ तत्पूर्वकं त्रिविधमनुमानं पूर्ववच्छे-
षवत्सामान्यतो दृष्टं च ॥ ५ ॥ प्रसिद्धसाधर्म्यात्साध्यसा-
धनमुपमानम् ॥ ६ ॥ आप्तोपदेशः शब्दः ॥ ७ ॥ स द्वि-
विधो दृष्टादृष्टार्थत्वात् ॥ ८ ॥ आत्मशरीरेन्द्रियार्थबुद्धि-
मनःप्रवृत्तिदोषप्रेत्यभावफलदुःखापवर्गास्तु प्रमेयम् ॥ ९ ॥
इच्छाद्वेषप्रयत्नसुखदुःखज्ञानान्यात्मनो लिङ्गमिति ॥ १० ॥

The major contribution of the Nyāya-śāstra is its profound approach to establishing valid knowledge. Using 528 sūtras, Nyāya discusses various aspects related to rules of reason, logic, epistemology and metaphysics. As per some Tibetan records Buddhist scholars spent years with Nyāya scholars to master the art of reasoning and logic. The picture shows the first ten sūtras from Nyāya-śāstra.

Source: https://upload.wikimedia.org/wikipedia/commons/thumb/6/65/Ancient_Nyayasutras_first_ten_sutras_in_Sanskrit.jpg/260px-Ancient_Nyayasutras_first_ten_sutras_in_Sanskrit.jpg

IKS IN ACTION 7.1

Vāda: The Indian Art of Debate

The debate between great intellectuals is a part of the Indian tradition and it has played a very significant role in advancing knowledge. One of the often-cited famous debates was between Ādi Śaṅkarācārya, the founder of Advaita philosophy and Maṇḍana Miśra, who eventually became his disciple. Maṇḍana Miśra's wife Ubhaya Bhāratī was the moderator of the discussion. The method of debate adopted by the Indians is unique and was highly structured and objective. There was no room for dogma or blind belief in arguments.

The ancient Indian scholars adapted a structured set of argumentative procedures for learning, where using a debate in which two opposing sides presented their views and finally concluded in a mutually acceptable manner by directing the flow of arguments with less deviation from the topic of discussion. The body of knowledge that enabled this is known as Nyāya, which means that by which one is led to a conclusion or a clarified knowledge. This is achieved through discussion or vāda, which is primarily meant for the discernment of the real nature of the thing under investigation and imparting the truth, as one understands it to the other party.

According to Nyāya, a scholastic debate involves two interactive participants–the proponent (one who holds the initial position) and a respondent. The goal of the discussion is the adoption of one of two opposing sides and there is no consideration of victory or defeat. What is adopted is defended using means of right knowledge and its opposite is assailed by confutation, without deviation from the established tenets.

The formal procedure adopted in a vāda can be briefly summarised as follows:

1. The initial proponent puts forward his thesis.
2. If the thesis is thought to be erroneous the respondent may refute it immediately (using step 6). On the other hand, if the thesis is accepted, then the respondent asks the proponent to outline the reason for accepting the thesis.
3. The proponent then offers proof outlining the reasons why the thesis should be accepted.
4. The respondent asks if the proof offered is based on a sound inference and a debate ensues.
5. The proponent negates the faulty relations and erroneous reasoning that may have occurred in the outlining of the proof of his thesis during the debate

This concludes the first part of the debate. If the respondent accepts both the thesis and the proof then the debate concludes. However, if the proof is deemed erroneous for some reason, then the second part of the debate is required.

6. The respondent offers a statement of refutation of the proponent's thesis, which thereby constitutes the initial starting point of his exposition. The refutation that follows aims to demonstrate the errors and inconsistencies of the proponent's position based upon the reasoning and evidence provided by the proponent.
7. The proponent responds with a rejoinder if he thinks that his critic's refutation is in some way erroneous. However, if the proponent accepts the soundness of the refutation, the respondent is asked to state formal proof of the refutation in a positive form.
8. Finally, the respondent offers formal proof of the refutation in inferential form.

Procedural argumentation adopted by Indians tries to resolve knowledge inconsistencies by analysing arguments and providing suitable counter-arguments. Indian logic is based on the conviction that logic is an instrument for the discovery and understanding of reality, which is different from the Western counterpart. Western philosophical systems treat argumentation differently. Western logic does not inquire whether the premises and conclusion of an argument are true. It is solely concerned with determining whether a given argument is valid.

Source: Adopted from **Mahalakshmi, G. and Geetha, T.V. (2009).** "An Indian logic-based argument representation formalism for knowledge sharing", **17**(1), *Logic: Journal of IGPL*, pp. 55–74.

We have made tremendous progress in the last 150 years with technological advances, scientific discoveries, and new inventions. There is a widespread feeling that we have developed capabilities to find new knowledge in hitherto unknown domains, however, this is not a new-found capability for the human race. Human beings armed with the thinking faculty have constantly strived to make innovations. Making such innovations follow a structured process, which culminates in resolving ambiguities in our understanding and finding new knowledge. The entire process requires that we have a robust framework to establish new knowledge. This has been the constant preoccupation of human beings. Ancient Indians too have developed a methodology for establishing new knowledge.

There is a feeling that the Indian thought process is driven by blind beliefs and dogmas rooted in a religious context. It is generally believed that much of the ideas developed and propagated down the ages have been on faith and were left unquestioned. If we explore the culture of developing new knowledge systems in India, we discover something on the contrary. A case in point is the development of Darśanas (Philosophical schools of thought) in India (in Chapter 3 we have introduced Darśanas and discussed some of the unique aspects). There are multiple schools of thought and there has been a healthy tradition of arguing and counter-arguing against the tenets of each school of thought in a highly professional, rational, and objective manner. In fact, one of the Darśanas, known as Nyāya, has specifically focused on how one can systematically inquire into a problem and establish new knowledge in a structured process. Using the tools and techniques developed by Nyāya researchers over 1000 years, a sound framework for logic and epistemology has been developed which has come in handy to critically analyse established tenets and bring in new knowledge. We shall see some aspects of these in this chapter.

7.1 THE KNOWLEDGE TRIANGLE

Knowledge (Jñāna) is defined as apprehension or consciousness. All knowledge is a revelation or manifestation of objects, just as a lamp manifests physical things placed before it. Knowledge may be valid or invalid. Valid knowledge (Pramā) is defined as the right apprehension of an object or knowing an object as it is. A valid knowledge corresponds to the reality and not anything other than that and is indeed produced by some valid means. Invalid knowledge includes memory, doubt, error, and hypothetical reasoning. Memory is not valid because the object remembered is not directly presented to the ātman, but only indirectly recalled. Doubt is uncertainty in cognition. Error is misapprehension as it does not correspond to the real object. Hypothetical reasoning is no real knowledge. When we see a rope as a rope, we have obtained the right knowledge. If we are uncertain whether it is a rope or a snake, we have doubt. We may need some means by which the doubt is unambiguously resolved. If we merely recall the rope that we have seen, it is memory. If we mistake the rope for a snake, we are erroneous.

Valid knowledge is that which has been established rigorously and systematically so that it is non-refutable, true, and reliable in the context and the conditions in which the knowledge is supposed to be applicable. Typically, in the Indian tradition, the aspects of valid knowledge could be best understood from the notion of a knowledge triangle. Figure 7.1 pictorially represents the knowledge triangle, which consists of three components and provides the overall context for obtaining the right knowledge.

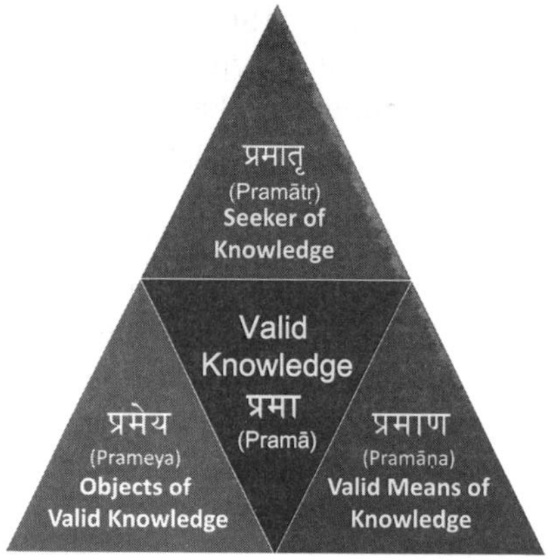

FIGURE 7.1 The Knowledge Triangle

- Firstly, the knowledge is obtained ultimately only by a seeker of the knowledge. Viewed in this manner all forms of knowledge manifest eventually as 'implicit' knowledge in the mind of the knowledge seeker. Whether it is a Nobel Laureate in Physics or a Scientist working in a laboratory developing a new formulation for medicine, or a spiritual seeker wanting to know who he/she is, the knowledge accrues in him/her only. Therefore, the first component of the knowledge triangle is the seeker of knowledge. The seeker of the knowledge is referred to as **Pramātṛ**. The knowledge seeker gets mentally involved in the process and commits physical and other resources, and time to the process. Once the knowledge seeker obtain the implicit knowledge he/she may choose to make it available to the others by way of tacit knowledge.
- The efforts of the knowledge seeker are directed towards some object or entity, as it becomes the context for seeking knowledge. This is the second aspect of knowledge, known as **Prameya**. For instance, a scientist developing a new vaccine will direct his/her entire focus in the knowledge creation process on the virus and all the aspects related to it. A physicist wanting to understand the world as a physical entity will direct the attention towards various aspects of the physical reality. The basic elements that constitute the physical reality, the processes that they employ to combine and regenerate into multiple forms and shapes may become the objects of knowledge that are being sought by the researcher. The term object does not have a mere 'physical' orientation. It includes all that towards which effort is directed to obtain the knowledge, which may include physical, meta-physical, virtual, and other entities.
- New knowledge creation is invariably a process of starting from known ideas and forms and progressing into the domain of the unknown. In this journey, since the destination is somewhat unknown, we need valid means to assure ourselves that the journey indeed is fruitful, and the process is robust and flawless. Therefore, the third component is to have valid means of obtaining the knowledge. In the Indian tradition, this is known as **Pramāṇa**.

Two schools of Indian thought, Nyāya, and Vaiśeṣika address the issue of knowledge of the reality, i.e. of the physical world of several entities and their interactions. Vaiśeṣika mainly confines itself to 'the exposition of reality' and Nyāya focuses on the issue of 'right knowledge of reality'. To put it in another way, Vaiśeṣika deals with metaphysics and ontology, and Nyāya deals with logic and epistemology. The key contribution of Nyāya-śāstra lies in providing a robust framework for establishing the right knowledge. The very first sūtra in the text begins with identifying sixteen factors that one needs to take into consideration while establishing the right knowledge. A correct understanding of these and appropriate use of them is critical in the process[1]. These sixteen factors provide a comprehensive set of concepts that help one to establish knowledge using a structured approach.

The principles laid out in Nyāya are fundamental and have had the widest appeal as a general framework for creating new knowledge and advancing our thoughts through logic and argumentation. The Nyāya principles have become a very useful tool for experts in all other fields (literature, grammar, philosophical studies, and other fields of knowledge). There have been numerous attempts both in India and outside to write commentaries on Nyāya.

Relating to our knowledge triangle, Vaiśeṣika deals with prameya of the knowledge triangle, whereas Nyāya mainly deals with pramāṇas, valid means of knowledge. However, in Nyāya-śāstra at a generalised level of abstraction 12 objects of knowledge (prameya) have been identified[2]. In this chapter, we shall use the Vaiśeṣika framework to understand the physical reality that presents itself as the object for knowledge creation. On the other hand, we shall use the Nyāya framework to understand various aspects of establishing valid knowledge.

7.2 PRAMEYA – A VAIŚEṢIKAN APPROACH TO PHYSICAL REALITY

Physics is one of the foundational aspects of Science and Technology, like Mathematics. It is a study of the 'reality' as we know and is concerned with space, time, matter, and energy. It provides a set of ideas, concepts, and theoretical backgrounds weaved into relevant frameworks so that we can make sense of things around us and understand the myriad combination of things that exist in the Universe and their inter-relationships. In the Indian tradition, these issues have been addressed in varying levels of abstraction in different works including some of the darśanas. In the course of this inquiry, it became necessary to understand the Universe in terms of its origin, its multi-dimensional nature, the constituents and their properties, and their inter-relationships. These topics form part of the modern-day branch of scientific study known as 'physics'. While modern science has provided much greater depth and detail into specific aspects of the physical reality, Vaiśeṣika has provided an overall framework and a higher level of abstraction of the physical reality. Vaiśeṣika is an effort to describe all the 'nameable and knowable' entities. The term Vaiśeṣika is derived from the term 'Viśeṣa', meaning an individual or special characteristic. The categorisation of the Universe is attempted using individual or atomic elementals and their association with one another, hence the term 'Vaiśeṣika'. This was propounded by Kaṇāda, a sage who is variously estimated to have lived between the 6th–2nd century BCE.

Kaṇāda composed about 370 sūtras (short aphorisms) using which he described the physical reality using a classification framework. Vaiśeṣika-darśana categorises substances that form the physical world, their attributes, and interactions. The system formally included space, time and matter, and other factors to establish the basic premises that govern the physical entities. The text establishes certain fundamental properties of physical entities linking them to basic atoms and paves the way for using these ideas in applied sciences such as Āyurveda and Śilpa-śāstra.

In the Indian tradition, the origin and dissolution of the Universe and its relation to God are assumed to be a two-stage process: primary or original and secondary or derivative. The primary stage of the evolution process is assumed to result in creating ether, time, space, and the ultimate atoms or elementals (Earth, Water, Fire, and Air). In the second stage, the focus is on how the multitude of physical entities are created once the Universe manifests through the primary stage of evolution. Kaṇāda's approach in Vaiśeṣika is directed towards the second stage of the creation. While Kaṇāda believed that an infinite intelligence created the Universe, he adopted a three-pronged approach in Vaiśeṣika to study the physical properties of the created Universe:

- Vaiśeṣikā is an effort to present a systematic framework to describe all the 'nameable and knowable' entities.
- Vaiśeṣikā establishes fundamental properties of physical entities linking them to basic atoms and paves the way for using these ideas in applied sciences such as Āyurveda and Śilpa śāstra.

- Uddeśa - A classification framework to enumerate 'knowable and nameable,' things of the physical reality
- Lakṣaṇa - Definitions to establish certain properties
- Parīkṣā - Examine using perception and inference to establish the knowledge

Vaiśeṣika proposed a framework to categorise and explain the entire gamut of the physical reality (predicable). Figure 7.2 illustrates the framework in simple terms. Predicable could be either existence or non-existence. Six sub-categories constitute existence and knowledge of them is considered the essence of the supreme good[3]. The first three, Dravya - substance, Guṇa - attributes, and Karma - action are objective aspects. The other three categories include Sāmānya - Universality, Viśeṣa - Specialty, and Samavāya - inherence or combination. These three are outcomes of intellectual discrimination.

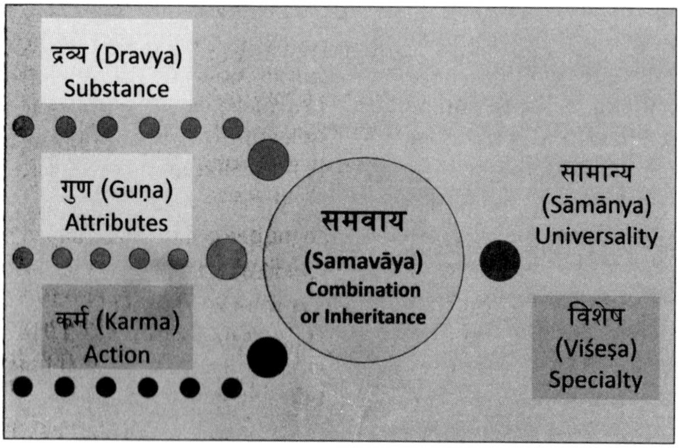

FIGURE 7.2 The Vaiśeṣika Framework

Physical reality invariably could be classified under 'Universality' (or Genus) and 'Specialty'(or Species). Universality is the principle of assimilation that presents the substratum of the observed substances, i.e. it presents the generic properties. On the other hand, 'Specialty' is the principle of differentiation within observed classes and patterns of existence. Vaiśeṣika presents certain principles pertaining to Universality and Specialty that characterise physical reality[4]. Substances (Dravya), attributes (Guṇa), and action (Karma) have both generalisation and specialisation. This enables us to make sense of the physical reality in terms of class – sub-class hierarchical structure. A combination is not a simple idea such as conjunction or addition of different aspects into one. It is a property, by which one develops the intuition, 'this is here' with respect to effect and cause. It is the relation of things and their attributes, inseparably associated by nature. Therefore, it indicates the property of inheritance.

Using this classification framework, Kaṇāda explained the physical reality from the concrete to abstract that forms part of the Universe. In the six-part classification, the first three (substance, attributes, and action) are objective and concrete. The other three belong to the domain of intellectual determination. This implies the role of the mind and the thinking faculty of the observer through a certain process of assimilation. Therefore the 'observer' is also part of the 'observed' in the final analysis and sense-making. The Vaiśeṣika-sūtras are organised under ten chapters, with two sections in each chapter. The first chapter presents the proposed classification framework and enumerates the components that form the classification. The second chapter describes the nine substances. The subsequent chapters deal with other topics related to the framework.

7.2.1 Dravyas – The Constituents of the Physical Reality

Dravya constitutes the basic building block of the physical reality. It is difficult to translate 'dravya' exactly into English. The possible words are realities, entities, or substances. For the sake of equivalence and familiarity let us use the word 'substance'. Substance, as we know is nothing but matter with certain characteristics. The notion of 'substance' in Vaiśeṣika is much broader than the modern-day definition. In Vaiśeṣika, non-corporal or imponderables such as time, space, ātman, and mind are also included. We need to be aware of these aspects when we use the word 'substance' in place of dravya. All the translated terms in the chapter have a similar limitation. These translated terms should not be confused with the modern-day perspectives of Science. For example, in Vaiśeṣika, Earth, Water, and Air are fundamental and homogenous kind of matter characterised by its own specific quality or attribute. On the other hand, for a modern scientist, these are compounds (or mixtures) each of which could be reduced to its parts.

There are certain fundamental characteristics of a substance as per Vaiśeṣika. Substances possess karma (action) and guṇa (attributes) in them[5] and are fundamentally a combinative cause. In other words, the basic atomic nature of the category of the substance allows the creating of many different things using the principle of Samavāya[6]. In this progressive process of samavāya, the entire physical reality could be visualised. One can see this at several levels of abstraction. For example, the threads of different colours can combine to form a beautiful fabric. Viewed in this fashion, the substances are the combinative causes as well as the effects of the causes. Further, they provide the Universality and the Specialty.

There are nine types of substances identified in Vaiśeṣika. These include Pṛthivī – Earth, Ap – Water, Tejas – Fire, Vāyu – Air, Ākāśa – Ether, Dik – Space, Kāla – Time, Ātmā – Ātman (or Self), and Manas – Mind[7]. Figure 7.3 has details of these and their sub-classifications. There are five physical substances (Earth, Water, Fire, Air,

- In the six-part classification proposed by Kaṇāda, the first three (substance, attributes, and action) are objective and concrete.
- The basic building blocks of the physical reality are the Dravya in the framework of Vaiśeṣika.

and Ether). Each of them has some specific quality perceptible to an external sense. Of the five physical substances, Ākāśa differs from the other four. Ākāśa is a non-corporeal substance of unlimited magnitude. The other four physical substances are capable of producing composite substances out of themselves, which is not the case in Ākāśa.

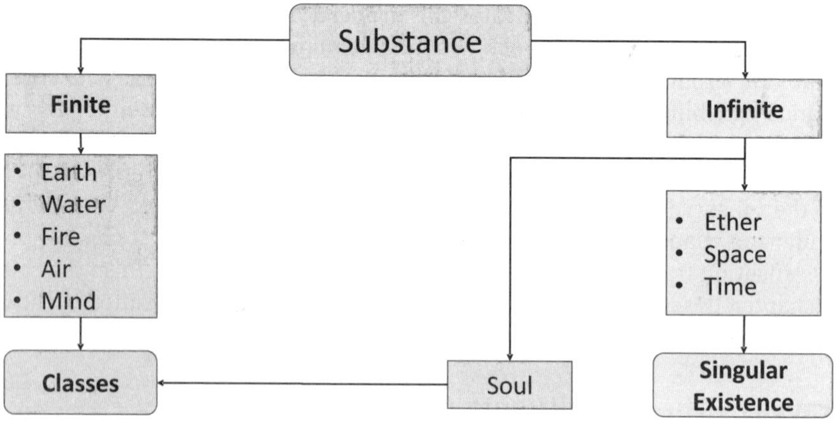

FIGURE 7.3 Nine Categories of Substances

Substances can be finite or infinite. Ether, Space, Time, and Ātman are considered infinite. On the other hand, Earth, Water, Fire, Air, and Mind are considered finite. Another useful dimension is whether the substances or unitary are of different classes. As indicated in Figure 7.3, Ether, Space, and Time are considered unitary and eternal. Since Vaiśeṣika has taken an atomic approach to understand and describe the physical reality, it is useful to know this dimension of the substances also. The eternal and unitary substances viz., ether, space, and time are the non-atomic substratum. On the other hand, the elementals earth, water, fire, and air are composed of atomic substances. According to Vaiśeṣika, two basic atoms (aṇus) can combine to become dvyaṇuka (dyads), and three such dyads are the constituent of a tryaṇuka (triad), which is the smallest form of a gross body visible to the naked eye. The notion of 'atom' (Aṇu) as referred to in Vaiśeṣika should not be confused with the current-day understanding of 'atom'. An Aṇu in Vaiśeṣika is merely the 'minima' and changeless entity. They are considered absolutely without any magnitude nor do they have a spatial dimension. On the other hand, in modern conception atoms have definite magnitude.

It is important to know that the notion of Ether (Ākāśa) is different from space as presented in Vaiśeṣika. Ākāśa is the typical ether in which all the substances exist. Moreover, it is the medium for the sound to travel. On the other hand, space is merely a dimension providing direction. Its role is to provide a spatial and locational reference for the entities that exist in physical reality. Mere space does not guarantee the propagation of sound particles.

The status of 'self' and 'mind' also needs to be clearly understood. Self is pure awareness and provides the capability to observe. Therefore, it serves the role of the 'observer' of physical reality and events in the Universe. However, to make use of this capability, a mechanism to observe and process the information obtained out of the observation is required. This is the role of the 'mind'. When one is in deep sleep, the 'self' exists but the mind is switched off. Therefore, no observations of the physical reality could be made during this time. Classical physics did not accept the 'self' or 'mind' as valid entities, primarily because there was no need for these entities. However, in quantum physics, the role of the 'observer' was as important as the role of 'observed'.

The four elementals Earth, Water, Fire, and Air are not to be understood merely in the sense of the term. For example, earth indicates the world of 'solid' substances, and water the world of 'liquids' and so on. Each one of them has some unique attributes that differentiate them from the other. Air possesses touch, and its touch is neither hot nor cold. The air exists as ultimate atoms and as their products. On the other hand, fire possesses a color and touch and it also exists as simple atoms and compounds. The color is white and luminous, and the touch is in the form of heat. Progressing further, water possesses taste in addition to touch and color. Viscidity and fluidity belong to water. Earth possesses smell in addition to the three properties of taste, color, and touch. Ether has the attribute of sound and not the others listed above. It is considered a substance, eternal, and universally present. Table 7.1 lists these differences. As seen in the table, Earth possesses 'smell', which others do not. Similarly, Ether possesses 'sound' which others do not. As we proceed downwards from Earth to Air, the number of attributes progressively come down.

In the Vaiśeṣikan system, 'time' is also considered a substance, unitary, and eternal. It provides a static background against which events happen. Our ability to sense prior, posterior, simultaneous, slowness, and quickness are the marks of the existence of time. Space is recognized from two simultaneously existing bodies fixed in direction and place. It is eternal and all-pervading. Although it is a single entity, the diversity is experienced on account of different effects. For example, using the reference of the Sun, space is perceived as North, East, South, and West, although it is one single entity. In the same manner, by constructing some partitions we recognise the unitary space as a bedroom, kitchen, classroom, etc. Ether, space, and time are not objects of external perception and are inferred in a relative sense from the effects.

TABLE 7.1 Some Attributes of Five Dravyas

Category	*Smell*	*Taste*	*Colour*	*Touch*	*Sound*
Earth	✓	✓	✓	✓	
Water		✓	✓	✓	
Fire			✓	✓	
Air				✓	
Ether					✓

According to Kaṇāda, the ātman (Self) is not an empty idea but a real existence. It is neither a by-product of physiological processes nor identical with different stages of consciousness. As proposed in the spiritual traditions, only such persons who have a certain level of spiritual development can experience its existence in conjunction with the mind. The ātman is considered as the substratum of attributes and is eternal. The mind is the internal organ of sense and like the ātman is not an object of ordinary perception. The mind is the organ of internal perception, which helps in producing the perception in the ātman through the contact of the senses and objects. In Vaiśeṣika, the methods of perception, and the conditions for this to happen have been laid out in the sūtras. Substances that are beyond the reach of the senses under ordinary circumstances are the ultimate atoms, air, space, time, ether, mind, and ātman.

7.2.2 Attributes – The Properties of Substances

Attributes are the inherent properties associated with the substances described above. Since they depend on the substances, they do not have an independent existence, nor can they be the independent cause for conjunction or disjunction process[8]. Seventeen attributes have been enumerated in Vaiśeṣika[9] and the subsequent works expanded it to twenty-four. These are either physical or psychological. Broadly these could be classified under three major heads (Figure 7.4):

- One set of attributes relate to finite substances. These include color, taste, smell, touch, prior and posterior aspects, fluidity, viscidity, and velocity. We have already seen how some of the attributes are related to Earth, Water, Fire, and Air.
- The second set of attributes are related to infinite substances (Ether and Sound for example) or with imponderable substances (such as Mind and Ātman). These include cognition, pleasure, pain, desire, aversion, volition, merit, demerit, impression, and sound. As evident from the list, these are soft attributes.
- The third set of attributes could be associated with both finite and infinite or imponderable substances. These include number, quantity, separateness, conjunction, and disjunction.

Not all attributes could be associated with every member of the group. For example, fluidity and viscidity are related to water. Similarly, number, quantity, separateness, etc. are not relevant in the case of substances of singular existence. Kaṇāda provides a detailed account of the production and destruction of these attributes in Chapters 5–10 in Vaiśeṣika through several sūtras.

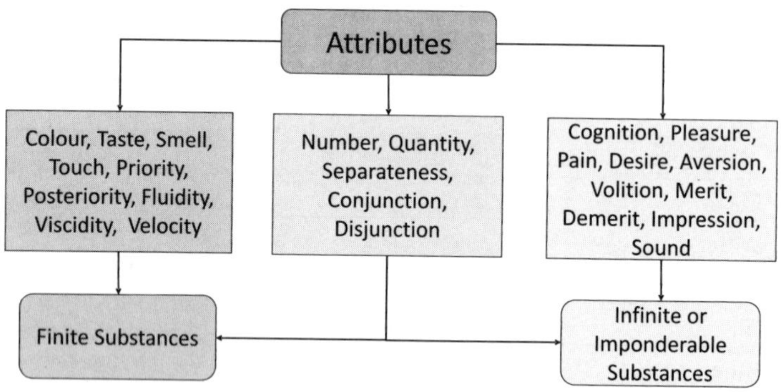

FIGURE 7.4 Attributes and Their Relationship to Substances

7.2.3 Action – The Driver of Conjunction and Disjunction

Action by definition is associated with one substance only. Unlike substances, it does not possess any attribute. However, it drives the process of conjunction and disjunction that occur with respect to the substances which are either single or in a combined form. On account of these properties, the action is not associated with ether, time, space, and ātman as these are infinite substances. Actions are classified into five types: throwing upwards, throwing downwards, contraction, expansion, and being in motion[10]. Although 'motion' is a generic term for action, its inclusion as one type is meant to include several other types of motions that we commonly experience. This includes rotation, harmonic motion, evacuation, percolation, etc.

IKS IN ACTION 7.2

Vaiśeṣikan Worldview of 'Existence'

Modern physics proposes a view where the atomic and the sub-atomic particles are the basic raw material and building block for the multitude of forms, names, and shapes that make up our objective reality. We find ideas in Vaiśeṣika, which resembles this approach. Kaṇāda visualised the world that we perceive to be of a combination of the basic substances and qualities that he identified. He established this basic feature of creation at the very beginning of the text using two sūtras:

द्रव्यगुणयोः सजातीयारम्भकत्वं साधर्म्यम् ॥
dravyaguṇayoḥ sajātīyārambhakatvaṃ sādharmyam || 1.1.6

It is common for the basic substances and the attributes to initiate or create entities and qualities of their own class.

द्रव्याणि द्रव्यान्तरमारभन्ते गुणाश्च गुणान्तरम् ॥
dravyāṇi dravyāntaram-ārabhante guṇāśca guṇāntaram || 1.1.7

Basic substances create other material substances and classes other than their own and so do the attributes.

While these two sūtras set the basic feature to explain the physical reality, Kaṇāda provided more details on how this happens at the atomic level. According to him, the ultimate atoms combine variously to form new substances that inherit unique attributes. In Chapter 4 of Vaiśeṣika, Kaṇāda discussed the issue of perception, in which he stated some rules that help us understand the objective reality that we perceive. The relevant sūtras are listed below:

सत् अकारणवत् नित्यम् ॥
sat akāraṇavat nityam || 4.1.1

Existence is uncaused and eternal (nitya).

By this, it is established that there are ultimate atoms eternally present which play the role in creating what we call 'existence'.

कारणाभावात् कार्याभावः ॥
kāraṇābhāvāt kāryābhāvaḥ || 4.1.3

In the absence of a cause, there is an absence of effect.

Essentially by this, it is inferred that different properties of existence, such as colour, taste, etc. are the fallout (effect) of these in the 'original'cause that brings these properties. These original causes are eternal and are the ultimate atoms as explained in the previous sūtra.

महत्यनेकद्रव्यवत्त्वात् रूपाच्चोपलब्धिः ॥
mahaty-anekadravyavattvāt rūpāccopalabdhiḥ || 4.1.6

The actual manifestation of a substantive object that we are able to perceive in terms of the properties of colour etc. is on account of the composition of several 'substances' (Dravya in the Vaiśeṣika framework).

Essentially this sūtra substantiates the previous one by explaining how the ultimate atoms combine variously to form the objects of our perception. There is one more sūtra that explains how the objects imbibe their attributes. It is stated below:

अनेकद्रव्य-समवायात् रूपविशेषाच्च रूपोपलब्धिः ॥
anekadravya-samavāyāt rūpaviśeṣācca rūpopalabdhiḥ || 4.1.8

Several 'substances' (Dravyas) in a combinative process ultimately provide the unique attributes to the object attained out of the combination.

प्रत्यक्षाप्रत्यक्षाणां संयोगस्य अप्रत्यक्षत्वात् पञ्चात्मकं न विद्यते ॥ 4.2.2
pratyakṣāpratyakṣāṇāṃ saṃyogasya apratyakṣatvāt pañcātmakaṃ na vidyate

Due to our inability to observe the conjunction processes of the five basic substances (elementals), we are unable to perceive the basic substances in the observable and unobservable physical substances.

अणुसंयोगः तु अप्रतिषिद्धः ॥
aṇusaṃyogaḥ tu apratiṣiddhaḥ || 4.2.4

Despite our inability to observe, the conjunction of basic atoms cannot be ruled out.

These sūtras establish the fact that a myriad combination of basic atoms eventually creates the physical world of differing names, shapes, and attributes.

Vaiśeṣika spells out the properties that are unique to actions and also the causes of the actions. Actions occur only in finite (corporeal) substances. This excludes other substances such as ether, time, space, self, and mind. Further, actions cannot initiate another action, nor can create a substance. The effects of the actions are felt in its own substratum or other places. For example, a falling ball will have an effect both on the ball and on the surface where it is falling or hitting. Several causes have been identified for action. Gravity is the foremost cause of several actions to happen. Other physical properties such as fluidity, impact, and impulse also cause action. Conjunction (concurrent events at a time or space) is also an important cause of action. However, Vaiśeṣika also includes another cause Adṛṣṭam, that which is not perceived by our senses, for action. Further, conjunction (by means of some impediment) can destroy action. Figure 7.5 summarises these aspects pertaining to action as enumerated in Vaiśeṣika.

Vaiśeṣika has discussed several observed phenomena in the physical world concerning action and motions and establishes certain properties of motion. Moreover, the critical role gravity plays in issues of motion and action have also been identified in the form of several sūtras. Classical physics has articulated them and developed mathematical constructs to model them, which we are aware of. However, knowledge of these was available to ancient Indians, as evident from the astronomical and mathematical concepts developed in India that pre-dates modern-day developments.

- Attributes are the inherent properties associated with the substances. Seventeen attributes have been enumerated in Vaiśeṣika.
- Vaiśeṣika has discussed several observed phenomena in the physical world concerning action and motions and establishes certain properties of motion.

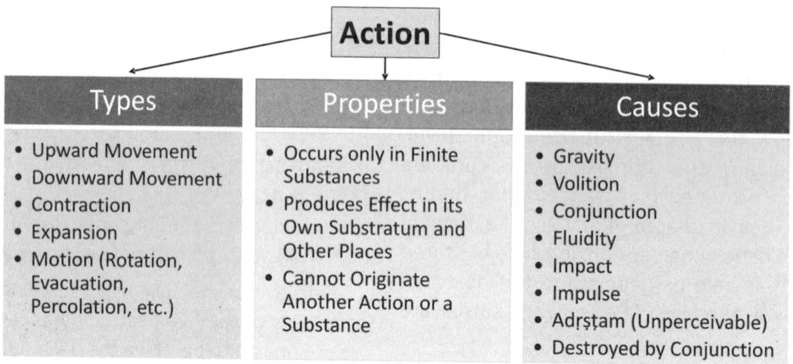

FIGURE 7.5 Action – Types, Properties and Causes

Gravitational Pull

In Vaiśeṣika we find the recognition of gravity causing the actions of throwing up and throwing down. This has been brought out explicitly in the form of some sūtras. These are enumerated below:

- गुरुत्व-प्रयत्न-संयोगानाम् उत्क्षेपणम् ॥ *gurutva-prayatna-saṃyogānām utkṣepaṇam* || 1.1.29
  Throwing up (or upward motion) is caused by the conjunction of effort and gravity.

  This sūtra establishes that unless there is an effort exerted to counteract the gravitational pull, upward movement of substances may not be possible. In continuation of the sūtra, the next two sūtras establish the free fall of substances on account of the force of gravity that governs upward and downward motions.

- संयोगाभावे गुरुत्वात् पतनम् ॥ *saṃyogābhāve gurutvāt patanam* || 5.1.7
  In the absence of conjunction (implying some form of impediment), objects freely fall on account of gravity.
- संस्काराभावे गुरुत्वात् पतनम् ॥ *saṃskārābhāve gurutvāt patanam* || 5.1.18
  In the absence of some propulsive energy generated by the action, objects fall freely on account of gravity.

Properites of Objects in Motion

In modern Physics, we are aware of Laws of motion which defines the notion of inertia. It is because of the inertial force that an object continues to move or loses its momentum. These ideas are articulated in Vaiśeṣika-sūtras[11]. The relevant sūtras are given below:

- नोदनविशेषाभावान्नोर्ध्वं न तिर्यग्गमनम् ॥
  nodanaviśeṣā-bhāvānnordhvaṃ na tiryag-gamanam || 5.1.8
  When there is no particular molecular movement or impulse (implying no external force is applied), there is no upward motion or sideward motion.
- नोदनादाद्यमिषोः कर्म तत्कर्मकारिताच्च संस्कारादुत्तरं तथोत्तरमुत्तरञ्च ॥ 5.1.17.
  nodanādādyamiṣoḥ karma tat-karma-kāritācca saṃskārād-uttaraṃ tathottaram-uttarañca ||

  The initial action of exerting a certain force causes the motion of an arrow released from a bow; from that momentum follows the next action (of arrow's motion) and so on.
- कार्यविरोधि कर्म ॥ *kārya-virodhi karma* || 1.1.14
  The effect of action (kārya) works against the action.

  Since it opposes the original cause that created the motion, perpetual motion is not possible. In classical physics, we recognise this as inertial property. Although the magnitude is not indicated, this sūtra also implies that every action has an opposite reaction.

7.3 PRAMĀṆA – THE MEANS OF VALID KNOWLEDGE

One of the components in the knowledge triangle is the means of obtaining valid knowledge. According to the Indian tradition, knowledge is ultimately produced in the 'ātman' and if the generating conditions are sound, knowledge is valid, otherwise, it is defective. Different schools of thought have come up with alternative means of obtaining valid knowledge. According to Nyāya-śāstra, there are four means of obtaining valid knowledge: Pratyakṣa, Anumāna, Upamāna, and Śabda[12].

Pratyakṣa may be defined as perception and is the primary means for acquiring knowledge. Pratyakṣa is nothing but direct perception, which enables one to obtain knowledge from the contact of a sense organ with its object. It is a direct experience of reality by eyes, ears, nose, touch, and taste. This is why the sensory organs are referred to in the Indian tradition as '*Jñānedriyas*' (organs of knowledge). Pratyakṣa can be thought of as the ultimate gate to ferry knowledge into the subtle organs of mind, intellect, and memory for anyone for further processing and internalisation. This process of internalisation will eventually convert the tacit knowledge obtained from various external sources through the sense organs into implicit. All other means of knowledge (Pramāṇas) eventually work in conjunction with Pratyakṣa to generate the right means of knowledge.

The knowledge so obtained is determinate and non-erratic[13]. Suppose there is an auditorium in which a program is being organised. If we are interested in answering the question, "How many people are attending the program?" The most authentic way to answer this would be to do an actual headcount and come to a conclusion, say 392 people are attending the program. In this example, we use our sense organs and count to arrive at an unambiguous answer.

Anumāna is inferential knowledge and is preceded by perception. 'Anu' in Sanskrit means 'follows' and 'māna' is knowledge. Therefore, anumāna points to the knowledge that follows something pre-existing and arrived at in a structured manner by relating to reasons and logic. There are two aspects involved in inferential knowledge. However, there is a concomitance (of the reason (hetu)) which makes it possible to make the inference as the hetu connects the other two aspects. We shall take a simple but a classical example to understand inference. We know that smoke is invariably associated with fire (the concomitance of smoke and fire is the key aspect for inference). If we see smoke on a hill, we conclude that there must be fire on that hill. From the presence of smoke in the hill as qualified by the knowledge that wherever there is smoke there is fire, we proceed to infer the presence of fire in the hill. Suppose if we want to know as to what to infer if we put butter on a gas oven. We can use our repository of inference to answer this question. We know that a gas oven always generates 'heat' by burning the cooking gas. Further, we also know that when butter is heated up it becomes ghee. Therefore, we infer that butter will become ghee when placed on a gas oven. The 'hetu' in this example is the heat which has the concomitance with the gas oven and it connects the butter and the gas oven.

Three types of inferences are proposed in Nyāya[14] (a priori, a posteriori, and commonly seen). In the case of a priori, previous knowledge of the cause will help us arrive at the knowledge. For instance, the moment we see heat being applied, we will be able to infer that ghee will be obtained from butter. In the case of a posteriori, knowledge is derived from the perception of the effect. For example, if we see warm ghee and a hotplate alongside, we will infer that there was butter that has been transformed into ghee. The commonly seen inference is similar to seeing many people walking on a wet road with an umbrella in their hands and inferring that it ought to have been raining in the area.

These are simple examples to illustrate anumāna. However, in reality, a structured logical framework is employed to validate knowledge using either a deductive or an inductive approach. The hetu (reason) is the prime driver of the inferential knowledge and in Nyāya, we have a reasoning logic (called avayava) to generate inferential knowledge, which we will discuss in Section 7.5.1.

Upamāna may be defined as a comparison or analogy. Since new knowledge generation is one of treading the path from 'the unknown' to the known, prior knowledge of related things plays a role in the process. In Nyāya, comparison and analogy obtained on account of the similarity of the unknown to another thing previously well-known[15] are known as Upamāna. It is produced by the knowledge of resemblance or similarity-based on certain attributes. There are two entities involved in analogy—the subject for the analogy (which relates to the unknown), the object for analogy (the known). Using one or more attributes, an axis for making the comparison is established through which new knowledge is developed. The knowledge developed is conditional to the choice of attributes but helps greatly in developing a better understanding of the unknown.

For example, with the knowledge of a cow and its physical and behavioral peculiarities, one can tread into the forest and develop a certain understanding of a hitherto unknown animal using attributes derived from the cow and projecting it on to the new animal. Hearing that a wild ox is like a cow the person who does not know about the wild ox infers that the animal which looks like a cow is a wild ox. In the Indian tradition, Upamāna is of three types:

- Sādṛśya-upamāna – In this type of upamāna, the similarity is the important source of knowledge. For example, Rakesh does not know about baseball and asks Ramesh who is aware of it. Ramesh tells him that baseball is similar to cricket. Later when Rakesh watches baseball, he gets to know that the sport he is watching over TV is baseball because he remembers the obsevsation by Ramesh that baseball resembles cricket. In this example, upamāna is based on similarity.
- Vaidharmya-upamāna – In this case of upamāna, the dissimilarity plays an important role in the establishment of knowledge. For example, Devikā is new to the field of engineering. She does not know what a spanner is. But she knows what a screwdriver is. She asks her colleague David, what the spanner looks like. David gives her a description of the spanner as follows, "it does not look like a screwdriver; spanner is a typical screw head that has two holes drilled in it, and the bit that mates with it has two pins that are set the same distance apart as the two holes in the screw head; there are three or four sizes, etc." Remembering this and seeing a tool dissimilar to the screwdriver with attributes similar to what was being mentioned she concludes that the tool is a spanner.
- Asādhāraṇa-dharma-upamāna – In this type of upamāna, the knowledge is established based on special quality present in the object or knowledge base. For example, the rhinoceros bears a horn on its nose is a peculiar sign which helps in its recognition and differentiation from the elephant.

There are multiple dimensions on which analogy can be drawn to develop a better understanding of the unknown. In a chemical process, suppose there is a description that the new material after the completion of the chemical process will be like 'rubber'. It helps the scientist to understand that perhaps the new material will have certain physical attributes such as elasticity. Similarly, if there is a discussion about a rebellion that broke out in the city and the police restoring law and order. If the question was, 'how did the police quell the rebellion?'. One way to communicate it unambiguously is perhaps to say, "the police dealt with the situation with an iron hand". In this case, the use of the word "iron hand" does not mean the police were all having special hands made of 'iron'. Rather the comparison and analogy to 'iron' are in the context of some attribute of iron such as 'firmness, hardness, not easy to bend, etc.' and relates it to the approach taken by the police to put the rebellion to an end.

From these discussions, it is evident that the efficacy of finding the new knowledge depends on the appropriate choice of the attributes to establish the similarity or otherwise between the unknown and the known aspect of knowledge.

Śabda is the verbal testimony of an authoritative expert in the subject. It also includes the authoritative texts of various śāstras composed by reliable experts. There are several situations in real-life where correct knowledge is obtained only from a reliable source such as an expert. This is simply because the subject matter is not known to many and it may require specialised expertise to analyse and find answers to the questions. Such knowledge is implicitly accumulated over long years of experience by a person of high standing and character. We encounter this often in our daily life. For example, if one develops chest pain and wants to know if there is a heart problem, the only way to find an answer to this is to go to a cardiologist. By virtue of extensive training, specialised knowledge, and deep experiences, the cardiologist will be in a position to resolve this question.

In the Indian tradition, this is accepted as a very important and valuable means of obtaining the right knowledge. In the Nyāya-śāstra, this is referred to as śabda, which is defined as the assertion of a reliable person (expert in the field)[16]. In the Indian tradition, the injunctions of the Vedas are considered as śabda-pramāṇa, as it provides instructions on several things that are 'not seen' or ordinarily known to human beings.

Figure 7.6 briefly summarises these discussions in a pictorial form. Direct perception is the basic means of acquiring knowledge and therefore all other pramāṇas will make use of it. The inference will make use of perception and analogy and finally, verbal testimony will make use of all the other three pramāṇas.

प्रत्यक्ष (Pratyakṣa) - Perception

- *How many people are attending this seminar? – 392*

अनुमान (Anumāna) - Inference

- *What will happen if you put butter on a hot plate? – It will become "ghee"*

उपमान (Upamāna) - Comparison and analogy

- *How did the police quell the rebellion? – They dealt with an "iron hand"*

शब्द (Śabda) - Word, testimony of past/present reliable experts

- *Is my heart working normally? – Cardiologist advice*

FIGURE 7.6 The Four Pramāṇas – An Illustration

7.4 SAṂŚAYA – AMBIGUITIES IN EXISTING KNOWLEDGE

The search for new knowledge, or for establishing a new tenet begins with the recognition of ambiguity in the knowledge that currently exists, which needs clarification. If a certain issue is well settled and there are no doubts, where is the scope for new knowledge? Similarly, if the issue is totally outside the domain of cognition there is no scope for new knowledge. In the Nyāya framework, the ambiguity is referred to as 'Saṃśaya'. This marks the beginning of the exercise for new knowledge creation. For instance, there is a lack of understanding of the properties of a certain material thereby leaving ambiguity of its potential applications. This may call for research and development of new knowledge. According to Nyāya ambiguity in knowledge occurs on account of five reasons[17]:

- When an observed phenomenon resembles several commonly known properties, there is ambiguity in understanding the phenomenon. Since peculiar aspects causing the phenomenon are not distilled, it clutters one's understanding and begs for further inquiry so that the distinctive aspects that make it unique could be identified. For example, in medical research, a new disease caused by a virus may resemble several known medical conditions in terms of symptoms. However, further research may be required to isolate the causes and distill certain unique aspects leading to the medical condition the virus causes. This provides the basis for removing the ambiguity in knowledge and is critical for developing medicines and effective treatment protocols for the disease caused by the new virus.

> - Pramāṇa could be defined as the means of obtaining the right knowledge. According to Nyāya Śāstra, there are four means of obtaining the right knowledge.
> - In the Indian tradition, verbal testimony is accepted as a very important and valuable means of obtaining the right knowledge.

- The other reason that causes ambiguity in knowledge is the recognition of properties that have no relation to anything known so far. While it definitely points to a new phenomenon hitherto unknown, it may not provide any explanatory power to articulate what causes the phenomenon and what are the implications of it. In several R & D applications, we encounter such examples. A new chemical process to extract some element may throw results that may not make sense at the outset. Some more research and deep inquiry will help us establish the causes and applications of the new process.
- The third reason for the ambiguity is conflicting findings from a study. Initial studies may exhibit certain properties. However, further studies may provide results that may conflict with the earlier findings, necessitating further inquiry into the process and searching for new knowledge and insights.
- Another variation of the conflicting results is considerable variations in the results. An observed phenomenon in a certain condition may get repeated in certain other situations, where we might not have anticipated. This creates a certain inconsistency in our understanding and results in the ambiguity of knowledge. In several studies involving behaviour of people, such results are expected. It merely points to more such number of studies involving greater number of samples and situations to identify underlying patterns and insights.
- There are also occasions when there is a total absence of our understanding of the problem itself when we study them. We may expect outcomes in certain situations but may not see it. In the extreme, we may not even have an idea of what to look for and where. All these will lead to ambiguity of knowledge on account of total non-apprehension of the issue that we are studying. This often happens in the initial stages of any exploratory study in unknown domains of research.

Figure 7.7 enumerates the reasons. In most of our attempts in discovering new knowledge, we encounter such reasons for ambiguity in knowledge. Identifying the nature of ambiguity that we face and establishing the need and the purpose for removing these ambiguities sets the knowledge seeker on the right path.

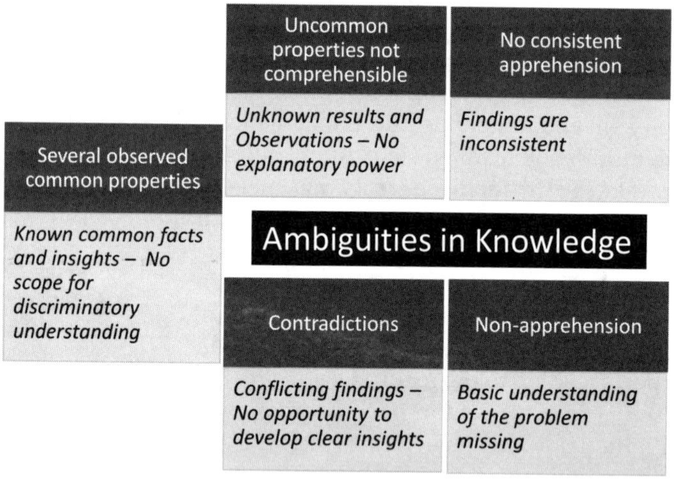

FIGURE 7.7 Causes for Ambiguities in Knowledge

7.5 FRAMEWORK FOR ESTABLISHING VALID KNOWLEDGE

The key contribution of Nyāya-śāstra lies in providing a robust framework for establishing the right knowledge. The very first sūtra in the text begins with identifying all the relevant factors that one needs to take into consideration while establishing the right knowledge. Sixteen categories have been identified and a correct understanding of these and appropriate use of them is critical in the process[18]. These sixteen categories provide a comprehensive set of concepts that help one to establish knowledge using a structured approach. Unless the right means of finding knowledge are directed to the object of the study, there is neither a context nor a possibility for a fruitful outcome. Therefore, the governing factors for conducting the study are 'Pramāṇa' and 'Prameya'. We have already discussed these aspects.

Nyāya prescriptions to establish valid knowledge has two principal components:

1. A tradition of debate with a mechanism to navigate through arguments and counter-arguments finally leading to the selection of the most acceptable one to resolve the ambiguity.
2. A methodology for those who want to engage in an honest, friendly, fair, and balanced debate for valid means of exploring new knowledge through a five-step reasoning logic.

Nyāya utilises two key aspects to establish new knowledge; One is a deductive/inductive reasoning framework and the other is a structured approach to debate (argumentation). In an oral tradition, new tenets, and knowledge are established after prolonged discussion (known as Tarka) on matters pertaining to the subject. Usually, there is a vādin (who seeks to advance a new knowledge) and a prativādin, an opponent who challenges the proposer of new knowledge. Based on a series of discussions in which arguments are placed by both sides a conclusion is finally reached on the basis of one convincing the other. This paves the way for advancing the knowledge into new domains. Therefore, logic and argumentation are considered as two main drivers of new knowledge. In the Nyāya-śāstra, these two aspects are well established through a series of sūtras (aphorisms).

On account of the prevalence of an oral tradition to establish knowledge in earlier days, factors 8 to 16 provide mechanisms to engage in the oral discussion. Figure 7.8 depicts a generalised framework for establishing valid knowledge using the categories identified by Nyāya. There are three components in this framework:

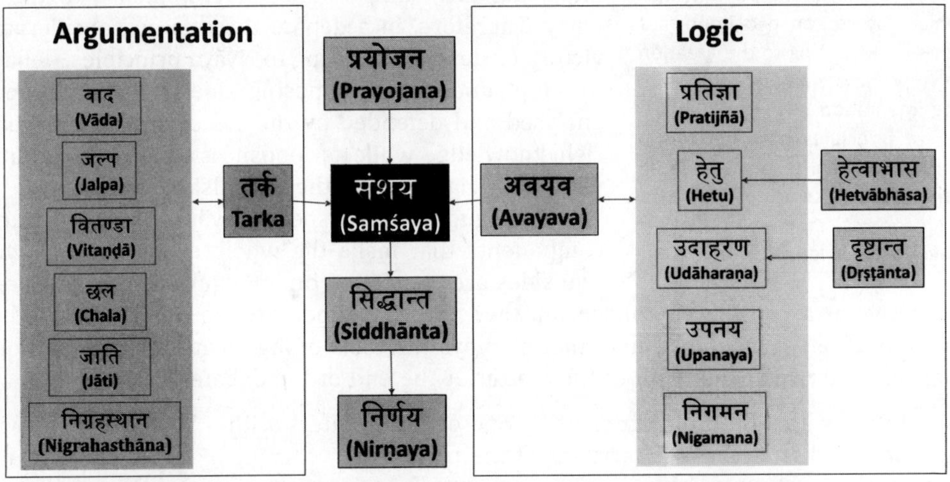

FIGURE 7.8 A Nyāya Based Framework for Establishing Right Knowledge

The middle part sets out the purpose, context, and outcome related to establishing new knowledge. The removal of ambiguity in existing knowledge is the fundamental driver that leads to the creation of new knowledge. Once the issue to be studied is identified, then the purpose of undertaking the exercise, the objectives to be met through the study and the benefits expected to accrue out of the exercise are to be clearly established before the commencement of the study. This is indicated by the factor 'Prayojana'. This helps to direct the efforts in a focused fashion, establish the reason for addressing the ambiguity on hand. The ambiguity is resolved using structured steps identified on the left and the right side of the framework. On account of this, a new tenet (Siddhānta) is established, which is basically, an outcome and verdict of the process adopted. This is finally confirmed and accepted by others (Nirṇaya) and it eventually becomes established knowledge.

The process adopted to arrive at valid new knowledge consists of two broad components, logic and argumentation, indicated on either side in Figure 7.8. The logical part (Avayava) consists of a structured and sequential five-step process that provides a reasoning logic to make valid inferences which provide the ammunition for the debate. The proposition (Pratijñā) is the starting statement which needs to be established to the satisfaction of the opponent. It gives rise to an inquiry necessitating the mention of the reason (Hetu). In the process of using the structured inquiry process, it is important to make use of familiar instances (Dṛṣṭānta) pertaining to the area of study. This will help find familiar situations and examples (Udāharaṇa) that can build support for the proposition. An example of the application of the propostion is 'Upanaya'. The final conclusion of the proposition is done on the basis of the above, known as *Nigamana*. Moreover, one must constantly endeavor to make sure that the logic advanced does not suffer from some of the potential fallacies (Hetvābhāsa).

On the other hand, the other component debate (Tarka) can be defined as reasoning in the presence of ambiguity of knowledge using some predefined methods for constructing and weighing up arguments through a structured process. It is primarily meant for the discernment of the real nature of the thing under investigation and imparting the truth, as one understands it, to the other party. Therefore, in a debate there is no consideration of victory or defeat. According to Nyāya principles, debate has to adopt one of two opposing sides. What is adopted is analysed and defended by the aid of any of the means of right knowledge, while its opposite is assailed by confutation, without deviation from the established tenets.

- Nyāya is one of the six Darśanas which focuses on two important aspects pertaining to the creation of new knowledge: Logic and Argumentation.
- Nyāya principles have become a very useful tool for experts in all other fields (literature, grammar, philosophical studies, and other fields of knowledge).

The argumentation aspect of Nyāya has highly structured components that make the whole exercise worthwhile. If two sides are contesting on ambiguity and have to resolve it and arrive at a considered conclusion, then these methods are variously employed. Vāda pertains to a constructive discourse that employs the tools of logic with the sole aim of both the parties to arrive at the truth of the matter at the end of the debate.

- **Jalpa** is a method of argument by one of the debaters with a keen desire to win the debate than necessarily arriving at the truth.
- **Vitaṇḍā** is another form of debate, wherein the debater is interested only in picking errors in the opponent's argument. It is akin to saying, "I know that you are wrong, but I do not know what is right".
- **Chala** is a method of arguing by simply picking up the loopholes in the opponent's argument and harping on it and thereby deflecting the argument away from the original objective. It is a tactic of distraction and derailment of the process and the outcomes.
- **Jāti** is a method of engaging in the debate by highlighting the internal inconsistency in the opponent's arguments and highlighting the contradictions.
- **Nigrahasthāna** is an occasion to raise objections and get the objection sustained by the mediator (as we witness in the modern judicial arguments). It is also an occasion to convince the moderator of a debate that the opponent is wrong and get the opponent out of the debate.

Based on these, it will be possible to deduce an acceptable tenet (Siddhānta) and reach a finality to the study (Nirṇaya) and present the results in such a manner that some useful new addition could be made to the existing body of knowledge. The above framework provides a generic structure for establishing right as well as new knowledge in any chosen field of study. The individual components of the framework warrant more explanation. We shall see some of the components in greater detail.

It is important to know some of the salient aspects of this methodology of new knowledge creation. Firstly, although the above discussion of the Nyāya methodology appears to give an oral slant, the same set of ideas could be used to argue in a written mode as well, although in a written format the arguments may happen in an asynchronous fashion. In this sense, this can be viewed as a generic framework for resolving the ambiguities and creating new knowledge. As is evident from the above discussion, the most appropriate and accepted form of tarka is vāda. However, a vādin must be aware of the alternative tactics that are likely to be deployed by the opponent and guard against these in not getting trapped into an argument. In the absence of this, the discussion may not yield the final result of establishing the new knowledge. Unlike other methods, the debater cannot merely get away by saying, "This is my view" or "I don't accept your views" but has to substantiate the same using reasoning and logic. Therefore, a vādin will invariably have to fall back on the reason and logic framework to mount convincing arguments in support of his/her view and put forth inferential knowledge.

- The key contribution of Nyāya Śāstra lies in providing a robust framework for establishing the right knowledge.
- The elements of a formal inquiry process in search of the right knowledge are enlisted under 'Avayava'.

7.5.1 Deductive/Inductive Logic Framework

Central to the process of establishing new knowledge is a deductive/inductive reasoning framework proposed in Nyāya (known as Avayava). It has five steps in the process[19] as depicted in Figure 7.9.

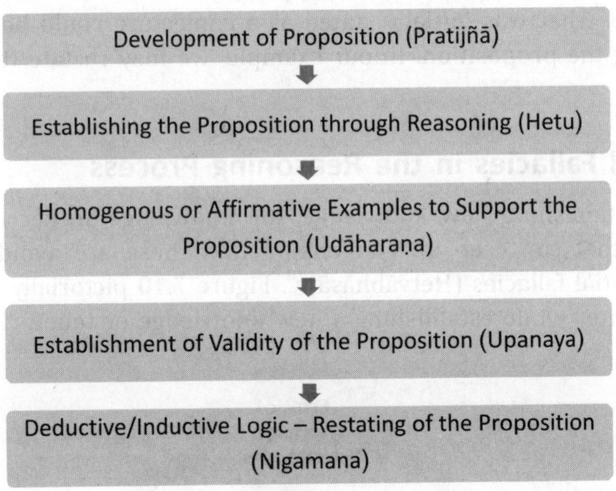

FIGURE 7.9 A 5-step Approach for Deductive/Inductive Logic

Step 1: Pratijñā (प्रतिज्ञा) – Development of a proposition

Deductive/inductive reasoning begins with the development of a proposition for the study. Based on the ambiguity to be resolved and the stated objectives, a proposition needs to be developed. For example, let us start with the proposition, 'Sound is non-eternal'.

Step 2: Hetu (हेतु) – Establishing the proposition through reasoning

Once the proposition is stated, one can draw upon the existing body of knowledge and argue the case for the proposition. In our example, we will reason 'sound is non-eternal' by arguing 'because it is produced'. Whatever is produced is non-eternal and therefore we seek to establish the proposition through this reasoning.

Step 3: Udhāharaṇa (उदाहरण) – Examples for supporting the proposition through reasoning

The theoretical support will merely help conjecture and state the proposition formally and logically. It requires support in terms of the observed phenomenon. Therefore, to establish the proposition, familiar instances need to be drawn from actual practice. In our example, the supportive observed phenomenon is, 'whatever is produced is non-eternal, as a pot'. In this case, the property of non-eternality is borrowed from the pot, which is a familiar substance. One can also use a counter-example (negative property). For example, whatever is 'not' non-eternal is 'not' produced as in the case of the soul (ātman).

Step 4: Upanaya (उपनय) – Establishment of the validity of the proposition

While the previous step may show support for the proposition based on the examples and counter-examples analysed, the validity of the proposition requires that it is sufficiently general and robust to state. This may also require checking for the logical and internal validity of the results. In the Nyāya-śāstra, some potential fallacies arising out of the work have been identified. Checking for these fallacies could help establish the logical and face validity of the propositions.

Step 5: Nigamana (निगमन) – Restating of the proposition

Once the above steps are completed, the deductive/inductive reasoning of the proposition is complete. Therefore, what was initially stated as a conjecture could be stated in a grounded fashion by restating the proposition. In our example, we may restate the proposition, 'Sound is non-eternal'.

7.5.2 Potential Fallacies in the Reasoning Process

In the course of establishing new knowledge, the knowledge seeker must be aware of the potential fallacies that can crop up and ensure that these are avoided. Nyāya-śāstra has identified five potential fallacies (Hetvābhāsā)[20]. Figure 7.10 pictorially depicts these fallacies that one can encounter while establishing a new knowledge or tenet.

FIGURE 7.10 Potential Fallacies in Reasoning

Savyabhicāra: Erratic conclusions: When it is possible to draw more than one conclusion from a deductive/inductive reasoning approach, it points to an erratic result. This is referred to as Avyabhicāra in Nyāya. Suppose we start with a proposition 'Sound is eternal' because it is intangible. Using the example that whatever is intangible is eternal, as atoms, we may conclude that sound is eternal. Alternatively we may start with the proposition that 'Sound is non-eternal' because it is intangible. Using the example that whatever is intangible is non-eternal, as intellect, we may conclude that sound is non-eternal. Here we have drawn two opposite conlcusions primarily because there is no universal connection between 'intangible', 'eternal' and 'non-eternal'. Such results of the study will make it erratic and we may not be in a position to make any concrete statement.

Viruddha: Contradictory conclusions: Another possibility is that starting with a proposition A, we proceed in the study and end up with a conclusion (Proposition 'A') which runs contrary to Proposition A. This means our proposition did not have enough support, or that the means that we have adopted were not sound enough. It can also point to the fact that the proposition was incorrectly stated while beginning the study.

Prakaraṇasama: The process goes back to the proposition: There may be occasions when the study did not yield anything substantial concerning the proposition. In the process of building support through the reasoning process, it may end up provoking the very proposition for which we are seeking support. Suppose we develop a proposition, 'Sound is non-eternal' and support this with a reason, 'because it is not possessed of the attribute of eternality'. In this case, 'non-eternal' and 'not being possessed of the attribute of eternality' means the same, and the reason begs the question.

Sādhyasama: Unproven proposition: Another fallacy that one can suffer from is the inability to prove the propositions despite the best of the efforts. Unlike in the previous cases, there is no directional sense in the findings in this case. The proposition is left wide open for further inquiry sometime in the future. For example, if the proposition 'Shadow is a substance' is argued with a reason, 'because it possesses motion' then this will stand only when we are able to prove that the shadow has motion. The motion may belong to a person obstructed by light.

Kālātīta: Mis-timed or Contextually irrelevant findings: Another fallacy of the process is that findings are not 'in sync' with the time. This is especially true when the tenet is to be established through an oral discussion between a proponent and an opponent of the idea. In other cases, the mistiming of the results can point to the contextual relevance of the findings. If the findings are not relevant because the context in which the results are being applied has changed, then it points to the fallacy of the entire effort.

7.5.3 Established Tenets in a Field of Study

The process of new knowledge creation does not happen in a vacuum. In most cases, it is an incremental process wherein based on the understanding of earlier results and established principles one will be able to hypothesize new conjectures to advance the knowledge. Moreover, deriving some insights from earlier studies become a key component in the process as it will help resolve some of the ambiguities. Therefore, one of the important elements in the field of research and new knowledge creation is the need to be aware of established tenets that already exist. Furthermore, using the logic and argumentation methods, it will be possible to establish a tenet. In Nyāya, four types of siddhāntas (established tenets) have been proposed[21]:

- Some tenets are accepted by every school of thought as they are fundamental in nature. This is referred to as '*Sarva-tantra-siddhānta*' in Nyāya. For example, in the Indian philosophical schools, the existence of the five elements (ether, air, fire, water, and earth) and the five sense organs are accepted by all. Similarly, in the scientific world, the existence of the force of gravity in the earth's atmosphere is accepted by everyone notwithstanding differences between the Newtonian (Classical) Physicists and Quantum Physicists on other tenets.
- Several principles are peculiar to different schools of thought, referred to as '*Prati-tantra-siddhānta' in Nyāya*. This is very common. For example, if we take two schools of health, Āyurveda, and Allopathy, they have different tenets concerning what causes disease in a human being. Even the disease management approach between the two schools differs on account of these differences in the siddhānta.
- The third variety of tenets, known as *Adhikaraṇa-siddhānta* points to a tenet governing over several other tenets if accepted. Suppose the hypothesis that the existence of water bodies on the Mars planet can support living beings is established through a proper study and accepted, then it means that other tenets related to this are already accepted. These include, for example, living beings need water to survive and several living beings exist when there is water, etc.
- There is a fourth type of siddhānta, known as *Abhyupagama-siddhānta*, which gets established in an implied fashion, on account of the details that we analyze about an issue. For example, when we accept phenomena such as eclipses, solar and lunar months, etc. it is already implied that the Sun and the Moon have a relative circular orbital motion with respect to the Earth.

Awareness of these concepts is very important for the development of the right knowledge. They help in positioning the knowledge in the right context and ensure that unintended fallacies do not hinder the process. Moreover, it helps to take adequate support from the existing body of knowledge in building several aspects of the new knowledge.

SUMMARY

- A valid knowledge corresponds to the reality and not anything other than that and is indeed produced by some valid means.
- Typically, in the Indian tradition, the aspects of valid knowledge could be best understood from the notion of a knowledge triangle, consisting of three components (Pramātṛ, Prameya, and Pramāṇa)
- Vaiśeṣika mainly confines itself to 'the exposition of reality' and Nyāya focuses on the issue of 'right knowledge of reality'.
- Vaiśeṣika presents a systematic framework to describe all the 'nameable and knowable' entities.
- Six sub-categories constitute existence. The first three, Dravya, Guṇa, and Karma are objective aspects. The other three categories Sāmānya, Viśeṣa, and Samavāya are outcomes of intellectual discrimination.
- The basic building blocks of the physical reality are the Dravya in the framework of Vaiśeṣika. The basic atomic nature of the substance allows the creating of many different things using the principle of Samavāya.
- There are nine types of substances identified in Vaiśeṣika. These include Pṛthivī, Ap, Tejas, Vāyu, Ākāśa, Dik, Kāla, Ātman, and Manas.

- Attributes are the inherent properties associated with the substances. Seventeen attributes have been enumerated in Vaiśeṣika and the subsequent works expanded it to twenty-four. These are either physical or psychological.
- Vaiśeṣika has discussed several observed phenomena in the physical world concerning action and motions and establishes certain properties of motion.
- Sixteen factors have been identified in Nyāya and a correct understanding of these and appropriate use of them is critical in the process of establishing knowledge.
- According to Nyāya-śāstra, there are four means available for obtaining the right knowledge (Pratyakṣa, Anumāna, Upamāna, and Śabda).
- Prtayakṣa is nothing but direct perception, which enables one to obtain knowledge from the contact of a sense organ with its object.
- Anumāna points to the knowledge that follows something pre-existing and arrived at in a structured manner by relating to reasons and logic.
- In Nyāya, comparison and analogy obtained on account of the similarity of the unknown to another thing previously well-known are known as Upamāna.
- Śabda is the verbal testimony of an authoritative expert in the subject.
- The quest for new knowledge invariably begins with the need to seek clarity on certain issues. According to Nyāya ambiguity in knowledge occurs on account of five reasons.
- Nyāya utilises two key aspects to establish new knowledge; One is a deductive/inductive reasoning framework and the other is a structured approach to debate (argumentation).
- Central to the process of establishing new knowledge is a five-step deductive/inductive reasoning framework proposed in Nyāya (known as Avayava).
- In the course of establishing new knowledge, the knowledge seeker must be aware of the potential fallacies that can crop up and ensure that these are avoided. Nyāya-śāstra has identified five potential fallacies (Hetvābhāsā).

REVIEW QUESTIONS

1. What do you understand by the term 'Knowledge Triangle'?
2. Given below are statements pertaining to establishing the right knowledge. Identify the pramāṇa applicable in each of these:
   (a) The constitutional experts advised the Governor on deciding if the Chief Minister enjoyed a majority in the Legislative Assembly.
   (b) Unexpectedly, chemical A turned blue upon mixing with another chemical B.
   (c) Alphonso is the king of mangoes.
   (d) The movie had a bad ending as many viewers came out of the Cinema hall, wiping their tears.
   (e) The housing colony in the evening glittered like Amarāvatī, the capital of Indraloka.
   (f) The yoga master asked the person to perform the āsana for 15 days to get rid of his pain.
   (g) There must have been a big fire as three fire engines have rushed into the factory just now.
3. Comment on the statement, "Vaiśeṣika is an ancient Indian approach to physics".
4. Briefly describe the overall framework of Vaiśeṣika for defining physical entities.
5. Do you see any similarity between concepts of current-day physics and those presented in Vaiśeṣika? What are the points of departure?
6. Enumerate the following as defined in Vaiśeṣika:
   (a) Substances
   (b) Attributes
   (c) Action

7. How are attributes related to the substances?
8. Briefly explain the notion of samavāya? What role does it play in creating the world of multiplicities?
9. What are the factors that cause different types of actions as per Vaiśeśīka?
10. What are the valid means suggested for attaining the right knowledge in Nyāya?
11. Comment on the statement, "Pratyakṣa is the fundamental means for valid knowledge".
12. What is the difference between anumāna and upamāna? Explain with some examples.
13. Why is śabda as a pramāṇa important? Under what conditions do we need śabda-pramāṇa?
14. Briefly describe the components of the Nyāya based framework for establishing the right knowledge.
15. Why do we experience ambiguity in knowledge?
16. What is the relevance of the principles laid out in Nyāya-śāstra in the current context?
17. Explain the steps involved in deductive/inductive reasoning as proposed in Nyāya.
18. What are the reasons for fallacies in establishing the right knowledge?
19. What do you understand by the term 'Siddhānta'? What is the relevance of it in the context of developing a new knowledge?

DISCOVER IKS

1. Indians have developed a method for structured debate through which meaningful discussions were held to address ambiguities in knowledge. Watch this video of Prof. V N Jha which gives a panoramic view of this tradition https://youtu.be/WinIPNgiUZ8. After watching the video, prepare a three-page report to answer the following questions:
   (a) What is the role of logic in the development of good knowledge? How does Nyāya contribute to this?
   (b) What are the methods used for debate and argumentation in the Indian tradition? Under what conditions are they used?
   (c) What are the unique features of the long Indian tradition of debate and dialogue?
2. Some of the foundations of physics find their expression in Vaiśeṣika, which was primarily focused on explaining the physical reality. Watch this video in which Prof. S C Kak brings out some of these ideas: https://youtu.be/GlxddwPwh5E. After watching the video, prepare a two-page note to address the following questions:
   (a) How is Vaiśeṣika related to physics? What are some of the components of physics that Vaiśeṣika seems to also address?
   (b) Can you identify some of the principles of physics that Vaiśeṣika has dealt with?

SUGGESTED READINGS

Bhaduri, S. (1975). "Studies in Nyāya – Vaiśeṣika Metaphysics", Bhandarkar Oriental Research Institute, Pune.

Chakrabarti, K. (1975). "The Nyāya-Vaiśeṣika Theory of Universals", *Journal of Indian Philosophy*, 3(3/4), pp. 363–382.

Chattopadhyaya, J-C. (1912). "The Hindu Realism", The Indian Press, Allahabad.

Kak, S. (2003). "Indian Physics: Outline of Early History", https://arxiv.org/abs/physics/0310001. Last accessed on Oct. 1, 2021.

Kak, S. (2016). *Matter and Mind: The Vaiśeṣika Sūtra of Kaṇāda*, Mount Meru Publishing, Ontario, Canada.

Matilal, B.K. (1985). *Logic, Language, and Reality*, Motilal Banarsidass Publishing, Delhi.

Oetke, C. (1996). "Ancient Indian Logic as a theory of non-monotonic reasoning", *Journal of Indian Philosophy*, **24**(5), pp. 447–539.

Prabhu, C.S.R. (2014). "The Physics of Vaiśeṣika", Sri Venkateswara Vedic University, Tirupati.

Sen, S. (1966). "The Impetus Theory of the Vaisesikas," *Indian Journal of History of Science*, **1**(1), pp. 34–45.

Shaw, J. (2010). "The Nyāya on Sources of Knowledge", in: P. Ghose (Ed.), Materialism and Immaterialism in Indian and the West: Varying Vistas (Centre for Studies in Civilization, Ministry of Human Resources, Government of India, New Delhi), pp. 117–152.

Simonson, S. (1946). "The Categories of Proof in Indian Logic", Philosophy and Phenomenological Research, **6**(3) pp. 400–409.

Sinha, N. (1990). *The Nyāya Sūtras of Gotama*, Motilal Banarsidass Publishers, Delhi.

Sinha, N. (2008). *The Vaiśeṣika Sūtras of Kaṇāda*, Cosmos Publications, New Delhi.

Sarma, V.V.S. (1994). "A Survey of Indian Logic From the Point of View of Computer Science", *Sadhana*, **19**(6), pp. 971–983.

Sharma, C. (1962). "Indian Philosophy: A critical survey", Barnes and Noble, pp. 163–198.

Tachikawa, M. (1971). "A sixth-century manual of Indian Logic (A Translation of the Nyāyapraveśa)", *Journal of Indian Philosophy*, **1**(2), pp. 111–145.

ENDNOTES

1. For details see **Sinha, N. (1990).** *The Nyāya Sūtras of Gotama*, Motilal Banrsidass Publishers. New Delhi. प्रमाण-प्रमेय-संशय-प्रयोजन-दृष्टान्त-सिद्धान्त-अवयव-तर्क-निर्णय-वाद-जल्प-वितण्डा-हेत्वाभास-छल-जाति-निग्रहस्थानानां तत्त्वज्ञानात् निःश्रेयसाधिगमः । pramāṇa-prameya-saṃśaya-prayojana-dṛṣṭānta-siddhānta-avayava-tarka-nirṇaya-vāda-jalpa-vitaṇḍā-hetvābhāsa-chala-jāti-nigrahasthānānāṃ tattvajñānāt niḥśreyasādhigamaḥ | Nyāya Sūtra 1.1.1. Sūtras 3–8 provide definitions and sub-classifications of the pramāṇas.
2. *Ibid.* pp. 6–9. The 12 objects of knowledge are given by the sūtra, आत्म-शरीर-इन्द्रिय-अर्थ-बुद्धि-मनः-प्रवृत्ति-दोष-प्रेत्यभाव-फल-दुःख-अपवर्गास्तु प्रमेयम् । ātma-śarīra-indriya-artha-buddhi-manaḥ-pravṛtti-doṣa-pretyabhāva-phala-duḥkha-apavargāstu prameyam | Nyāya Sūtra 1.1.9. Sūtras 1.1.10–1.1.22 provide definitions of the 12 objects.
3. For details see **Sinha, N. (2008).** *The Vaiśeṣika Sūtras of Kaṇāda*, Cosmo Publications, New Delhi, pp. 8–17. This is formally stated in the Vaiśeṣika sūtra 1.1.4 as, धर्मविशेषप्रसूताद् द्रव्य-गुण-कर्म-सामान्य-विशेष-समवायानां पदार्थानां साधर्म्य-वैधर्म्याभ्यां तत्त्वज्ञानात् निःश्रेयसम् । dharmaviśeṣaprasūtād dravya-guṇa-karma-sāmānya-viśeṣa-samavāyānāṃ padārthānāṃ sādharmya-vaidharmyābhyāṃ tattvajñānāt niḥśreyasam |
4. *Ibid.* pp. 37–50. Part 2 of Chapter 1 discusses these in several sūtras. See for example, सामान्यं विशेष इति बुद्ध्यपेक्षम् । sāmānyaṃ viśeṣa iti buddhyapekṣam | Vaiśeṣika Sūtra (1.2.3); द्रव्यत्वं गुणत्वं कर्मत्वञ्च सामान्यानि विशेषाश्च । dravyatvaṃ guṇatvaṃ karmatvañca sāmānyāni viśeṣāśca | Vaiśeṣika Sūtra (1.2.5).
5. *Ibid.* pp. 27–28. क्रिया-गुणवत् समवायिकारणमिति द्रव्यलक्षणम् । kriyā-guṇavat samavāyikāraṇamiti dravyalakṣaṇam | Vaiśeṣika Sūtra (1.1.15).
6. *Ibid.* pp. 33–333. कारणमिति द्रव्ये कार्यसमवायात् । kāraṇamiti dravye kāryasamavāyāt | Vaiśeṣika Sūtra (10.2.1).
7. *Ibid.* pp. 51–144. Books 2 and 3 of Vaiśeṣika provide the definitions and lay down the key characteristics of the nine substances.

8. *Ibid.* pp. 28–29. द्रव्याश्रयगुणवान् संयोगविभागेष्वकारणमनपेक्ष इति गुणलक्षणम् । dravyāśrayaguṇavān saṃyog avibhāgeṣvakāraṇamanapekṣa iti guṇalakṣaṇam | Vaiśeṣika Sūtra (1.1.16).

9. *Ibid.* pp. 19–20. रूपरसगन्धस्पर्शाः संख्याः परिमाणानि पृथक्त्वं संयोगविभागौ परत्वापरत्वे बुद्धयः सुखदुःखे इच्छाद्वेषौ प्रयत्नाश्च गुणाः । rūparasagandhasparśāḥ saṃkhyā: parimāṇāni pṛthaktvaṃ saṃyogavibhāgau paratvāparatve buddhayaḥ sukhaduḥkhe icchādveṣau prayatnāśca guṇāḥ | Vaiśeṣika Sūtra (1.1.6).

10. *Ibid.* pp. 20–23. उत्क्षेपणमवक्षेपणमाकुञ्चनं प्रसारणं गमनमिति कर्माणि । utkṣepaṇamavakṣepaṇamākuñcanaṃ prasāraṇaṃ gamanamiti karmāṇi | Vaiśeṣika Sūtra (1.1.7).

11. For more details please see, **Kak, S. (2016).** *Matter and Mind: The Vaiśeṣika Sūtra of Kaṇāda*, Mount Meru Publishing, Ontario, Canada.

12. For details see, **Sinha, N. (1990).** *The Nyāya Sūtras of Gotama*, Motilal Banarsidass Publishers, Delhi, p. 3. Nyāya Sūtra 1.1.3 defines the four pramāṇas, प्रत्यक्ष-अनुमान-उपमान-शब्दाः प्रमाणानि । pratyakṣa-anumāna-upamāna-śabdāḥ pramāṇāni |

13. *Ibid.* p. 3. इन्द्रियार्थ-सन्निकर्षोत्पन्नं ज्ञानम् अव्यपदेश्यम् अव्यभिचारि व्यवसायात्मकं प्रत्यक्षम् । indriyārtha-sannikarṣotpannaṃ jñānam avyapadeśyam avyabhicāri vyavasāyātmakaṃ pratyakṣam | Nyāya Sūtra 1.1.4.

14. *Ibid.* p. 4. अथ तत्पूर्वकं त्रिविधमनुमानं पूर्ववत्-शेषवत्-सामान्यतो दृष्टं च । atha tatpūrvakaṃ trividhamanumānaṃ pūrvavat-śeṣavat-sāmānyato dṛṣṭaṃ ca | Nyāya Sūtra 1.1.5.

15. *Ibid.* p. 4. प्रसिद्ध-साधर्म्यात् साध्य-साधनम् उपमानम् । prasiddha-sādharmyāt sādhya-sādhanam upamānam | Nyāya Sūtra 1.1.6.

16. *Ibid.* p. 5. आप्तोपदेशः शब्दः । āptopadeśaḥ śabdaḥ | Nyāya Sūtra 1.1.7.

17. *Ibid.* p. 10. The five sources of saṃśaya are enumerated in the sūtra as, समानानेकधर्मोत्पत्तेः विप्रतिपत्तेः उपलब्धि-अनुपलब्धि-अव्यवस्थाश्च विशेषापेक्षो विमर्शः संशयः । samānānekadharmotpatteḥ vipratipatteḥ upalabdhi-anupalabdhi-avyavasthāśca viśeṣāpekṣo vimarśaḥ saṃśayaḥ | Nyāya Sūtra 1.1.23.

18. *Ibid.* pp. 1–5. प्रमाण-प्रमेय-संशय-प्रयोजन-दृष्टान्त-सिद्धान्त-अवयव-तर्क-निर्णय-वाद-जल्प-वितण्डा-हेत्वाभास-छल-जाति-निग्रहस्थानानां तत्त्वज्ञानात् निःश्रेयसाधिगमः । pramāṇa-prameya-saṃśaya-prayojana-dṛṣṭānta-siddhānta-avayava-tarka-nirṇaya-vāda-jalpa-vitaṇḍā-hetvābhāsa-chala-jāti-nigrahasthānānāṃ tattvajñānāt niḥśreyasādhigamaḥ | Nyāya Sūtra 1.1.1. Sūtras 1.1.3 – 1.1.8 provide definitions and sub-classifications of the pramāṇas.

19. *Ibid.* pp. 13–14. In Nyāya sūtra 1.1.32, the elements of a deductive/inductive reasoning logic are given as, प्रतिज्ञा-हेतु-उदाहरण-उपनय-निगमनानि अवयवाः । pratijñā-hetu-udāharaṇa-upanaya-nigamanāni avayavāḥ | Further explanations of each of these are given in sūtras 33–39.

20. *Ibid.* pp. 21–23. These are stated in the Nyāya in the sūtra 1.2.4 as, सव्यभिचार-विरुद्ध-प्रकरणसम-साध्यसम-कालातीताः हेत्वाभासाः । savyabhicāra-viruddha-prakaraṇasama-sādhyasama-kālātītāḥ hetvābhāsāḥ | Each of these are further explained in sūtras 1.2.5 to 1.2.9.

21. *Ibid.* pp. 11–13. The basic definition of an established tenet is stated in the Sūtra, तन्त्राधिकरणाभ्युपगम-संस्थितिः सिद्धान्तः । tantrādhikaraṇābhyupagama-saṃsthitiḥ siddhāntaḥ | Nyāya Sūtra 1.1.26. Sūtras 27–31 state the four types and their respective definitions.

Part 3

SCIENCE, ENGINEERING AND TECHNOLOGY IN IKS

CHAPTER

8 Mathematics

LEARNING OUTCOMES

After finishing this chapter, you will be able to:

- Understand the nature of contributions made by ancient Indian Mathematicians
- Appreciate the unique nature of the Indian Mathematical contributions
- Understand specific contributions in Arithmetic, Geometry and Trigonometry
- Know the key role played by Āryabhaṭa and others in the field of Mathematics
- Appreciate the role of Piṅgala in the field of Binary Mathematics

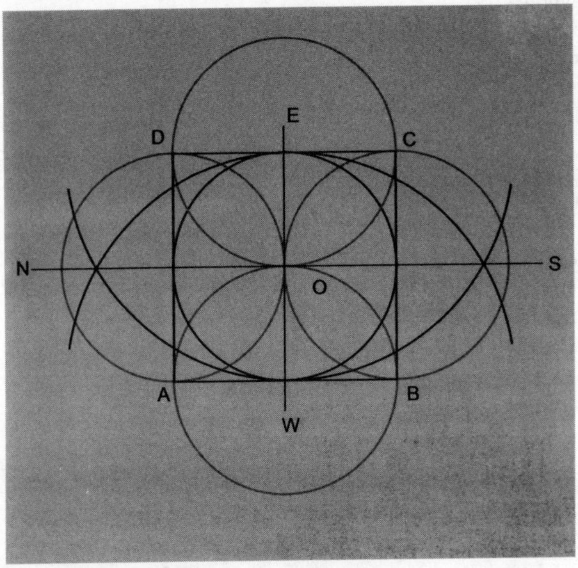

Geometry is an ancient Science in India. Just with a pole anchored on the ground and a thread attached to it, Indians were able to generate complex geometrical shapes. What we see here is a procedure for construction of a square mentioned in Baudhāyana-śulba-sūtra, an ancient mathematical text taught in the Department of Mathematics in some Universities in the West as 'Rope Geometry'.

IKS IN ACTION 8.1

Ancient Indians: Tryst with Mathematics

Mathematics on the Indian subcontinent has a rich and long history going back over 5,000 years and thrived for centuries before advances were made in Europe. Its influence spread to China, Southeast Asia, the Middle East, and Europe. Apart from introducing the concept of zero, Indian mathematicians made seminal contributions to the study of geometry, arithmetic, binary mathematics, the notion of negative numbers, algebra, trigonometry, and calculus among other areas. The decimal place value system that is employed worldwide today was first developed in India.

For a large part, European mathematicians were reluctant to accept negative numbers as meaningful. Many took the view that negative numbers were absurd. This reluctance to adopt negative numbers, and indeed zero, held European mathematics back for many years. Although the reputation of Indian mathematics continues to suffer from the Eurocentric bias. Notwithstanding this, the Indian subcontinent has a strong and continuing mathematical heritage, the pinnacle of its expression being the Mathematical genius in Ramanujan who lived in the 20th Century CE.

Let us look at the genius of the mathematicians in the Vedic age by taking an example. It is a regular feature for the Vedic people to make several offerings to the Gods in a sacrifice. The sacrificial altars were not a standard shape such as a square or a rectangle. There were more than 70 different shapes of altars used in various sacrifices. These include shapes such as tortoise, falcon, and chariot wheels. The construction of these involved several complex shapes including, the isosceles triangle, rhombus, and circle.

The construction complexity had certain other dimensions. For example, there were a fixed number of bricks of certain shapes to be used. The number of layers in the altar was also fixed. Furthermore, there were other constraints such as the area of certain altars to be equal to other altars. Among other things, this is a tough mathematical problem to solve.

The problem is best demonstrated by taking the example Śyena-citi, a falcon-shaped structure. Figure 8.1 is the schematic representation of a Śyena-citi (Falcon bird-shaped altar).

As seen from Figure 8.1, the falcon has five components: the head, the body, the tail, and the two wings. Five differently shaped bricks have been employed to construct this structure. These include a square, a right-angled triangle, an equilateral triangle, and two compound shapes made from a rectangle, triangle, and a trapezium. Table 8.1 has the details on the number of bricks used in the construction of the Śyena-citi.

FIGURE 8.1 The Śyena-citi

As evident from the table there are strict constraints in terms of the number of bricks of each type to be used with respect to each part of the Śyena-citi. Construction of such a shape with these differently shaped bricks by satisfying the constraints is no easy task. It calls for a good understanding of several branches of mathematics including geometry, arithmetic, and algebra.

TABLE 8.1 Type and the Number of Bricks Used

	B1	B2	B3	B4	B5	
Head	1		6	6	1	14
Body	30	6	10			46
Wings	30	62	16			108
Tail	8	4	20			32
Total	69	72	52	6	1	200

Mathematics, referred to as Gaṇita is an integral part of Indians from very ancient times[1]. Ancient Indians developed several concepts of mathematics primarily in the course of solving several real-life problems that they were facing. For example, one of the Vedāṅgas is Jyotiṣa, which confines itself with required mathematical concepts, measurements, and approximation techniques to make certain predictions of the movement of celestial bodies in the sky. Similarly, another Vedāṅga, Kalpa required principles of geometry to construct Vedic altars of many different shapes. Buddhists and Jains were also deeply interested in mathematical concepts and produced canonical works discussing some useful mathematical concepts.

What started as a requirement at an early stage continued through the post-Vedic period as an uninterrupted tradition and many useful contributions to the field of mathematics were made throughout Indian history. Over the years, the contributions spanned almost all modern areas of mathematics. These include algebra, geometry, trigonometry, indeterminate equations, square, cube, and their roots, permutation and combinations, numerical methods and approximations, infinite series, to name a few.

8.1 UNIQUE ASPECTS OF INDIAN MATHEMATICS

Indian mathematics is unique, and it vastly differs from the modern approaches. A few of them are worthy of mention:

(a) There is a popular thinking among many that the world is divided into those who know and love mathematics and those who don't. This separation was primarily because we are educated that mathematics works with left-brain functionalities and literature works with right-brain functionalities. Therefore, the design of the pedagogy and delivery have kept these away in two separate compartments. In contrast, Indian mathematics is a seamless blend of poetry, literature, logic, and mathematical thinking weaved into a single work. All great works of mathematics are invariably great literary works too and they appeal to everyone on account of this natural blend. Therefore, there is no fear or stress in learning Mathematics.

> - Indian Mathematics is a seamless blend of poetry, literature, logic, and mathematical thinking weaved into a single work.
> - The use of sūtras and pithy verses is characteristic in the Indian Mathematical tradition to convey complex ideas and concepts.

(b) Mathematics was considered as a part of life. This is why Mathematics could be found in temple inscriptions, literary work addressing issues of life, and in a discussion on religion or spirituality. Bhāskarācārya in his Līlāvatī, for example, brings interesting mathematical concepts by posing interesting riddles to a student and solving them. Similarly, in Brahmasūtrabhāṣya of Śaṇkarācārya and the Vyāsabhāṣya on Patañjali's Yogasūtras refers to the decimal place value system as an example while discussing a philosophical issue.

(c) The use of sūtras is characteristic in the Indian tradition to convey ideas and concepts. These improve the retention of complex ideas and details very easily. Indian mathematics uses these mechanisms as much as possible. For example, there is a simple sūtra for remembering the pattern that constitutes a binary cycle of length three based on eight groups defined by Piṅgala in his Chandaḥ-śāstra. Āryabhaṭa's uniqueness lies in his ability to use sūtras to bring the utmost simplicity to describe complex information. A shining example is his description of the entire sine table using a verse of just two lines (see the last Endnote in Chapter 1). One who is familiar with Āryabhaṭa's system of numeration will be able to remember the sine table with very little effort.

(d) There has been an uninterrupted tradition of mathematical thinking and it has widely spread across the length and breadth of India. Mathematical concepts were developed by those from Gāndhāra (modern-day Afghanistan) to those in Bengal, as well as by those from Kashmir to Kerala.

(e) One of the important characteristics of Indian Mathematics is its algorithmic approach and the fact that it allows for approximate solutions based on the needs of real-life situations. Indian mathematicians also adopted what is today referred to as the constructive approach where the emphasis is on finding a procedure to solve a problem rather than merely seeking proofs of existence of a solution.

8.2 GREAT MATHEMATICIANS AND THEIR CONTRIBUTIONS

Before we discuss some key mathematical concepts developed by Indians, it will be interesting to have a bird's eye view of some of the great mathematicians and their contributions. The number of mathematicians India has produced, and their contributions are significant. Table 8.2 illustrates the major contributions. A closer look at the table reveals a continuous stream of mathematical theory building. Several areas of modern mathematics have been addressed. These include for example Fibonacci series, Pell's equation, and Pascal's triangle. In this chapter, we shall take a few examples from different areas of mathematics and briefly sketch the contributions made by Indians in the past and illustrate a few methods. This is to provide a curtain riser view of the Indian mathematical tradition and its contribution to the development of science, engineering, and technology.

- Indian Mathematics dealt with almost all areas of modern mathematics for more than 1000 years.
- Mathematical concepts found in Vedic texts, Buddhist and Jain works suggest that this culture is several thousand years old.

8.3 ARITHMETIC

Arithmetic is the branch of mathematics dealing with computation using numbers. Arithmetic is part of the daily life of any society. Commercial operations and trade require handling numbers and basic mathematical operators such as addition, subtraction, multiplication, and division. Indian arithmetic was quite sophisticated by the time of Āryabhaṭa in the 5th century CE. This is primarily because of a fully developed decimal place value system employing 0 to 9. In the 7th century, Brahmagupta established the use of 0 as a number and gave rules for multiplication, division, addition, and subtraction of zero. Indian decimal place value system was well known by the middle of 7th century as evident from the following observation of the Syriac bishop Severus Sebokht, "Indians possess a method of calculation that no word can praise enough. Their rational system of mathematics, or their method of calculation. I mean the system using nine symbols.[2]"

TABLE 8.2 Ancient Indian Mathematicians and Their Salient Contributions

Sl. No.	*Details of the Work/Mathematician*	*Period, Location*	*Salient Contributions*
1	**Vedic Texts**	3000 BCE or earlier	The earliest recorded mathematical knowledge, number system, Pythagorean type triplets; Decimal system of naming numbers, the concept of infinity;
2	**Lagadha** – Vedāṅga-jyotiṣa	~ 1300 BCE	Astronomical concepts; a mathematical model for sun-moon movement in time; equinoxes and solstices;
3	**Śulba-sūtras** (Baudhāyana, Āpasthamba, Kātyāyana and Mānava Śulba-sūtras)	800–600 BCE	Earliest Texts of Geometry; Approximate value of the square root of 2, and π. Exact procedures for the construction and transformations of squares, rectangles, trapezia, etc.
4	**Pāṇini** – *Aṣṭādhyāyī*	500 BCE; Śalātura (in Khyber province in Pakistan)	Algorithmic approaches; Originator of the Backus–Naur Form (BNF), used in programming languages today, Context-sensitive rules, Arrays, inheritance, polymorphism;
5	**Piṅgala** – *Chandaḥ-śāstra*	300 BCE	Binary sequences; Conversion of Binary to Decimal system and vice versa; 'Meru Prastara' (Pascal's triangle); Optimal Algorithms to calculate powers; Zero as a Symbol;
6	**Bauddha Mathematical Works**	about 500 BCE to 500 CE	Multi-valued logic, Discussion of indeterminate and infinite numbers;
7	**Jaina Mathematical Works** – *Sūrya-Prajñapti, Jambūdvīpa-prajñapti, Bhagavatī-sūtra, Sthānāṅga-sūtra, Uttarādhyayana-sūtra, Tiloyapannati, Anuyoga-dvāra-sūtra*	200 BCE to 300 CE	Concept of logarithms, large numbers; algorithms for raising a number by a power; the arc of a circle; combinatorics; mensuration; Decimal system; Approximation of π;
8	**Āryabhaṭa** – *Āryabhaṭīyam*	476–550 CE; Kusumapura, near Pataliputra, Bihar	Concise verses; Algorithm for square root, cube root, Place value system; Sine table; geometry; quadratic equations; Linear indeterminate equations; Sums of squares and cubes of numbers; Planetary astronomy; Plane and spherical trigonometry;
9	**Varāha Mihira** – *Bṛhat Saṃhitā, Bṛhat-jātaka, Pañca-siddhāntikā*	482–565 CE, Ujjain, Madhya Pradesh	Summary of five ancient siddhāntas; Sine table, trigonometric identities; ($\sin^2 + \cos^2$); combinatorics; Magic squares;
10	**Bhāskara I** – *Commentary on Āryabhaṭīya, Laghu-bhāskarīyam and Mahā-bhāskarīyam*	600–680 CE; Vallabhi region, Saurashtra, Gujarat	Expanded Āryabhaṭa's work on Integer solution for indeterminate equations; Approximate formula for the sine function, Planetary Astronomy;

(*Contd.*)

TABLE 8.2 Ancient Indian Mathematicians and Their Salient Contributions (*Contd.*)

Sl. No.	*Details of the Work/Mathematician*	*Period, Location*	*Salient Contributions*
11	**Brahmagupta** - *Brāhmasphuṭa-siddhānta; Khaṇḍakhādyaka*	598–668 CE; Bhillamala in Rajasthan	Rules of arithmetic operations with zero and negative numbers, Algebra (Bījagaṇita); linear and quadratic indeterminate equations; Pythagorean triplets, Formula for the diagonals and area of a cyclic quadrilateral; notion of arithmetic mean.
12	**Virahāṅka** - *Vṛttajātisamuccaya (in Prākṛt)*	~600 CE	Fibonacci numbers; Moric metres.
13	**Śrīdharācārya** - *Triśatikā and Pāṭīgaṇita*	870–930 CE; Bhūrisṛṣṭi or Bhurshut village, Hugli, West Bengal	Arithmetic, Algebra, and Commercial Mathematics; Approximation of square root of a non-square number; Quadratic equations; Practical applications of algebra.
14	**Mahāvīrācārya** - *Gaṇita-sāra-saṅgraha*	800–870 CE; Gulbarga Karnataka;	A comprehensive, exclusive textbook on mathematics covering arithmetic-geometry- algebra. Continuing the ancient Jaina mathematics tradition; permutations and combinations; arithmetic and geometric series; the sum of squares and cubes of numbers in arithmetic progression.
15	**Jayadeva**	10th Century CE or earlier	Cakravāla (cyclic) method for solution of the second-order indeterminate equation.
16	**Shrīpati** - *Gaṇita-tilaka, Siddhānta-śekhara, Dhikoṭidākaraṇa, etc.*	1019–1066 CE; Rohiṇīkhaṇḍa, Maharashtra	Planetary Astronomy
17	**Bhāskarācārya (Bhaskara-II)** - *Līlāvatī on arithmetic and geometry; Bījagaṇita on algebra; Siddhānta-śiromaṇi on astronomy; Vāsanābhāṣya on Siddhānta-śiromaṇi.*	1114–1185 CE; Hailed from Bijjadavīda	Canonical textbooks used all over India, Detailed explanations including Upapatti (demonstration or proof); addition formula for sine function. Surds; permutations, and combinations; Solution of indeterminate equations, Ideas of calculus, including mean value theorem, planetary astronomy; construction of several instruments.
18	**Nārāyaṇa Paṇḍita** - *Gaṇita-kaumudī – a treatise on arithmetic and Bījagaṇita Vataṁśa – a treatise on algebra.*	1325–1400 CE;	Advanced textbooks taking forward the works of Bhāskarācārya, further properties of cyclical quadrilaterals, summation and repeated summations of arithmetic series, theory and construction of Magic squares, further developments in combinatorics.

Sl. No.	Details of the Work/Mathematician	Period, Location	Salient Contributions
19	**Mādhava of Saṅgamagrāma**	1340–1425 CE; Sangama Grama, in Kerala	Founder of Kerala School of Mathematics – a pioneer in the development of calculus; Infinite series and approximations for π, Infinite series and approximations for cosine and sine functions.
20	**Parameśvara,** – *Dṛgganita, Siddhāntadīpikā;* Commentaries on Āryabhaṭīyam, Mahā-bhāskarīya; Laghu-bhāskarīya, Līlāvatī, and Sūryasiddhānta	1360–1460 CE; Alathiyur, (near Tirur), Kerala	Properties of Cyclic quadrilateral; iterative techniques.
21	**Nīlakaṇṭha Somayājī,** *Tantra-saṅgraha; Āryabhaṭīya-bhāṣya, Siddhānta-darpaṇa*	1444–1544 CE; Near Tirur, Kerala	Irrationality of π, basic ideas of calculus; revised planetary theory, which is a close approximation to Kepler's model; Exact results in spherical astronomy.
22	**Jyeṣṭhadeva** – *Yukti-bhāṣā*	1500–1575 CE; Kerala	Hailed as the first textbook of Calculus; detailed explanations and proofs of the infinite series given by Mādhava.
23	**Śaṅkaravāriyar** – *Kriyākramakarī commentary on Līlāvātī and commentary on Tantra-saṅgraha*	1500–1569 CE Kerala	Explanations and Proofs of the results and procedures given in Līlāvatī.
24	**Gaṇeśa Daivajña** – *Buddhi-vilāsinī (commentary on Līlāvatī);*	1504 CE; Nandi Grama, Nadod, Gujarat	Explanations and Proofs of the results and procedures given in Līlāvatī.
25	**Kṛṣṇa Daivajña** – *Bījapallva-Commentary on Bījagaṇita of Bhāskarācārya*	1600 CE Delhi	Explanations and Proofs of results and procedures given in Bījagaṇita.
26	**Munīśvara** – *Siddhānta-sārvabhauma, commentary on Līlāvatī; Pātisāra;*	17th Century CE; Varanasi	Explanations and Proofs of the results and procedures given in Līlāvatī; trigonometric identities.
27	**Kamalākara** – *Siddhānta-tattva-viveka*	1616–1700 CE; Varanasi, Uttara Pradesh	Addition and subtraction theorems for the sine and the cosine; Sines and cosines of double, triple, etc., angles.

8.3.1 Square of a Number

Āryabhaṭa identifies the geometric aspect and the mathematical operation of the square in his definition. As per his definition[3] varga is a square and is also a geometric object whose sides are equal. Bhāskara–I (629 CE) in his commentary on Āryabhaṭīya provides an algorithm for finding the square of any number. The verse that explains the algorithm is given below:

- An algorithm for computation of cube root was given for the first time by Āryabhaṭa.
- Brāhma-sphuṭa-Siddhānta, gives a good description of calculations with positive and negative numbers, Zero, fractions and surds.

अन्त्यपदस्य वर्गं कृत्वा द्विगुणं तदेव चान्त्यपदम् ।
शेषपदैराहन्यात् उत्सार्योत्सार्य वर्गविधौ ।।

antyapadasya vargaṃ kṛtvā dviguṇaṃ tadeva cāntyapadam |
śeṣapadairāhanyāt utsāryotsārya vargavidhau ||

The steps of the algorithm as provided by the above verse can be enumerated as follows:

Step 1: अन्त्यपदस्य वर्गं कृत्वा (*antyapadasya vargaṃ kṛtvā*).

Square the last digit (most significant digit) first. Place it in a new row (two places to the right of digits in the previous row).

Step 2: द्विगुणं तदेव चान्त्यपदम् (*dviguṇaṃ tadeva cāntyapadam*).

Multiply the last digit with two and each of the remaining digits to the right and place them to the right in the same row.

Step 3: शेषपदैराहन्यात् उत्सार्योत्सार्य (*śeṣapadairāhanyāt utsāryotsārya*).

Remove the current most significant digit.

If there are no more digits to operate go to step 5.

Step 4: The next digit becomes the last digit now. Go to step 1.

Step 5: Perform the final addition to get the square of the number.

It may be noted that this ancient rule for squaring, uses $\frac{n(n-1)}{2}$ multiplications for squaring an n-digit number.

EXAMPLE 8.1: Find the square of the number 1638.

Final Digit Position	7	6	5	4	3	2	1
Row 1: Last digit = 1 Multiplication with other digits Digit '1' is removed	$(1^2) = 1$	(2*1*6) = 12	(2*1*3) = 6	(2*1*8) = 16			
Row 2: Last digit = 6 Multiplication with other digits Digit '6' is removed			$(6^2) = 36$	(2*6*3) = 36	(2*6*8) = 96		
Row 3: Last digit = 3 Multiplication with other digits Digit '3' is removed					$(3^2) = 9$	(2*3*8) = 48	
Row 4: Last digit = 8 Digit '8' is removed No more digits available							$(8^2) = 64$

Final Addition of the numbers:

Digit Position						
7	**6**	**5**	**4**	**3**	**2**	**1**
1						
1	2					
		6				
		1	6			
	3	6				
		3	6			
			9	6		
				9		
				4	8	
					6	4
2	6	8	3	0	4	4

The answer is 26,83,044.

8.3.2 Square Root

Āryabhaṭa presents an algorithm for determining the square root of a number. The methodology of finding the square root revolves around the concept of splitting the digits into pairs, starting from the least significant digit, the odd place being designated as varga (V), and the even place as avarga (A). For example, if we take two numbers 55225 and 205209 the split will look as follows:

V	A	V	A	V		A	V	A	V	A	V
5	5	2	2	5		2	0	5	2	0	9

Āryabhaṭa then provides a simple recursive algorithm that starts from the most significant varga place and progressively introduces one digit at a time until all digits are exhausted. The algorithm for square root is provided in the following verse:

भागं हरेत् अवर्गात् नित्यं द्विगुणेन वर्गमूलेन ।
वर्गाद्वर्गे शुद्धे लब्धं स्थानान्तरे मूलम् ।।
bhāgaṃ haret avargāt nityaṃ dviguṇena vargamūlena |
vargādvarge śuddhe labdhaṃ sthānāntare mūlam ||

The above verse has indicated the following steps to the algorithm:

1. Designate the varga and avarga digits starting from right to left (least significant to most significant).
2. Take the first varga sthāna at the leftmost (most significant) along with an avarga digit at its left (if any).
3. Remove (subtract) the maximum possible square from this number, and the square root of the square that we can remove will be added to the square root line (this is a place where we are accumulating the answer as we perform the operation).
4. Along with the remainder of the previous operation, bring the next digit down. The next digit is avarga digit (as we have just completed the varga digit related operation). भागं हरेत् अवर्गात् नित्यं द्विगुणेन वर्गमूलेन (*bhāgaṃ haret avargāt nityaṃ dviguṇena vargamūlena*)

   Whenever we operate at avarga digit, we need to divide the number by two times the current value of the square root that we have stored in the square root line. The quotient obtained in this division will be added to the square root line.
5. Along with the remainder of the previous operation, we will bring the next digit down. The next digit is varga digit (as we have just completed the avarga digit related operation).
   वर्गाद्वर्गे शुद्धे (*vargādvarge śuddhe*)

   Whenever we operate at a varga digit, we need to remove the square of the quotient obtained in the previous step.
6. If some more digits are remaining go to step 4, else go to step 7.
7. लब्धं स्थानान्तरे मूलम् (*labdhaṃ sthānāntare mūlam*).
   The final result in the square root line is the answer.

As evident from the above algorithm, this procedure runs in a recursive fashion bringing down one digit at a time. When the procedure stops after exhausting all the digits the final value accumulated in the square root line is the answer. If the number is not a perfect square this procedure aborts mid-way or leaves a remainder at the end of the operation.

EXAMPLE 8.2: Find the square root of the number 19,881.

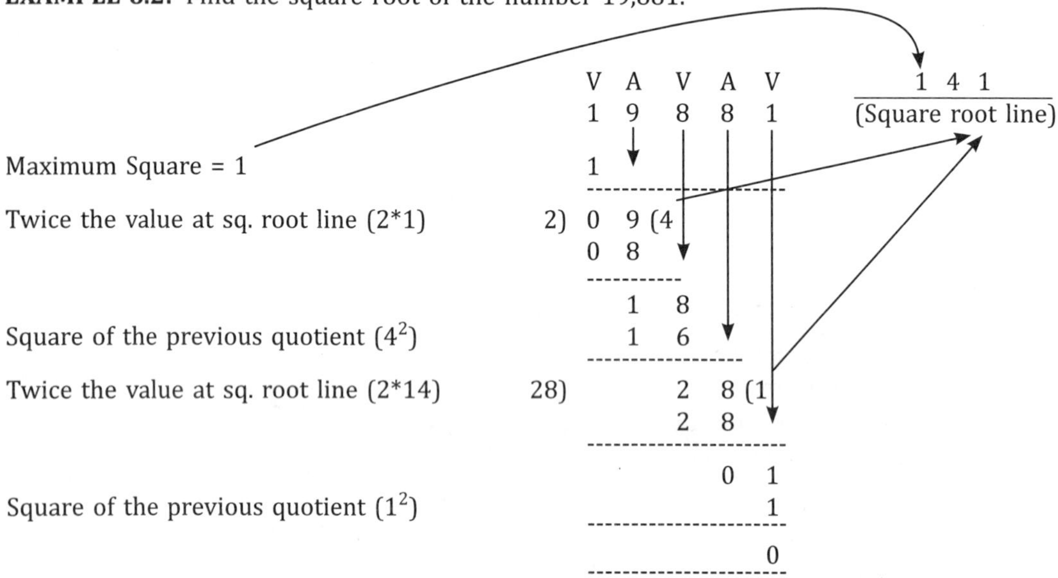

The final answer = 141.

Cube Root

Āryabhaṭa was the first to establish a procedure to calculate the cube root (*Ghana-mūla*) of a number. Āryabhaṭa provided an algorithm for determining the cube root of a number that is strikingly similar to that of the algorithm for square root. If a number has '*n*' digits, the number of digits of the cube of that number will be $\geq 3n - 2$ and $\leq 3n$. With this in mind, Āryabhaṭa designates the digits in a number using three different nomenclatures. Āryabhaṭa then provides a simple recursive algorithm that starts from the most significant place and progressively introduces one digit at a time until all digits are exhausted. An illustration of this is available in Shukla and Sarma's critical edition of Āryabhaṭīya[4].

8.3.3 Series and Progressions

A sense of the knowledge of progressions of ancient Indians is evident in the Vedic text. For example, in the *Camaka-praśna* (Taittirīya-saṃhitā 4.5.11) there is a mention of an arithmetic progression of odd numbers starting from one to 33 followed by another arithmetic progression of even numbers starting from four to 48 in steps of four[5]. In the *Vājasaneyī-saṃhitā* we have the *yugma* (even) and *ayugma* (odd) series: 4, 8, 12, 16, 48 and 1, 3, 5, 7, ... Similarly, there is a geometric series (12, 24, 48, 96, 196608, ..., 393216) mentioned in the *Pañcaviṃśa-brāhmaṇa*. This indicates an appreciation of odd and even numbers as distinct entities in the Vedic period and also the formation of different types of series. Āryabhaṭa I (499 CE), and Brahmagupta (628 CE) considered the cases of the sums of the sums, the squares and the cubes of the natural numbers. Mahāvīra (850 CE) gave a rule for the summation of an interesting geometric series. Nārāyaṇa (1356 CE) provided a more generalised method for repeated summation of partial series. We shall see some of them in this section[6].

IKS IN ACTION 8.2

Square Root of Imperfect Squares

Several sources in the ancient Indian texts point to multiple attempts and methods to obtain the square root of an imperfect square. Let us look at some of them here.

Śulba-sūtra Formula for $\sqrt{2}$

The square root of two has been explored as early as the Vedic times (Śulba-sūtra). According to Bodhāyana-śulba-sūtra (BSS 2.12), the value of the square root of two can be obtained using the following sūtra:

प्रमाणं तृतीयेन वर्धयेत् तच्चतुर्थेन आत्मचतुस्त्रिंशेनोनेन सविशेषः ।

pramāṇaṃ tṛtīyena vardhayet taccaturthena ātmacatustriṃśenonena saviśeṣaḥ |

प्रमाणं तृतीयेन वर्धयेत् *(pramāṇaṃ tṛtīyena vardhayet) essentially meaning, add 1/3 to 1;* तच्चतुर्थेन *(taccaturthena), meaning add one fourth of this* $\frac{1}{4}*\frac{1}{3}$*, however,* आत्मचतुस्त्रिंशेनोनेन *(ātma-catustriṃśenonena) take out* $\frac{1}{34}$ *from this.* सविशेषः *(saviśeṣaḥ)* essentially denotes it is a special case (meaning an approximate number). Using this sūtra,

The value of $\sqrt{2} \approx 1+\frac{1}{3}+\frac{1}{3\times 4}\left(1-\frac{1}{34}\right)$

$$=\frac{577}{408} \approx 1.4142156863$$

Bakshali Method

Bakshali manuscripts (estimated to have been written sometime during 300–600 CE) discusses several unique mathematical issues, including that of finding the square root of an imperfect square. Any imperfect square N may be expressed as $\sqrt{A^2+b}$. According to Bakshali Manuscript, the square root of the number N may be expressed as:

$$\sqrt{N}=\sqrt{A^2+b} \approx A+\frac{b}{2A}-\frac{\left(\frac{b}{2A}\right)^2}{2\left(A+\frac{b}{2A}\right)}$$

Approximations to the Square Root of a Non-square Number

Śrīdhara (850 CE) in his Triśatikā, has explained how Āryabhaṭa method can be used to get better approximations to the square root of a non-square number. For instance, if D is a non-square number, we can use the expression below to calculate $\sqrt{D}$ to any desired accuracy:

$$\sqrt{D}=\frac{\sqrt{D*10^{2n}}}{10^n}$$

Source: Based on the NPTEL Video on "Mathematics in India – From Vedic period to modern times. https://nptel.ac.in/courses/111/101/111101080/. Last accessed on Oct. 1, 2021.

The term *upaciti* or *citi* is used to indicate a series in general. For example, the series $1 + 2 + 3 + \cdots + n$, which starts with '1' and has a common difference of '1' is called एकोत्तराद्युपचिति (*ekottarādi-upaciti*). Let us consider the case of the following the series of natural numbers up to n terms: 1, (1 + 2), (1 + 2 + 3), …, $(1 + 2 + 3 + \cdots + n)$. The term चितिघनः(*citighana*) is used to denote the sum of this series. Āryabhaṭa gives the formula for computing the sum in the following verse:

एकोत्तराद्युपचितेर्गच्छाद्यकोत्तरत्रिसंवर्गः ।
षड्भक्तः स चितिघनः सैकपदघनो विमूलो वा ॥ २१ ॥
ekottarādy-upaciter-gacchādyakottara-trisaṃvargaḥ |
ṣaḍbhaktaḥ sa citighanaḥ saikapada-ghano vimūlo vā || Gaṇita-pāda 21 ||

Of the series (*upaciti*) which begins with the term '1' and has a common difference '1', (*ekottarādi*), take three terms in continuation, of which the first is equal to the given number of terms, and find their continued product (*gacchādyakottara-trisaṃvargaḥ*). That (product) divided by 6 gives the citighana (*ṣaḍbhaktaḥ sa citighanaḥ*). Alternatively (*vā*), it can be obtained by the number of terms plus one subtracted from the cube of that (*saikapada-ghano vimūlo*), divided by 6 (*ṣaḍbhaktaḥ*). This can be expressed in the following notation:

$$1+(1+2)+(1+2+3)+\cdots+(1+2+3+\cdots+n)=\frac{n(n+1)(n+2)}{6} \text{ or } \frac{(n+1)^3-(n+1)}{6}$$

The use of the term citighana is indeed interesting. It literally means the contents of a pile (of say balls) in the shape of a pyramid on a triangular base. The equivalent physical interpretation of the above series is that the pyramid is so constructed that there is 1 ball in the topmost layer, 1 + 2 balls in the next lower layer, 1 + 2 + 3 balls in the further next lower layer, and so on.

Sum of the Series of Squares and Cubes

Let us now consider the summation of the series of squares and the summation of the series of cubes, of the first n natural numbers indicated as ΣN^2 and ΣN^3 respectively.

$$\Sigma N^2 = 1^2 + 2^2 + 3^2 + \cdots + n^2$$

and

$$\Sigma N^3 = 1^3 + 2^3 + 3^3 + \cdots + n^3$$

Āryabhaṭa designated ΣN^2 as वर्गचितिघनः (*varga-citighanaḥ*) and ΣN^3 as घनचितिघनः (*ghana-citighanaḥ*). The formulae for the summation of the series of squares and cubes is provided in the following verse:

सैक-सगच्छ-पदानां क्रमात् त्रिसंवर्गितस्य षष्ठोंऽशः ।
वर्गचितिघनः स भवेत् चितिवर्गो घनचितिघनश्च ॥ २२ ॥
saika-sagaccha-padānāṃ kramāt trisaṃvargitasya ṣaṣṭhoṃ'śaḥ |
varga-citighanaḥ sa bhavet, citivargo ghana-citighanaśca || Gaṇita-pāda 22 ||

The product of the three quantities (*trisaṃvargitasya*), viz., the number of terms (n), the number of terms plus one ($n+1$), and the same increased by the number of terms ($n+1+n$), (*saika-sagaccha-padānāṃ kramāt*), when divided by 6 (*ṣaṣṭhoṃ'śaḥ*) gives the sum of the series of squares of natural numbers (*varga-citighanaḥ sa bhavet*). The square of the sum of the series of natural numbers (*citivargo ghana*) gives the sum of the series of cubes of natural numbers (*ghana-citighanaśca*).

According to the above verse,

$$\Sigma N^2 = 1^2 + 2^2 + 3^2 + \cdots + n^2 = \frac{n(n+1)(2n+1)}{6}$$

and

$$\Sigma N^3 = 1^3 + 2^3 + 3^3 + \cdots + n^3 = \left(\frac{n(n+1)}{2}\right)^2$$

Repeated Summation of Series (Vārasaṅkalita)

In the previous case we saw the *saṅkalita* (sum of a series) for the first n natural numbers. A more general case of the repeated summation of such series was provided by Nārāyaṇa. Vāra means repeated (or again). Therefore, this method of repeated summation of a series can be designated as *vārasaṅkalita*.

Let the symbol ${}^{n}V_1$ denote the arithmetic series of the first 'n' natural numbers. This can be expressed as: ${}^{n}V_1 = 1 + 2 + 3 + \cdots + n$. Similarly, let ${}^{n}V_2$ denote the series of the partial sums of the series ${}^{n}V_1$. If we use 'r' to denote the number of terms up to which we want to sum up, then we can create partial sums of ${}^{n}V_1$. For example,

For $n = 1$, ${}^{1}V_1 = 1$; $n = 2$, ${}^{2}V_1 = 1 + 2$; $n = 3$, ${}^{3}V_1 = 1 + 2 + 3$; and so on. Using this we can write the second order sum of partial series as:

$${}^{n}V_2 = \sum_{r=1}^{r=n} {}^{r}V_1$$

$${}^{n}V_2 = 1 + (1 + 2) + (1 + 2 + 3) + \cdots + (1 + 2 + 3 + \cdots + n)$$

Similarly one can write the other sums of the partial series for the higher order in the following fashion:

$${}^{n}V_3 = \sum_{r=1}^{r=n} {}^{r}V_2;$$

In general, the *m*th order sum of partial series of number can be represented as:

$${}^{n}V_m = \sum_{r=1}^{r=n} {}^{r}V_{m-1}$$

Nārāyaṇa provided a formula for the computation of the *m*th order *vāra-saṅkalita*. It denotes in this case the operation of forming a new series by taking the sums of the previous series. He provided an expression to calculate the sum using the following verse:

एकाधिकवारमिताः पदादिरूपोत्तरा पृथक् तेंऽशाः ।
एकाद्येकचयहरास्तद्घातो वारसङ्कलितम् ॥
ekādhika-vāramitāḥ padādi-rūpottarā pṛthak teṃ'śā: |
ekādy-ekacayaharās-tadghāto vārasaṅkalitam ||

- Āryabhaṭīya gives a good indication of the knowledge of algorithmic approach and use of recursive algorithms for problem-solving.
- The birth of Indian geometry could be traced to the Vedic time.

The terms of the sequence beginning with the pada (number of terms, i.e. *n*) and increasing by 1 (पदादिरूपोत्तरा:) taken up to *(m+1)* times (*ekādhika-vāramitāḥ*) are successively the numerators (*pṛthak teṃ'śā:*) and the terms of the sequence beginning with unity and increasing by 1 (*ekādy-ekacayaharāḥ*) are respectively the denominators. The continued products of these (fractions) (*tadghātaḥ*) gives the vāra-saṅkalita (*vāra-saṅkalitam*). According to the above, since *n* is the number of terms and *m* the order, we get the following sequence of fractions:

$$\frac{n}{1}, \frac{n+1}{2}, \frac{n+2}{3}, \ldots, \frac{n+m}{m+1}$$

and the sum of the series is the product of this sequence of fractions, given by:

$${}^{n}V_m = \frac{n\,(n+1)(n+2)\ldots(n+m)}{1.2.3\ldots(m+1)}$$

This expression is the most generalised form for summation of series. Substituting $m = 1$ above will fetch us the sum of the first 'n' natural numbers. Substituting $m = 2$ in the above equation will give us one of the two formulae of Āryabhaṭa that we derived in the previous section.

8.4 GEOMETRY

As we saw in Chapter 2, Yajñas formed a very important part of Vedic life. The performance of Vedic rituals involved the construction of a variety of Vedic altars (*yajña-vedīs*) as per certain specifications. The Brāhmaṇa portion of the Yajurveda contains details about the arrangement of the sacrificial ground and the construction of the altars. For this purpose, one first fixes cardinal directions and then goes on to construct altars of different shapes and dimensions using prefabricated bricks. Śulba-sūtras give exact methods for the construction of such altars. On account of these, the birth of Indian geometry can clearly be traced to the Vedic times.

Śulba-sūtras are a part of the Vedāṅga. These sūtras give information on the methods of the layout of the vedī, citi, and the maṇḍapa. Śulba means a thread and with the help of a thread and a pin (or a pole), construction methods for various shapes have been described in the śulba-sūtras. The four śulba-sūtras attributed to Bodhāyana, Manu, Āpastambha, and Kātyāyana are the predominant ones. We also have śulba-sūtras due to Maitrāyaṇa, Varāha, and Vādhūla. The oldest of the śulba-sūtras attributed to Bodhāyana is estimated to have been written prior to 800 BCE. Śulba-sūtras are manuals for the construction of Vedic altars. Therefore, Śulba-sūtras also discuss the manufacturing of bricks, thereby indicating a good knowledge of the materials and manufacturing processes which seems to have prevailed in those times.

8.4.1 Property of Right-angled Triangle in Śulba-sūtras

In current mathematical texts, we are taught an important theorem being attributed to Pythagoras (570–495 BCE). However, the first general statements of this so-called Pythagoras theorem is actually found in the Śulba-sūtras. In the Baudhāyana-śulba-sūtra (prior to 800 BCE), the theorem is stated in the following form (this result is known by name *Bhuja-koṭi-karṇa-nyāya*): "The sum of the areas of the squares formed by the length and breadth of a rectangle equals the area produced by the diagonal of the rectangle"[7]. Let us consider a rectangle of length a and breadth b. Let the diagonal of the rectangle be c. According to the Baudhāyana formula, $a^2 + b^2 = c^2$. This is graphically illustrated in Figure 8.2.

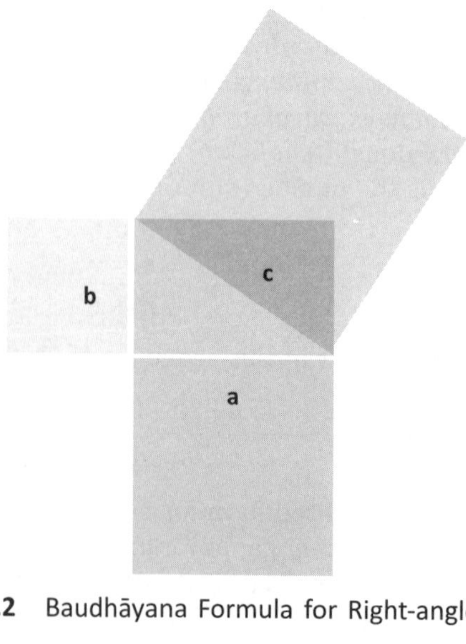

FIGURE 8.2 Baudhāyana Formula for Right-angled Triangle

IKS IN ACTION 8.3

Fun and Practicality in Indian Mathematics

Indian mathematical works have several interesting problems of day-to-day importance solved using mathematical principles. Further, it is described as a game in poetic verses. Āryabhaṭīya, Līlāvatī and other Indian mathematical works describe several such entertaining and interesting problems. Let us see an example.

Ancient Indians used *śaṅku*, a gnomon, as a device for measurement to fix the direction and coordinates, to measure the length scale and host of such applications. One verse poses a problem related to gnomonic shadow.

शङ्कुगुणं शङ्कुभुजाविवरं शङ्कुभुजयोर्विशेषहृतम् ।
यल्लब्धं सा छाया ज्ञेया शङ्कोः स्वमूलाद्धि ॥
śaṅku-guṇaṃ śaṅku-bhujāvivaraṃ śaṅku-bhujayor-viśeṣahṛtam |
yal-labdhaṃ sā chāyā jñeyā śaṅkoḥ sva-mūlāddhi ||

According to this verse, the height of the gnomon multiplied by the distance between gnomon and the lamp post is to be divided by the difference between the lamp post and the gnomon. The quotient thus obtained, should be known as the length of the shadow measured from the foot of the gnomon. This can be better understood graphically using in Figure 8.3.

In Figure 8.4, there is a lamp post (denoted by AC) and a śaṅku (denoted by EF). The objective is to find the length of the shadow cast by the śaṅku on account of the lamp post.

Triangles FEB and CDF are similar.

Therefore, $EB = \frac{EF * DF}{DC} = \frac{EF * AE}{AC - EF}$

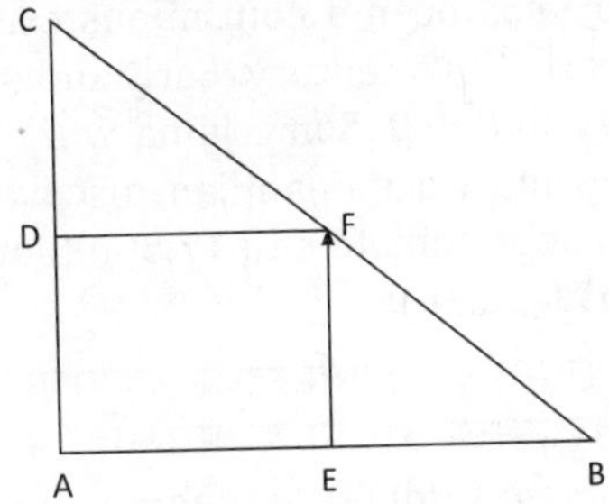

FIGURE 8.3 The Shadow Problem – An Illustration

Such formulations are used in the design and construction of temples when the architect has to design for the sunlight to fall on the shrine. Bhāskara defines an inverse problem and provides the solution, i.e., how to calculate the height of the lamp post knowing the height of śaṅku and the length of the shadow. The same principle is applied to several astronomical calculations including calculation of eclipses as shown in Figure 8.5.

Considering half symmetry, of sun, earth, moon along the horizontal axis in the orbital system, and getting the equivalence of the triangle ABC above, we have:

EF is the half diameter of the earth (śaṅku)

AC is the half diameter of the sun (lamp post)

AE is the distance between sun and earth

EB is the earth's shadow (to be determined)

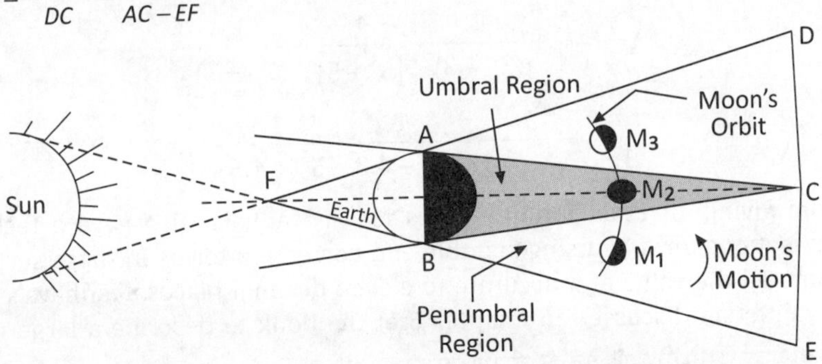

FIGURE 8.4 Shadow Problem – An Astronomy Example

8.4.2 The Value of π

Right from the Vedic period, Indians understood that the ratio of the area of a circle to its perimeter was a constant and that this value can only be approximated. What is interesting is that there has been a continuous evolution of the approximation of π from the Vedic period till the early 20th century. Early Indian works took the value of π as 3. Śulba-sūtras estimated π as close to 3.088. Some Jaina works approximated π as the square root of 10. In early 20th century Srinivasa Ramanujan discovered an infinite series which was later used during the 1980s to approximate π to 17.5 million places. Āryabhaṭīyam discusses π in the following verse (Gaṇitapāda, 2.10):

चतुरधिकं शतमष्टगुणं द्वाषष्टिस्तथा सहस्राणाम् ।
अयुतद्वयविष्कम्भस्य आसन्नो वृत्तपरिणाहः ॥

caturadhikaṃ śatam-aṣṭaguṇaṃ dvāṣaṣṭis-tathā sahasrāṇām /
ayuta-dvaya-viṣkambhasya āsanno vṛttapariṇāhaḥ //

According to the verse, the circumference of a circle is eight times (100 + 4) added to 62,000 (*caturadhikaṃ śatam-aṣṭaguṇaṃ-dvāṣaṣṭis-tathā-sahasrāṇām*). The diameter of the circle is 20,000 (*ayuta-dvaya).* Since the ratio of the circumference to the diameter yields the value of π, we can compute π to be:

$$\pi = \frac{(100+4)\times 8+62000}{20000} = \frac{62832}{20000} = 3.1416$$

Bhāskarācārya in his work Līlāvatī, (verse 199), derives the value of π using a different approach and arrives at the same value.[8] According to this,

$$\pi = \frac{3927}{1250} = 3.1416$$

In the 14th century CE, the celebrated astronomer mathematician Mādhava discovered several infinite series for π which were later re-discovered in Europe by James Gregory (1671), Gottfried Leibniz (1674), and Abraham Sharp (1699). Some of Mādhava's series are given below:

$$\frac{\pi}{4} = 1 - \frac{1}{3} + \frac{1}{5} - \frac{1}{7} + \cdots$$

$$\frac{\pi}{\sqrt{12}} = 1 - \frac{1}{3.3} + \frac{1}{3^2 \times 5} - \frac{1}{3^3 \times 7} + \cdots$$

$$\frac{\pi}{4} = \frac{3}{4} + \frac{1}{(3^3-3)} - \frac{1}{\left(5^3-5\right)} + \frac{1}{(7^3-7)} - \cdots$$

$$\frac{\pi}{16} = \frac{1}{1^5+4.1} - \frac{1}{3^5+4.3} + \frac{1}{5^5+4.5} - \cdots$$

Apart from giving an exact infinite series for π, Mādhava also gave a technique finding better approximations for π by using suitable end correction terms in these series. In this way Mādhava estimated the value of π accurate to eleven decimal places. Mādhava's verse uses the bhūta-saṃkhyā system discussed in Chapter 6 of the book to describe a large number:

विबुध-नेत्र-गज-अहि-हुताशन-त्रि-गुण-वेद-भ-वारण-बाहवः ।
नव-निखर्व-मिते वृत्तिविस्तरे परिधिमानमिदं जगदुर्बुधाः ॥

vibudha-netra-gaja-ahi-hutāśana-tri-guṇa-veda-bha-vāraṇa-bāhavaḥ /
nava-nikharva-mite vṛttivistare paridhimānam-idaṃ jagadurbudhāḥ //

In this verse, vibudha means devas, who are 33 in number. Using the bhūta-saṃkhyā system discussed in Chapter 6 of the book, we can compute π as follows:

$$\pi = \frac{2827433388233}{9*10^{11}} = 3.141592653592...$$

The following Table 8.3 summarises the different approximations to value of π that were discovered in the Indian tradition right from the Vedic times to the time of Ramanujan.

TABLE 8.3 History of Approximations to π by Indian Mathematicians

	Value of π	*Accuracy (Decimal places)*	*Method*
Śulba-sūtras (around 800 BCE)	3.08888	1	Geometrical
Jaina texts (500 BCE)	$\sqrt{10}$ =3.1623	1	Geometrical
Āryabhaṭa (499 CE)	$\frac{62832}{20000}$ = 3.1416	4	Polygon doubling (4.2^8 = 1024 sides)
Bhāskarācārya (Līlāvatī)	3927/1250 = 3.1416	4	Polygon doubling
Mādhava (1375 CE)	$\frac{2827433388233}{9*10^{11}}$ = 3.141592653592...	11	Infinite series with end corrections
Ramanujan (1914 CE)		17 million	Modular equation

8.5 TRIGONOMETRY

Trigonometry is called *jyotpatti,* the science of computation of chords in Indian mathematics. Consider a circle of radius R as shown in Figure 8.5. DA = $R\theta$, is an arc. AB = $R \sin \theta$, is called the jyā corresponding to the arc, $R\theta$. Earlier it was called jyārdha or 'half a bow-string', but later it was just called 'jyā'. R is taken to be $\frac{21600}{2\pi} \approx 3438$, where the circumference of the circle is 21,600 units (number of 'minutes' in a radian). OB = $R \cos \theta$ is called the koṭijyā or kojyā or cojyā. Jīvā was another name for jyā. When this was transmitted to Arab countries,

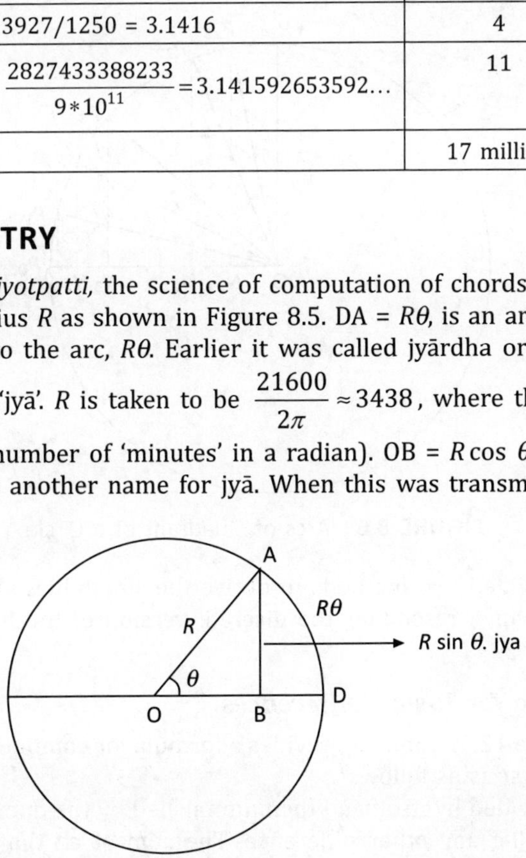

FIGURE 8.5 The Jyā and Cojyā in Indian Mathematics

jīvā became jība, which was read as 'jayb' in Arabic. In common parlance, 'jayb' was 'pocket' or 'fold' in Arabic. When many mathematical texts in Arabic were translated into Latin, 'jayb' got translated as 'Sinus', which means 'pocket' or 'fold' in Latin. This eventually became 'sine'. Since the complement of jyā is known as kojyā, the complement of sine became 'cosine' in Europe.

Computation of the 'R sines'

R sine values are required if we need to compute any quantity which involves the sine and cosine functions. Let us consider the first quadrant in a circle, since the values of the *R* sines and *R* cosines in other quadrants can be related in a simple manner to the values in the first quadrant. One way to obtain these values is to divide the quadrant into certain parts and determine *R* sine values for these. Once these values are available, one can compute the other values of the angles by interpolation. Let us consider one quadrant of a circle as shown in Figure 8.6. Let this quadrant be divided into 24 parts (P_0, P_1, P_2, ..., P_{24}). Each arc measures $\frac{90}{24} = 3°45'$. In the figure, we are interested in the chord lengths P_iN_i, which is $R \sin i\theta$. If we see the triangle P_2N_2O, the chord length P_2N_2 is the sine of the right-angle triangle of angle θ_2. For any chord in between these 24 chords, 1st order or 2nd order interpolation is used to get the intermediate values.

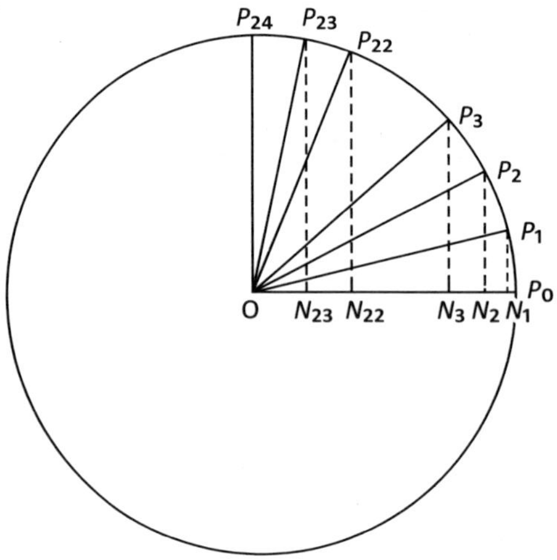

FIGURE 8.6 Arcs of Quadrant of a Circle

Āryabhaṭa has provided two methods to derive the sine tables. A geometric method, and an analytical method which resembles the discrete version of the harmonic equation as we know today.

Āryabhaṭa's Formula for R sine Differences

In the Gaṇita-pāda verse 12, Āryabhaṭa provides a formula for computing the *r* sine differences. The meaning of the verse is as follows[9]:

"The first *R* sine divided by itself and then diminished by the quotient will give the second difference. For computing any other difference. The sum of all the preceding differences is divided by the first *R* sine and the quotient is subtracted from the preceding difference. Thus, all the remaining differences (can be calculated)."

Let R_1, R_2, R_3,..., R_{24}, denote the 24 R sines and δ_1 (=R_1), δ_2, δ_3, δ_4, ..., δ_{24} denote the 24 R sine-differences. Then, according to the above verse,

$$\delta_2 = R_1 - \frac{R_1}{R_1}$$

$$\delta_{n+1} = \delta_n - \frac{\delta_1 + \delta_2 + \delta_3 + \cdots + \delta_n}{R_1} = \delta_n - \frac{R_n}{R_1}$$

Nīlakaṇṭha's Formula for R sine Differences

Nīlakaṇṭha (1500 CE) provided a much accurate estimate of R sine differences by providing a correction to the above formula. According to him, "the first R sine divided by itself and then diminished by the quotient gives the second R sine difference. To obtain the other R sine differences, divide the preceding R sine by the first R sine and multiply the quotient by the difference between the first and second R sine differences and subtract the resulting product form the preceding R sine difference". It is stated mathematically below:

$$\delta_{n+1} = \delta_n - \left(\frac{R_n}{R_1}\right)(\delta_1 - \delta_2)$$

This is the same as the relation

$$2 \sin nx - \sin\{(n+1)x\} - \sin\{(n-1)x\} = (\sin nx/\sin x)\ (2 \sin x - \sin 2x)$$

This approach to computing R sine difference was extraordinary to be conceived at the time of Āryabhaṭa. French mathematician and astronomer Jean Baptiste Joseph Delambre, (1749–1822) paid rich tributes to Āryabhaṭa's analytical method, "the method is curious: it indicates a method of calculating the table of sines using their second differences ..., the differential process has not up to now been employed except for Briggs, who himself did not know that the constant factor was the square of the chord. Here then is a method, the Indians possessed, and which is found neither among Greeks not amongst the Arabs[10]".

8.6 ALGEBRA

Algebra became the core of mathematics in the 20th century, which many historians termed as 'algebraization of mathematics'. Algebra provided simplicity, clarity, and precision, to the mathematicians. Algebra is one of the main areas of contribution by ancient Indians as they considered it as a subject of great utility. The ancient Indian name for the science of algebra is bīja-gaṇita. Bīja means 'element' or 'analysis' and gaṇita 'the science of calculation.'

The science of algebra is broadly divided by the Indian mathematicians into two major parts. Of these, one deals with analysis (bīja). The other part discusses concepts that are essential for analysis. It includes the laws of signs, the arithmetic of zero (and infinity), operations with unknowns, surds, and the linear indeterminate equation (known as kuṭṭaka or pulveriser). Therefore, one is able to identify several important features of algebra in the Indian mathematical works. For instance, symbols were used for unknowns by Brahmagupta (it is also indicated in Āryabhaṭīya and its commentary by Bhāskara). Operations with negative numbers were also introduced by Brahmagupta. Furthermore, we can notice that linear and quadratic equations were solved by Āryabhaṭa and Brahmagupta. The Indian mathematicians recognised few fundamental operations in algebra, viz, addition, subtraction, multiplication, division, squaring and the extraction of the square-root.

Recognition of negative numbers and their treatment in mathematical operations were evident very early. Brahmagupta (628 CE) says: "The sum of two positive numbers is positive, of two negative numbers is negative; of a positive and a negative number is their difference." The use of symbols to denote unknowns, working on equations to solve unknown were the key contributions of ancient Indians. Indians discovered the usage of fundamental arithmetic operators *'yu'* for *yuta* – addition; *'kṣa'* for *kṣaya* – subtraction; *'gu'* for *guṇa* – multiplication; *'bhā'* for *bhāga* – division. They used *'mū'* or *'ka'* for *mūla* or *karaṇi* denoting root. They used the first letter of the names of different colors to denote different unknown variables. With these, Indians analysed and classified equations – called *samīkaraṇa.*

Āryabhaṭa in his work *Āryabhaṭīya (Gaṇitapāda,* Verse 24), discusses an interesting case of how to determine the two numbers whose product and difference are known[11]. The procedure explained in the verse can be algebraically stated as follows. Let us consider two numbers x and y. Let $x - y = a$ and $xy = b$. Then as per the verse, x and y can be determined as follows:

$$x = \frac{\sqrt{4b+a^2}+a}{2};\ y = \frac{\sqrt{4b+a^2}-a}{2}$$

8.7 BINARY MATHEMATICS AND COMBINATORIAL PROBLEMS IN CHANDAḤ-ŚĀSTRA OF PIṄGALA (300 BCE)

We saw the work of Piṅgala known as Chandaḥ-śāstra in Chapter 2 and Chapter 6. In this section, we shall study his contributions in the field of binary mathematics and combinatorial problems. The metres in Sanskrit (chandas) were analysed by means of two kinds of syllables, laghu (L), and guru (G) from a prosody perspective. Any syllabic metre (varṇa-vṛtta) is therefore characterized as a sequence of L and G. If we replace laghu (L) with '1' and guru (G) with '0', then it transforms the metrical pattern into a binary sequence. A binary sequence is a sequence composed of '1's or '0's. For example, 1,0,0,1 and 0,1,0,0 or examples of binary sequences of length 4. In Chapter 8 of Chandaḥ-śāstra, Piṅgala has analysed a host of combinatorial problems relating to metrical patterns or, equivalently, binary sequences. These are as follows:

- *Prastāra:* A procedure by which all possible metrical patterns or, equivalently, binary sequences of a given length are generated sequentially as an array (prastāra).
- *Saṃkhyā:* The process of finding the total number of binary sequences (rows) in the prastāra (array).
- *Naṣṭa:* For a given row number in an array, the process of identifying the corresponding binary sequence directly.
- *Uddiṣṭa:* Given a binary sequence, the process of identifying the corresponding row number in the array (prastāra), directly.
- *Lagakriyā:* The process of finding the number of binary sequences in the array with a given number of '1's or '0's.
- *Adhvayoga:* The process of finding the space occupied by the array (prastāra) (to determine the floor area needed).

These concepts introduced by Piṅgala around 300 BCE are perhaps the earliest instance of computations involving binary numbers. The basic building block of this binary analysis is the generation of the prastāras. Sūtras 20–23 in Chapter 8 of Chandaḥ-śāstra provide the details for the generation of the binary arrays[12]. The method indicated in the sūtras is as follows:

To begin with, the array of binary sequences of length one is a table of two rows (see Figure 8.7), row 1 = 0, row 2 = 1. Every new array of increasing length is generated using a simple procedure. This is demonstrated in Figure 8.7. At every iteration, the existing array is replicated, followed by adding one more column to the replicated array. In the first half of the new column '0' is inserted and in the second half, '1'. The procedure can be repeated in this manner to generate array of binary sequences of length. As seen in Figure 8.7, in the second array, column 1 in rows 1 to 2 and rows 3 to 4 are merely the repeat of the one-syllable array of length 1. In the second column '0' and '1'. are repeated. Similarly, in the third array, the first two columns in rows 1 to 4 and rows 5 to 8 are the repeat of the array of sequences of length two. In the last array, the first three columns in rows 1 to 8 and rows 9 to 16 are the repeat of the array of sequences of length three. One can proceed in this manner to generate arrays of binary sequences of any length in a recursive fashion.

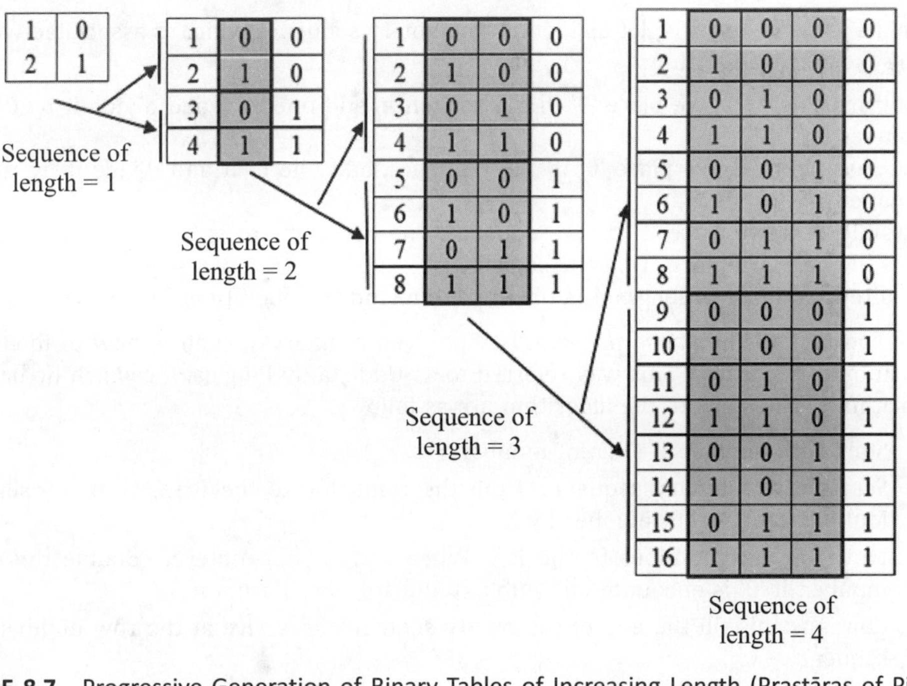

FIGURE 8.7 Progressive Generation of Binary Tables of Increasing Length (Prastāras of Piṅgala)

As we can see, the number of binary sequences of length n is 2^n. This is called the Saṃkhyā or the total number of rows in the array (prastāra). We can also see that there is a relation between a binary sequence and the row in which it occurs in the array. This relation is the following:

Row-number of a binary sequence = Mirror image of the binary sequence (viewed as a binary number) +1

For example, the fifth row in the array of sequences of length 3 is 001. The mirror image of this is 100. This, when viewed as binary number, is: $2^2.1 + 2.0 + 1.0 = 4$; and $4 + 1 = 5$.

Similarly, the 14th row of array of sequences of length 4 is 1011. The mirror image of this is 1101. This, when viewed as a binary number, is: $2^3 .1 + 2^2.1 + 2.0 + 1.1 = 13$; and $13 + 1 = 14$.

We shall now describe the methods given by Piṅgala to identify the binary sequence that is associated with a particular row number in a given array and vice-versa. Piṅgala called these processes *Naṣṭa and Uddiṣṭa.*

Let us consider the array of length 4 (the last array in Figure 8.7). There are some interesting mathematical problems that one encounters. For example, given a row-number, how can we identify the binary sequence associated with that row in the array. Piṅgala provides a simple algorithm to identify the binary sequence, known as Naṣṭa[13]. The steps are as follows:

- Start with the desired row number.
- Divide it by 2. If it is perfectly divisible, place '1' in the sequence.
- If the number is not divisible by two, place '0' in the sequence, add one to the number and divide.
- Repeat the process until the binary sequence of required length is obtained.

For the array of length 4, let us identify the binary sequence which is associated with row 13 using the above algorithm:

13 is not divisible by 2, we place '0' in the sequence, add one to it and divide it by two. The new number is '7'.
7 is not divisible by 2, we place '0' in the sequence, add one to it and divide it by two. The new number is '4'.
4 is divisible by 2, we place '1' in the sequence. The new number is 2.
2 is divisible by 2, we place '1' in the sequence.
The required sequence of length 4 is obtained. We stop the operation. The answer is '0011'.

The converse of the above problem is that given a binary sequence, how to identify the row position in the array? This was referred to as Uddiṣṭa by Piṅgala, for which he has given an algorithm[14]. The steps of the algorithm are as follows:

1. Start with number 1. Current number = 1.
2. Scanning the binary sequence from the right, locate the first '1' in the sequence. Multiply the current number by 2.
3. Move the next number to the left. When a '1' is encountered, double the current number. If '0' is encountered double it and subtract 1 from it.
4. Continue this till the end of the binary sequence to arrive at the row number of the sequence.

Let us consider the following sequence from the array of length 4–'0111'. Applying the above algorithm, we get the following:

Start with the current number = 1.
The first number is '1'. Therefore, the current number is $1 \times 2 = 2$.
The second and third numbers are also '1'. So, we double the earlier result twice. $2 \times 2 \times 2 = 8$.
The last number is '0'. So, we double the last result and then subtract 1 from it. $8 \times 2 - 1 = 15$.
Therefore, the row position is 15.

EXAMPLE 8.4: Find the binary sequence associated with 37th array of length of 6.

Using the Naṣta algorithm of Piṅgala, we obtain the sequence as follows:

37 is not divisible by 2, we place '0' in the sequence, add one to it and divide it by two. The new number is '19'.
19 is not divisible by 2, we place '0' in the sequence, add one to it and divide it by two. The new number is '10'.
10 is divisible by 2, we place '1' in the sequence. The new number is '5'.
5 is not divisible by 2, we place "0" in the sequence, add one to it and divide it by two. The new number is '3'.
3 is not divisible by 2, we place '0' in the sequence, add one to it and divide it by two. The new number is '2'.
2 is divisible by 2, we place '1' in the sequence.

The required binary sequence of length 6 is obtained. We stop the operation. The answer is '001001'.

EXAMPLE 8.5: Identify the row number of the binary sequence '1010100' array of length 7.

We use the Uddiṣṭa algorithm to arrive at the position in the array.
We start with number 1.
The first letter '1' is found in the third position from the left. Therefore, the number is 1 × 2 = 2.
The next letter is '0'. So, we double the last result and then subtract 1 from it. 2 × 2 - 1 = 3.
The next letter is '1'. We double the last result. The number is 3 × 2 = 6.
The next letter is '0'. We double the last result and then subtract 1 from it. 2 × 6 – 1 = 11.
The last number is '1'. We double the last result. 11 × 2 = 22.

Therefore, the row position is 22.

Another interesting problem is to find out how many binary sequences of length '*n*' there are, that contain '*r*' number of '1's. Essentially, this boils down to the combinatorial problem of choosing '*r* out of *n*', which leads to the binomial co-efficient ^nC_r. Piṅgala refers to this problem as lagakriyā. The procedure is best explained by the "Varṇa-Meru" of Piṅgala. Figure 8.8 has a schematic representation of Varṇa-Meru. The construction follows a simple algorithm. Start with a single square. Place '1' in the square. Below this are successive rows with one more number of square compared to the previous row in the top. In each square, numbers have to be placed as per the following rule:

The number to be placed inside any square is the sum of the numbers contiguous to this square in the row above. Continue this procedure to construct the triangle shown in Figure 8.8. For the two squares in the second row, there is only one square above. Therefore, we place the number 1 in each of the squares. In the third row, the second square is contiguous to both the squares above. Therefore, we place the number 1 + 1 = 2. In the same way we get 1,3,3,1 in the fourth row. In this manner we may proceed to construct the Varṇa-Meru shown in Figure 8.8. The same triangle of binomial coefficients was re-discovered by the French mathematician Blaise Pascal (1655 CE). It is therefore generally referred to as Pascal's triangle in modern books.

1

1 1

1 2 1

1 3 3 1

1 4 6 4 1

1 5 10 10 5 1

FIGURE 8.8 Varṇa Meru (due to Piṅgala)

Thus, we can see that Piṅgala laid a solid foundation for combinatorial theory and binary mathematics, which was followed by later-day Mathematicians in the country. *Varāhamihira's Bṛhatsaṃhitā* (550 CE), mentions 1820 different combinations that can be obtained by choosing 4 perfumes from a set of 16 basic perfumes. ($^{16}C_4$ = 1820). Varāhamihira also discusses the construction of a *Meru* - tabular form for arriving at this binomial coefficient.

8.8 MAGIC SQUARES IN INDIA

Mathematics can be utilitarian and at the same time a tool for human curiosity, fun, and a matter for mind-expanding exercise. One such example is the magic square problems in arithmetic. An array of numbers containing an equal number of rows and columns is called a magic square when the summation of the numbers in every column and those in every row and each diagonal happens to be the same. On the other hand, if the summation of the 'other' diagonals also is the same, then such a square is called a pan-diagonal magic square. Figure 8.9 is a simple illustration of the magic squares. In the first square [Figure 8.9(a)] the row sums, column sums, and the sum of the main diagonals are 34. However, the other diagonals do not add up to 34. For example, the shaded diagonal (6, 5, 7, and 8) does not add up to 34. Please note that the matrix structure is to be imagined like a torus (a rolled version like a cylinder). Therefore, after 7, the next number in the diagonal will be 8. On the other hand, in Figure 8.9(b), all the diagonals add up to 34. This is a pan-diagonal magic square.

12	3	**6**	13
14	**5**	4	11
7	16	9	2
1	10	15	**8**

(a) Magic square

10	3	**13**	8
5	**16**	2	11
4	9	7	14
15	6	12	**1**

(b) Pan-diagonal magic square

FIGURE 8.9 Magic and Pan-diagonal Magic Square; Magic Sum = 34

The construction of such magic squares has been known in India from very early times. Indian mathematicians specialized in the construction of a special class of magic squares called *sarvatobhadra* or *pan-diagonal magic squares*. Bhadra-gaṇita is the name for the study of magic squares in the Indian mathematical tradition. Various 3 × 3 magic squares are attributed to the ancient astronomer Garga (100 BCE). The work Kakṣapuṭa of Nāgārjuna (100 CE) describes a method for constructing 4 × 4 magic squares, which gives pan-diagonal squares when the magic sum is even. Magic squares have been used to address certain combinatorial issues involving arithmetic. For example, Varāramihira (587 CE) made use of a 4 × 4 magic square to specify varying proportions of four ingredients that can be mixed to prepare different perfumes. 4 × 4 magic squares are found on the gates of buildings, on the walls where shopkeepers transact their business, and on the covers of calendars used by astrologers even to this day. A 4 × 4 square occurs in a Jaina inscription of the 11th century CE, found in the ancient town of Khajuraho. A similar square dated to 1480 CE was found in the Gwalior fort. The first chapter of the notebooks of Srinivasa Ramanujan is on magic squares.

Construction of a 4 × 4 Pan-diagonal Magic Square

Perhaps the easiest magic square to understand is the one presented by Nāgārjuna in his work Kakṣapuṭa, where he provided a mnemonic and a formula to construct a generic magic square. The mnemonic is as follows: "अर्क इन्दुनिधा नारी तेन लग्न विनासनम्" (*arka indunidhā nārī tena lagna vināsanam*). There are 16 syllables in this mnemonic. Using the katapayādi formula (see Chapter 6 for details), the above mnemonic can be converted into 16 numbers. All vowels are zero in kaṭapyādi. Therefore, in the mnemonic, the first number is zero. Populating the 4 × 4 magic square with this number gives us the first template. See Figure 8.10(a) for details. Let the desired sum of the magic square be $2n$. Then using the formula given in Figure 8.10(b) the individual cells of the magic square can be arrived at. For example, if the desired sum is 100, then n = 50. The magic square for this is available in Figure 8.10(c).

0	1	0	8
0	9	0	2
6	0	3	0
4	0	7	0

(a)

n-3	1	n-6	8
n-7	9	n-4	2
6	n-8	3	n-1
4	n-2	7	n-9

(b)

47	1	44	8
43	9	46	2
6	42	3	49
4	48	7	41

(c)

FIGURE 8.10 Nāgārjuna's Scheme for Generating a 4 × 4 Pan-diagonal Magic Square with an Illustration for n = 50

The mathematics involved in the construction of magic squares and other magic figures was first systematically and elaborately discussed by the mathematician Nārāyaṇa (1356 CE) in his Ganiṭakaumudī. Some of his methods were unknown in the west and were recently discovered by the efforts of several scholars. The last chapter of the seminal work Gaṇitakaumudī is devoted to Bhadragaṇita, which contains 55 verses giving rules and 17 verses giving examples. Consider an nth order magic square. Nārāyaṇa classifies the magic squares into three categories:

(a) if $n = 4m$, then it is called *Samagarbha* (double even squares)

(b) if $n = 4m + 2$, then it is *Viṣamagarbha,* and (semi-double even squares)

(c) if $n = 2m + 1$, then it is *Viṣama* (odd squares), where $m = 1, 2, 3,...$

After classifying the squares into samagarbha, viṣamagarbha, and viṣama, Nārāyaṇa discusses a generalized method to construct the magic square for the desired sum, S and the size of the square, n. The n^2 numbers to be filled in an $n \times n$ magic square follow an arithmetic progression with an initial number a, and the common difference d. Nārāyaṇa presented a general method for filling up a 4 × 4 pan-diagonal magic square by an arithmetic sequence using a succession of *turaga-gati* (horse movements)[15]. Positioning 1 at the top left corner he generated 24 such squares and concluded that in all there are 384 such pan-diagonal squares. In the 20th century a few mathematicians studied the 4 × 4 pan-diagonal magic squares and also concluded that only 384 such pan-diagonal squares are possible. Further, using some more analysis they came up with certain properties of the 4 × 4 pan-diagonal magic squares. These properties are useful for constructing a 4 × 4 pan-diagonal magic square.

	3		
5	16	2	
	9		
			1

(a)

10	3	13	
5	16	2	
4	9	7	
			1

(b)

10	3	13	8
5	16	2	
4	9	7	14
15		12	1

(c)

10	3	13	8
5	16	2	11
4	9	7	14
15	6	12	1

(d)

FIGURE 8.11 Construction of a 4 x 4 Pan-diagonal Magic Square

Property 1: In a pan-diagonal 4 × 4 magic, the entries of any 2 × 2 sub-square formed by consecutive rows and columns add up to the magic.

Property 2: In a pan-diagonal 4 × 4 magic square the sum of an entry with another which is two squares away from it along a diagonal is always half the magic sum.

Property 3: In a 4 × 4 magic square with entries, 1, 2, ..., 16, each element has the same set of neighbours in each of the 384 pan-diagonal versions. In particular, the element 16 has as neighbours 2, 3, 5, 9.

Using these properties, we can construct a 4 × 4 pan-diagonal matrix using numbers 1, 2, 3, ..., 16 as follows:

Step 1: First we place 1 in any of the cells and place 16 two cells diagonally away from 1.

Step 2: Using property 3, we generate the first few elements of the 4 × 4 pan-diagonal magic square by placing 2, 3, 5, and 9 in any order as neighbours of 16 (Figure 8.11(a)).

Step 3: The magic sum for a 4 × 4 pan-diagonal magic square for numbers 1 to 16 is 34. We use the above properties to fill all the remaining cells in the magic square. For example,

(a) Using property 1 we fill in some of the adjacent cells as shown in Figure 8.11(b).

(b) Using property 2 to fill in some more cells (highlighted cells in Figure 8.11(c)).

(c) Finally using property 1, we fill the balance cells (highlighted cells in Figure 8.11(d)).

SUMMARY

- Ancient Indians developed several concepts of mathematics primarily because they needed to solve a lot of real-life problems that they were facing even during the Vedic period.
- Mathematical concepts were developed by those living from Gāndhāra (modern-day Afghanistan) to those in Bengal, as well as by those from Kashmir to Kerala.
- Brahmagupta, in his work Brahma-sphuṭa-siddhānta, gives a good description of working with fractions, calculations with positive, negative numbers, and with zero.
- Several sources in the ancient Indian texts point to multiple attempts and methods to obtain the square root of an imperfect square.
- *Āryabhaṭīya* gives a good indication that ancient Indians had a well-developed algorithmic approach to problem-solving and were able to utilise recursive algorithms.
- Śulba-sūtras, a section of the Kalpa part of the Vedāṅga, have dealt with the subject of geometry in detail.
- Āryabhaṭīya, Līlāvatī and other Indian mathematical works have several interesting problems of day-to-day importance described as a game in poetic verses.
- Right from the Vedic period, there has been a continuous evolution of the approximation of π till the 20th century. Mādhava estimates π using an infinite series with end correction, which is accurate to eleven decimal places.
- Āryabhaṭa has provided two methods to derive the sine tables. A geometric method, and an analytical method which is very unique and not found in works of any other mathematician, till about 15th century.
- In Chapter 8 of Chandaḥ-śāstra, Piṅgala has analysed a host of problems related to handling binary sequences. The concepts developed during 200–300 BCE are relevant for the modern-day computations involving binary numbers.
- The construction of magic squares has been known in India from very early times. Bhadra-gaṇita is the name for the study of magic squares in the Indian mathematical tradition.

REVIEW QUESTIONS

1. Enumerate the unique aspects of Indian Mathematics. Can you illustrate them with some examples?
2. Name three great Indian mathematicians and enumerate their key contributions.
3. Identify three major areas of contribution of Ancient Indian mathematicians.
4. What key inferences can we draw by an examination of the key contributions of Indian mathematicians?

5. Why was knowledge of geometry important for the ancient Indians?
6. Briefly explain about the knowledge of π that Indian Mathematicians possessed. How does it compare with that of other Mathematicians?
7. What was the motivation for Indians to study the right-angled triangle and other geometric shapes?
8. Comment on the statement, "Indian Mathematics is a blend of poetry, literature, and mathematics". Give examples to support your views.
9. What are the key contributions of Piṅgala to modern mathematical thinking?
10. What is the difference between a magic square and a pan-diagonal matrix square?
11. Construct a 4 × 4 pan-diagonal magic square. What is the magic sum for the square that you constructed?

EXERCISE PROBLEMS

1. Calculate the square of the following numbers using the Āryabhaṭa method:
   (a) 149 (b) 2347 (c) 642 (d) 369 (e) 1777
2. Find the varga and avarga sthānas of the following numbers:
   (a) 17,342 (b) 1,23,456 (c) 69,900,342
3. Verify the answers that you obtained in question (1) by deriving the square root of the answers using the Āryabhaṭa method.
4. Find the square root of the following numbers using the Āryabhaṭa method:
   (a) 21,609 (b) 2,85,156 (c) 56,644 (d) 9,27,369 (e) 31,329
5. Verify the answers that you obtained in question (4) by finding the square of the answers using the Āryabhaṭa method.
6. Use the Bhakshali method to find the square root of the following numbers:
   (a) 126 (b) 912 (c) 8,174 (d) 21,924 (e) 83,369
   (*Hint*: Find the nearest perfect square of the number to identify A and b).
7. Construct a table of five-letter binary word (Piṅgala's prastāra).
   (*Hint*: Use the four-letter binary table to construct the five-letter binary table).
8. Consider a five-letter binary table and identify the pattern for the following rows using the naṣṭa algorithm of Piṅgala:
   (a) Row 7 (b) Row 16 (c) Row 25 (d) Row 31
9. For a five-letter binary table identify the row number corresponding to each of the binary word using Piṅgala's uddiṣṭa algorithm:
   (a) 10101 (b) 00101 (c) 10010 (d) 11011 (e) 01010

DISCOVER IKS

1. Indian's contribution to the field of Mathematics is not known as not much effort has been made to bring it to the attention of interested people. Go through the video available at the following site to further reinforce your understanding of the contribution of Ancient Indians in the area of Mathematics: https://www.youtube.com/watch?v=huJDNh0G3kw. After watching the video, develop a three-page note to answer the following questions:
   (a) What are the key contributions of Indians in the area of Mathematics?
   (b) Identify two Mathematicians and their key contributions?
   (c) How do the Indian Mathematicians' contributions compare with that of the Western counterpart?

2. Bhāskarācārya II, one of the greatest Mathematicians of India produced great mathematical works including mathematics, number system, and astronomy. Watch the video on Bhāskarācārya II available in the following site: https://www.youtube.com/watch?v=WoJGJeOyLEc. After watching the video, develop a three-page note to answer the following questions:
   (a) Briefly sketch the biographical sketch of Bhāskarācārya II. When did he compose the Līlāvatī?
   (b) What are the major works of Bhāskarācārya II? Explain what the issues are covered in his works.
   (c) How did Bhāskarācārya II deploy unique real-life examples to introduce interesting mathematical concepts and problems? Give some examples.

SUGGESTED READINGS

Datta, B. and Singh, A.N. (1962). *History of Hindu Mathematics: Parts I and II*, Asia Publishing House, Mumbai.

Datta, B. (1932). *Ancient Hindu Geometry: The Science of Sulbas*, Calcutta University Press, Reprinted Cosmos Pug., New Delhi, 1993.

Divakaran, P.P. (2018). *The Mathematics of India: Concepts, Methods, Connections*. Springer (Hindustan Book Agency), New Delhi.

Joseph, G.G. (1990). *The Crest of the Peacock: Non-European Roots of Mathematics*, Penguin, London.

Joseph, G.G. (2009). *A Passage to Infinity: Medieval Indian Mathematics from Keralà and Its impact*. Sage, New Delhi.

Joseph, G.G. (2016). *Indian Mathematics Engaging with the World from Ancient to Modern Times*, World Scientific, London.

Kedarnath (1938). *Chandaḥ śāstra of Piṅgala with Commentary Mṛtasañjīvanī of Halāyudha Bhaṭṭa*, 3rd ed., Mumbayyāṃ: Nirṇayasāgarakhyamudraṇāye, Mumbai.

Kolachana, A., Mahesh K. and Ramasubramanian, K. (2019). "Hindu mathematics in the seventh century as found in Bhāskara I's commentary on the *Āryabhaṭīya* (IV)", In: Kolachana A., Mahesh K., Ramasubramanian K. (Eds), Studies in Indian Mathematics and Astronomy, *Sources and Studies in the History of Mathematics and Physical Sciences*, Springer, Singapore.

Madhusudana Sastri (1994). *Vṛttaratnākara of Kedāra with Commentaries Nārāyaṇī and Setu*", Chaukhamba, Varanasi.

Parameswaran, S. (1998). *The Golden Age of Indian Mathematics*, Swadeshi Science Movement, Kerala.

Plofker, K. (1963). *Mathematics in India*, Princeton University Press, New Jersey, USA.

Ramasubramanaian, K. (2015). "Gaṇitānada": Selected works of Radha Charan Gupta on History of Mathematics, *Indian Society for History of Mathematics*, New Delhi.

Sarasvati Amma, T.A. (1999). *Geometry in Ancient and Medieval India*, Motilal Banarsidass, 2nd ed., New Delhi.

Sen, S.N. and Bag, A.K. (1983). "The Sulbasutras", INSA, New Delhi.

Shukla, K.S. and Sharma, K.V. (1976). "Āryabhaṭīya of Āryabhaṭa", Indian National Science Academy, New Delhi.

Sridharan, R. (2005). "Sanskrit Prosody: Piṅgala Śāstra and binary arithmetic", in G.G. Emch et al. (Eds.), *Contributions to the History of Indian Mathematics*, Hindustan Book Agency, Delhi, pp. 33–62.

Srinivasiengar, C.N. (1988). *The History of Ancient Indian Mathematics*. The World Press, Kolkata.

Van Nooten, B. (1993). "Binary numbers in Indian antiquity", *Journal of Indian Philosophy*, **21**, pp. 31–50.

ENDNOTES

1. We gratefully acknowledge the support provided by Professors K Ramasubramanian, M D Srinivas and M S Sriram in helping us develop this chapter. We have drawn several examples and ideas from their NPTEL Lecture series on "Mathematics in India: From Vedic period to Modern times". For more details on the NPTEL Lecture series, visit https://nptel.ac.in/courses/111/101/111101080/. Last accessed on Oct. 1, 2021.
2. For more details see, https://en.wikipedia.org/wiki/Arithmetic. Last accessed on Oct. 1, 2021.
3. वर्गः समचतुरश्रः फलं च सदृशद्वयस्य संवर्गः। vargaḥ samacaturaśraḥ phalaṃ ca sadṛśadvayasya saṃvargaḥ |
4. For more details see, **Shukla, K.S. and Sharma, K.V. (1976).** "Āryabhaṭīya of Āryabhaṭa", Indian National Science Academy, pp. 37–38.
5. एका च मे तिस्रश्च मे पञ्च च मे सप्त च मे नव च म एकादश च मे त्रयोदश च मे पञ्चदश च मे सप्तदश च मे नवदश च म एकविंशतिश्च मे त्रयोविंशतिश्च मे पञ्चविंशतिश्च मे सप्तविंशतिश्च मे नवविंशतिश्च म एकत्रिंशच्च मे त्रयस्त्रिंशच्च मे चतस्रश्च मेऽष्टौ च मे द्वादश च मे षोडश च मे विंशतिश्च मे चतुर्विंशतिश्च मेऽष्टाविंशतिश्च मे द्वात्रिंशच्च मे षट्त्रिंशच्च मे चत्वारिंशच्च मे चतुश्चत्वारिंशच्च मेऽष्टाचत्वारिंशच्च मे ... ekā ca me tisraśca me pañca ca me sapta ca me nava ca ma ekādaśa ca me trayodaśa ca me pañcadaśa ca me saptadaśa ca me navadaśa ca ma ekaviṃśatiśca me trayoviṃśatiśca me pañcaviṃśatiśca me saptaviṃśatiśca me navaviṃśatiśca ma ekatriṃśacca me trayastriṃśacca me catasraśca me'ṣṭau ca me dvādaśa ca me ṣoḍaśa ca me viṃśatiśca me caturviṃśatiśca me'ṣṭāviṃśatiśca me dvātriṃśacca me ṣaṭriṃśacca me catvāriṃśacca me catuścatvāriṃśacca me'ṣṭācatvāriṃśacca me ... **R.L. Kashyap (2005).** "Yajur Veda Taitirīya Samhita" Sri Aurobindo Kapāli Sāstry Institute of Vedic Culture, Banglore, Volume 3, Kāṇḍa 4, pp. 448.
6. For a discussion on this, see, **Dutta, B. and Singh, A.N. (1993).** "Use of Series in India: Revised by Kripa Shankar Shukla" *Indian Journal of History of Science,* **28**(2), pp. 103–129.
7. दीर्घचतुरश्रस्याक्ष्णया रज्जुः पार्श्वमानी तिर्यङ्मानी च यत् पृथग्भूते कुरुतस्तदुभयं करोति dīrghacaturaśrasyākṣṇayā rajjuḥ pārśvamānī tiryaṅmānī ca yat pṛthagbhūte kurutastadubhayaṃ karoti | Baudhāyana Śulba Sūtra (1.12).
8. व्यासे भनन्दाग्निहते विभक्ते खबाणसूर्यैः परिधिः सुसूक्ष्मः । द्वाविंशतिघ्ने विहृतेऽथ शैलैः स्थूलोऽथवा स्याद् व्यवहारयोग्यः ॥

   vyāse bhanandāgnihate vibhakte khabāṇasūryaiḥ paridhiḥ susūkṣmaḥ | dvāviṃśatighne vihṛte'tha śailaiḥ sthūlo'thavā syād vyavahārayogyaḥ || **Padmanabha Rao, A.B. (2014).** "Bhāskarācārya's Līlāvatī" Part–II, Chinamaya International Foundation Shodha Sansthan Ernakulam, Kerala, India, Verse 199.
9. प्रथमाच्चापज्यार्धाद्यैरूनं खण्डितं द्वितीयार्धम् । तत्प्रथमज्यार्धांशैस्तैस्तैरूनानि शेषाणि ॥

   prathamāccāpajyārdhādyairūnaṃ khaṇḍitaṃ dvitīyārdham | tatprathamajyārdhāṃśaistaistairūnāni śeṣāṇi || (Gaṇita-pāda, Verse 12).
10. **Datta, B. and Singh, A.N. (1983).** "Hindu Trigonometry", *Indian Journal of History of Science,* **18**, p. 77.
11. द्विकृतिगुणात् संवर्गाद् द्व्यन्तरवर्गेण संयुतान्मूलम् । अन्तरयुक्तं हीनं तद् गुणकारद्वयं दलितम् ॥

    dvikṛtiguṇāt saṃvargād dvyantaravargeṇa saṃyutānmūlam | antarayuktaṃ hīnaṃ tad guṇakāradvayaṃ dalitam || For details see, **Sharma, K.V. (1976).** "Āryabhaṭīya of Āryabhaṭa", Indian National Science Academy, New Delhi, Verse 24.
12. द्विकौ ग्लौ । मिश्रौ च । पृथग्लामिश्राः । वसवस्त्रिकाः । dvikau glau | miśrau ca | pṛthaglāmiśrāḥ | vasavastrikāḥ | Chandaḥ Śāstra 8.20–8.23. See for details, **Paṇḍit Kedārnāth and Vāsudev Laxmaṇ Shāstrī Paṇaśikar (1927).** *Chandaḥ Shāstra*, Nirṇaya Sāgar Press, Mumbai.
13. *Ibid.* लर्धे । सैके ग् । lardhe | saike g | 8.24–8.25.
14. *Ibid.* प्रतिलोमगणं द्विर्लाद्यम् । ततोग्येकं जह्यात् । pratilomagaṇaṃ dvirlādyam | tatogyekaṃjahyāt | Chandas Śāstra, 8.26–8.27.
15. For a discussion on the magic squares, see for example, **Dutta, B. and Singh, A.N. (1976).** "Magic Squares in India: revised by Shukla K.S.", *Indian Journal of History of Science,* **27**(1), pp. 51–120.

CHAPTER

9 Astronomy

LEARNING OUTCOMES

After finishing this chapter, you will be able to:

- Develop some awareness about the contributions made by several Indian Astronomers
- Develop a basic understanding of the celestial coordinate system
- Understand the basic elements of the Indian calendar
- Develop a basic understanding of the components of Indian Pañcāṅga
- Become aware of the instruments used for study of astronomy in India

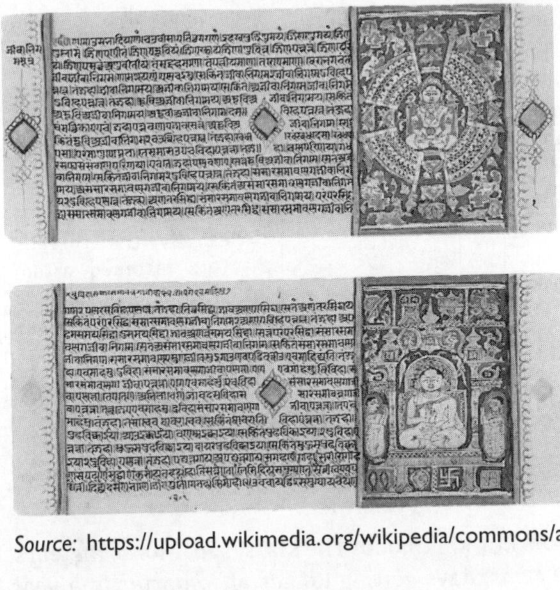

This picture is a manuscript of Sūryaprajñapti-sūtra in Jain prākṛt on paper found in Western India dating to 1500 CE. It is a Śvetāmbara Jain text belonging to 3rd–4th BCE and is a part of the Jain canonical literature called Upāṅga Āgamas. It is an astronomical work that gives information on the sun, moon and planets and their motions.

Source: https://upload.wikimedia.org/wikipedia/commons/a/a3/Suryaprajnapati_Sutra.jpg

IKS IN ACTION 9.1

Astronomical Gifts from the King of Benares to the Prince of Wales in 1876

The Prince of Wales toured India during 1875–76. During this time, the Mahārāja of Benares, Sir Ishwari Prasad Narayan Singh gifted some astronomical instruments to the Prince. The astronomical instruments were prepared according to the descriptions in *Siddhānta-śiromaṇi* (Bhāskarācārya, 1150 CE). Table 9.1 has the details of the instruments presented by the Mahārāja.

The Mahārāja also presented scale models of the masonry astronomical instruments found in the 'Hindu Observatory' at Benares. This observatory was one of the five set up by Sawai Jai Singh II, the king of Jaipur, in the 1720s at Jaipur, Delhi, Mathura, Ujjain and Benares. The model of the observatory of Benares was made in silver and sissoo-wood.

The Mahārāja presented a clock, showing the signs of the Zodiac, phases of the moon, dates of the month, days of the week, hours, and minutes. It is supposed to be a model of a larger clock of the same kind in the inner court of the Ramnagar Fort. This unique clock was made by Mulchand, the State clock maker and it provides several chronological and astronomical information. It shows not only correct time of day but also the position of the sun and phases of the moon together with the signs of zodiac and the date of the month and year. To get this information otherwise, several scientific instruments would be necessary.

These presents were historically very significant. They also bear testimony to the status of Astronomy as practiced in India. Unfortunately, none of these items are found in any of the museums in Britain. One hopes that they are still in some private royal collection. For the history of scientific instruments in India, it is important that this collection is located, brought to the knowledge of the people and experts in the field of astronomy.

Source: **S.R. Sarma (2014).** "Astronomical Instruments presented by the Mahārāja of Benares to the Prince of Wales in 1876", *Bulletin of the Scientific Instrument Society*, **122**, pp. 12–15.

TABLE 9.1 List of Instruments Gifted

1	Digaṃśa-Yantra	For finding the degrees of Azimuth of a planet/star
2	Dhruva-Prota-Cakra-Yantra	For finding the degrees of declination of a planet/star.
3	Yantra-Samrāṭ (King of instruments)	For finding the distance (in time) from the meridian and the declination of a planet/star, and of the Sun.
4	Bhitti-Yantra (A Mural Quadrant)	For finding the Sun's greatest declination and the latitude of the place.
5	Viṣuvad-Yantra (Equinoctial Circle)	For ascertaining the distance in time) of the Sun, or of any star from the meridian.
6	Phalaka-Yantra	An instrument for finding the time after sunrise (invented by Bhāskarācārya).
7	Cakra-Yantra	For finding the altitude and zenith distance of the Sun, the longitude of planets.
8	Cāpa-Yantra and Turīya-Yantra	For finding the zenith distance and altitude of the sun.
9	Śaṅku (Gnomon)	To ascertain the points of the compass, the place of the observer (latitude, and time).
10	Armillary Sphere	Represents several celestial circles. By the threads that are fastened within the globe it is possible to determine the parts of any spherical triangle on the globe.

We are all fascinated by the sky above, right from our childhood. The stars, sun, moon, meteors, phenomena such as sunrise, sunset, monsoon rains, days getting longer and shorter in a year, and so on, trigger several questions in our mind. Ancient Indians like others elsewhere in the world were drawn deep into these same questions. They dwelt deeply on the issues of celestial objects and heavenly bodies and their relationship with the Earth, human beings, and human life. Indians from time immemorial developed this intense sense of inquiry, and a natural curiosity. Driven by a fertile imaginative mind they deeply observed and analyzed the natural phenomena and sought answers rationally in the process of developing an ecosystem of knowledge in allied areas in Mathematics.

Astronomy is a branch of science that studies celestial objects, space, and the physical universe as a whole using concepts mainly from Mathematics. The sky and outer space have been a great source of curiosity for mankind forever. Therefore, Astronomy has been a branch of study right from pre-historic times. All ancient civilizations have made a methodical observation of the night sky and developed their understanding of the celestial phenomena. Intense observation of the patterns in the sky using some instruments designed for this purpose has contributed to a better understanding of astronomy. Ancient Indians have also contributed to the study of astronomy in very significant ways and have laid some foundations for its growth in the days to come. We shall see some of the salient aspects of Indian astronomy[1].

9.1 UNIQUE ASPECTS OF INDIAN ASTRONOMY

In modern science, the study of astronomy, space, and celestial objects are considered different and alien entities vis-à-vis human beings. Space is 'out there' and it is a matter of curiosity to understand what it consists of and how it can be studied using mathematical constructs and models. The celestial bodies have no relationship to the society or the culture, except that they influence climatic conditions, rainfall, etc. The Sun, for instance, is imagined as nothing but an inert matter consisting of material and radiating a large amount of energy. How can an entity that brings life to the entire living beings on the Earth be considered lifeless and inert? In contrast, in several cultures including the Indian, the Sun is considered as a living entity and a deity.

Knowledge of astronomy was widely used by all sections of Indian society, not just by the subject matter experts. Villagers, farmers, arts, and craftsmen and householders would know the meaning of Rāśi (Zodiac Sign), Nakṣatra (Star), and months of the calendar, and certain details about every day (Pañcāṅga). They developed some understanding of how seasons are formed, as seasonal changes affect economic activities such as farming, and the general health of individuals. This is common knowledge, even if the citizens are not educated in modern methods. There are cultural-religious aspects too. In India, it is a common tradition for the newly married couple to be shown the pair of stars known as Vasiṣṭha and Arundhatī (corresponding to the visual binary system in the constellation of 'Ursa Major') as part of the marriage ceremony by the priest or the family elders. The celestial binary is held up as a model for the married life of the couple.

> ♦ Knowledge of astronomy was widely used by all sections of Indian society.
> ♦ The observations and computations of measures related to celestial bodies such as stars and planets have been consistently accurate and ahead of the times as observed by many.

There are several differences between the current (predominantly Western) approach and that of the Indian and other ancient civilizations to astronomy. These are summarized below:

- The celestial entities are an integral part of all living beings on Earth. There is a strong sense of mutual dependence between the earthly and the celestial entities.

- Astronomy, a study of the celestial entities and phenomena is interconnected with cultural practices and daily life. Many events and planning of activities are done with reference to the celestial entities. 'Kālanirṇaya' (Determination of time) was considered the purpose of astronomy. Therefore, consulting the Pañcāṅga is a daily necessity and often the day begins with this activity.
- Astronomy is not a study of some alien entities but an integral and important aspect of one's life. Several concepts and models were developed by ancient Indians to address this requirement. This partly explains the growth and maturity of astronomical thinking in Indian society right from early times.
- Indians developed a systematic procedure of study, observation, data collection, codification, pattern recognition, and analysis. This led to the development of advanced mathematics including arithmetic, geometry, algebra, trigonometry, and the basics of calculus.

9.2 HISTORICAL DEVELOPMENT OF ASTRONOMY IN INDIA

The development of astronomy as a systematic area of study was driven by multiple considerations by civilizations world over. They looked up to the sky for a few important reasons. First, it helped them develop a calendar to predict seasons. The stars in the sky served as a reliable aid for directional and navigational purposes and for astrological predictions. Moreover, the sky was always considered as an abode of gods. Above all, observing the sky has been a fascinating experience for many. The Indian civilization was driven by similar considerations:

- The Vedic living required methods for fixing an auspicious time for performing various rituals and activities. Knowledge of the cardinal directions is also required to set the Vedic altars.
- In ancient times, there were two major wealth-generating activities: Agriculture and Trade. Agriculture was the main driver of the economy and wealth creation. Ability to predict the weather conditions and understand the mechanism of season formation was a key requirement to improve agricultural productivity.
- The other economic activity is related to international trade. This required a good understanding of the latitude, longitude, and loxodrome. Moreover, there was a need for a calendar as well.
- Astronomy plays a crucial role in navigation. The basic idea in navigation is to follow rising or setting location of a particular constellation for a certain period and then turn towards another constellation. Therefore, knowledge of astronomy becomes crucial. Indians have been regularly using astronomy for maritime purposes and some studies suggest that details of this have been well documented[2].

All these called for a calculation of time and timekeeping, which resulted in the formation and continuous evolution of the Indian calendar system Pañcāṅga, which is an accurate system of calendar even today. Indian Astronomy developed continuously in time right from the Vedic period. For example, in the Atharvaveda-saṃhitā, we are able to see references to stars, planets, and comets[3]. Parāśara-tantra consists of planet and comet observations made during the 2nd Millennium BCE[4]. The text provides details of the sidereal periods of Jupiter and Saturn.

Furthermore, the movement of Mars has been described qualitatively. The text also contains a list of twenty-six comets. In the Vedāṅga Jyotiṣa, the concept of Yuga consisting of 5 solar years of 62 lunar months has been defined. The concept of Yuga was introduced to synchronize the Solar and Lunar calendars. Two intercalary months Aṁhaspati and Saṁsarpa were added to complete a Yuga[5]. Recent studies use astronomy software for simulating the Vedic sky. Using this they analyze the statements in Śatapatha-brāhmaṇa about the Kṛttikā (Pleiades) and other such astronomical events mentioned in the Ṛgveda and conjecture that these could have been observed around 3000 BCE[6]. Table 9.2 has a list of references in the Vedic text and the possible dates for the text based on this. Another area of study in astronomy relates to cosmology (mainly the origin of the Universe). The Nāsadīya-sūkta of Ṛgveda (RV 10.129) speculates on the origin of the Universe. The Yajurveda and the Atharvaveda gives a full list of 27 stars commencing from Kṛttikā. A study of Itihāsas (Rāmāyaṇa and Mahābhārata) reveals numerous references to stars, planets, and their position in the sky. This indicates a good understanding by the society of celestial entities and their movement in the sky.

- In the Vedāṅga Jyotiṣa the concept of Yuga consisting of 5 solar years, 67 lunar sidereal cycles, 1830 days, and 1835 sidereal days has been defined.
- Nīlakaṇṭha Somayājī in 1500 CE proposed a revised model of planetary motion which closely approximates the mathematical model of planetary motion given by Kepler over a hundred years later in 1609.

The basic canonical texts of the Jains are around 45, besides a large number of subsidiary texts. Of these, Sthānāṅga and Bhagavatī-sūtra contain information on mathematics and astronomy. Two other individual texts, the Nandī-sūtra and Anuyogadvāra-sūtra deal with numerous topics, including topics on astronomy and mathematics, which a Jaina monk was supposed to know. Tattvārthādhigama-sūtra of Umāsvāti (185–219 CE) discusses cosmology and astronomy. Among later Jaina works on astronomy, a noteworthy addition is Jyotissāra by Thakker Pheru (14th century CE) in 238 verses, divided into four chapters[7].

TABLE 9.2 Astronomy References in Vedic Texts and Their Dating

No.	*Vedic Reference*	*Statement*	*Date*
1	Śatapatha-brāhmaṇa (2.1.2.3)	Kṛttikā (Pleiades) never swerve from the east.	2950 BCE
2	Maitrāyanīya-brāhmaṇa Upaniṣad (6.14)	Winter solstice at the mid-point of the Śraviṣṭhā segment and the summer solstice at the beginning of Māgha.	1660 BCE
3	Vedāṅga-jyotiṣa	Winter solstice at the beginning of Śraviṣṭhā and the summer solstice at the mid-point of Āśleṣā.	1300 BCE
4	Taittirīya-āraṇyaka (II.19.1)	Abhaya Dhruva (currently identified as alpha-Draconis) in the Śiṃśumāra Constellation is the pole star.	2800 BCE*

Source: Adopted from **Subhash Kak (2010).** "Archeoastronomy in India", arXiv. https://arxiv.org/abs/1002.4513.
Source:* **Iyengar, R.N. (2011). "Dhruva the Ancient Indian Pole Star – Fixity, Rotation and Movement", *Indian Journal of History of Science*, **46**, pp. 26–39.

In the discipline of Indian astronomy, a class of texts called 'siddhānta' was developed beginning 5th century CE. The siddhānta astronomy adopted more sophisticated mathematics, incorporated the planets in the system, devised a system of coordinates for the determination of the periods of planetary revolutions and the relative sizes of the Earth, the Sun, and the Moon. Āryabhaṭa developed a mathematical approach to astronomy in his work Āryabhaṭīyam, (499 CE). This is considered the first full-fledged treatise on mathematical astronomy in India. *Varāhamihira* in his *Pañca-siddhāntikā* (~530 CE) refers to five different astronomical texts which were prevalent during his time. They include Paitāmaha, Vasiṣṭha, Romaka, Pauliṣa, and Saura Siddhāntas. They deal with the true motion of the moon, sun, diurnal problems, lunar and solar eclipses, movement of planets such as Mercury, Venus, Jupiter, and Saturn.

Jyotirmīmāṃsā of Nīlakaṇṭha Somayājī, written in 1504 CE, is another important work in Indian astronomy. Nīlakaṇṭha stressed the importance of astronomical observation and argued for the necessity of correcting parameters periodically based on observations of eclipses, the Sun, Moon, and the planets. In 1500 CE, Nīlakaṇṭha revised the prevalent method for calculating planetary positions and proposed a planetary model in which the planets orbit around the Sun, which itself moves around the Earth. This is essentially the modern picture in a geocentric framework. Table 9.3 lists the major contributions of Indian astronomers from ancient time till the 19th century CE.

TABLE 9.3 Indian Astronomers and Their Seminal Contributions

Sl. No.	*Details of the Work/Mathematician*	*Period, Location*	*Salient Contributions*
1	***Author not known*** – *Sūrya-siddhānta*	Prior to 6th century CE	It appears that there are several versions of it available. An ancient version is summarised in Varāhamihira's Pañcasiddhāntikā. A modern version is very popular even now among the traditional scholars and calendar makers.
2	**Varāhamihira** – *Pañca-siddhāntikā*	6th century CE	Presents an updated summary of five ancient siddhāntas.
3	**Āryabhaṭa** – *Āryabhaṭīya*	Born 476 CE; Kusumapura, near Pataliputra, Bihar	A section on mathematics, foundations of trigonometry; Calculation of the sine function; Rotation of the Earth; Accurate algorithms for the positions of the Sun, the Moon and planets; Earth in the cosmos; Eclipses.
4	**Bhāskara I** – *Āryabhaṭīya-bhāṣya, Mahā-bhāskarīya*	7th century CE	Commentary on Āryabhaṭīya explaining its mathematics and astronomy in detail; Develops the Āryabhaṭan system in his own texts.
5	**Brahmagupta** – *Brāhmasphuṭa-siddhānta, Khaṇḍakhādyaka*	7th century CE	A detailed system of calculations pertaining to the Sun, the Moon, and planets; Many new algorithms and explanations in astronomy. path-breaking results in mathematics like vargaprakṛti (quadratic indeterminate equations), and cyclic quadrilaterals; Khaṇḍakhādyaka, a practical manual of Indian astronomy.
6	**Lalla** – *Śiṣyadhīvṛddhida-tantra*	8th–9th century CE	A textbook which expounds on the Āryabhaṭan system, with new algorithms.
7	**Mañjulācārya** – *Laghumānasa*	10th century CE	An explicit expression for the 'second correction' to the longitude of the Moon; Derivative of sine function and instantaneous velocity of the Sun and the Moon.

(*Contd.*)

Sl. No.	*Details of the Work/Mathematician*	*Period, Location*	*Salient Contributions*
8	**Śrīpati** – Siddhānta-śekhara	11th century CE	An important text quoted by the later astronomers.
9	**Bhāskarācārya II** – *Siddhānta-śiromaṇi, Vāsanābhāṣya, Karaṇakutūhala*	Born 1114 CE	Most of the standard calculations and algorithms in Indian astronomy included, mistakes rectified, generalizations made, a calculation-manual using ready-made tables, and arithmetical simplifications.
10	**Kerala School** **Mādhava of Saṅgamagrāma**– *Veṇvāroha, Sphuṭacandrāpti* **Parameśvara of Vaṭasseri**– *Dṛggaṇita, Bhaṭadīpika, Siddhānta-dīpikā* **Nīlakaṇṭha Somayājī or Somasutvan of Trikkantiyur**– *Tantra-saṅgraha, Āryabhaṭīya-bhāṣya* **Jyeṣṭhadeva** – *Gaṇita-yuktibhāṣā* Acuyta Pisaroti– *Sphuṭanirṇayatantra* Śankaravarman– *Ṣaḍratnamāla*	**14th–19th Century** 1340–1425 CE 1360–1455 CE 1444–1550 CE 1500–1610 CE 16th century CE 19th century CE	This school made important contributions to mathematical analysis – Derivation of infinite series for π, sine and cosine functions, much before the subject developed in Europe. A major revision of the traditional planetary theory in 1500 CE. Innovations in astronomical computations; Systematisation of the applications of spherical trigonometry to astronomy; Improved theory of eclipses.
11	**Gaṇeśa Daivajña** – *Grahalāghava*	Born 1507 CE	Simplified procedures for calculation of planetary positions, used for preparing almanacs or Pañcāṅgas even now.
12	**Kamalākara**–*Siddhānta-tattva-viveka*	Born 1616 CE	Elaborate work mostly based on Indian concepts and parameters but incorporates elements of the Greek astronomer Ptolemy's system.
13	**Candraśekhara Sāmanta** – *Siddhānta-darpaṇa*	Born 1835 CE	Important modifications in planetary parameters revised the lunar theory, designed simple instruments, reformed the traditional calendar of Odisha.
14	**Raja Sawai Jai Singh** – *Yantrarāja-racanā, Zij Muhammad Shahi*	1688–1743 CE	Built famous observatories in several parts of North India.

9.3 THE CELESTIAL COORDINATE SYSTEM

A celestial coordinate system is a mechanism for specifying positions of planets, stars, and other celestial objects relative to physical reference points available to an observer situated on the earth. It helps to locate a celestial luminary using a three-dimensional or a spherical plane. Figure 9.1 provides the rudimentary elements of the celestial coordinate system. The celestial pole is an imaginary large sphere concentric to the earth. All the objects in the sky can be projected on the celestial sphere. The celestial north pole and celestial south pole are analogous to the north pole and south pole of the earth respectively. Similarly, the celestial equator is concentric to the earth's equator. Likewise, all elements of the celestial sphere such as the celestial hemisphere are analogous to the earth's hemisphere. The only difference is that it is projected on to a great circle so that all celestial luminaries can be mapped on to the sphere using some celestial coordinates.

Another useful coordinate system is the ecliptic. **Ecliptic** is an imaginary line depicting the Sun's movement around earth. When an observer sees the Sun's path from Earth's reference frame, it appears to move around the Earth in a path that is tilted with respect to the spin axis at 23.5°. Figure 9.1(a) graphically illustrates the ecliptic. The Ecliptic plane sets the reference plane and is the basis for the ecliptic coordinate system which is used for all further astronomical calculations. A third aspect of the coordinate system is the horizontal coordinate system from the perspective and the position of an observer. The horizontal plane with the Zenith and the Nadir are the two poles of this system. The **zenith** is an imaginary point directly 'above' a particular location (highest point), on the celestial sphere and **nadir** is 180° from zenith (the lowest point). It helps to position the stars and other luminaries in the sky relative to the observer's horizon. As we know by experience, the position of the celestial luminaries varies with time and is useful for the observer to locate and track these celestial entities. Figure 9.1(b) is a simple illustration of this. The vector from an observer to a point of interest (such as a star or any celestial body) is projected perpendicularly onto a reference plane. The angle between the projected vector and a reference vector on the reference plane is called the **azimuth**.

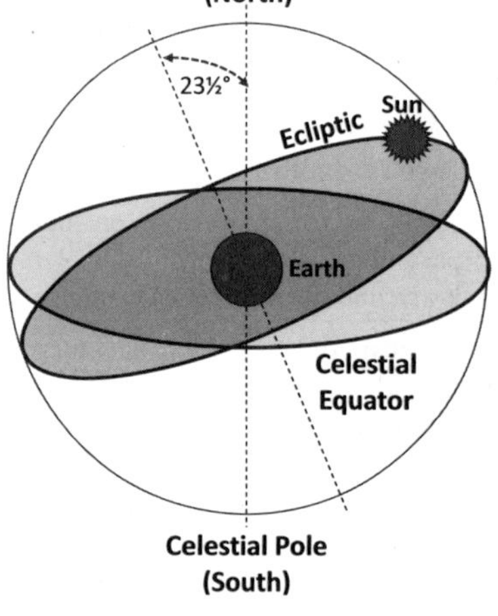

FIGURE 9.1 (a) Illustration of an Ecliptic

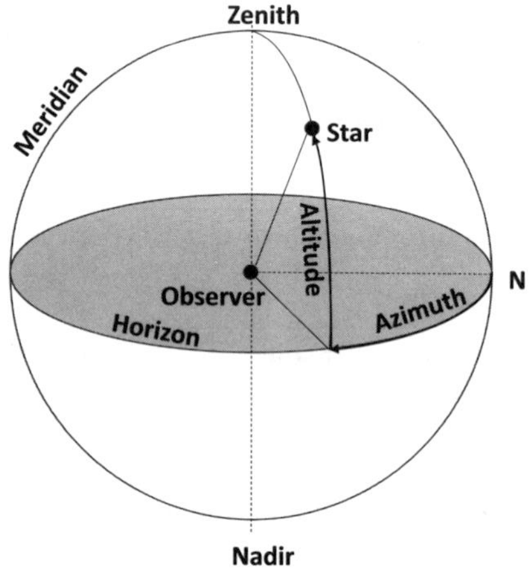

FIGURE 9.1 (b) Azimuth and Zenith

Every celestial object appears to rise in the eastern part of the sky, travel up along a 'diurnal path', and set in the western part. The relative positions of the stars are fixed, and they do not appear to move with respect to each other. However, the Sun, the Moon, and the planets seem to move eastwards in the background of stars (apart from their daily motion from east to west). Let us consider the movement of the Sun, as viewed from the Earth, as shown in Figure 9.2. At the time of Sunrise on a particular day, the Sun is at A which is in the same direction as a star S1. After one day, at the next Sunrise, it would be at B, which is in the same direction as a star S2, east of S1. This change in the position of the celestial objects forms the backbone of computation of various time units such as year, month, and day. The calendaring and other astronomical calculations are done using these basic principles. Further, some of the elements of the seasons are also inferred from analyzing the trajectory of the Sun in the ecliptic.

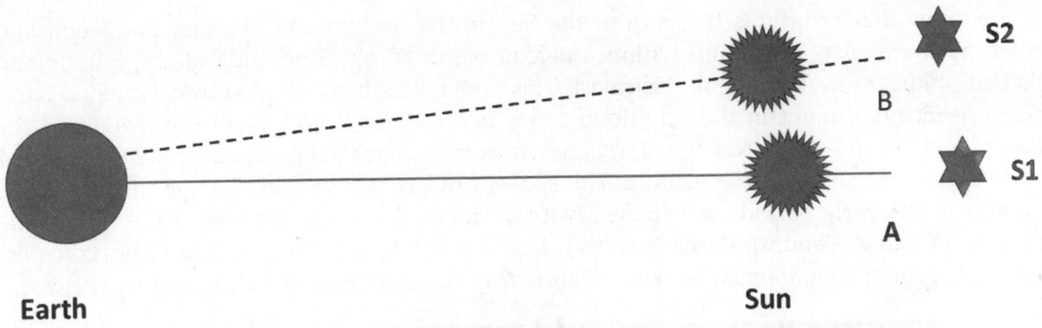

FIGURE 9.2 Motion of the Sun in the Background of Stars

A solstice occurs when the sun's path crosses the extreme north or south points on the earth's equator. In Figure 9.3, when the Sun is at S4, it is at the southernmost point with respect to the equator. This is known as the *winter solstice*. Similarly, when the Sun is at S2, it is at the northernmost point with respect to the equator. That point is known as the *summer solstice*. At S1 and S3, the ecliptic intersects the equator, and these points are known as the *equinoxes*. When the Sun is at equinox, it is an equinoctial day. Between S4 and S2, the Sun moves northwards (known in the Indian system as Uttarāyaṇa), and between S2 and S4, the Sun moves southwards (known as Dakṣiṇāyana). We find Vedic references to these important astronomical concepts. Taittirīya-saṃhitā (6.5.3) observes[8], 'Thus the Sun moves southwards for six months and northwards for six months'. The equinoctial day (viṣuvat) is mentioned in Aitareya-brāhmaṇa (18.4)[9]. Methods to compute the *n*th equinox is available in Vedāṅga-jyotiṣa[10].

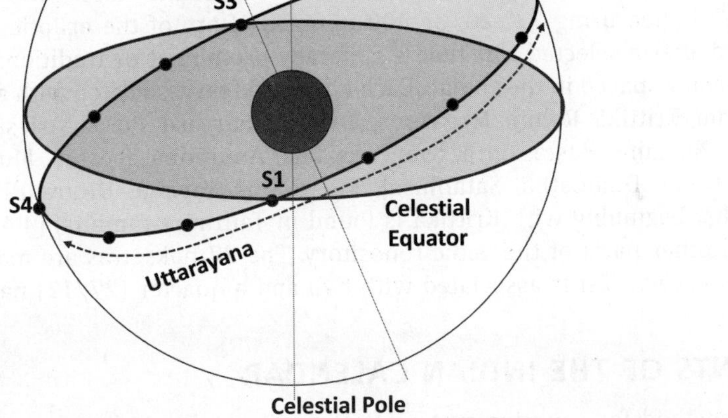

FIGURE 9.3 Illustration of Solstices and Equinoxes

We know that ecliptic is the path of the Sun in the background of stars. The Moon and almost all the planets are found within a belt of width 8° on either side of the ecliptic and this belt is known as 'zodiac', or Rāśicakra. This zodiac has been divided into 12 equal parts from a fixed initial point in the ecliptic in order to trace the trajectory of the Moon and the other planets in the context of the stars, known as rāśis. Each segment spans 30°. Figure 9.4 shows the zodiac signs on the ecliptic. The concept of the twelve-fold division of the ecliptic is traced to the early periods of Vedāṅga Jyotiṣa. The twelve Rāśis are Meṣa (Aries), Vṛṣabha (Taurus), Mithuna (Gemini), Karka (Cancer), Siṃha (Leo), Kanyā (Virgo), Tulā (Libra), Vṛścika (Scorpio), Dhanus (Sagittarius), Makara (Capricorn), Kumbha (Aquarius), and Mīna (Pisces).

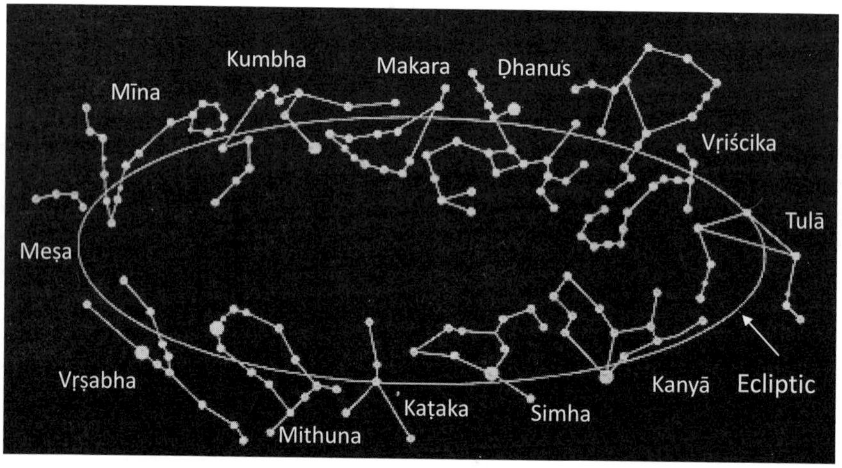

FIGURE 9.4 The Zodiac Signs on the Ecliptic

The 'sidereal period' of an object is the time taken by it to complete one revolution in the background of stars. The Moon moves in an orbit around the Earth, which is slightly inclined to the ecliptic. The Moon's sidereal period is close to 27.32 days, which constitutes one lunar cycle. In other words, the Moon covers nearly 1/27th part of the ecliptic per day. In the Indian system, the zodiac has been divided into 27 equal parts from a fixed initial point in the ecliptic in order to trace the trajectory of the Moon in the context of the stars. Each such division is known as nakṣatra, measuring 13° 20′ or 800 minutes of arc of the ecliptic (360/27). Each division is named after a selected star that is generally prominent or traditionally well-known and is broadly equally spaced in the zodiac. Each day would be associated with a nakṣatra. They are Aśvinī, Bharaṇī, Kṛittikā, Rohiṇī, Mṛgaśiras, Ārdrā, Punarvasu, Puṣya, Āśleṣā, Maghā, Pūrvā Phalgunī, Uttarā Phalgunī, Hasta, Citrā, Svātī, Viśākhā, Anurādhā, Jyeṣṭhā, Mūla, Pūrvāṣāḍhā, Uttarāṣāḍhā, Śravaṇa, Dhaniṣṭhā, Śatabhiṣaj, Pūrvā Bhādrapadā, Uttarā Bhādrapadā, and Revatī. The full list beginning with Kṛittikā is found in Taittirīya-saṃhitā, (4.4.10.1-3), and in Atharvaveda and other parts of the Vedic repository. The 27 nakṣatras are mapped on to the 12 rāśis. Therefore, each rāśi is associated with two and a quarter (27/12) nakṣatra.

9.4 ELEMENTS OF THE INDIAN CALENDAR

A calendar is a method for counting systematically and continuously the successive days by cyclic periods such as year and month. Calendaring is done based on two luminaries in the sky, namely the Sun and the Moon[11]. One of the critical uses of astronomy is to calendar the time and identify seasons, which are related to the relative movement of the Sun with respect to the Earth. The Indian system of astronomy has dealt with this elaborately as evidenced in the Vedic texts. While in the Vedic texts we merely find the description of these elements, the computation details of these are found in Vedāṅga Jyotiṣa, though approximate and simple. Later day mathematicians introduced formal methods for mathematical computations of the calendar. There are three clear time-markers as even casual observations of the sky reveal, viz., day, month, and year. All major civilizations grappled with the problem of how these time units are related to each other.

The Notion of a Year

The elapsed time for the Sun to return to the same star, which is nothing but the period of the revolution of the Earth around the Sun, is the solar year. Solar calendars are based on this. On the other hand, the period of the return of the Moon in opposition (Full moon) or conjunction to the Sun (New moon) in relation to the Earth, is a lunar month. 12 such successive months form the lunar year, and this is the basis of all lunar calendars. Both the calendars are in use today. The solar-year calendar is followed for indicating the dates of the year in the states of Tripura, Assam, Bengal, Odisha, Tamil Nadu, Kerala, and partly Punjab and Haryana while the lunar calendar is followed in other states for the same purpose. But in all states for fixing the dates of religious festivals and for selecting the auspicious time for undertaking many socio-religious activities, the lunar calendar is used. The lunar calendar, however, is pegged to the solar calendar, or more precisely, the lunar months are linked with the solar months to prevent these from getting delinked with the seasons. Therefore, it is more apt to say that in the Indian tradition a luni-solar calendar is followed. Knowledge of the components of the solar and lunar calendar is critical.

What constitutes a year has been variously defined in the Indian system[12]. The Vedic year consisted of 12 months, each of 30 days (known as Sāvana). This results in a year consisting of 360 days. This was synchronized to the seasons by adding 5 days to the calendar. The Ṛgveda (1.164.11) depicts this as, "The wheel (of time), formed with twelve spokes, revolves around the heavens, without wearing out. O Agni, on it, are 720 sons (viz., days and nights)"[13]. According to this verse, a year has 12 months and 360 days. 5 or even 5.25 days were added later. In Yajurveda, there are references to 12 lunar months (known as Vatsara) amounting to 354 days. The synchronization was done using the Ekādaśarātra ceremony to account for 365 days in a year leaving an error of 0.25 days. Subbarayappa and Sarma have compiled the list of astronomical references in various ancient Indian literature pertaining to these aspects[14]. It appears that the authors of the Ṛgveda were aware of the discrepancies between the duration of the Lunar year and Solar year and the need to add an intercalary month for synchronizing the two. Table 9.4 summarises the differences between alternative year cycles in the Indian tradition.

A solar year is approximately 365.2564 days (modern value) and 12 lunar months is approximately 354.3671 days. There is a gap of nearly 10.89 days between the two. Therefore, to align the lunar year with a solar year which is connected with the seasons, an additional lunar month has to be introduced in some years. In Yajurveda (4.4.11) there is a mention of the need to add 2 intercalary months in 5 years[15]. It had been noticed that the Sun and the Moon return together to nearly the same position in the framework of stars, after five years. The concept of Yuga was introduced to synchronize the Solar and Lunar calendars. Two intercalary

TABLE 9.4 Alternative Classification of Year Cycles in the Indian Tradition

	Solar Year	*Sāvana Year*	*Lunar Year*
Year	12 Solar Months (366 days)	12 Sāvana Months (360 days)	12 Lunar Months (354 days)
Month	30.5 days	30 days	29.5 days
Yuga	60 Solar Months	61 Sāvana Months	62 Lunar Months
Remarks	Known as Saṃvatsara	Known as Idāvatsara	Known as Anuvatsara

months Aṁhaspati and Saṁsarpa were added to complete a Yuga.[16] A 5-year yuga cycle is mentioned in Taittirīya and Vājasaneyī saṃhitās.

Solar and Lunar Months

The Sūrya-siddhānta defines the solar month as the time taken by the Sun to traverse a rāśi, which is the difference between the time of ingress of the Sun to one rāśi (saṅkrānti) and to the time of ingress to the next rāśi. As Meṣa is the first rāśi, the time taken by the Sun to traverse this rāśi forms the 1st solar month, and it is called Meṣa in Kerala, and Vaiśākha in Tripura, Bengal, Odisha, Haryana, and Punjab, Bahag in Assam, and Citra in Tamil Nadu. The second solar month is similarly formed by the time taken by the Sun to traverse the second rāśi, which is Vṛṣabha, and so on for other months. The names of the twelve months and six seasons are given in the Taittirīya-saṃhitā (1.4.14; 4.4.11)[17] as "madhu, mādhava (vasanta ṛtu), śukra, śuci (grīṣma ṛtu), nabha, nabhasya (varṣa), īśa, ūrja (śarad ṛtu), sahas, sahasya (hemanta ṛtu), tapas, tapasya (śiśira ṛtu)".

A lunar month (cāndra-māsa) is the time interval between two new Moons (Amāvāsyā), or two full Moons (Pūrṇimā). It consists of two pakṣas; Śukla-pakṣa (bright fortnight) from Amāvāsyā to Pūrṇimā, and the Kṛṣṇa-pakṣa (dark fortnight) from Pūrṇimā to Amāvāsyā. Figure 9.5 illustrates this idea pictorially. A normal lunar year has twelve lunar months. The names of the twelve lunar months are Caitra, Vaiśākha, Jyeṣṭha, Āṣādha, Śrāvaṇa, Bhādrapada,

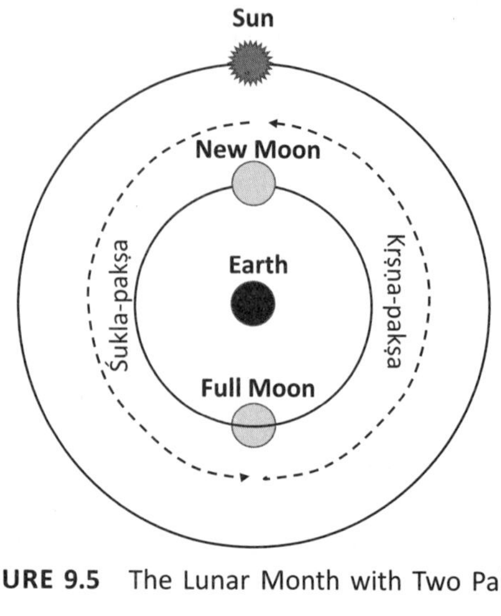

FIGURE 9.5 The Lunar Month with Two Pakṣas

Āśvayuja, Kārtika, Mārgaśira, Puṣya, Māgha, and Phālguna. During most Caitra months the Moon will be close to the star Citrā (Spica), on the full moon day of the month. Similarly, during most Vaiśākha months, the Moon will be near the star Viśākhā on the full moon day in that month. This is the logic behind the nomenclature of the months.

The Notion of Tithi

The Indian calendar is based on the true longitudes of the Sun and the Moon and is related to the actual happenings in the sky. The relative position of the Moon and the Sun on a day determines the tithi. In the Stellar background, the Sun moves at the rate of 1° per day eastwards and Moon moves at the rate of 13° per day. Tithi captures the angular separation between the Sun and the Moon. It is a time-unit during which the angle between the Sun and the Moon (as viewed from the Earth) increases by 12° or its integral multiples. At Amāvāsyā, the Sun and the Moon are perfectly aligned (that is why we are not able to sight the moon) and the angle is 0°, and the first tithi begins. At the end of the first tithi, the angular separation is 12° and the second tithi begins. The second tithi ends, and the third begins when the angle is 24°, and so on. Figure 9.6 illustrates this pictorially. The 12° is merely an average. In reality, the actual duration of the tithis will vary on account of the perturbing forces affecting the true motions of the Moon and the Sun. The actual duration of a particular tithi may vary from 26 hours 47 minutes to 19 hours 59 minutes.

As we saw, there are rudiments of a calendar with adhikamāsas (intercalary months), and 27 nakṣatras as markers of the Moon's movement. However, the descriptions in the saṃhitās are qualitative. A definite quantitative calendrical system is described in Vedāṅga Jyotiṣa. It is available in Ṛgvedic and Yajurvedic versions. Vedāṅga-jyotiṣa is the first text in India to give mathematical algorithms in astronomy. There is nothing on planetary motion in this work. There are short algorithms for finding tithi, nakṣatra, Sun's position in the sky, etc. The Vedāṅga-jyotiṣa gives rules for calculating some astronomical quantities related to the Sun and the Moon. The motions of the Sun and the Moon (in the background of stars) are assumed to be uniform.

New Moon
Sun = S_0, Moon = M_0; $M_0\hat{E}S_o = 0°$
After 1st tithi
Sun = S_1, Moon = M_1; $M_1\hat{E}S_1 = 12°$
After 2nd tithi
Sun = S_2, Moon = M_2; $M_2\hat{E}S_2 = 24°$
Full Moon
Sun = S_{15}, Moon = M_{15}; $M_{15}\hat{E}S_{15} = 180°$

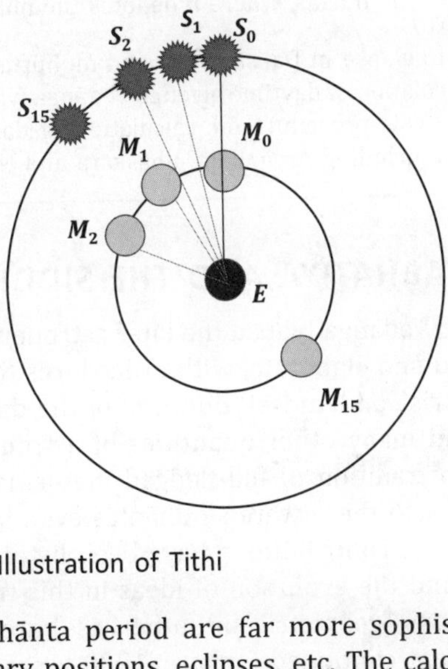

FIGURE 9.6 An Illustration of Tithi

The later astronomical texts of the siddhānta period are far more sophisticated and elaborate, with procedures to calculate planetary positions, eclipses, etc. The calculations are more accurate, and the calendar is well-defined and more advanced. All the texts adhere to the sūtra format, and there is continuity in the tradition. Precise trigonometric measurements were developed and used to achieve accurate time measurement. This development and refinement continued over 1000 years. Trigonometric functions accurate to the first minute was developed and used in Sūrya-siddhānta, and Āryabhaṭīya. This was improvised to seconds (vikalā) by Vaṭeśvara (904 CE), and thirds (tatparā) by Mādhava (14th century CE).

A calendar system was developed using the prime meridian of Ujjayinī (current Ujjain, also known as Avantipura in Indian literature). Methods were developed for recalibration of the calendar for any local place.

How to Compute the Quantum of Daytime on any Day?

It is a common observation that sunrise and sunset times are different each day in a year. Hence the quantum of the daytime is not constant. Ancient Indian astronomers had formulated procedures to calculate the daytime on any given day, using their observations. Basically, they had observed that on the equinoctial day, when the sun is on the equator, the durations of daytime, and night-time are equal. On the other hand, the quantum of daytime is the highest at the summer solstice and least at the winter solstice.

Vedāṅga-jyotiṣa, gives a simple arithmetical rule for the duration of the daytime (in verse 22 of the *Ṛgveda* version and verse 40 of *Yajurveda* version):

यदुत्तरस्यायनतो गतं स्याच्छेषं तथा दक्षिणोऽयनस्य ।
तदेकषष्ट्या द्विगुणं विभक्तं सद्वादशं स्याद्दिवसप्रमाणम् ॥
yaduttarasyāyanato gataṃ syāccheṣaṃ tathā dakṣiṇo'yanasya |
tad-ekaṣaṣṭyā dviguṇaṃ vibhaktaṃ sadvādaśaṃ syād-divasapramāṇam ||

According to this, the number of days which have elapsed in the northward course of the Sun (Uttarāyaṇa) or the remaining days in the southward course (Dakṣiṇāyana) doubled and divided by 61, plus 12, is the daytime (in muhūrtas) of the day taken. Hence, the duration of the daytime D_t is given by:

$D_t = \left(12 + \frac{2n}{61}\right)$ muhūrtas, where n denotes the number of days elapsed in Uttarāyaṇa, or the number of days yet to elapse in Dakṣiṇāyana and muhūrta = 1/30th of a civil day (of 24 hours).

The calculation of daytime given in Vedāṅga Jyotiṣa, matches closely with contemporary astronomy calculations. This procedure for calculation of daytime was further improvised by later day Indian astronomers including Āryabhaṭa, Bhāskara and Nīlakaṇṭha Somayājī.

9.5 ĀRYABHAṬĪYA AND THE SIDDHĀNTIC TRADITION

Compared to Vedāṅga Jyotiṣa the later astronomical texts of the siddhāntic period are far more sophisticated and elaborate, with procedures to calculate the planetary positions, eclipses, the times of sunrise and sunset, duration of the day, relation between the shadow of the Sun and the time, and many other quantities of astronomical interest. Āryabhaṭa was the pioneering figure in the tradition of full-fledged mathematical astronomy in India called the siddhāntic tradition, though there were siddhāntas even before Āryabhaṭīya. The major astronomers and their distinctive contributions have been listed in Table 9.2. There were many innovations in techniques and the evolution of ideas in this tradition. Detailed explanations, derivations and demonstrations (geometric and otherwise) are found in the commentaries. This development and refinement continued for over 1200 years.

Āryabhaṭīya is the first extant (discovered and available) text on mathematical astronomy in India. It was a pioneering work that established the framework for mathematical astronomy in India. It contains a systematic treatment of all the traditional astronomical problems. The text was composed in 499 CE. Āryabhaṭīya describes the procedures for calculating the positions of the Sun, the Moon, and the planets, astronomical variables associated with the daily paths of these objects in the sky, especially the Sun, and the important phenomena of eclipses. In all these calculations, trigonometry plays a crucial role. The 'jyārdha', or the half-chord introduced in Āryabhaṭīya is far more convenient than the Greek chord for astronomical computations. This jyārdha is nothing but the modern sine function (apart from a constant factor). A very important recursion relation for the sines and an explicit table of sines are given in Āryabhaṭīya. Āryabhaṭīya has only 121 verses and is organized into 4 parts, namely: Gītikāpāda, Gaṇitapāda, Kālakriyāpāda, and Golapāda. Table 9.5 has a summary of the issues discussed in Āryabhaṭīya.

TABLE 9.5 Organisation of Āryabhaṭīya and the Issues Covered

No.	*Section Name*	*No. of Verses*	*Issues Covered*
1	Gītikāpāda	13 verses	The letter-numeral notation for numbers, concepts of Kalpa and Mahāyuga, revolution numbers of planets and parameters associated with them.
2	Gaṇitapāda	33 verses	Square, Square root, cube, cube root, areas of a triangle, a trapezium, and general plane figures, volumes of right pyramids and spheres, value of π, methods for computing Sines geometrically, constructing a sine table, arithmetic progression, summation of first n natural numbers, the sum of their sums, sums of squares and cubes of first n natural numbers, Kuṭṭaka procedure to solve linear indeterminate equations, relative velocities of moving objects, and a problem related to interest calculation.
3	Kālakriyāpāda	25 verses	Reckoning of time, calendrical concepts, planetary models (epicycle and eccentric circle theories), and explicit procedures for calculation of planetary positions, etc.
4	Golapāda	50 verses	Problems of spherical astronomy as seen at different latitudes, diurnal problems associated with the motion of the Sun, Moon, and planets on the celestial sphere, situation of the earth and its shape, brightness/darkness of planets, parallax, lunar eclipses, solar eclipses and so on.

According to Āryabhaṭa, the Earth is a sphere at the center of the framework in which stars and planets move. According to a verse in Āryabhaṭīya[18]:

> *"The globe of the Earth stands (supportless) at the center of the circular frame of asterisms surrounded by the orbits (of the planets); it is made up of water, earth, fire, and air and is spherical."*

According to the earlier astronomical works in India and elsewhere, the Earth was stationary, and all the celestial objects rotated in the sky, completing one rotation in one day. The celestial objects rise in the eastern part of the sky, and set in the western part, and are visible when they are above the horizon. However, Āryabhaṭīya presented a different view. The diurnal motion of the celestial objects was mentioned in one of the verses as follows[19]:

"Just as a man in a boat moving forward sees the stationary objects as moving backward, just so are the stationary stars seen by people at Lanka (on the equator), as moving exactly towards the west."

What is implied in this verse is that though it appears that objects in the sky are moving from east to west, they are indeed stationary, and it is the Earth which is moving (rotating) along with the entities situated on it.

As we had seen earlier, a yuga of five years was in vogue at the time of Vedāṅga Jyotiṣa. On the other hand, in smṛtis and the old Sūrya-siddhānta (before Āryabhaṭīya), we have the concept of a mahā-yuga of 43,20,000 years. A similar approach is taken in Āryabhaṭīya. Moreover, a mahā-yuga is taken to be composed of 4 sub-yugas, namely Kṛta, Tretā, Dvāpara, and Kali. In other texts, the durations of these sub-yugas are in the ratio, 4:3:2:1 respectively. In Āryabhaṭīya they are all of the equal duration, namely, 10,80,000 years. It can be inferred that the beginning of the current Kali-yuga is on February 18, 3102 BCE which was a Friday.

- Āryabhaṭīya is the first pioneering work that established the framework for mathematical astronomy in India.
- Āryabhaṭīya describes the procedures for calculating the positions of the Sun, the Moon and the planets, astronomical variables associated with the daily paths of these objects in the sky.

In a mahā-yuga, all the planets and the auxiliary quantities associated with them make integral numbers of revolutions. The number of revolutions in the stellar background made by the planets in a mahā-yuga as per Āryabhaṭīya is given in Table 9.6. As evident from the table, Āryabhaṭa's computations were amazingly close to the modern value, considering that it was estimated in the 5th century CE. The number of civil days in a mahā-yuga known as the Yuga-sāvana-dina (D_Y) is also specified. As per Āryabhaṭīya D_Y = 15,77,917,500.

TABLE 9.6 Planetary Revolutions in a Mahā-yuga and the Inferred Sidereal Periods in Āryabhaṭīya

Planet	*No. of Revolutions*	*Sidereal Period*	*Modern Value*
Sun	43,20,000	365.25868	365.25636
Moon	5,77,53,336	27.32167	27.32166
Moon's Apogee	4,88,219	3231.98708	3232.37543
Moon's Nodes	2,32,226	6974.7491	6793.39108
Mecury*	1,79,37,020	87.96988	87.96930
Venus*	70,22,388	224.69814	224.70080
Mars	22,96,824	686.99974	686.97970
Jupiter	3,64,224	4332.27217	4332.58870
Saturn	1,46,564	10766.06465	10759.20100

* For Mercury and Venus: Śīghroccas (Heliocentric revolution numbers). Yuga-sāvana-dina (D_Y) = 15,77,917,500. The knowledge of the revolution numbers and the yuga-sāvana-dina is necessary to compute the positions of the planets on any day.

The angular position of the planet at any time with respect to a reference line is called its longitude (θ). In all the Indian texts, the reference line is the direction of the beginning point of the Meṣa-rāśi in the stellar background. The Moon and the planets move in planes that are slightly inclined to the ecliptic. We ignore this inclination for the time being. Assume for a moment that the planets move uniformly (with constant speed) in circles with the earth as the center in the plane of the ecliptic only. Then this angle is termed the 'mean longitude' and denoted by θ_0. This varies at a uniform rate. According to Āryabhaṭīya, the mean longitudes of all the planets are zero at the beginning of Kali-yuga.

The apparent motions of the Sun, Moon, and planets in the background of stars are not uniform. Two corrections have to be applied to the 'mean lingitude' θ_0 to obtain the 'true' (geocentric) longitude. These are:

1. *Manda-saṃskāra:* This is due to the non-uniformity of motion due to the eccentricity of the planet's orbit. It is called the 'Equation of centre' in modern astronomy. This is the only correction to the Sun and the Moon (for Moon, there are some other minor corrections specified in later texts). In the case of the actual planets called tārāgrahas in India (traditionally only Mercury, Venus, Mars, Jupiter, and Saturn), we obtain the true heliocentric longitude, that is, the longitudes with respect to the Sun after the manda-saṃskāra.
2. *Śīghra-saṃskāra:* This correction is done only for the planets. This converts their heliocentric longitudes (as observed from the Sun) to the geocentric longitudes (as observed from the Earth).

Āryabhaṭīya discussed the above two corrections for the first time in the Indian tradition. In the early 17th century CE, the European astronomer, Kepler had given the correct model for the motion of the planets. According to him, planets move in elliptical orbits around the Sun. The planetary model which is implicit in Āryabhaṭīya and also in the later Indian texts is broadly equivalent to the Kepler model and gives approximately the same final results for the longitudes. There were some problems with the interior planets, Mercury and Venus which were resolved by Nīlakaṇṭha Somayājī in 1500 CE in his Tantra-saṅgraha.

There were some problems associated with the calculational procedure for Mercury and Venus in the traditional Indian model. Around 1500 CE, Nīlakaṇṭha Somayājī modified the procedure based on a detailed scientific analysis and proposed a revised planetary model in his Āryabhaṭīya-bhāṣya and other works. According to this model, the planets move in eccentric orbits around the Sun (that is, the centers of the orbits will be slightly displaced from the Sun), which itself orbits around the Earth. See Figure 9.7 for a simple schematic representation of this model. This was before the well-known heliocentric model of Copernicus in 1542 CE, which has some shortcomings. These are absent in the revised model of Nīlakaṇṭha, which is in a geocentric framework. It is essentially the same as Tycho Brahe's model for the planetary motion proposed around 1580 CE, where the planets move around the Sun, with the Sun itself orbiting the Earth.

9.6 PAÑCĀṄGA – THE INDIAN CALENDAR SYSTEM

Pañcāṅga is the backbone of Indian astrology and calendar. All festivals and local events are based on pañcāṅga, and it plays a vital role in the daily lives of the vast majority of the people in the country. Therefore, there are several regional versions of Pañcāṅga throughout the country to cater to the local requirements. The development of calendric astronomy in India in a systematic manner can be traced back to the Vedāṅga Jyotiṣa, wherein the calendar was formulated based on cycle-periods of five years, called yuga. However, it was the Siddhānta Jyotiṣa, which employed scientific computations that ushered in the refined calendar and Pañcāṅga system that is in vogue today.

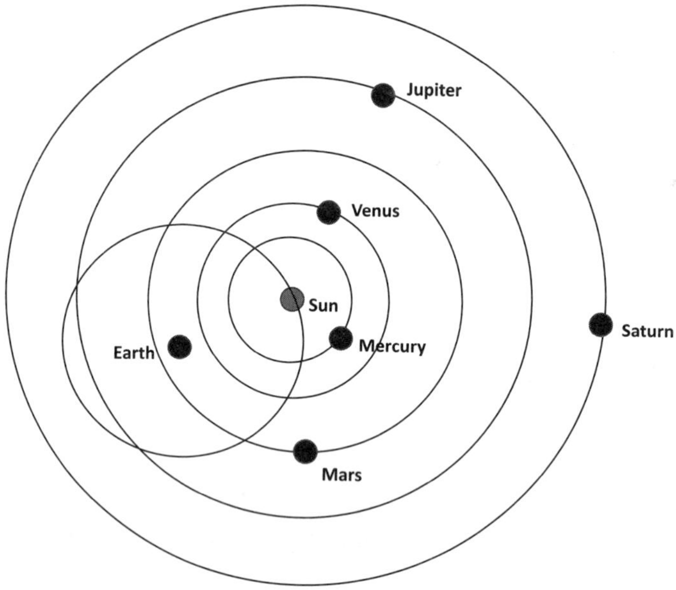

FIGURE 9.7 Nīlakaṇṭha Somayāji's Planetary Model

Many of them employ Sūrya-siddhānta and also use some texts popular in specific regions. In recent times, Graha-lāghava of Gaṇeśa Daivajña and Siddhānta-darpaṇa of Candraśekhara Sāmanta are used to make pañcāṅga calculations. Pañcāṅga, as the name suggests has five components: tithi, nakṣatra, vāra, karaṇa, and yoga. Based on the astronomy concepts developed, pañcāṅga computes the five components at any given instant. In the pañcāṅgas used by the Indian society, such calculations are done, and the results are made available in a ready reckoner format for direct reading and interpretation. The calculations are based on the true longitudes of the Sun and the Moon.

Tithi

Tithi plays a very important role as it determines most of the religious festivals and functions in India. The computational aspects of tithi in a pañcāṅga follows our earlier discussion on tithi. Let θ_M and θ_S be the true Longitudes of the Moon and the Sun in degrees respectively at any instant. Let $\frac{(\theta_M - \theta_S)}{12} = I + f$, where I is an integer, and f is the fractional part. Then I is the number of tithis that have elapsed in the lunar month.

Let us consider two cases for the true longitudes of the Moon (θ_M) and the Sun (θ_S) at any instant. The calculation of tithi is done as follows:

Case 1: (θ_M) = 60° 12′ and (θ_S)= 19° 7′, the computations are as follows:

$$\frac{(\theta_M - \theta_S)}{12} = \frac{(60^\circ\, 12' - 19^\circ\, 7')}{12} = \frac{41^\circ\, 5'}{12} = 3\frac{61}{144}$$

Therefore 3 tithis have elapsed, and the current tithi is 4th, which is Caturthī in Śukla-pakṣa.

Case 2: (θ_M) = 201° 2′ and (θ_S)= 219° 17′. Since $\theta_M < \theta_S$ we add 360° before performing the subtraction.

$$\frac{(\theta_M - \theta_S)}{12} = \frac{(561°\,2' - 219°\,17°)}{12} = \frac{341°\,45'}{12} = 28\frac{23}{48}$$

Therefore 28 tithis have elapsed, and the current tithi is 29th, which is Caturdaśī in Kṛṣṇa-pakṣa.

Karaṇa

This is half of a tithi. In one karaṇa, the angular separation between the Moon and the Sun increases by 6°. There are 60 karaṇas in a lunar month. At any instant, the procedure to find the number of karaṇas elapsed, and the time elapsed in the current karaṇa is the same as for the tithi except that 12 should be replaced by 6.

To find the karaṇa, we repeat the above calculation with 6 as the divisor.

Case 1:

$$\frac{(\theta_M - \theta_S)}{6} = \frac{(60°\,12' - 19°\,7')}{6} = \frac{41°\,5'}{6} = 6\frac{61}{72}$$

So, 6 karaṇs have elapsed, and the current karaṇā is 7th.

Case 2:

$$\frac{(\theta_M - \theta_S)}{6} = \frac{(561°\,2' - 219°\,17')}{6} = \frac{341°\,45'}{6} = 56\frac{23}{24}$$

So, 56 karaṇs have elapsed, and the current karaṇa is 57th.

Nakṣatra

As we have already seen, the ecliptic is divided into 27 equal parts called nakṣatras beginning with Aśvinī and ending with Revatī. Hence each nakṣatra corresponds to $\frac{360 \times 60}{27} = 800$ minutes along the ecliptic. The nakṣatra at any instant refers to the particular portion of the ecliptic in which the Moon is situated. When the longitude of the Moon in minutes is divided by 800 the quotient gives the number of nakṣatras that have elapsed, and the remainder corresponds to the minutes covered by the Moon in the present nakṣatra.

If we consider the above example, the nakṣatras for the two longitudes of the Moon will be as follows:

To find the nakṣatra, we first convert θ_M into minutes and divide it by 800.

Case 1: (θ_M) = 60° 12′ =3612′.

Nakṣatra = $\frac{3{,}612}{800} = 4\frac{103}{200}$. Therefore 4 nakṣatras have elapsed, and the current nakṣatra is 5th (Mṛgaśiras).

Case 2: (θ_M) = 201° 2′ = 12,062′.

Nakṣatra = $\frac{12{,}062}{800} = 15\frac{31}{400}$. Therefore 15 nakṣatras have elapsed, and the current nakṣatra is 16th (Viśākhā).

Yoga

Yoga means addition, which indicates the period during which the sum (yoga) of the nirayaṇa longitudes of the Sun and the Moon amounts to 13° 20′ or integral multiples. The yoga at any moment can be found by dividing the sum of the longitudes of the sun and moon at the moment by 13° 20′. The quotient will give the serial number of the yoga expired and the remainder will give the elapsed part of the current yoga in units of arc. It pertains to a 27-fold division of the ecliptic. $\theta_S + \theta_M$ in degrees is found (if the number exceeds 360, we can subtract 360 from it). It is converted into minutes (by multiplying this by 60) and then divided by 800. The quotient (Q) represents the number of yogas completed, and Q + 1 represents the current Yoga. We shall compute the yoga for the two cases:

Case 1: $(\theta_M) = 60° \ 12'$; $(\theta_S) = 19° \ 7'$; $\theta_M + \theta_S = 60° \ 12' + 19° \ 7' = 79° \ 19'$. Converting it to minutes, we get 4759′. Dividing this by 800, we get,

$$\frac{4759}{800} = 5\frac{759}{800}.$$

Therefore, five yogas have elapsed and the current one is 6th.

Case 2: $(\theta_M) = 201° \ 2'$; $(\theta_S) = 219° \ 17'$; $\theta_M + \theta_S = 201° \ 2' + 219° \ 17' = 420° \ 19'$. After subtracting 360° from this and converting it to minutes, we get 3619′. Dividing this by 800, we get,

$$\frac{3619}{800} = 4\frac{419}{800}.$$

Therefore, four yogas have elapsed and the current one is 5th.

Vāra

Vāra is the day of the week as we know it in modern parlance. It is interesting to note that this calendar term is used by all calendars of the world maintaining the same name sequence and serial order of the days in the cycle of 7 days which comprise the week. There is a practical issue in determining the day of the week. Events are recorded as per the calendar in use at a region at a point in time. The method of recording might have changed over time and there are variable lengths of the year in the luni-solar calendar. Therefore, it becomes difficult to compute the exact time gap between the occurrence of two events which is essential for astronomical and calendrical analysis. This problem is tackled by the method of counting continuously the number of mean solar days from a fixed distant epoch.

Āryabhaṭa I was the first astronomer in the world to conceive the brilliant idea of counting continuously the days without the involvement of months and years which varies from one calendar to another. This system is named 'ahargaṇa', literally meaning 'count of days'. As per Āryabhaṭīya, the beginning of the current Kali-yuga was on a Friday (18 February, 3102 BCE). Let A be the ahargaṇa corresponding to a given day. Divide A by 7. If the remainder is 0, the given day is a Friday, if it is 1, it is a Saturday, etc. This is used to check the computed value of A. Suppose the ahargaṇa of the day for which we need to compute the pañcāṅga is 18,70,348. The day of the week is determined as follows:

The ahargaṇa of the day is 18,70,348. Dividing this by 7 we get a quotient of 2,67,192 and a reminder of 4. For reminder 4, the day of the week is Tuesday.

9.7 ASTRONOMICAL INSTRUMENTS (YANTRAS)

Astronomy, unlike certain other disciplines like architecture and metallurgy, is more a science of observation and computation. The positions and movements of heavenly bodies have to be observed and recorded very accurately before a theory to explain their motions can be propounded. The theories have to be revised if their predictions do not tally with the observations. Also, for several astronomical computations, the time since the sunrise has to be measured precisely. In the case of motions of stars, planets and other luminaries in the sky visual observations are not likely to be accurate. Therefore, it is necessary to devise instruments to ascertain the positions and motions of heavenly bodies and measure the duration of time. Ancient Indians recognized this and developed a variety of instruments, (yantras). The instruments that were used in Indian astronomy include the water clock (ghaṭī-yantra), gnomon (śaṅku), cross-staff (yaṣṭi-yantra), armillary sphere (gola-yantra), board for the sun's altitude (phalaka-yantra), sundial (kapāla-yantra) to name a few. We shall see a brief description of some of these.

Gnomon

The simplest instrument is just a straight, vertical stick (śaṅku) which has a pointed tip. This is called a 'gnomon' in astronomy, which along with a rope formed a basic but very powerful instrument in Indian astronomy. The Indian texts describe methods to find various astronomical quantities like the directions, time of the day, the latitude of the place, and the longitude of the Sun, etc. from the shadow of the gnomon. When the shadow of the pillar was at its shortest length on any day, the gnomon indicated midday. To obtain some idea of the time of the morning or afternoon, an observer can assess the length and direction of the shadow. The first use of the gnomon in India appears to be to fix the cardinal points.

Procedure for Fixing the Cardinal Direction

One of the first requirements is to fix the cardinal directions (east, west, south, north) in a given location. Kātyāyana-śulba-sūtra gives the following method to draw the east-west line. Figure 9.8 depicts this procedure pictorially. Place a gnomon *OX*, exactly vertically on level ground, so that its tip at *X* points at the zenith *Z*. Draw a circle with a suitable radius with the base of the gnomon *O* as the Centre. Mark the points *W′* and *E′* where the tips of the shadow of the gnomon are on the circle, in the forenoon and the afternoon respectively. Then, *E′W′* is in the east-west direction. This is because the shadows of the same length would be

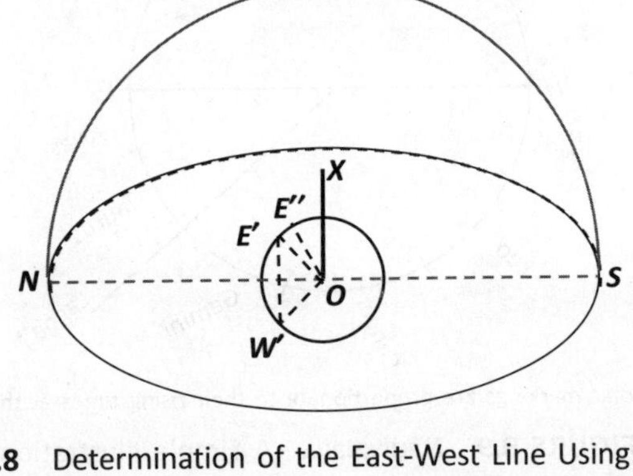

FIGURE 9.8 Determination of the East-West Line Using a Gnomon

symmetrically situated with respect to the north-south line, and $E'W'$ is perpendicular to that and is the east-west line. One can draw a line parallel to this passing through O, for locating the east and west points at O. There is a slight inaccuracy in the above procedure, due to the change in the declination of the Sun during the day. So, if W' and E'' are the points at which the shadows touch the circle, the east-west line is $E'W'$, where E' is slightly displaced from E''. The expression for the amount of displacement is discussed in a commentary on Brāhmasphuṭa-siddhānta of Brahmagupta and many texts which followed it.

Armillary Sphere

In its simplest form, an armillary sphere is a ring of bronze fixed in the plane of the equator. When the shadow of the upper half exactly covered the lower half, it indicates an equinox. However, the armillary sphere described by Bhāskara II and other Indian astronomers is elaborate. It consisted of three spheres; a bhagola or the sphere of the fixed stars, a khagola or the sphere of the sky outside the bhagola, and a third sphere known as dṛggola outside the khagola, in which the circles forming both the spheres khagola and bhagola are mixed. This can be used to find the time since sunrise and the 'lagna' at that time[20].

Nāḍīvalaya-yantra

The Nāḍīvalaya is a large circular wooden disc with an axis in the centre (see Figure 9.9). It is divided into 60 ghaṭikās and also into 12 signs of the ecliptic with variable arcs corresponding to the periods of their risings in the place of observation. The twelve periods thus marked are further subdivided into finer units. The signs are inscribed in the reverse order of the signs, that is first Aries, then Taurus to the West of Aries, and so on. It is placed in a plane parallel to the equinoctial so that the axis in the centre points towards the north pole. The longitude of the Sun in signs, and degrees for the sunrise of the given day is obtained by calculation and that position is marked on the circle. The disc is rotated around the axis so that the shadow of the axis falls on the mark made for the position of the Sun at sunrise. Using this method, the disc is fixed in position. As the sun rises, the shadow of the axis advances from the marks made for the point of the sunrise to the nadir. The number of ghaṭis will be seen between the point of sunrise to the position of the shadow which will also indicate the lagna or the sign on the eastern horizon.

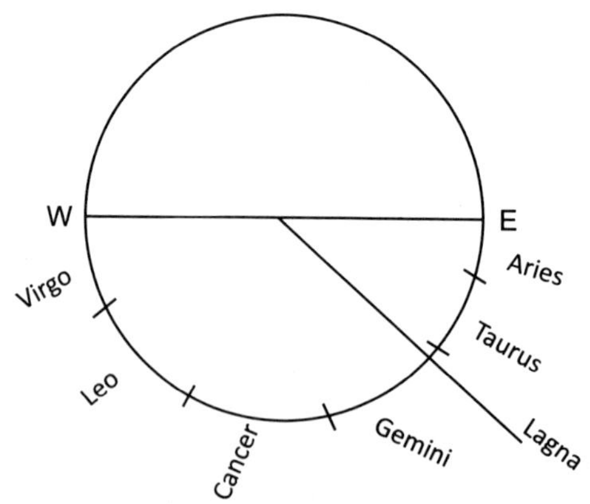

* The spacing of zodiac markings are proportionate to their rising times at the place installation.

FIGURE 9.9 Nāḍīvalaya – A Simple Illustration

Gola-yantra

The armillary sphere and the Nāḍīvalaya described above will measure the time and indicate only the lagna after sunrise on a day when the sky is clear. However, it is necessary to measure time during the night also and to determine the lagna at a particular instant. For this purpose, Indian astronomers had devised other instruments. One of these was Gola-Yantra, which is made of wood, perfectly spherical, and uniformly dense all around but light (in weight). According to Āryabhaṭa, it should be made to rotate keeping with time with the help of mercury, oil, and water and by proper application of astronomical principles. Sūrydeva a commentator of his work provides details of how to construct this instrument.

Cakra-yantra

Cakra-Yantra is made in the form of a plate of metal or seasoned wood and a needle is fixed at the center. When the instrument is illuminated by the rays of the Sun on both sides, the shadow of the needle will give the angular height of the Sun. For his observatories at Jaipur and Varanasi, Raja Sawai Jai Sıngh built Cakra Yantras which are very large instruments. There are two instruments at Jaipur each 6 feet in diameter and one at Varanasi, 3 feet 7 inches in diameter, one inch thick and two inches broad, faced with brass, on which degrees and minutes are marked They are mounted on pillars and fixed so as to revolve around an axis parallel to the earth's axis. The axis carries a pointer, which indicates the hour angle on the fixed circle. Figure 9.10 is a simple representation of a Cakra-Yantra.

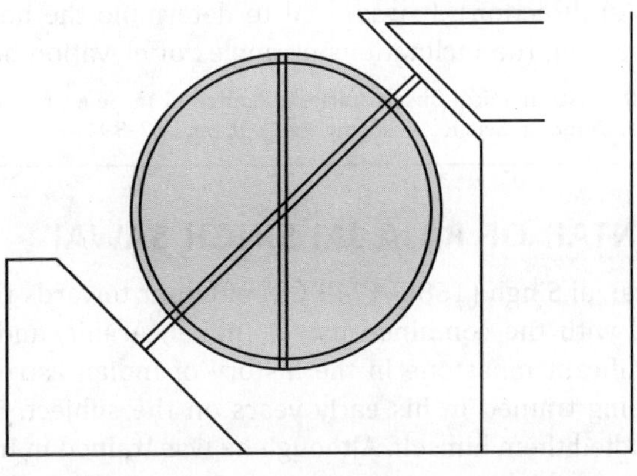

FIGURE 9.10 A Simple Representation of Cakra-Yantra

Astronomical Instruments Described in Siddhānta-śiromaṇi

Astronomy is an observational science. The positions and movements of heavenly bodies need to be continuously observed and recorded with good accuracy before a theory to explain their motions can be propounded. Therefore, astronomical instruments play a crucial role in advancement of theory in astronomy. Indian astronomers have devised several instruments over time. All these instruments constituted astronomical observatories in older times. In Siddhānta-śiromaṇi of Bhāskarācārya, we find a detailed description of several astronomical instruments. Some of them with brief descriptions are given below:

1. **Gola-yantra, the armillary sphere:** A sphere in which all movable and fixed circles are designed. An observer can perform observations sitting within the sphere itself. This instrument serves the purpose of an astrolabe. Bhāskarācārya performed lunar observations with the help of this instrument and compiled his Bījopanaya.
2. **Cakra-yantra:** A wooden or metallic wheel-like structure with an axis fixed in a hole at its centre. It was used to determine longitudes and latitudes of planets.
3. **Cāpa-yantra:** This is half the structure of Cakra yantra.
4. **Turīya-yantra:** This constitutes one quadrant of the Cakra yantra with a stick or nālika (tube) for observing celestial bodies in order to determine their zenith distances and altitudes.
5. **Nāḍīvalaya:** A cakra in the plane of the equator used to determine directly timings of rising and setting of signs.
6. **Ghatī-yantra:** A droṇa (bowl-shaped) water-clock with a hole of standardized size at its bottom.
7. **Nara or Śaṅku:** (the gnomon) made of Ivory or a metal.
8. **Phalaka-yantra:** A plank with a circle of radius 30 aṅgulas drawn on it. The circle graduated in ghaṭīs and degrees. When hanging in the plane of vertical circle, the instrument can read zenith distance directly through graduations on the circle.
9. **Dhī-yantra:** It is a simple stick instrument (augmented by a plumbline-like device to assign vertical direction). It was used to determine the heights and distances of objects by measuring the inclinations or angles of elevation at different points.

Source: **Sharma, S.D. (2000).** "Astronomical Observatories", Chapter 11 in, Sen, S.N and Shukla, K.S. (Eds.), *History of Astronomy in India*, Indian National Science Academy, 2nd ed., pp. 367–394.

9.8 JANTAR MANTAR OF RĀJĀ JAI SINGH SAWAI

The efforts of Raja Sawai Jai Singh (1686–1743 CE) of Jaipur towards the fostering of scientific observational practices with the combined use of Indian, Arabic, and European advances in astronomy mark a significant milestone in the history of Indian astronomy. Jai Singh was an astronomer himself, being trained in his early years on the subject. He collected texts of all three traditions and studied them himself. Although he was trained in Indian astronomy he was not satisfied with the use of old data on astronomical constants. Therefore, he prepared metallic instruments after making some improvements but found that the accuracy of his instruments was still not satisfactory. He felt the need for making gigantic instruments that are sturdy, free from wear and tear, and not affected by changing weather conditions, etc. Further, Jai Singh believed that European instruments were not large and suffered from errors. Therefore, he took a different approach to build astronomical instruments and created a versatile astronomical observatory by setting up a group of essential instruments in one place. These observatories known as Jantar Mantar were built in five places: Delhi, Jaipur, Varanasi, Ujjain, and Mathura during his regime. The observatory at Mathura was located on the banks of Yamuna in a fort built by Raja Man Singh of Amber. Due to multiple attacks, there are not even debris of the observatory anymore. The last remains of the Mathura Jantar Mantra were demolished in 1857 CE when a contractor, Jyoti Prasad bought the fort.

When Jai Singh constructed the Delhi observatory, he ensured that the rules of geometry are properly adhered to, for adjustments to meridians and latitudes of the place so that all errors were rectified. The Delhi observatory was completed by 1724 CE. He erected instruments of his own invention, such as Jai Prakāśa-yantra, Rāma-yantra, and Samrāṭ-yantra. Samrāṭ-yantra had semidiameters of 18 cubits (1 cubit = 20 inches) with a thin layer of lime and stone which provided good stability to the instruments. The observations obtained from the instruments erected at the Delhi Jantar Mantar were found to give measurements that tallied well with other known results. To confirm the reliability of the results obtained through observations, he constructed similar observatories in Jaipur, Ujjain, Varanasi, and Mathura.

The Jantar Mantar deployed a variety of known astronomical instruments such as Cakra Yantra, Nāḍīvalaya, and Kapāla-yantra. When Jai Singh constructed the Jantar Mantar, he not only utilized the tradition of astronomical techniques in India but also made some innovations by bringing in the astronomical techniques he learnt from the Arabic and European astronomers. Let us see some details about Samrāṭ-yantra.

Yantra Samrāṭ is a big instrument with a gnomon wall which is set in the precisely determined north-south direction and is made up of lime brick stones. This is 3 hastas (1 hasta = 18 inches) in thickness and 24 hastas in length. Its upper part is slanting, made up of stones, and points towards Polaris. In the south the root side is 4.5 hastas in height and on the front side in the north it is three inches short of 15 hastas. It has ladders in between. On the eastern and western sides, there are two stony arcs with radii 6 hastas and are more than a quadrant each in size. These have a breadth of about 4 hastas and a thickness of 9.5 aṅgulas. Their centres lie on the śankuppālīs (the edges of the gnomon wall). On each arc, there are quadrants which on edges have divisions in 15 equal ghatis. Each division is further graduated in uniform six subdivisions which are a little less than an aṅgula in measure. On the śaṅkupālīs (on both edges of the gnomon wall) there are two small iron rings, whose centers coincide with the centers of the cāpa-pālīs (the edges of the arcs). Figure 9.11 is a simple sketch of the instrument.

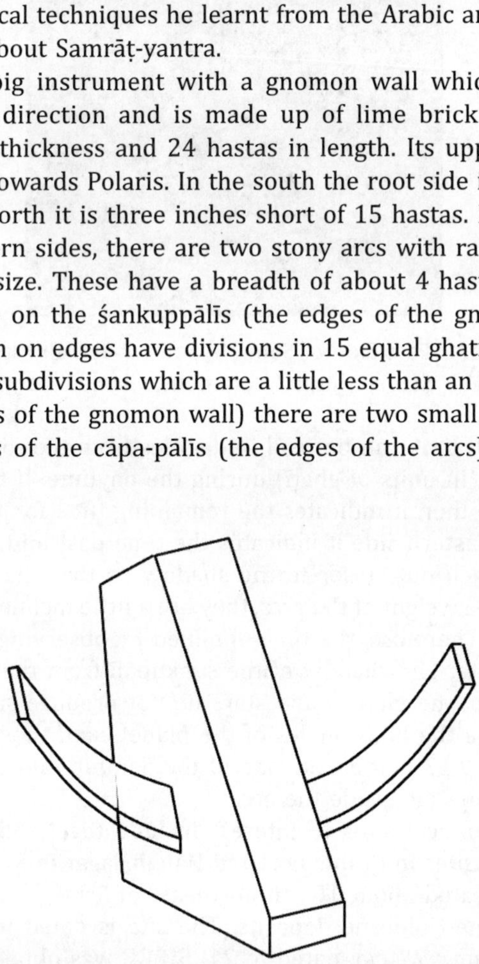

FIGURE 9.11 A Simple Sketch of Samrāṭ-yantra

FIGURE 9.12 East Side of Samrāṭ-yantra 2 at Delhi Jantar Mantar
Source: https://commons.wikimedia.org/

Using this instrument, time can be read in degrees from the top of the gnomon up to the shadow on the cāpa-pālī (in units of ghaṭī) during the daytime. If the shadow of the gnomon falls on the western side, then it indicates the remaining time for midday. On the contrary, if the shadow falls on the eastern side it indicates the time past midday. Steps built by the side of arcs enable one to reach out closer to the shadow so that the measurement is accurate. However, on account of the weight of the arcs, they are a little inclined from the actual positions near the terminal edges. Therefore, the time obtained by observing the shadows at the edges is likely to be less accurate. The shadow of the śaṅkupālī from the Moon is not so clear as it is from the Sun. Moreover, the planets and stars do not produce shadow at all. Nevertheless, it is possible to determine the hour angles of the planets and those of stars with the help of this instrument[21]. Figure 9.12 is the east side of the Samrāt Yantra 2 in Delhi Jantar Mantar, where we can see the stairs alongside the arc.

Other archaeoastronomical sites of interest include the Neolithic site of Burzahom (in Kashmir), and megalithic sites in Brahmagiri and Hanamsagar in Karnataka in the South India. The Burzahom, site is located about 10 km northeast of Srinagar in the Kashmir Valley on a terrace of Late Pleistocene-Holocene deposits. The site is dated to around 3000–1500 BCE. A stone slab of size 48 cm × 27 cm, dated to 2125 BCE was obtained that shows two bright objects in the sky with a hunting scene in the foreground.

The megalithic stone circles of Brahmagiri in the Chitradurga district of Karnataka in South India, have been dated to 900 BCE. Studies pertaining to this suggest that site lines from the center of a circle to an outer tangent of another circle point to the directions of the sunrise and full moon rise at the time of the solar and lunar solstices and equinox. Hanamsagar in Karnataka is another megalithic site with stone alignments pointing to cardinal directions. There are 2,500 stones arranged in a square of side 600 meters with 50 rows and 50 columns with a separation between stones of about 12 m. The lines are oriented in cardinal directions. It has been argued that the directions of summer and winter solstice can be fixed in relation to the outer and the inner squares. Other astronomical observations such as estimating the time of the day using the shadow, the prediction of months, seasons, and passage of the year could have been possible using the structure in the site[22].

SUMMARY

- Knowledge of astronomy was widely used by all sections of Indian society, not just by the subject matter experts.
- Good knowledge of astronomical concepts of ancient Hindus is also evident from several Vedic texts.
- Āryabhaṭīya is the first text on astronomy in India. It established the framework for mathematical astronomy in India. It contains a systematic treatment of all the traditional astronomical problems.
- The elapsed time for the Sun to the same star, which is nothing but the period of the revolution of the Earth around the Sun, is the solar year.
- It had been noticed that the Sun and the Moon return together to nearly the same position in the framework of stars, after five years. The concept of Yuga was introduced to synchronize the Solar and Lunar calendars.
- A lunar month (cāndra-māsa) is the time interval between two new Moons (Amāvāsyā), or two full Moons (Pūrṇimā). It consists of two pakṣas; Śukla-pakṣa (bright fortnight) from Amāvāsyā to Pūrṇimā, and the Kṛṣṇa-pakṣa (dark fortnight) from Pūrṇimā to Amāvāsyā.
- In the Indian system, the ecliptic is divided into 27 equal divisions, known as nakṣatras. Each day would be associated with a nakṣatra.
- Tithi captures the angular separation between the Sun and the Moon. It is a time-unit during which the angle between the Sun and the Moon (as viewed from the Earth) increases precisely by 12°.
- The Vedāṅga Jyotiṣa gives rules for calculating some astronomical quantities related to the Sun and the Moon. The astronomical texts of the siddhānta period are far more sophisticated and elaborate, with procedures to calculate planetary positions, eclipses, etc.
- Around 1500 CE, Nīlakaṇṭha Somayājī modified the procedure based on a detailed scientific analysis and proposed a revised planetary model in his Āryabhaṭīya-bhāṣya and other works.
- There are several regional versions of Pañcāṅga throughout the country to cater to the local requirements. They all employ Sūrya-siddhānta to compute the Pañcāṅga. In recent times, Graha-lāghava of Gaṇeśa Daivajña is used to make pañcāṅga calculations.
- The instruments used in Indian astronomy include the water clock (ghaṭi-yantra), gnomon (śaṅku), cross-staff (yaṣṭi yantra), armillary sphere (gola-yantra), board for the sun's altitude (phalaka-yantra), and sundial (kapāla-yantra).
- The simplest astronomical instrument is just a straight, vertical stick (śaṅku) which has a pointed tip. This is called a 'gnomon' in astronomy.
- In its simplest form, an armillary sphere is a ring of bronze fixed in the plane of the equator.
- The efforts of Raja Sawai Jai Singh of Jaipur towards the fostering of scientific observational practices with the combined use of Indian, Arabic, and European advances in astronomy mark a significant milestone in the history of Indian astronomy.

REVIEW QUESTIONS

1. Why is the study of Astronomy important for mankind?
2. How does the Indian approach to astronomy differ from the other approaches?
3. Prepare a one-page note outlining the salient contributions of Indians to the field of astronomy. In what way did it contribute to the development of astronomical thought in other parts of the world?
4. Briefly describe the contents of the text, Āryabhaṭīya.
5. What are the key contributions of Āryabhaṭīya to the field of astronomy?
6. What is the concept of a 'yuga' in a calendrical system? Why is it required?
7. Briefly describe various aspects of the Indian conceptualization of a 'year'.
8. Explain the notion of the ecliptic and the 'northern' and 'southern' motions of the Sun.
9. Explain the solar and a lunar month in an Indian calendar. Explain the term 'adhikamāsa'.
10. What do you understand by the following terms:
    (a) Nakṣatra and Tithi
    (b) Rāśi
    (c) Vatsara
    (d) Saṃvatsara
    (e) Uttarāyaṇa
    (f) Idāvatsara and Anuvatsara
    (g) Dakṣiṇāyana
11. Why do we need to perform Manda-saṃskāra and Śīghra-saṃskāra?
12. What is a Mahā-yuga? What is Ahargaṇa?
13. Discuss the basic features of Nīlakaṇtha Somayājī's planetary model.
14. Why is Pañcāṅga called so? What are the elements of a Pañcāṅga?
15. Comment on the statement, "Raja Sawai Jai Singh's contributions to the field of astronomy is timely and significant".
16. Briefly describe the instruments used by ancient Indians for astronomical purposes.

DISCOVER IKS

1. A calendar is an important requirement for us as it helps us keep track of time, agree to commitments, and finish tasks as per plan. Calendaring is one of the core issues addressed by astronomers. Listen to the video available in the link, https://www.youtube.com/watch?time_continue=661&v=MvpuC7Dg4e0&feature=emb_logo. After watching the video, prepare a note to answer the following questions provided below:
    (a) What is your understanding of the Western approach to calendaring? What are some of its limitations?
    (b) Compare and contrast the Indian Calendar and Western Calendar. Which one is more scientific? Why?
2. Several ancient Indian texts deal with various aspects of astronomy. The video link provides an interesting conversation on "Ancient Indian astronomy and Engineering". https://www.youtube.com/watch?v=UeaXs1O9qGw. After watching the video, prepare a two-page note to answer the following questions:
    (a) Identify four key observations made by Prof. R N Iyengar with respect to the astronomical thought and developments in Ancient India.
    (b) Who are Parāśara and Vṛddha-Garga? What are his major contributions in the area of Astronomy?
    (c) What is the connection between Vedic practices and astronomy?

3. Sūrya-siddhānta is one of the frequently quoted texts among the Pañca-siddhāntas in India. In this video, there is a discussion on Sūrya-siddhānta: https://www.youtube.com/watch?v=bQNhQ7wxOvA. After watching the video, prepare a note that answers the following questions:
   (a) What is Sūrya-siddhānta? What does it deal with?
   (b) How old is this text and what is its importance?

SUGGESTED READINGS

Abhyankar, K.D. (2005). "Earliest Vedic Calendar", *Indian Journal of History of Science*, 40, pp. 1–7.

Balachandra Rao, S. and Uma, S.K. (2006). "Grahalaghavam of Ganesha Daivagna", Indian National Science Academy.

Balachandra Rao, S. and Venugopaly, P. (2008). "Eclipses in Indian Astronomy", Bharatiya Vidya Bhavan, Bengaluru.

Iyengar, R. N. (2003). "Internal Consistency of Eclipses and Planetary Positions in Mahabharata", *Indian Journal of History of Science*, **38**, pp. 77–115.

Iyengar, R.N. (2005). "Eclipse Period Number 3339 in Rigveda", *Indian Journal of History of Science*, **40**(2) June, pp. 139–152.

Iyengar, R.N. (2008). "Archaic Astronomy of Parashara and Vridha Garga", *Indian Journal of History of Science*, **43**, 1.

Iyengar, R.N. (2010). "Comets and Meteoritic Showers in the Rig Veda and their Significance", *Indian Journal of History of Science*, **45**, 1.

Kak, S.C. (1987). "On Astronomy in Ancient India", *Indian Journal of History of Science*, **22**(3), pp. 205–221.

Kak, S. (2010). "India", Chapter 6 in Heritage Sites of Astronomy and Archaeoastronomy in the context of the UNESCO World Heritage Convention A Thematic Study, Clive Ruggles and Michel Cotte (Eds.), Published by ICOMOS, pp. 99–112.

Narahari Achar, B.N. (1977). "A Note on The Five-Year Yuga of the Vedanga Jyotisa", *Electronic Journal of Vedic Studies*, 3–4, pp. 21–28.

Ramasubramanian, K. and Sriram, M.S. (2011). *Tarkasaṅgraha of Nīlakaṇṭha Somayājī*, Hindustan Book Agency, New Delhi.

Sastry, T. and Kuppanna, S. (1985). "Vedanga Jyotisa by Lagadha (edited and translated)", critically edited by K V Sarma, Indian National Science Academy, New Delhi. India.

Sewell, R. and Dikshit, S.B. (1995). *The Indian Calendar*, Motilal Banarsidass, Delhi.

Shyalaja, B.S. (2007). "The zodiacal pillars of Sringeri", *Current Science*, 92, pp. 846–849.

Sen, S.N. and Shukla, K.S. (2000). "History of Astronomy in India", Indian National Science Academy, 2nd ed., New Delhi.

Subbarayappa, B.V. and Sarma, K.V. (1985). *Indian Astronomy: A Source Book*, Nehru Centre, Mumbai.

Sarma, S.R. (2015). "Astronomical Instruments in India", *Handbook of Archaeoastronomy and Ethnoastronomy*, Springer Science Business Media, New York.

Sarma, S.R. (2018). "Water Clock and Steelyard in the Jyotiṣkaraṇḍaka", *International Journal of Jaina Studies (Online)*, **14**(2), pp. 1–49.

Vahia, M.N. and Yadav, N. (2011). "The Origin and Growth of Astronomy, as Viewed from an Indian Context", In: Orchiston W., Nakamura T., Strom R. (Eds.), Highlighting the History of Astronomy in the Asia-Pacific Region, *Astrophysics and Space Science Proceedings*, Springer, New York, pp. 61–84.

ENDNOTES

1. We gratefully acknowledge that a substantive part of the material in this chapter is based on a write-up on Astronomy provided by Prof. M S Sriram, Professor, K V Sarma Research Foundation, Chennai.
2. **Vahia, M.N. and Yadav, N. (2011).** "The Origin and Growth of Astronomy, as Viewed from an Indian Context", In: Orchiston W., Nakamura T., Strom R. (Eds.), Highlighting the History of Astronomy in the Asia-Pacific Region. *Astrophysics and Space Science Proceedings*, Springer, New York, pp. 61–84.
3. We see the following prayer in the 19th Kāṇḍa, Sūkta 9:

   शं नों मि॒त्रः शं वरुं॒णः शं वि॒वस्वां॒छमन्तंकः । उत्पाता॒ः पार्थिवा॒न्तरिंक्षा॒ः शं नों दि॒विचंरा॒ ग्रहांः॥७॥
   शं नो॒ भूमिर्वेप्यमा॒ना शमु॒ल्का निर्हतं च॒ यत् । शं गावो॒ लोहितक्षीरा॒ः शं भूमि॒रवं तीर्य॒तीः॥८

   नक्षंत्रमु॒ल्काभिहंतं॒ शमंस्तु न॒ः शं नोंऽभिचा॒राः शमुं सन्तु कृत्याः । शं नो॒ निखांता व॒ल्गाः शमु॒ल्का देंशोपस॒र्गाः शमुं नो भवन्तु ॥९॥
   शं नो॒ ग्रहांश्चान्द्रम॒साः शमांदि॒त्यश्चं राहु॒णा। शं नों मृत्युर्धूमकेंतु॒ः शं रु॒द्रास्ति॒ग्मतेंजसः ॥१०॥

   śaṃ nó mi̱traḥ śaṃ varúṇaẖ śaṃ vi̱vasvā̱ṃchamantákaḥ | utpātā̱ḥ pārthivā̱ntarikṣā̱ḥ śaṃ nó di̱vicárā̱ grahā̱ḥ́ ||7|| śaṃ no̱ bhūmirvepyamā̱nā śamu̱lkā nirhátaṃ ca̱ yat | śaṃ gāvo̱ lohitakṣīrā̱ḥ śaṃ bhūmi̱ravá tīrya̱tīḥ||8 nakṣátramu̱lkābhihátaṃ̱ śamástu naẖ śaṃ nó'bhicā̱rāḥ śamú santu kr̥̱tyāḥ | śaṃ no̱ nikhātā va̱lgāḥ śamu̱lkā déśopasa̱rgāḥ śamú no bhavantu ||9|| śaṃ no̱ grahāścāndrama̱sāḥ śamādi̱tyaśca rāhu̱ṇā| śaṃ nó mr̥tyurdhū̱makétuẖ śaṃ ru̱drāsti̱gmatéjasaḥ ||10||

   Favour us Mitra, Varuṇa, Vivasvān, and the God of Death, Portents on earth and in the air, and planets wandering in heaven! 19.9.7

   Gracious to us be trembling earth, gracious the flaming meteor stroke! Gracious be kine who yield red milk, gracious be earth when sinking down! 19.9.8

   Gracious be meteor-stricken constellation, gracious to us be magic spells and witchcraft! Gracious to us be buried charms, and gracious the meteors and the portents of the region! 19.9.9

   Kind be the Powers who seize the Moon, with Rāhu be Ādityas kind! Favour us Death and Cornet, and Rudras with penetrating might! 19.9.10
4. See for example, **Iyengar, R.N. (2008).** Archaic Astronomy of Parashara and Vriddha Garga, *Indian Journal of History of Science*, **43**, 1, pp. 1–28.
5. See **Narahari Achar, B.N. (1998).** "On the enigma of the five-year yuga of Vedāṅga Jyotiṣa", *Indian Journal of History of Science*, **33**(2), pp. 101–109, for a discussion of the yuga system and several Vedic and Purāṇic references.
6. See for more details, **Narahari Achar, B.N. (1999).** "On Exploring the Vedic Sky with Modern Computer Software", *Electronic Journal of Vedic Studies*, **5**, Issue 2 (December), p. 3.
7. For some more details, see, **Sarma, K.V. (2000).** "A Survey of Source Materials". Chapter 1 in Sen, S.N and Shukla, K.S. (Eds.), History of Astronomy in India, Indian National Science Academy, 2nd ed., pp. 1–23.
8. तस्माद् आदित्यः षण्मासो दक्षिणेनैति षडुत्तरेण । tasmād ādityaḥ ṣaṇmāso dakṣiṇenaiti ṣaḍuttareṇa |
9. एकविंशं एतद् अहर् उपयन्ति विषुवत्तं मध्ये संवत्सरस्य । ekaviṃśaṃ etad ahar upayanti viṣuvattaṃ madhye saṃvatsarasya | Aitareya Brāhmaṇa 18.4 The ekaviṃśa [rite] is performed on the [equinox] day, occuring in the middle of the year.
10. विषुवं तद्गुणं द्वाभ्यां रूपहीनं तु षड्गुणम् । यल्लब्धं तानि पर्वाणि तथार्धं सा तिथिर्भवेत् ॥ viṣuvaṃ tadguṇaṃ dvābhyāṃ rūpahīnaṃ tu ṣaḍguṇam | yallabdhaṃ tāni parvāṇi tathārdhaṃ sā tithirbhavet || R̥gveda Vedāṅga Jyotiṣa, Verse 31.
11. For a good discussion on the Indian calendar please see, **Chatterjee, A.K. and Chakravarty, A.K. (2000).** "Indian Calendar from Post-Vedic period to AD 1900, Chapter 9 in Sen, S.N and Shukla, K.S. (Eds.), History of Astronomy in India, Indian National Science Academy, 2nd ed., pp. 276–335.

12. Five systems of year are mentioned in the Vedic corpus. Saṃvatsara is the solar year, the time taken by the Sun to pass through the twelve zodiacs. Idāvatsara is the name of the Sāvana year, consisting of 12 months, each of 30 days duration. Anuvatsara is the lunar year, each month of which ends on the amāvāsyā day. Vatsara is the name of the year with 12 lunar cycles, each of about 27.32 days, the time taken by the moon to pass through the Nakṣatras. Parivatsara is the time taken by the Jupiter to pass from one sign of ẓodiac to another.
13. द्वादशारं नहि तज्जराय वर्वर्ति चक्रं परि द्यामृतस्य । आ पुत्रा अग्ने मिथुनासो अत्र सप्त शतानि विंशतिश्च तस्थुः ॥ dvādaśāraṃ nahi tajjarāya varvarti cakraṃ pari dyāmṛtasya | ā putrā agne mithunāso atra sapta śatāni viṃśatiśca tasthuḥ: || (Ṛgveda 1.164.11).
14. **Subbarayappa, B.V. and Sarma, K.V. (1985).** *Indian Astronomy: A Source Book*, Nehru Centre, Mumbai.
15. *Ibid* p. 50.
16. See **Narahari Achar, B.N (1998).** "On the Enigma of the five-year yuga of Vedāṅga Jyotiṣa", *Indian Journal of History of Science*, **33**(2), pp. 101–109, for a discussion of the yuga system and several Vedic and Purāṇic references.
17. मधुश्च माधवश्च वासन्तिकावृतू, शुक्रश्च शुचिश्च ग्रैष्मावृतू, नभश्च नभस्यश्च वार्षिकावृतू, इषश्च ऊर्जश्च शारदावृतू, सहश्च सहस्यश्च हैमन्तिकावृतू, तपश्च तपस्यश्च शैशिरावृतू ।। शतपथब्राह्मणम्/काण्डम् ८/अध्यायः ३/ब्राह्मणम् २ ॥ maduśca mādhavaśca vāsantikāvṛtū, śukraśca śuciśca graiṣmāvṛtū, nabhaśca nabhasyaśca vārṣikāvṛtū, iṣaśca ūrjaśca śāradāvṛtū, sahaśca sahasyaśca haimantikāvṛtū, tapaśca tapasyaśca śaiśirāvṛtū || Śatapatha-brāhmaṇa 8/3/2.
18. वृत्तभपञ्चरमध्ये कक्ष्यापरिवेष्टितः स्वमध्यगतः । मृज्जमशिखिवायुमयो भूगोलः सर्वतोवृत्तः ॥ vṛttabhapañcaramadhye kakṣyāpariveṣṭitaḥ svamadhyagataḥ | mṛjjamaśikhivāyumayo bhūgolaḥ sarvatovṛttaḥ || Āryabhaṭīya, Golapāda, Verse 6.
19. अनुलोमगतिर्नौस्थः पश्यत्यचलं विलोमगं यद्वत् । अचलानि भानि तद्वत् समपश्चिमगानि लङ्कायाम् ॥ anulomagatirnausthaḥ paśyatyacalaṃ vilomagaṃ yadvat | acalāni bhāni tadvat samapaścimagāni laṅkāyām || Āryabhaṭīya, Golapāda, Verse 9.
20. For an explanation of this, please see, **Rai, R.N. (2000).** "Astronomical Instruments *Sharma",* Chapter 10, In, Sen, S.N. and Shukla, K.S. (Eds.), History of Astronomy in India, Indian National Science Academy, 2nd ed., pp. 337–366.
21. For an explanation of this, please see, **Sharma, S.D. (2000).** "Astronomical Observatories", Chapter 11, In, Sen, S.N. and Shukla, K.S. (Eds.), History of Astronomy in India, Indian National Science Academy, 2nd ed., pp. 367–394.
22. For some more details on these, see **Kak, S. (2010).** "Archeoastronomy in India", arXiv. https://arxiv.org/abs/1002.4513. Last accessed on Oct. 1, 2021. Also see, **Vahia M.N. and Yadav N. (2011).** "The Origin and Growth of Astronomy, as Viewed from an Indian Context", In: Orchiston W., Nakamura T., Strom R. (Eds.), Highlighting the History of Astronomy in the Asia-Pacific Region, *Astrophysics and Space Science Proceedings. Springer*, New York, pp. 61–84.

CHAPTER

10 Engineering and Technology: Metals and Metalworking

LEARNING OUTCOMES

After finishing this chapter, you will be able to:

- Develop familiarity with the Science, Engineering and Technology (S & T) heritage of ancient and medieval India
- Understand the role of archaeological, and other evidence in assessing the S & T heritage of India
- Understand about ancient Indian pursuits in the area of metal and metalworking and idol making

This is a gold coin of the Gupta king Samudragupta (330–376 CE) kept at the British Museum. Samudragupta, with halo, standing facing left, wearing cap, decorated coat and trousers and earrings, holding a spear in left hand and making an offering with right hand over altar. In left field is a Garuda standard with ribbons and crescent above.

Stamping such coins with intricate details requires knowledge of mining, extraction of Gold and further metal forming techniques such as die casting.

Source: https://commons.wikimedia.org/wiki/File:SamudraguptaCoin.jpg

IKS IN ACTION 10.1

Wootz Steel: The Rise and Fall of a Great Indian Technology

Wootz steel is a unique Indian process of making steel, which was widely sought for manufacture of swords in the eastern Mediterranean region as such a quality of steel was simply not available elsewhere. Wootz steel was produced as early as 700 BCE in several places in India and exported to and traded throughout ancient Europe and the Arab world. It became particularly famous in the Middle east for manufacture of Damascus swords.

This sword is known for its edge sharpness, flexibility, strength and the typical surface structure on the sword's surface. The word 'wootz' might have originally been wook; the Tamil language root word for the alloy is urukku. Other Dravidian languages have similar-sounding words for steel. Ukku, is the word for steel in Kannada and Telugu languages.

In the 12th century CE, the Arab Edrisi mentioned that the Hindus excelled in the manufacture of iron and that it was impossible to find anything to surpass the edge of Indian steel. Arabs took ingots of wootz steel to Damascus following which a thriving industry developed there for making weapons and armour.

By the end of the 17th century CE, shipments running into tens of thousands of wootz ingots were traded from the Coromandel coast to Persia. This indicates that the production of wootz steel was almost on an industrial scale predating the Industrial Revolution in Europe. In India till the 19th century CE, swords and daggers of wootz steel were made at centres including Lahore, Amritsar, Agra, Jaipur, Gwalior, Tanjavur, Mysore, and Golconda. Unfortunately, none of these centres survive today.

Because of the oral tradition followed in India, not too many records were available documenting the process of wootz production. It is mainly the European travelers such as Buchanan in 1807, Benjmain Heyne in 1818, H W Voysey in 1832, and Josiah Marshall Heath in 1840 who left detailed accounts. However, Wootz steel caught the attention of scientists in England, France, Russia, and Sweden.

As England had colonized India, there was considerable interest in studying wootz steel. Michael Faraday, the inventor of electricity was fascinated by wootz steel and enthusiastically studied it. Along with the cutler Stodart, Faraday attempted to study how to make Damascus steel and they incorrectly concluded that aluminium oxide and silica additions contributed to the properties of the steel and published their results in 1820. Though Faraday could not replicate the wootz steel, he is hailed as the father of alloy steel.

After careful studies, the Scientists in the West discovered the composition and microstructure of the steel and their relation to mechanical properties. This laid the foundations of modern materials science and the investigations on wootz steel continue to inspire researchers to this day.

Recent investigations on the properties of the ultra-high carbon wootz steel such as super-plasticity justify it being called an advanced material of the ancient world with not merely a past but also perhaps a future.

Unfortunately, on the other hand, the British Raj introduced production taxes and mining bans. This must have been partly responsible for the disruption of mining and steel manufacture. Eventually, it may have contributed to the loss of know-how in India in the last 150 years.

Source: Based on **Ranganathan, S. and Srinivasan, S. (2006).** "A Tale of Wootz steel" Resonance (June 2006), pp. 67–77 and **Prakash, B. (1977).** "Metals and Metallurgy", Chapter 3 [3.2] in History of Technology in India Vol. I, A.K Bag (Ed.), Indian National Science Academy, New Delhi, pp. 80–174, and other published material.

Science, Engineering, and Technology (S & T) are the buzzwords of the last two centuries. Everything we learn in our educational system today on S & T has its origin in the West. Furthermore, the technological innovations that we benefit from and the products that we use are also of Western origin. Of late, there have been more activities in the Indian S & T sector, and we seem to be catching up with the cutting-edge technology of the world with great difficulty. The pace and spread of current technological developments have overshadowed our history and make us believe that the ancestral societies other than the West have very little to contribute to the S & T ecosystem. We also tend to believe that all that we have in the last 250–500 years are the only living example of S & T. On account of these, an average well-educated Indian believes that Science and Technology are of Western origin and India has very little to contribute. If we go by this logic, then the medieval society ought to have had a very simple and primitive living in which S & T played practically no role. In this chapter and a few that follow, we focus on this aspect.

10.1 THE INDIAN S & T HERITAGE

Indian civilisation has a rich heritage spanning over the recorded history of the last two millennia. Several archaeological and literary resources point to the existence of a vibrant community in antiquity. Several artifacts, relics, and physical structures dot the length and breadth of the country that bears testimony to this. In the earlier chapters of the book, we discussed in some detail the scientific heritage of ancient Indians and their contributions. We were able to see how with a well-developed body of knowledge in Mathematics, Astronomy, and other fields several theoretical foundations were laid for the benefit of the scientific community. However, we are not aware if Engineering and Technology were part of ancient Indian society. Did our ancestors have any knowledge of S & T? If so, how did they employ the knowledge and how did they benefit from it?

We can approach this question by exploring if there is any evidence of the use of S & T in ancient times. The evidence could be from any of the following three:

(a) The archaeological evidence that shows how S & T ought to have been deployed in ancient times

(b) Living examples of physical entities that point to the use of good S & T practices

(c) Literary resources pointing to the use of S & T in ancient times

In this section of the book, we deploy all three of these approaches to understand the S & T heritage of ancient Indians. We explore a variety of available evidence to make sense of the S & T heritage that could have existed in ancient and medieval times in India. Furthermore, we explore the available knowledge sources in the domain of S & T. Using these we discuss various aspects of S & T in Chapters 11 and 12. A related field of Town Planning and Architecture is separately discussed in Chapter 12.

Archaeological excavations in the last 150 years in India have fetched a rich collection of artefacts that reveal several clues to the Engineering and Technology heritage of ancient Indians. Amongst other things, the excavations of Kalibangan in Rajasthan and Lothal in Gujarat have revealed functional town-planning, dwelling houses built of different sizes of burnt bricks but of a standard proportion of dimensions, tiled flooring, well-developed drainage systems, and pottery which are both utilitarian as well as decorated. Furthermore, it also points to agricultural operations, shipbuilding, metalworking, especially iron and steel, copper, and zinc, and a host of other skilled crafts such as intricate bead-making, and jewellery among the people. Excavations show that Lothal had a man-made dock for berthing boats (or smaller size ships)—a trapezoid basin measuring about 214 m × 36 m. What appears to be the largest maritime structure of the times, had facilities for loading and unloading[1]. There are studies on town planning and architectural developments, skill, and developments.

Archaeological and literary evidence suggests that using the technology they were making military equipment, agricultural implements, ornamental pieces (see Figure 10.1 for an example from 1st century BCE), idols, coins, and a host of other items for medical applications. They were also adept at extracting Mercury and using it in alchemy. Several travellers' accounts (such as that of Herodotus, Ktesias, Hiuen Tsang, and Pliny) reveal how famous the Indian metalsmiths were for their quality and workmanship. We shall see some details of these aspects in Chapter 11.

10.2 MINING AND ORE EXTRACTION

In technology, ancient Indian's contributions to metalworking, metallurgy, and material science are noteworthy. Indian metalsmiths had notable achievements, particularly developing unique skills in mining and working with iron and steel, copper and its alloys, and zinc. Archaeological examination of the vicinity of the major mining centres of modern India viz., Zawar, Rajpura-Dariba, Rampura-Agucha, Khetri, Chamba, Deri-Ambaji, Singhbhum, Chitradurga, and Cudappah presents an interesting picture. These locations have extensive ancient mine workings and debris, and heaps of slags and retorts. Furthermore, a study of the ruins of temples and townships in various places in India corroborates the degree to which mining, and metalworking have developed in ancient India. From an analysis of the archaeological specimens obtained one can infer the skill and technical superiority exhibited by our ancestors in metallurgical works, which we seem to be largely unaware of.

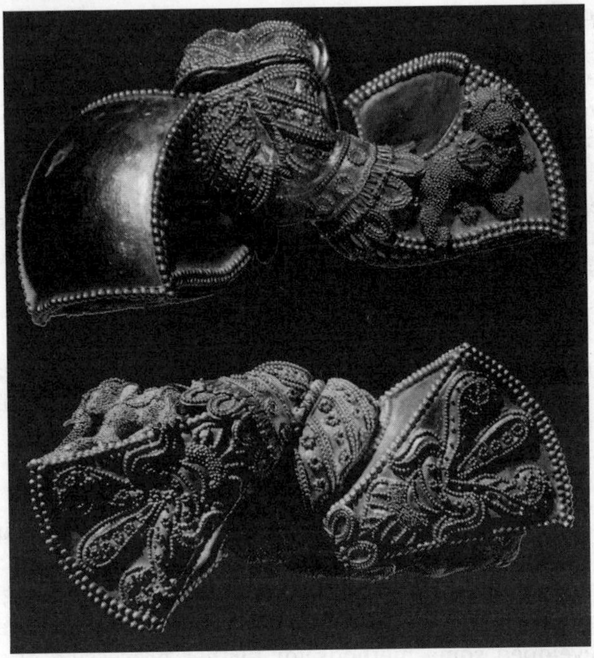

FIGURE 10.1 Andhra Pradesh Royal Earrings (1st Century BCE)
Source: PHGCOM, Photographed at the Metropolitan Museum of Art.

In April 1980, the Hindustan Zinc Limited (HZL) sponsored a three-year research project on the recovery of zinc from the ancient slags which was successfully conducted at the Indian Institute of Technology Kanpur. In 1982, HZL collaborated with the British Museum Research Laboratory and the Department of Archaeology, M.S. University of Baroda on archaeological investigations which led to the discovery of the zinc distillation outfit, including furnaces and retorts showing the production strategy. Extensive archaeological and archaeo-material investigations followed[2]. These investigations suggest that the tradition of underground mining in India goes back to the thirteenth century BCE. For example, excavations in the South Lode (100 m depth) of Rajpura-Dariba mine (80 km north-east of Udaipur) have been C-14 dated as 1260 BCE, 1130 BCE, and 1050 BCE. In Zawar, about 30 km south-west from Udaipur, the ancient mines (earliest C-14 date obtained is 430 BCE) are found, both opencast and underground confirming widespread underground mining of lead-zinc ores in the southern Rajasthan during the 5th-4th centuries BCE onwards. The art of smelting zinc ore and recovery of zinc metal by distillation must have been discovered before the 4th century BCE. This is supported by the discovery of a brass vase containing 34.34% zinc in Takṣaśilā[3].

The excavation of an early historic iron smelting site near Dhatwa in the Tapti valley in Surat District of Gujarat led to an investigation into the metallurgical technology of the site. This revealed evidence of an iron-smelting industry in the form of iron objects, heaps of iron metallurgical tap slag, and pieces of iron ore in layers 3, 2, 1, and on the surface of the mound. Large amounts of iron objects and slag were recovered from layers 1 and 2 suggesting that the industry must have developed over time[4].

Radiocarbon dating of mine timber and charcoal shows that large-scale mining took place between Chalcolithic (early Iron Age in India) and 1800 years ago. Hindustan Zinc Limited has given information on the history of mining and smelting at Zawar, Rajpura-Dariba, and Rampura-Agucha. Several remains of the mining process are available to date in the Zawar mines. Samples have been taken from here for carbon dating. Radiocarbon dating of the remains of wooden stairways, haulage scaffolds, and other artefacts retrieved from the ancient mines indicate that lead-zinc mining existed as far back as 500 BCE. Similarly, radiocarbon of pieces of mine timber, rope, and bamboo basket from the ancient workings indicated that these mines were in production approximately 3000 years ago. Furthermore, a piece of lead metal recovered from the slag dump was found to contain 97.5% lead and 150 ppm silver indicating thereby that the smelting technology was advanced.

The ancient mining and ore extraction process across metals followed a somewhat similar methodology. This involved fire-setting and quenching with water to create cracks in the rock followed by extraction of ore with hammering, chiselling, and scrapping tools. The mining operations left a series of open cut at the top and the excavations were supported by timber. Earthen pitchers were used for carrying water and wooden trolleys for transportation of the ore.

Ancient Indians developed skills to explore ore deposits and mine portions that are remunerative. There are studies providing intricate details on how the ancient Indians were performing the copper mining in the Khetri mines in Rajasthan. Using an initial process of firing and quenching to crack the ore, they knew how to extract the ore using hammers and chisels. They also developed some methods for further processing of the ore to obtain the molten metal. Underground excavations up to nearly 500 feet are visible in several areas in these hill regions. There are several tunnels of dimension 5 ft × 4 ft and these have some provisions for improving the atmospheric conditions at that level. The mining process involved a few steps. First, the ore was dislodged by setting the fire. Further, it was quenched in water, and using a chisel and hammer it was extracted. These extracted ores were carried out of the mines by the labourers in their heads for further processing.

> - In the Vedic texts, we find ample references to Gold, Silver, iron, copper, and their alloys.
> - The Indian metal smiths were adept in alloy technology as they could produce alloys of controlled composition.

To crack the ore, large quantities (about 6–7 tonnes) are stacked and set fire. The workers immediately leave the mines only to return after three days. The studies on the mining process suggest that each worker proceeded into the mine with a chisel, lamp, and a small basket. The lamp placed on his head provided not only illumination but also helped him identify the glittering particles of ore. While at work the miner seated upon his heels with the lamp upon his head, the hammer in his right hand and the chisel in his left hand, and the small basket upon his knee in which he received all the fragments of ore that were struck off by the chisel. The basket was passed through a chain of workers, and in this manner, the ore was cascaded to the surface eventually[5].

Once the ore is brought to the surface, they were finely powdered using heavy hammers, so that they are conducive for the roasting process. The powdered ore was mixed with cow-dung and made into rolls about four inches long which were dried first in the Sun and then roasted in the open air in a fire of cow-dung cake. The ore was then ready for smelting. There is enough evidence to show the smelting of the ore using locally built furnaces, cemented with clay having nozzles and bellows. Using a pair of sticks, the bellow valve was built and they were opened when the bag was raised for blowing the air. Through a hole made at the bottom of the furnace, the molten metal was stirred and drawn out. After lighting the furnace, the ore was gradually introduced alternatively with charcoal and flux (refuse from the old iron furnaces).

10.3 METALS AND METALWORKING TECHNOLOGY

In modern times metals and alloys play an important role in our our lives. Several engineering gadgets and equipment make use of a variety of metals and their alloys. Ancient Indians have made use of base metals and alloys in multiple ways. This includes the making of military aids, ornaments, vessels, tools, artefacts, statues, and coins to name a few. In particular, Indians were known to be quite advanced in iron and steel, and zinc. The archaeological findings of Harappa and Mohenjo-Daro as well as in burial sites in South India have artefacts in Gold, Silver, Copper, and Bronze. The bronze statue of a dancing girl found at Mohenjo-Daro bears testimony to the technological skills of the ancient craftsmen. The statue is evidence of the skills of the craftsmen in drilling fine holes and casting the statue using the lost wax process (known in modern times as the cire-perdue process). In the Vedic texts, we find ample references to Gold, Silver, iron, copper, and their alloys[6]. Therefore, it is not surprising to find that mining, ore processing, extraction of metals, methods for metalworking, and alloying were known to them. In this section, we will see in more detail the above aspects of Indian metalworking and metallurgical principles[7].

A detailed study of the archaeological remains and the repository of literary works from 5th century CE to 12th century CE, in addition to numerous references in the Vedic texts, clearly point to certain unique aspects of metalworking and metallurgy in India during ancient and medieval times. Some of them include the following:

- The smelters and metalsmiths had gained a high degree of knowledge regarding furnace design, combustion of fuel, refractories, and skill in operating the furnaces at the desired temperature. They were also aware of the temperature specifications and other conditions required to carry out the process.
- The knowledge of the Indian metal smiths on Iron-Carbon alloy was clearly superior and ahead of the times. This is evident from their ability to produce the famous wootz steel, which was in demand in the West.
- Indians were the first to introduce Zinc to human civilisation and also to develop Cu-Zn alloys.
- The Indian metal smiths were also adept in alloy technology as they could produce alloys of controlled composition.
- Indians developed good skills in designing and casting a variety of artefacts and deployed good moulding and diecasting methods. The available evidence suggests that they were casting small as well as large objects in the country.

10.3.1 Gold Extraction Process

The process for the extraction of gold in ancient times has been analysed and documented by Bharat Gold Mines Ltd. According to the available information, mercury was added to a mixture of black sand and gold ore, and the mixture was rubbed with a little common salt. After some time, the gold and mercury would form an amalgam. The amalgam is separated by adding water to the mixture and agitating the mixture. The amalgam was rolled in a damp rag and the mercury was squeezed off. Finally, it was burnt in the fire to extract the gold. If the gold was found alloyed with silver or other base metals, it was hammered into a thin sheet between two stones. The plate was burnt in a two-layered cow-dung. The cow dung absorbed the base metals, and the pure gold was separated. This shows that the technique of Hg (Mercury) amalgamation and gilding were apparently known to the Indian craftsmen from the very early times[8].

- Hg, as well as HgS, have been extensively used in the preparation of Āyurvedic medicines.
- Hg is also used for the amalgamation process for the extraction of Silver and Gold.

However, the most common process of separation of native gold from the sand and quartz rock employs a simple gravity separation or panning. In this process, the quartz is first crushed to a fine size to extract the gold particle and then separate it by suspending the mixture in water. When the pan containing the suspension is agitated under water gold particles settle at the bottom of the pan and the sand and soil are washed away. In the case of a larger scale of operation, big pans were suspended in the river water through a sling so that the mixture could be easily agitated.

The main purpose of the extraction of gold was to make ornaments and coins. It was also used to prepare 'Suvarṇa Bhasma' in Āyurvedic preparations. Ancient craftsmen could subject the purified gold metal bar to metalworking techniques such as forging, punching, embossing, etc. so that ornaments of intricate designs could be produced. It appears that ancient craftsmen could also use other metal forming techniques such as rolling and wire drawing. Gold coins were either made by die casting in clay moulds or by the technique of punch marking[9].

10.3.2 Zinc Production

Of all the base metals, Zinc was perhaps the last to be discovered and used during the pre-Christian era. A study by Hegde, Craddock, and Sonawala in 1984 concluded that India was the first to introduce this metal to the rest of the world between 600 to 200 BCE[10]. Geographically Rajasthan stands out as a major mining province dating back to about 1000 BCE. Studies based on the debris and slag dump in the Zawar region of Rajasthan it is estimated that about 15,000 tonnes of Zinc ought to have been mined and processed. Zinc was exported to other countries even before the 11th century CE. Studies on the Zawar zinc mines provide rich insights into zinc mining in ancient India. According to some other estimate, about 250,000 tonnes of zinc concentrates ought to have been extracted from about 2.5 million tonnes of ore in the mined area, before modern mining operations commenced.

- Ancient Indians adopted a novel technique of downward drift reduction distillation process for Zinc production, which is a precursor to the modern processes adopted all over the world.
- The distillation apparatus found near the ancient Takṣaśilā (Needham) was unique and much advanced.

Extensive archaeological excavations conducted in several parts of Rajasthan clearly show that the earliest artefact containing an appreciable amount of zinc anywhere in the world is from India. A pioneering contribution of the Indian subcontinent in terms of the history of science and technology was the mastery of the metallurgy of zinc production. In the Aravalli ranges of northwestern India, amongst rich polymetallic mineral deposits one can see several mining galleries, slag heaps, and retorts which bear testimony to the early extraction of lead and silver at Dariba and Agucha and, most uniquely of zinc, at Zawar. The Zawar area near Udaipur has yielded unique evidence suggesting that the extraction of metallic zinc flourished on an almost semi-industrial scale by the mid-14th century CE as indicated by some carbon dates from the analysis of archaeo-metallurgical debris. Surveys and excavations revealed the extensive presence of furnaces or *koṣṭhis*, intended for an ingenious process of zinc extraction by downward distillation[11].

Zinc smelting was done during the 9th century CE, which was upgraded to an industrial scale around the 13th century CE. Several clay retorts, presumably used for zinc smelting are found scattered in the ruins of Zawar. At Zawar elements of transport systems are evident, with zig-zag paths on slopes with stone or wooden steps or ladders. The size of the passage on the main routes indicates that it would have been possible to carry loads in baskets on the head. Since India had plenty of iron from the early days, the tools used for extraction of the ore such as chisels, hammers, and scrappers were made of iron. For lighting, oil lamps were used, and the available evidence indicates that the lamps were placed at fixed positions in the mining areas.

Ancient Indians adopted a novel technique of downward drift reduction distillation process for Zinc production, which is a precursor to the modern processes adopted all over the world. The Zinc has a melting point of 410°C and a boiling point of 930°C. Moreover, in the open air at about 550°C Zinc gets oxidized to ZnO (Zinc Oxide). Therefore, the only viable method to extract Zinc is to rapidly cool the Zinc vapour to around 500°C to produce the liquid metal thereby preventing its reoxidation. The ingenuity of the ancient Indians was making this possible using the downward drift reduction distillation process. Some studies suggest that in India the process of distillation of water, wine, and probably mercury was known even in the Vedic period (Yajurveda). The distillation apparatus found near the ancient

Takṣaśilā (Needham) was unique and much advanced. This system uses a separate condenser tube fitted to the mouth of the distill to carry the vapours to the condenser vessel which was water-cooled. A similar apparatus called Ḍamru-yantra was used for the preparation of Āyurvedic formulations. We will discuss different yantras used for such extraction purposes in Section 10.5.

The principle of the downward drift distillation process can be explained in simple terms using the description of a yantra available in Rasa-Ratna-Samuccaya (**RRS**) for extraction of Zinc, Mercury, etc. Imagine having a pair of pots placed one over the other with their mouths aligned. What it means is that the lower pot is in the normal position and the upper pot is placed upside down over the lower pot. The upper inverted pot has a solid charge and is sealed with clay. It has a reed stick at the centre for the escape of gases. During the processing, the upper pot is heated by building a firing platform around it. Once it reaches 600°C, the reed is charred and burnt off, paving way for the reduced metal vapour to flow downwards. The lower pot acts as a condenser and it has mechanisms for rapid cooling so that it can convert the vapour into liquid metal. Figure 10.2 is a simple illustration of the yantra for the downward distillation process.

FIGURE 10.2 A Simple Illustration of the Yantra Used for Downward Drift Distillation Process

Source: **Sen, S.N. and Subbarayappa, B.V. (2009).** "A Concise History of Science in India", Indian National Science Academy, 2nd ed., p. 429. Reproduced with permission.

A similar but advanced method was used by the Zawar smelters. The process used by them consisted of the following steps. Zinc is extracted by smelting Zinc Sulphide ore. After sizing, the ore was mixed with charcoal dust and fired into a heap to convert it into oxide by controlled roasting in an oxidizing atmosphere. During this process, ZnS (Zinc Sulphide), PbS (Lead-II Sulphide), and other sulphide minerals got converted to their respective oxide but no reduction of these oxides by carbon was permitted. The roasted ore was mixed with more charcoal powder, salt, and borax as flux and thoroughly mixed with cow dung and water, and then made into balls of 5 to 10 mm dia. by hand rolling. These pellets were dried in Sun and then filled into the brinjal-shaped retorts.

Many retorts have been found at Zawar. They are of two principal sizes with a capacity of 750 c.c. and 2000 c.c. As described in RRS, the main retort or crucible is in the shape of a brinjal. It was made from a clay mixture consisting of locally available refractory clay, rice husk, and sometimes mixed with iron ore dust. Many such compositions have been mentioned in ancient Indian literature.

10.3.3 Copper Mining and Extraction Process

India abounds in ancient specimens of copper. Utensils made of copper, owing to the peculiar sanctity attached to it by ancient Indians, have been used in India from ancient times in the performance of religious ceremonies. Wires made of copper and brass have from early times been used in the construction of stringed musical instruments for which India has been famous. A long and flourishing tradition of the copper industry in ancient India can be inferred from the following archaeological specimens[12]:

- As many as 424 copper implements and weapons and 102 pieces of thin silver plates were discovered in one place measuring about three feet in length, three feet in breadth, and four feet in depth in the village of Gungeria in the Nagpur Division of the erstwhile Central Provinces in 1870 CE. The copper implements were mostly celts, shovels, axe-blades, spades, manufactured for warlike, domestic, and agricultural purposes. These have been kept in the Archaeological Department of the Indian Museum, Kolkata. Figure 10.3 is a sample of copper swords belonging to pre-historic times.

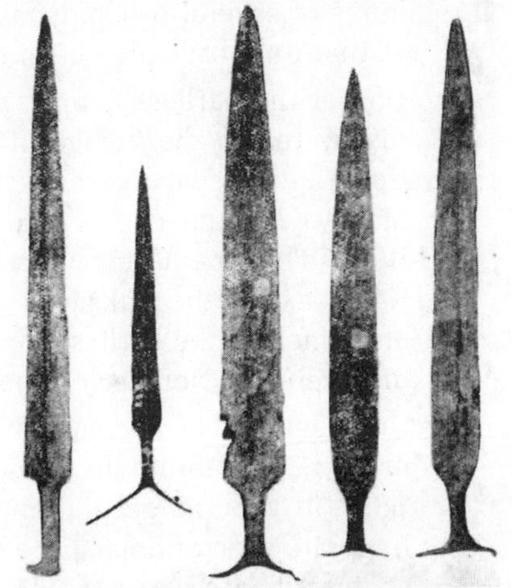

FIGURE 10.3 Prehistoric Copper Swords Discovered at Fatehgarh (Indian Antiquary, 1905, p. 236).

- A big solid copper bolt was found in the Rampurwa Ashoka pillar near the frontiers of the Kingdom of Nepal. The bolt is barrel shaped in appearance, slightly tapering at the two ends. It is 242 inches long, with a circumference of 14 inches at the centre. The two ends of the bolt have a circumference of about 12 inches. This bolt ought to have been fabricated as early as the third century BCE, thereby testifying to the high metallurgical skill of ancient Indians.
- A colossal copper statue of Buddha was discovered at Sultanganj in the district of Bhagalpur in the ruins of an old Buddhist monastery (see Figure 10.4). It was 7 feet 6 inches high and weighed nearly 1 ton. Based on the mode of its construction and discovery of a coin belonging to the time of Candragupta II in the vicinity of the monastery, its date is estimated to be 5th century CE. Lumps of copper ore were also found suggesting that the smelting and casting operations were done on the spot. It has been taken away and preserved in the Birmingham Museum.

> - The Chinese traveller Hiuen Tsang has written that Indians knew the method of preparing brass from a mixture of Copper and Calamine.
> - Rasa-ratna-samuccaya gives a vivid account of the processes of extraction of copper and its use in the preparation of various Āyurvedic formulations.

- Another statue, 80 feet high was reportedly witnessed by the Chinese traveler Hieun-Tsang near Nālandā in a Vihāra of brass constructed by King Śilāditya (also known as King Harṣavardhan who ruled from 606–647 CE). According to Hiuen-Tsang, "A pavilion of six stages is required to cover this gigantic copper. It was the work of Rājā Pūrṇavarman, the last descendant of King Aśoka, a king of 7th century CE".
- The principal use of copper was to mint coins by the kings. The punch-marked copper coins of many kings of Northern India (1st century BCE), copper coins of the Kanishka

(1st century CE), and the Gupta kings have been found in many places. In Central and Southern India copper coins of the Andhra and Khatrapa dynasties are amongst the earliest. Brass coins of kings of several other dynasties living at that time have also been collected.

- One of the earliest copper plates was discovered in the village of Sohgaura, Gorakhpur district, with inscriptions of Maurya Brāhmī characters (320–230 BCE). The Sue Vihār inscriptions of Kaniṣka and the Takṣaśilā plates are also amongst the earliest copper plates discovered in Northern India.
- One remarkable ancient copper *ghoṭi or loṭa,* was found by Major Hay in 1857 in the village of Kundla in the Kangra district, Punjab. The vessel has an inscribed scene running around it which represents Prince Siddhārtha (later to become Buddha) going in a royal procession in a chariot drawn by four horses. From the mode of the inscription use of copper in the utensil, making is dated anywhere between 1st century BCE and 3rd century CE.

FIGURE 10.4 Copper Statue of Buddha Discovered at Sultangunj (**Neogi, P. (1918).** "Copper in Ancient India", Indian Association for the Cultivation of Science, Special Publication No. 1, Kolkata, p. 21).

Extraction of Copper for Āyurvedic Purposes

RRS gives a vivid account of the processes of extraction of copper and its use in the preparation of various Āyurvedic formulations. According to RRS, there are two varieties of copper: pure (red) copper from Nepal and the other impure or black copper (Mleccha). The Nepal copper was of high purity (99.5%), brick red, and very ductile whereas the impure copper contained copper oxides and other impurities like Pb, Sn, As, Zn, etc. which made it hard and brittle. RRS has mentioned many processes for obtaining pure copper from chalcopyrite by smelting small charges in crucibles. In Āyurveda, copper in the powder form (bhasma) is used for therapeutic formulations. Therefore, certain methods are specified for the extraction of copper using lemon juice as a reductant. In this process, after purification of chalcopyrite, the ore mineral is roasted at 750–900°C adding lemon juice at the rate of 100 cc per 100 g of the ore during the process. During the roasting process, Cu (Copper) and Fe (Iron) get converted to citrates. The roasted ore is mixed with 25% Borax (flux) and more lemon juice, and the mixture is pressed into 20–30 mm balls. After drying these balls in Sunlight, they are melted at 1250°C using a 4-stage process. At the end of the process sulphide free copper is extracted.

Another process mentioned in the RRS is based on the precipitation of Cu from blue, vitriol solution ($CuSO_4$) by the process of cementation. In this process, concentrated $CuSO_4$ solution was kept in an iron bowl, where on the surface a copper layer was formed following the reaction given below:

$$CuSO_4 + Fe = FeSO_4 + Cu$$

After some time, a thick layer of pure copper particles gets deposited on the iron surface. After removing the solution, the copper is extracted by scraping it away from the iron surface. This pure copper is finally washed with water free from Sulphur oxide and used for the preparation of copper bhasma.

10.3.4 Copper Alloys

Brass articles of the 1st century BCE to 1st century CE have been found on the excavation of some ancient stupas at Māṇikyālaya in 1830. Another inscribed brass urn of the same date as the former has been discovered in a tope about 30 miles west of Kabul in Wardak district. This urn, which in shape and size approaches closely the ordinary water vessels in use in India to this day, was originally thickly gilt and its surface has, in consequence, remained well-preserved[13]. Lothal (2200–1500 BCE) showed one highly oxidized antiquity, another Harappan site of Ross (also in Gujarat), has yielded a few samples of the chisel, cell, rod, and bangle, made of brass and assaying up to 1.54% zinc. The Chinese traveller Hiuen Tsang has written that Indians knew the method of preparing brass from a mixture of copper and Calamine. Hiuen Tsang has described a brass 'vihār' (unfinished convent) at Nālandā during the period of Śilāditya (Harṣavardhana) whose walls, doors, and windowsill all were covered with brass. On the other hand, huge brass guns and cannons of the Moghul period bear testimony to the skills of the medieval metalworkers. Table 10.1 has a list of brass objects discovered during various archaeological excavations in different parts of the country.

TABLE 10.1 Brass Objects in Ancient India

Sl. No.	*Date and Site*	*Archaeological Specimen*
1	~ 1500 BCE Lothal	No. 4189 Copper Object
2	Harappan Rosdi	Chisel, Celtrod, Bangle
3	4th Century BCE, Takṣaśilā	Vase Bm
4	2nd Century BCE, Takṣaśilā	Bangle
5	2nd Century CE, Gujarat	Female Figure Carrying Flower-Container Indo-Parthian
6	5th Century CE	Gāndhāra Buddha
7	6th Century CE, Akota	Ambikā
8	7th Century CE, Mahudi	Ṛśabhanātha
9	8th Century CE, Kashmir	Śiva
10	9th Century CE, Nalanda	Buddha
11	11th Century CE, W. Tibet	Mañjuśri
12	Gujarat 1350 CE	Ambikā
13	~ 1480 CE	Model Temple with Four Doors 10 × 24.5 cm
14	~ 1485 CE	Viṣṇu-Nārāyaṇa
15	Rajasthan 15th–16th Centuries CE	Rajput Prince on Horse
16	Gujarat 1554 CE	Kal Bhairava

Adopted from **Biswas, A.K. (1993).** "The Primacy of India in Ancient Brass and Zinc Metallurgy", *Indian Journal of History of Science*, **28**(4) pp. 309–330.

Ancient Indian's supremacy in bronze technology is baffling. One can infer the existence of Copper-bronze technology by examining the excavations of the Sindhu Saraswati civilisation. The use of bronze for ornamentation was evident from the discovery of ancient bronze articles at Tirunelveli in Tamil Nadu whilst excavating the ancient pre-historic burial sites. The bronze articles included ornamental vase stands, bowls, jars, and cups of different patterns with ornamental bowl lids. Bronze bangles, necklaces, ear ornaments, and diadems were also found. The Madras Government Museum has a very large collection of such pre-historic bronze objects from Adichanallur and Coimbatore probably dating from the 2nd or the 1st century BCE.[14] From the Bhir mound at Takṣaśilā, eight bronze vessels and mirrors with more than 20% tin, belonging to the 4th century BCE to the 1st century CE were reported. Similarly, a few vessels from the Nilgiris cairns of southern India with around 20–30% tin-bronze were also excavated.

Copper-bronze iron-casting, is a living tradition in India, having an active application in artistic and decorative item manufacture and in the field of idol making. With the advent of firearms in the Mughal period, guns of huge dimensions were cast with artistic designs. A wide variety of efficient distillation practices and extraction processes were in vogue in India for more than a thousand years[15]. There is a continuing Indian tradition of high-tin beta bronze vessel making by a traditional community of Kammalar (or bronzesmiths) in Kerala. Some studies show evidence for the use of high tin bronze in the fabrication of a wafer-thin water clock from Kerala, which resembles a sinking water clock with a hole at its bottom. As it floats, water flows into the bowl and it sinks after a certain time interval. Water clocks and sand clocks were used in Kerala to measure the time for medication, particularly *dhāra*[16]. As we all know, brass along with bronze was very largely used in making statues of gods and goddesses in the Middle Ages. Another variation of this is the making of idols using pañca-loha (a combination of five metals—gold, silver, copper, zinc, and iron). Several pañca-loha idols from Tamil Nadu temples have been stolen or taken away in the last 300 years.

10.3.5 Mercury

- The technique of amalgamation and gilding of Mercury was known to the Indian craftsmen from the very early times.
- Geographically Rajasthan stands out as a major mining province dating back to about 1000 BCE.

The compound HgS (Mercury (II) sulphide) is referred to as Rasa-sindūra in the Āyurvedic text Caraka-saṃhitā. Hg (Mercury), as well as HgS, have been extensively used in the preparation of Āyurvedic medicines. In RRS and other literature on the Rasa-śāstra description of the use of Cinnebar (Hiṅgula) to produce Mercury (Hiṅgulākṛṣṭa-rasa) has been described. The apparatus to be used for distillation and extraction is also given in some detail. HgS gets decomposed easily on heating in air or with lime to red hot temperature. The reactions are as follows:

$$HgS + O_2 \longrightarrow Hg + SO_2$$

$$3HgS + 2CaO \longrightarrow 2Cas + 3Hg + SO_2$$

The Āyurvedic literature also describes the process of purification of this metal and its reconversion to HgS for Mercurial preparations. As we have already discussed, Hg is also used for the amalgamation process for the extraction of Silver and Gold.

10.3.6 Lead and Silver

PbS can be converted into PbO through a simple process of roasting and further reduced to molten lead at 500°C. The process used in ancient India in places such as Zawar uses this methodology using a process where the roasting and reduction is carried out in an open pit furnace. In such a furnace temperature as high as 1000°C can be easily obtained and the reduced metal flows out and gets collected in the front pit. At this temperature, Ag_2S (Silver Sulphide) also gets decomposed and reduced and forms an alloy with molten lead. Silver metal was subsequently obtained from the alloy of Pb-Ag by reheating it in a shallow hearth furnace with bone charcoal at the bottom. During remelting of the alloy, excess air is blown so that the lead gets oxidized into PbO. Since this has a very low melting point it is skimmed off, leaving behind the purified liquid silver. It is tapped out and cast into ingots[17].

Kauṭilya's Arthaśāstra describes Silver–Copper alloys and the techniques applied for minting silver coins of a definite weight. The study of the punch marks on ancient coins indicates a high level of craftsmanship in the preparation of punching dies.

10.4 IRON AND STEEL IN INDIA

Ancient specimens of iron are widely available in India which will convince anyone that India has always been a rich iron-producing country. A host of archaeological excavations firmly establish that ancient Indians were well aware of the use of iron. Indians manufactured massive iron objects during the Christian era, which were much ahead of the times. The iron pillar at Qutub Minar (in the Mehrauli area of South Delhi) which weighs about 6,000 kg is well known, but other iron pillars, viz the one at Dhar (12th century CE) which weighs 7,000 kg and is nearly twice as big as the Iron Pillar of Delhi in Qutub Complex, and another on Mount Abu, which is less known. Another notable example is the 29 iron beams in the Konark temple in Odisha. Moreover, the gigantic iron beams at Konark, which lay buried in the sea sand for a few centuries were recently unearthed. These as well as numerous iron beams at Puri and Bhubaneswar where as many as 239 pieces were found in the Puri Gunduchibari temple are shining evidence for the supremacy of iron in ancient Indian culture[18]. The corrosion resistance and sheer size of these artefacts point to highly evolved practices in metalworking involving iron. Moreover, the presence of numerous Aśoka pillars of stone cut out in a faultless manner from single pieces of stone of gigantic dimensions presupposes the use of the finest steel saws and steel chisels in India in the 4th century BCE.

A large number of prehistoric iron implements such as swords, daggers, tridents, spears, javelins, arrows, spades, hangers, saucer lamps, beam rods, and tripods have been unearthed during the excavation of numerous burial sites in the Tuticorin district of Tamil Nadu. The stūpa of Bodh Gaya belongs to the Aśokan times. The foundations of the stūpa, on excavation, have a piece of iron slag that has been preserved in the Kolkata Museum. This piece of iron stag dated the 3rd century BCE is believed to be the most ancient archaeological evidence of the manufacture of iron in India. Besides the iron slag, many iron clamps five or six inches long and about one inch broad have been found in the main temple and in various stupas in Bodh Gaya, which are preserved in the Indian Museum[19].

With the advent of the carburisation of iron, a special type of high carbon steel was produced in India from as early as the fourth century BCE, known as wootz steel, used for military applications for producing tough swords, helmets, and armour. The word 'wootz,' by which name Indian steel was and is still known in Europe seems to have been prepared from time immemorial and was the metal

- Indians manufactured massive iron objects during the Christian era, which were much ahead of the times.
- The Suśruta saṁhitā has mentioned the fabrication of more than 100 surgical tools made of iron-carbon alloys and the process of heat treatment to obtain a razor-sharp edge.

from which the famous Damascus blades were prepared. From the presence of numerous Aśoka pillars of stone neatly carved out of single pieces of stone of gigantic dimensions we can infer the use of saws and chisels made out of steel in India in the 4th century BCE. The use of a large number of surgical instruments by Suśruta (3rd century BCE), some of which 'could bisect a hair longitudinally,' also points to the use of steel in India in making cutlery and such medical instruments[20].

It appears that the ancient Indians, especially from the time of the Guptas, were aware of the technology of producing corrosion resistance iron. This calls for a good understanding of metallurgy and material science, to use modern terms. Dr. P.C. Ray in his History of Hindu Chemistry remarks, "...The wrought iron-pillar in Qutub complex, which weighs ten tons and is some 1500 years old, the huge iron girders at Puri, the ornamental gates of Somnath, and the feet wrought-iron gun at Nurvar are monuments of a by-gone art and bear silent but eloquent testimony to the marvellous metallurgical skill attained by the Hindus...." [21]. The high status of iron and steel technology in ancient and medieval India is reflected in the manufacture and use of numerous large iron objects, including forge-welded cannons. Such cannon, found at Nurwar, Mushirabad, Dhaka (in Bangladesh), Bishnupur, Bijapur, Gulbarga, and Thanjavur, bear ample evidence to the medieval Indian blacksmith's skill in the design, engineering, and construction of large forge-welded iron objects. Based on its weight and size, the cannon at Thanjavur (see Figure 10.5), must be regarded as one of the largest forge-welded iron cannons in the world. According to a recent authoritative history of the ancient city of Thanjavur, the cannon was manufactured in Thanjavur during the regime of Raghunatha Nayak (1600–1645 CE)[22].

FIGURE 10.5 The Thanjavur Cannon

Source: By P. Jambulingam—Own work, CC BY-SA 4.0, https://commons.wikimedia.org/w/index.php?curid=35437933

The Delhi iron pillar (5th century CE) is made of almost pure iron (99.72%) and is crowned with the decorative capital. The pillar is unrivalled for it is corrosion-free property for such a long time and having withstood the ravages of weather (see Figure 10.6). The Sanskrit inscriptions found on the pillar indicate that it was constructed during the reign of Candragupta Vikramāditya II (375–414 CE), of the Gupta dynasty. The total length of the pillar is 23 feet 6 inch, including the decorated capital which is 3 feet and 5 inches tall. It was reported that the top of the pillar was adorned originally with an idol of Garuda (eagle), which was removed when the pillar fell into the hands of the Muslim rulers. The pillar ought to have been manufactured using a forge welding process as it was the only technique known to ancient Indians. In contrast, the forging of such large iron objects began in the West in the 19th century[23].

FIGURE 10.6 The Delhi Iron Pillar
Source: https://commons.wikimedia.org/wiki/File:QtubIronPillar.JPG

A 12th century CE iron pillar at Dhar in Central India is much larger (44 feet 4 inches). On Mount Abu, situated in the lower part of Rajputana, stands the temple Acaleśwar, built in 1412 CE. An iron pillar, the third of its kind, about 12 feet high is situated in the courtyard of the temple with a Śaiva Triśūl or trident on its top. It is stated that just before the death of the Pathan Emperor Alauddin when a rebellion broke out in all parts of India, the Hindus defeated the invaders and built this pillar as a mark of victory with the molten implements of war[24]. A rather less known iron pillar is located in Adi-Mukambika temple at Kodachadri village about 40 km from Kollur, in a remote forest of the Western Ghats in Karnataka. The pillar is estimated to be 46 feet high with a rectangular cross-section of 8.5 cm × 5.8 cm weighing about 500 kg. Based on a series of modern metallographic tests conducted at the Indira Gandhi Centre for Atomic Research, Kalpakkam, it was found that the X-ray examination could not reveal the definite presence of any element or compound besides pure iron[25]. There are over 260 iron beams of the same period in the temples at Puri and Konark in Odisha, each about six meters long. These living specimens unambiguously point to a sophisticated and perhaps well-established iron and steel working skills of ancient Indians[26].

The Delhi iron pillar is famous for its corrosion resistance for nearly 1600 years and has widely attracted the attention of archaeologists and corrosion technologists. It bears testimony to the high level of skill achieved by the ancient Indian ironsmiths. One plausible argument for corrosion resistance has been the low relative humidity in Delhi. However, the material of construction is likely to play a significant factor in corrosion resistance. This is because, at several other places where the relative humidity is high, we find corrosion-resistant iron pillars and beams. As we have already seen, notable among them include the iron beams in the Sūrya temple at Konark and the iron pillar at Mūkāmbikā temple.

The Suśruta Saṁhitā has mentioned the fabrication of more than 100 surgical tools made of iron-carbon alloys and the process of heat treatment to obtain a razor-sharp edge that is capable of splitting a thin hair into two longitudinally. As per the text, applying a carbonaceous paste on the edge of a surgical knife, heating it red hot, and then subjecting it to hardening and tempering treatments will make the knife razor-sharp that it can split a hair longitudinally. Figure 10.7 has a sample set of such tools designed by Suśruta.

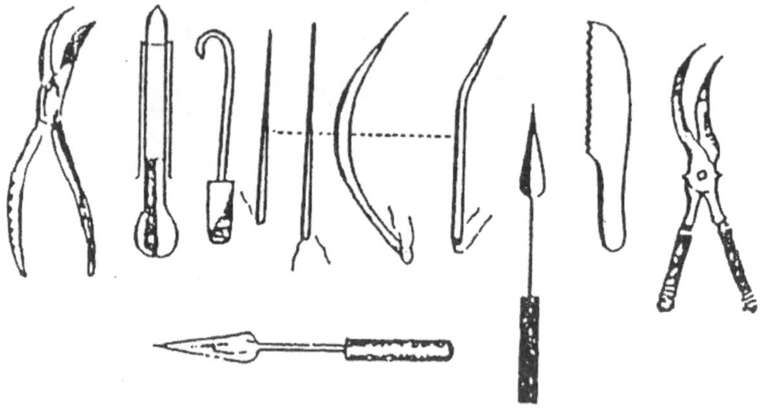

FIGURE 10.7 A Sample Set of Tools Designed by Suśruta Using Fe-C Alloy

Source: **Bag, A.K. (1977).** *History of Technology in India,* Vol. I, Indian National Science Academy, New Delhi, p. 102. Reproduced with permission.

Smelting of Iron

The reason for expertise in iron is attributed to the fact that India has one of the largest and richest deposits of Hematite, Magnetite, and their hydrated forms like limonite, etc. spread all over India. Ancient Indians have been extracting the iron from these sources and producing sponge iron or wrought iron using a variety of furnaces. Figure 10.8 is a sketch of a reconstructed furnace belonging to 700 BCE. The furnaces built by Indians differed from those in Western Asia and Europe of the same period. While the Indian furnaces were built using prefabricated clay blocks, the others were made by digging a hole in the earth and arranging the stone pieces to the required shape. The Indian furnaces could be reused after repair whereas the Western furnaces were to be discarded after one use.

The smelting of iron in the past was done by a special group (caste) of persons in each tribe and they all worshipped the God 'Asura'. In Bihar, Odisha, and the Eastern part of Karnataka the iron smelting was carried out, by these 'Asuras' and in Central India by 'Agarias'. There are studies on the 'Agaria' tribe, and their iron smelting practice spread through the central Indian regions including Rewa, Udaipur, Ranchi, and Koraput. Another group associated with ironworking is 'Lohārs' or 'Lohārins' (blacksmiths). While Agarias were involved in smelting the iron ore to produce wrought iron blooms, Lohārs worked on the wrought iron to manufacture various tools and objects using suitable heat treatment operations.

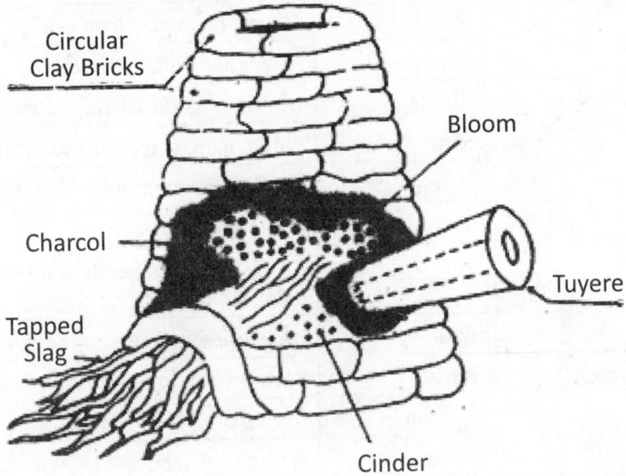

FIGURE 10.8 Sketch of a Reconstructed Furnace (700 BCE)

Source: Bag, **A.K. (1977).** *History of Technology in India,* Vol. I, Indian National Science Academy, New Delhi, p. 110. Reproduced with permission.

The ancient iron-making practice consisted of several operations and each one of them was performed under the strict supervision of the master craftsman, who is often an older person in the tribe. The operation of the furnace was considered to be a sacred ritual and they prayed to the tribal God 'Asura' for the success of the smelting. Many of the 'Mantras' and verses that they recite have been preserved as folk songs. The womenfolk also participated in this activity and there were certain tasks assigned to them. RRS describes in some detail the characteristics features of the women and their duties.

The advanced state of use of iron and its carbon alloys is evident from the ability to study the properties of iron and classify them accordingly into different types. Chapter 5 of RRS has about 15 verses where these classifications have been given in some detail. The iron-carbon alloy has been classified into three main groups[27]: Kānta-loha (soft iron), Tīkṣṇa-loha (High carbon steel), and Muṇḍa-loha (cast iron). Based on other characteristics of the iron-carbon alloy, these have been further classified into sub-groups as shown in Table 10.2. The classification of iron-carbon alloy indicates that the Indian iron technology was much advanced, and the ancient craftsmen were capable of selective use of various grades of this alloy. RRS is a text assigned to the 12th century CE. This type of classification was not known elsewhere. According to some studies, it was only in 1772 that the classification as grey, white and mottled cast iron fractures were attempted in the west[28].

TABLE 10.2 Classification of Fe-C Steel Alloys as per Rasa-ratna-samuccaya

Type of Iron	*Name*	*Properties*
Kānta-loha (Soft Iron)	Bhrāmaka	Very soft, magnetic iron
	Cumbaka	Mildly magnetic, sticks to iron pieces
	Karṣaka	Attracts iron objects
	Drāvaka	Strong magnetic iron
	Romaka	Permanent magnet, develops magnetic field around it
Tīkṣṇa-loha (Carbon Steel)	Khara	Develops good cutting edge, breaks on bonding
	Sāra	Softer iron, it has fibrous fracture
	Hṛnnāla	Hard and tough, has fibrous fracture
	Tārāvaṭṭa	Develops good cutting edge
	Vājira	Good hardening and tempering property, bluish in colour, hard cutting edge
	Kāla	Develops hard cutting edge after tempering
Muṇḍa-loha (Cast Iron)	Mṛdu	Soft, brittle, low melting point
	Kunṭha	Grey iron
	Kadāra	White cast iron

Source: **Bag, A.K. (1977).** *History of Technology in India,* Vol. I, Indian National Science Academy, New Delhi, p. 126.

10.4.1 Extraction of Iron from Biotite by Āyurvedic Method

In RRS and other Āyurvedic texts, several processes have been mentioned for the extraction of iron from Biotite and other iron-bearing minerals for the preparation of medicine. This process is known as Satvapātana, and it consists of three steps: Śodhana (purification), Bhāvana (Maceration and Trituration), and Damana (Heating and smelting). The first step involves purifying the mineral by heating and quenching it in some specified extracts seven times. The suggested liquids used for purification of biotite are[29]:

1. Kāñjī (Acidic fermentative liquid)
2. Triphalā (Decoction of a mixture of Terminienelia chebula, Terminalia belerica, and Emblica Officinalis)
3. Cow's urine
4. Cow's milk

This process is accompanied by intermediate Bhāvana and finally pelletisation of the treated mineral mixed with certain other ingredients. RRS specifies the following proportion of the charge materials: Mica (purified) 200 g, Borax (flux) 50 g, and Musali powder (reductant) 50 g. To this mixture, water is added, and the mixture is pelletised to 25–30 mm balls and dried in the Sun. The pellets are charged into a crucible and heated to high temperatures in a furnace. A furnace design that can attain a temperature of 1400°C, has been described in RRS text.

Varāhamihira (550 CE) in Bṛhat-saṃhitā discussed the following processes for carburisation and hardening of iron swords in Chapter 50 (verses 23–26)[30]:

(i) Make a paste of the gelatin from the sheep's horn and excreta of pigeon and mouse with the juice of the plant Arka (Caletropis Gigantica) and smear this to the steel after rubbing it with sesame oil. After heating the sword to red hot condition, sprinkle on it any of the following: water, milk of horse, camel or goat, ghee, blood, fat, or bile. Then sharpen the edge.

(ii) Plunge the steel, red hot into a solution of plantain ashes in whey, keep it for twenty hours, and then sharpen the edge.

10.4.2 Manufacture of Steel

Steel may be prepared in two ways, firstly, by removing part of the carbon of cast iron before it is converted into wrought iron; or secondly, by carburising wrought iron or making it combine with the requisite quantity of carbon. The second process is called the *process of cementation* and has been discovered in England only in the 18th century CE, while Indian iron was prepared by this process from time immemorial. The iron produced by the Indian method is always wrought iron, unlike cast-iron which is produced in modern blast furnaces. The main reason is that the modern blast furnaces reach a temperature of 1400°C or more. On the other hand, the temperature in the furnaces is low, sufficient enough to only soften the iron.

The process of Indian steelmaking may be summarised in the following way:

Wrought iron is first obtained by the direct method, viz. heating the ores of iron with charcoal in small blast furnaces (the blast being admitted employing hand bellows) without the intermediate formation of cast iron as has already been described. To convert the wrought iron into steel each piece is cut into three parts, each of which is put into a crucible, together with a handful of the dried branches of 'taṅgedu' (Cassia auriculata) and another of fresh leaves of 'vonaṅgady' (convolvulus laurifolia). The mouth of the crucible is then properly sealed with a handful of red mud and is arranged in circular order with their bottoms turned towards the centre in a hole made on the ground for the purpose. The hole is then filled up with charcoal and large bellows are kept blowing for six hours, by which time the operation is finished. The crucibles are then removed from the furnace, ranged in rows on moistened mud, and water is thrown on them whilst yet hot. The steel is found in conical pieces at the bottom of the crucibles, the form of which it has taken[31].

The chemical action that takes place, as evident from some studies is that during the application of heat to the closed crucible the dry wood and green leaves would yield charcoal as well as an abundant supply of hydrocarbons. The joint action of carbon and hydrocarbons on the iron greatly facilitates the formation of steel quickly. The European method of cementation using charcoal alone used to take six or seven days, and even fourteen to twenty days. while the Indian process takes only four to six hours.

The carburisation of hot sponge or bloom could be done by selective carburisation of cutting edge of the implements by the application of carburising paste and reheating to 950 to 1000°C. This method was followed by ancient Indians to develop sharp cutting edges such as surgical knives and swords as we have already seen. However, the iron treated by the carburising process suffered from structural heterogeneity and non-consistency of quality as well as low fracture strength. These difficulties were solved by

- Varāhamihira (550 CE) in Bṛhat-saṃhitā discussed the processes for carburisation and hardening of iron swords.
- The European method of cementation using charcoal used to take six or seven days, and even fourteen to twenty days, while the Indian process takes only four to six hours.

the introduction of the process of production of Wootz steel, which is a classic innovation of ancient Indians.

Yuktikalpataru an 11th century CE work mentions the relative properties of iron-carbon alloys produced in different regions of India and provides a comparative metric for their relative superiority as given below[32]:

- Krauñca—iron is supposed to be two times better than Sāmānya (probably Muṇḍa-loha).
- Kaliṅga (Odisha)—8 times better than Krouñca iron.
- Bhadra—100 times better than Kaliṅga iron.
- Vajra—1000 times better than Bhadra iron.
- Pāṇḍi—6 times better than Vajra iron.
- Niravi—10 times better than Pāṇḍi iron.
- Kānta—Ten billion times as good as Niravi iron.

10.5 LOST WAX CASTING OF IDOLS AND ARTEFACTS

This is a process adopted for producing metal sculptures based on an original sculpture. The metal sculpture is made of silver, gold, brass, or bronze. The beautiful idols and icons made during the Chola era were produced using bronze using this technique. Figure 10.9 is an example of a bronze idol of Śiva–Pārvatī, made using the lost wax casting process probably in the 11th century CE. In the Indian tradition, the lost-wax casting process is also used to produce idols using a combination of five metals (Gold, Silver, Lead, Copper, and Zinc) known as pañca-loha idols. Intricate works and a good post-casting finish can be achieved by this method. Such is the specialty, antiquity, and grandeur of these pañca-loha idols that a good number of them have been stolen, smuggled, or simply taken out of the country. The oldest known example of this technique is a 6,000 year-old amulet from the Sindhu Sarasvati Civilisation. There are other examples from somewhat later periods from Mesopotamia in the third millennium BCE.

In this process, a pattern of the desired shape is made out of beeswax first (therefore the process is known as *Madhūcchiṣṭa Vidhānam*—meaning process using remains of beehives). A mould is then prepared by applying coatings of prepared clay slurry on it. Later, when the refractory clay slurry has dried the wax is removed by baking the refractory shell. Before casting the molten metal, the refractory shell is generally embedded in a box filled with the sand-clay mixture for further support to the refractory wall. After pouring the metal and its solidification the mould is carefully dismantled to take out the metal casting. Further refinements and finishing touches are given by filing, chiselling, engraving, and polishing.

Several literary sources in India have documented the lost wax metal casting process in detail. Chapter 14 of Viṣṇu-saṃhitā, which is a part of Viṣṇu Purāṇa, dated to 5th century CE mentions the need to make a wax model first before making a metal replica. Similarly, Chapter 68 of Mānasāra has details on *Madhūcchiṣṭa Vidhānam*. Manasollāsa also known as Abhilaṣitārtha-cintāmaṇi of 12th century CE also provides a detailed account of the method of preparation of wax pattern and slurry coating. As per these texts, the composition of the clay slurry coating consisted of clay mixed with a certain proportion of finely ground charred rice husk and sodium chloride. Some studies point to the use of the lost-wax technique currently by the tribals of Bastar for producing various brass and bronze objects.

10.6 APPARATUSES USED FOR EXTRACTION OF METALLIC COMPONENTS

Three groups of experts were working on the extraction of metals from the ore. First are the metalworking engineers and metalsmiths. They mine the ore and extract the metal. The second group was the Āyurvedic practitioners. They also extracted metals and prepared therapeutic formulations using the powder form of the metals (bhasma). Thirdly the alchemists were involved in extracting metals and used them in their alchemical preparations. While the second and third groups might have worked on a smaller scale, the metalworking engineers engaged on a bigger scale. Nevertheless, the process and requirements were similar, and they had been adopting similar processes for extraction.

FIGURE 10.9 A 11th Century CE Bronze Idol of Śiva–Pārvatī
Source: https://upload.wikimedia.org/wikipedia/commons/9/94/Bronze_siva.png

A closer look at the alchemy works belonging to the early centuries of the Common Era gives us a good indication of how a laboratory of a scientist looked like and what type of apparatuses were fabricated by them and used for extraction of various chemicals and metallic components. Several apparatus and appliances called yantras were employed by the alchemists.

For instance, RRS has described a variety of crucibles (Mūṣā) and furnaces for the manufacture of Āyurvedic medicines. It also describes in total 51 kinds of metallurgical tools (upakaraṇas), 36 kinds of equipments (yantras), 17 types of crucibles, and 9 types of furnaces. The alchemist, the Āyurvedic expert, and the metalworking engineer seem to be benefiting from each other's know-how and used similar apparatuses, albeit in varying sizes to suit their requirements. We have already seen some of these in the previous sections. However, we shall discuss this aspect in a focused manner here.

The crucible, known as mūṣā-yantra is the frequently used apparatus by the alchemists as it serves as a generic container for a variety of operations. The mūṣā-yantra was predominantly earthen. To prepare the crucibles, the earth of anthill, rice husk, iron rust, chalk, human hair, and a few other ingredients are mixed in a certain proportion, and to this goat's milk is added. The mixture is kneaded into a dough-like mass which would then be shaped into the desired forms, and finally sun-dried[33]. There were various types of crucibles for different operations. For the extraction of zinc from calamine, a crucible of the shape of brinjal (Solanum melongena) is used. To this crucible, a tubular end is added, which expands towards its mouth like a flower.

We shall see in brief some of the apparatuses found in the ancient Indian laboratory[34]. The apparatus is simple and made of earthen material shaped into different forms. In all the cases the source of heat was either cow-dung or wood. As we know, this may not yield a high temperature, and therefore wherever required the substances are heated for a few days, in some cases, weeks.

Dolā-yantra uses a suspension mechanism using which the ingredient to be subjected to some treatment is kept in a piece of cloth, tied, and suspended using a rod. This is immersed in a pot half-filled with the desired liquid. The liquid is then heated from the outside. See Figure 10.10(a) for an illustrative sketch of Dolā-yantra.

(a) Dolā-yantra

(b) Pātana-yantra

FIGURE 10.10 Examples of Yantras Used by Alchemists in India

Source: **Sen, S.N. and Subbarayappa, B.V. (2009).** *A Concise History of Science in India,* Indian National Science Academy, 2nd ed., pp. 426–430. Reproduced with permission.

Svedanī-yantra is used for steaming purposes. The mouth of a pot is covered with a piece of cloth and the substance to be steamed is placed on the cloth. Water is kept in the pot and is covered using another pot. The substance is steamed by boiling the water.

Pātana-yantra is used whenever a sublimation or distillation process is involved. It consists of a well-baked pot of suitable dimensions. Water is filled in this pot to its neck and another bigger pot is kept over it in an inverted position. The two necks of the pots are firmly sealed using an adamantine made of lime, raw sugar, and buffalo's milk. A downward sublimation process is applied as follows. When the substance is smeared inside the upper vessel and heated from outside by burning cow-dung cakes, it vaporizes and collects in the water in the lower pot. On the other hand, the substance may also be heated in the lower pot and the upper one is kept cooled by applying wet rags. The sublimate deposits in the interior of the upper pot. Figure 10.10(b) is a schematic representation of a version of Pātana-yantra for the downward distillation process. In this version, the vapour passes through a pipe into another pot kept alongside. The second pot is cooled with water and the vapour condenses into liquid and is collected in the second pot.

Ḍhekī-yantra is also used for the distillation of substances such as mercury. A hole is made slightly below the neck of a pot and one end of a bamboo tube is introduced into it. The other end is fitted into a brass vessel that is filled with water. Mercury mixed with the other desired substances is subjected to distillation in this type of apparatus.

Vālukā-yantra is of a type of sand-bath for heating substances uniformly and for usually a long time. A long-necked bottle, containing the substance to be heated, is kept buried in sand in an earthen pot up to three-fourth of its height. The apparatus is heated from below for a long time till a straw when placed on the surface of the sand, catches fire. Instead of sand, sometimes salt is used.

- Yuktikalpataru an 11th century CE work mentions the relative properties of iron-carbon alloys produced in different regions of India.
- The alchemist, the Āyurvedic expert, and the metalworking engineer seem to be benefiting from each other's know-how.

Dhūpa-yantra is used for fumigation purposes. Fumigation of the leaves of gold or silver with the fumes of sulphur or arsenic substances is done in this apparatus. Two vessels are employed for this purpose. In the lower vessel, iron bars are placed in a slanting position below its mouth, and leaves of gold or silver are placed on them. Sulphur or the arsenic substance is placed in the lower vessel and the other vessel is used for covering purposes. On heating, the fumigation takes place.

For heating mercury and sulphur together, the following details of an mūṣā-yantra are given in the Rasārṇava[35]: The apparatus consists of two crucibles; one of them has a narrow orifice. In one crucible sulphur is taken and in the other mercury. Mercury and sulphur are to be moistened with filtered garlic juice. The crucible containing sulphur is to be inserted into the one having mercury, and the apparatus carefully lowered into an earthen pot over which another earthen pot is to be kept, and the rims luted with a cloth. It is heated from outside by cow-dung fire for three days.

SUMMARY

- Indian civilisation has a rich heritage spanning over the recorded history of the last two millennia. Moreover, there are several other archaeological and literary resources that point to the existence of a vibrant community in antiquity.
- In technology, ancient Indian's contributions to metalworking, and in more general terms to metallurgy and material science are noteworthy.
- The high status of iron and steel technology in ancient and medieval India is reflected in the manufacture and use of numerous large iron objects, including forge-welded cannons.
- A long and flourishing tradition of the copper industry in ancient India can be inferred from a number of archaeological specimens.
- Extensive archaeological excavations conducted in several parts of Rajasthan clearly show that the earliest artifact containing an appreciable amount of zinc anywhere in the world is from India.
- The Vedic corpus has multiple references to iron, copper, gold, silver, and tin. There are a few important treatises pertaining to the 5th century CE to 13th century CE that deal with metallurgy and metalworking.
- Ancient Indians developed skills to explore ore deposits and mine portions that are remunerative.
- Rasa-ratna-samuccaya has mentioned many processes for obtaining pure copper from chalcopyrite by smelting small charges in crucibles. In Āyurveda, copper in the powder form (bhasma) is used for therapeutic formulations.
- Ancient craftsmen could subject the purified gold metal bar to metalworking techniques such as forging, punching, embossing, etc. so that ornaments of intricate designs could be produced.
- Studies on the Zawar zinc mines provide rich insights into zinc mining in ancient India. Ancient Indians adopted a novel technique of downward drift reduction distillation process for zinc production, which is a precursor to the modern processes adopted all over the world.
- The compound HgS is referred to as Rasa-sindūra in the Āyurvedic text Caraka-saṃhitā. Hg, as well as HgS, have been extensively used in the preparation of Āyurvedic medicines.
- Ancient Indians have been extracting iron from ore deposits and producing sponge iron or wrought iron using a variety of furnaces.
- The carburisation of hot sponge or bloom could be done by selective carburisation of cutting edge of the implements by the application of carburising paste and reheating to 950°C to 1000°C. This method was followed by ancient Indians to develop sharp cutting edges such as surgical knives and swords.
- A closer look at the alchemy works belonging to the early centuries of the Common Era gives us a good indication of how a laboratory of a scientist looked like and what type of apparatuses were fabricated by them and used for extraction of various chemicals and metallic components.
- The beautiful idols and icons made during the Chola era were produced using bronze using the lost wax casting technique.

REVIEW QUESTIONS

1. What are the ways by which one can make an assessment of the S & T heritage of an ancient culture?
2. Do you think that ancient Indians had a good understanding of mining and metallurgy? Is there archaeological evidence to support this?
3. How can we infer ancient Indians' knowledge and use of iron and copper?
4. What is the status of zinc with respect to ancient Indian society?

5. What are the key features of iron pillars in India? What can we infer from these existing specimens?
6. What are the various uses that ancient Indians put copper and its alloys to?
7. Briefly comment on the mining operations done by ancient Indians.
8. Briefly describe the process and the steps involved in extracting copper from the mines.
9. What is the status of zinc with respect to ancient Indian society?
10. What do you understand by the term 'downward drift distillation process'? Why is it important in the extraction of zinc metal?
11. Explain the process of extraction of copper powder for use in Āyurvedic formulations.
12. What are the salient differences between the Indian method and the western method of extracting iron from iron ore?
13. Outline the salient aspects of the Indian iron and steel processes in ancient India.
14. What are the different varieties of iron listed by ancient Indians?
15. What are the alternative methods deployed for the sharpening of swords, knives, and sharp tools?
16. What do you understand by the term 'Yantra' in the context of metalworking? Describe four yantras used by alchemists in India.
17. Briefly describe the 'lost wax' casting technique.

DISCOVER IKS

1. Ancient Indian culture has contributed in myriad ways to the development of Science, Engineering, and Technology. This video captures 4 or 5 areas in which the ancient Indians have contributed to the development of Science, Engineering and Technology: https://www.youtube.com/watch?v=QxgK0dX872k. Watch the video carefully and prepare a one-page report each on the following questions:
   (a) How have the Indians promoted Scientific thought using Mathematics and Astronomy?
   (b) What was the status of metalworking in ancient India? How did the technology find use in Europe?
   (c) What is the status of healthcare in ancient India? How did it contribute to modern practice?
2. Indians were adept in casting idols that are rich and intricate in features. This has been mainly possible on account of the method of casting used by them. The practice which started 2000 years ago continues. Watch the video that demonstrates the making of an idol using the last wax casting process by clicking on this URL: https://www.youtube.com/watch?v=XTjmiAo8YJQ.

   After watching the video prepare a report that answers the following questions:
   (a) What do you understand by the Lost wax casting process of making an idol? Why is it called so?
   (b) What are the unique features of this process?
   (c) What are the critical success factors for this process to deliver a perfect idol?

SUGGESTED READINGS

Bag, A.K. (1997). *History of Technology in India*, Vol. I, Indian National Science Academy, New Delhi.

Balasubramaniam, R. (2000). "On the corrosion resistance of the Delhi iron pillar", *Corrosion Science*, **42**, pp. 2103–2129.

Balasubramaniam, R., Saxena, A., Anantharaman, T.R., Reguer, R., and Dillmann, P. (2004). "A Marvel of Medieval Indian Metallurgy: Thanjavur's Forge-Welded Iron Cannon", *JOM*, **56**(1), pp. 17–23.

Biswas, A.K. (1993). "The Primacy of India in Ancient Brass and Zinc Metallurgy", *Indian Journal of History of Science*, **28**(4) pp. 309–330.

Bose, D.M, Sen, S.N. and Subbarayappa, B.V. (2009). *A Concise History of Science in India*, Universities Press, 2nd ed., Hyderabad.

Chaudhuri, M. (1976). "Shipbuilding in the Yuktikalpataru and Samaraṅgana Sūtradhāra" *Indian Journal of Historical Studies*, **11**(2), pp. 137–147.

Craddock, P.T., Hegde, K.T.M., Gurjar, L.K., and Willies, L. (2017). *Early Indian Metallurgy: The Production of Lead, Silver and Zinc through Three Millennia in North West India*, Archetype Publications Ltd., London.

Deva, K. (1959). "Temples of Khajuraho in Central India", *Ancient India*, **15**, pp. 43–65.

Gandhi, S.M. (2014). "Copper, Zinc, Lead Ores—their Exploitation and Metal Extraction by the Ancients in the Northwestern India", *Journal Geological Society of India*, **84** (September 2014), pp. 253–266.

Hegde, K.T.M. (1973). "A Model for Understanding Ancient Indian Iron Metallurgy", *Man, New Series*, **8** (3), pp. 416–421.

Neogi, P. (1914). "Iron in Ancient India", Indian Association for the Cultivation of Science, *Bulletin No. 12*, Kolkata.

Neogi, P. (1918). "Copper in Ancient India", Indian Association for the Cultivation of Science, Special Publication No. 1, Kolkata.

Rajput, D.S., Gokarn, R., Jagtap, C.Y., Galib, R., Patgiri, B.J., and Prajapati, P.K. (2016). "Critical Review of *Rasaratna Samuccaya*: A Comprehensive Treatise of Indian Alchemy", *Anc. Sci. Life*, **36**(1) pp. 12–18.

Ray, P. (1967). "Origin and tradition of Alchemy", *Indian Journal of Historical Studies*, **2**(1), pp. 1–21.

Ray. P. (1956). "History of Chemistry in Ancient and Medeival India", Indian Chemical Society, Kolkata.

Singh, P.K., Dey, P., Jaina, S.K., and Mujumdar, P.P. (2020). "Hydrology and water resources management in ancient India", *Hydrol. Earth Syst. Sci.*, **24**, pp. 4691–4707.

Srinivasan, K.R. (1958). "The Pallava Architecture of South India", *Ancient India*, **14**, pp. 114–138.

Srinivasan, S. (2016). "Metallurgy of Zinc, High-tin Bronze and Gold in Indian Antiquity: Methodological Aspects", *Indian Journal of History of Science*, **51**(1), pp. 22–32.

Subbarayapa, B.V. (1982). "Glimpses of science and technology in ancient and mediaeval India", *Endeavour, New Series*, **6**(4), pp. 177–182.

Tripati, S. (2011). "Ancient maritime trade of the eastern Indian littoral", *Current Science*, Vol. **100**(7), pp. 1076–1086.

ENDNOTES

1. **Subbarayapa, B.V. (1982).** "Glimpses of Science and Technology in Ancient and Mediaeval India", *Endeavour, New Series*. **6**(4), pp. 177–182.
2. **Biswas, A.K. (1993).** "The Primacy of India in Ancient Brass and Zinc Metallurgy", *Indian Journal of History of Science*, **28**(4) pp. 309–330.
3. *Ibid*.
4. **Hegde, K.T.M. (1973).** "A Model for Understanding Ancient Indian Iron Metallurgy", *Man, New Series*, **8**(3), pp. 416–421.

5. See for details, **Singh, R.D. (1997).** "Material Technology—3.1 Mining", Chapter 3 in *History of Technology in India,* Vol I, A.K. Bag (Ed.), Indian National Science Academy, New Delhi, pp. 48–79.
6. These are numerous references in the Veda for the metals and alloys and their use. See for example the popular passage in Yajur Veda which mentions these metals: अश्मा च मे॒ मृत्तिका च मे गि॒रयश्च मे॒ पर्वताश्च मे॒ सिकताश्च मे॒ वन॒स्पतयश्च मे॒ हिरण्यं च॒ मेऽयश्च मे॒ सीसं च मे॒ त्रपुश्च मे श्या॒मं च मे लो॒हं च मे॒... aśmā ca me̱ mṛttikā ca me gi̱rayáśca me̱ parvatāśca me̱ sikátāśca me̱ vana̱spatáyaśca me̱ hiráṇyaṃ ca̱ me'yáśca me̱ sīsaṃ́ ca me̱ trapúśca me śyā̱mamaṃ cá me lo̱haṃ cá me̱... Vājasaneyī mādhyandina-saṃhitā 18/13. http://titus.uni-frankfurt.de/texte/etcd/ind/aind/ved/yvw/vs/vs.htm?vs018.htm Last accessed on October 1, 2021.
7. For an excellent treatment of this subject please refer to **Prakash, B. (1997).** "Metals and Metallurgy", Chapter 3 [3.2] in *History of Technology in India,* Vol. I, A.K. Bag (Ed.), Indian National Science Academy, New Delhi, pp. 80–174. We have substantially drawn material from this chapter to discuss the issues in this section.
8. *Ibid.*
9. *Ibid.*
10. **Hegde, K.T.M., Craddock, P.T., and Sonawana, V.H. (1984).** "Report on the Excavations of Ancient Zinc Smelting Furnace at Zawar, Rajasthan", as mentioned in **Prakash, B. (1997).** "Metals and Metallurgy", Chapter 3 [3.2] in *History of Technology in India,* Vol. I, A.K. Bag (Ed.), Indian National Science Academy, New Delhi, pp. 80–174.
11. **Srinivasan, S. (2019).** "Book Review: Early Indian metallurgy: The Production of Lead, Silver and Zinc through Three Millennia in North West", *Indian Journal of Historical Studies,* **54**(3), pp. 376–378.
12. **Neogi, P. (1918).** "Copper in Ancient India", Indian Association for the Cultivation of Science, Special Publication No. 1, Kolkata.
13. **Biswas, A.K. (1993).** "The Primacy of India in Ancient Brass and Zinc Metallurgy", *Indian Journal of History of Science,* **28**(4), pp. 309–330.
14. See for details, **Paramasivan, S. (1941).** "Investigations on Ancient Indian Metallurgy I: A Pre-historic Bronze Bowl", *Proceedings of the Indian Academy of Sciences, Section A,* **13**(2), pp. 87–93.
15. **Subbarayapa, B.V. (1982).** "Glimpses of science and technology in ancient and mediaeval India", *Endeavour, New Series.* **6**(4), pp. 177–182.
16. **Srinivasan, S. (2016).** "Metallurgy of Zinc, High-tin Bronze and Gold in Indian Antiquity: Methodological Aspects", *Indian Journal of History of Science,* **51**(1), pp. 22–32.
17. **Prakash, B. (1997).** "Metals and Metallurgy", Chapter 3 [3.2] in *History of Technology in India,* Vol. I, A.K. Bag (Ed.), Indian National Science Academy, New Delhi, pp. 80–174.
18. **Neogi, P. (1914).** "Iron in Ancient India", Indian Association for the Cultivation of Science, *Bulletin No. 12,* Kolkata.
19. *Ibid.*
20. *Ibid.*
21. **Neogi, P. (1914).** "Iron in Ancient India", Indian Association for the Cultivation of Science, *Bulletin No. 12,* Kolkata.
22. **Balasubramanian, R., Saxena, A., Anantharaman, T.R., Reguer, S., and Dillmann, P. (2004).** "A Marvel of Medieval Indian Metallurgy: Thanjavur's Forge-Welded Iron Cannon", *JOM,* **56**(1), pp. 17–23.
23. **Balasubramaniam, R. (1998).** "The decorative bell capital of the Delhi iron pillar", *JOM;* **50**(3), pp. 40–47.
24. **Neogi, P. (1914).** "Iron in Ancient India", Indian Association for the cultivation of Science, *Bulletin No. 12,* Kolkata.

25. For more details see, **Anantharaman, R. (1999).** "The iron pillar at Kodachadri in Karnataka", *Current Science*, **76**(11), pp. 1428–1430.

26. **Subbarayapa, B.V. (1982).** "Glimpses of science and technology in ancient and mediaeval India", *Endeavour, New Series*. **6**(4), pp. 177–182.

27. The classification of iron-carbon alloys is described in the following verses of Rasa-ratna-samuccaya:

    मुण्डं तीक्ष्णञ्च कान्तञ्च त्रिप्रकारमयः स्मृतम् ॥
    muṇḍaṃ tīkṣṇañca kāntañca triprakāramayaḥ smṛtam || RRS 5.67

    मृदु कुण्ठं कडारञ्च त्रिविधं मुण्डमुच्यते ॥
    mṛdu kuṇṭhaṃ kaḍārañca trividhaṃ muṇḍamucyate || RRS 5.68

    खरं सारञ्च हृन्नालं ताराचट्टञ्च वाजिरम् ।
    काललोहाभिधानञ्च षड्विधं तीक्ष्णमुच्यते ॥
    kharaṃ sārañca hṛnnālaṃ tārācaṭṭañca vājiram |
    kālalohābhidhānañca ṣaḍvidhaṃ tīkṣṇamucyate || RRS 5.74

    भ्रामकं चुम्बकञ्चैव कर्षकं द्रावकं तथा । एवं चतुर्विधं कान्तं रोमकान्तञ्च पञ्चमम् ॥
    bhrāmakaṃ cumbakañcaiva karṣakaṃ drāvakaṃ tathā | evaṃ caturvidhaṃ kāntaṃ romakāntañca pañcamam || RRS 5.83.

    See for details, **Vidyābhūṣaṇa, Āśubodha and Vidyāratna, Nityabodha (1927).** "Rasaratnasamuccayaḥ: (prācīna-rasa-granthaḥ)". Kalikātā.

28. See for details, **Singh, R.D. (1997).** "Material Technology–3.1 Mining", Chapter 3 in *History of Technology in India,* Vol. I, A.K. Bag (Ed.), Indian National Science Academy, New Delhi, p. 126.

29. The process is described in the following two verses of Rasa-ratna-samuccaya:

    प्रतप्तं सप्तवाराणि निक्षिप्तं काञ्जिकेऽभ्रकम् । निर्दोषं जायते नूनं प्रक्षिप्तं वाऽपि गोजले ॥
    prataptaṃ saptavārāṇi nikṣiptaṃ kāñjike'bhrakam | nirdoṣaṃ jāyate nūnaṃ prakṣiptaṃ vā'pi gojale || RRS 2.16

    त्रिफलक्वथिते चापि गवा दुग्धे विशेषतः । ततो धान्याभ्रकं कृत्वा पिष्ट्वा मत्स्याक्षिकारसैः॥
    triphalakvathite cāpi gavā dugdhe viśeṣataḥ | tato dhānyābhrakaṃ kṛtvā piṣṭvā matsyākṣikārasaiḥ|| RRS 2.17

30. The two processes for sharpening the sword are given in the following verses of Bṛhat-saṃhitā:

    इदमौशनसं च शस्त्रपानं रुधिरेण श्रियमिच्छतः प्रदीप्ताम् ।
    इविषा गुणवत्सुताभिलिप्सोः सलिलेनाक्षयमिच्छतश्च वित्तम् ॥२३॥
    वडवौष्ट्रकरेणुदुग्धपानं यदि पापेन समीहतेऽर्थसिद्धिम् ।
    झषपित्तमृगाश्वबस्तदुग्धैः करिहस्ताच्छिदये सतालगर्भैः ॥२४॥
    आर्कं पयो हुडुविषाणमर्षीसमेतं पारावताखुशकृता च युतः प्रलेपः ।
    शस्त्रस्य तैलमथितस्य ततोऽस्य पानं पश्चाच्छितस्य न शिलासु भवेद्विघातः ||२५||
    क्षारे कदल्या मथितेन युक्ते दिनोषिते पायितमायसं यत् ।
    सम्यक् शितं चाश्मनि नैति भङ्गं न चान्यलोहेष्वपि तस्य कौण्ठ्यम् ॥ २६ ॥

    idamauśanasaṃ ca śastrapānaṃ rudhireṇa śriyamicchataḥ pradīptām |
    iviṣā guṇavatsutābhilipsoḥ salilenākṣayamicchataśca vittam ||23||
    vaḍavauṣṭrakareṇudugdhapānaṃ yadi pāpena samīhate'rthasiddhim |
    jhaṣapittamṛgāśvabastadugdhaiḥ karihastācchidaye satālagarbhaiḥ ||24||
    ārkaṃ payo huḍuviṣāṇamarṣīsametaṃ pārāvatākhuśakṛtā ca yutaḥ pralepaḥ |
    śastrasya tailamathitasya tato'sya pānaṃ paścācchitasya na śilāsu bhavedvighātaḥ ||25||
    kṣāre kadalyā mathitena yukte dinoṣite pāyitamāyasaṃ yat |
    samyak śitaṃ cāśmani naiti bhaṅgaṃ na cānyaloheṣvapi tasya kauṇṭhyam || 26 ||

    See for details, **Sastri, V.S. and Bhat, M.R. (1946).** *Varahamihira's Brihatsamhita*, Soobbiah & Sons, Bengaluru.

31. **Neogi, P. (1914).** "Iron in Ancient India", Indian Association for the Cultivation of Science, *Bulletin No. 12*, pp. 65–76, Kolkata.

32. The relevant verses from Yuktikalpataru are reproduced here:
सामान्याद् द्विगुणश्चोक्तं कलिर्दशगुणस्ततः । कलेः शतगुणं भद्रं भद्राद्वज्रं सहस्रधा || 85.31
वज्रात् षष्टिगुणः पाण्डिर्निरविर्दशभिर्गुणैः । ततः कोटिसहस्रेण ह्ययस्कान्तः प्रशस्यते ॥ 85.32
sāmānyād dviguṇañcoktaṃ kalirdaśaguṇastataḥ | kaleḥ śataguṇaṃ bhadraṃ bhadrādvajraṃ sahasradhā || 85.31
vajrāt ṣaṣṭiguṇaḥ pāṇḍirniravirdaśabhirguṇaiḥ | tataḥ koṭisahasreṇa hyayaskāntaḥ praśasyate || 85.32

33. **Sen, S.N. and Subbarayappa, B.V. (2009).** *A Concise History of Science in India*, Indian National Science Academy, 2nd ed., p. 426.

34. *Ibid.* pp. 427–430.

35. The verses in Rasārṇavam that describes this process is found in Chapter 4 and is reproduced below:
लोहमूषाद्वयं कृत्वा द्वादशाङ्गुलमानतः । ईषच्छिद्रान्वितामेकां तत्र गन्धकसंयुताम् ॥ 4.8
मूषायां रसयुक्तायाम् अन्यस्यां तां प्रवेशयेत् । तोयं स्यात् सूतकस्याधः ऊर्ध्वाधो वह्निदोपनम् ॥ 4.9
रसोनकरसं भद्रे यत्नतो वस्त्रगालितम् । दापयेत्प्रचुरं यत्नात् आप्लाव्य रसगन्धकौ ॥ 4.10
स्थालिकायां निधायोर्ध्वं स्थालीमन्यां दृढां कुरु । सन्धिं विलेपयेद्यत्नात् मृदा वस्त्रेण चैव हि ॥ 4.11
स्थाल्यन्तरे कपोताख्यं पुटं कर्षाग्निना सदा । यन्त्रस्याधः करीषाग्निं दद्यात्तीव्राग्निमेव वा ॥ 4.12
एवन्तु त्रिदिनं कुर्य्यात् ततो यन्त्रं विमोचयेत् । तप्तोदके तप्तचुल्ल्यां न कुर्य्याच्छीतले क्रियाम् ॥ 4.13
न तत्र क्षीयते सूतो न च गच्छति कुत्रचित् । अनेन क्रमयोगेन कुर्य्याद्गन्धकजारणम् ॥ 4.14
ऊर्ध्वं वह्निरधश्चापो मध्ये तु रससङ्ग्रहः । मूषायन्त्रमिदं देव जारयेद्गगनादिकम् ॥ 4.15
lohamūṣādvayaṃ kṛtvā dvādaśāṅgulamānataḥ | īṣacchidrānvitāmekāṃ tatra gandhakasaṃyutām || 4.8
mūṣāyāṃ rasayuktāyām anyasyāṃ tāṃ praveśayet | toyaṃ syāt sūtakasyādhaḥ ūrdhvādho vahnidopanam || 4.9
rasonakarasaṃ bhadre yatnato vastragālitam | dāpayetpracuraṃ yatnāt āplāvya rasagandhakau || 4.10
sthālikāyāṃ nidhāyordhvaṃ sthālīmanyāṃ dṛḍhāṃ kuru | sandhiṃ vilepayedyatnāt mṛdā vastreṇa caiva hi || 4.11
sthālyantare kapotākhyaṃ puṭaṃ karṣāgninā sadā | yantrasyādhaḥ karīṣāgniṃ dadyāttīvrāgnimeva vā || 4.12
evantu tridinaṃ kuryyāt tato yantraṃ vimocayet | taptodake taptacullyāṃ na kuryyācchītale kriyām || 4.13
na tatra kṣīyate sūto na ca gacchati kutracit | anena kramayogena kuryyādgandhakajāraṇam || 4.14
ūrdhvaṃ vahniradhaścāpo madhye tu rasasaṅgrahaḥ | mūṣāyantramidaṃ deva jārayedgaganādikam || 4.15
For details see, **Ray, P.C. (1910).** *Rasarnavam.* Unknown publisher. https://ia801608.us.archive.org/23/items/in.ernet.dli.2015.312525/2015.312525.The-Rasarnavam.pdf Last accessed on Oct. 1, 2021.

CHAPTER 11 Engineering and Technology: Other Applications

LEARNING OUTCOMES

After finishing this chapter, you will be able to:

- Familiarise with some texts that discuss certain aspects of Engineering and Technology
- Develop awareness about use of dyes and paints and irrigation and water management, physical structures and temples in ancient India
- Familiarise with some aspects of shipbuilding
- Familiarise with a few surgical practices and tools prescribed in Āyurveda

Indian maritime has a long-standing history. According to Roman historians, Tamil Pandya embassies were received by Augustus Caesar. It is therefore not surprising that shipbuilding is an age-old industry in India. What we see here is a model of a Chola (200–848 CE) ship's hull, built by the ASI, displayed in a Museum in Tirunelveli, Tamil Nadu.

Source: Photo by Everdawn at English Wikipedia, http://creativecommons.org/licenses/by-sa/3.0

IKS IN ACTION 11.1

Irrigation Systems and Practices in South India

The irrigation system began in South India from the initial centuries of the first millennium CE. Kallanai built by Karikal Chola is one of the oldest living irrigation systems in the world today. Karikal Chola built flood banks and strengthened the embankment of river Cauvery thereby launching a culture of large irrigation works. The Pandyas of Madurai, the Pallavas of Kanchi, the Cholas of Thanjavur, the Hoysalas of Dwarasamudra, the Chalukyas of Kalyani, and the Kakatiyas of Warangal continued this legacy during the 7th–12th century CE and introduced good water management practices.

The Pallavas of Kanchi who ruled from the 4th century CE were great tank builders. They set up a good irrigation management system consisting of tanks, inundation canals, and channels and wells. The concentration of medium and small-scale tanks in the northern parts of Tamil Nadu (Chingleput and North Arcot districts, popularly known as Tank Districts) was on account of their initiatives. Some of the tanks constructed by the Pallavas continue even to this day to serve the purpose for which they were intended.

Management of Irrigation Systems

The Pallava records inform us that while most of the projects were initiated and funded by the State, private individuals were encouraged to construct irrigation works. However, the maintenance of the irrigation system was entrusted to the village assembly. The Chalukyans had a separate department of waterworks called Vāri-gṛha-karaṇa which was responsible for the execution and maintenance of irrigation works. The Mahājanas were responsible for collecting taxes and land revenue and utilising the same for proper maintenance of irrigation works.

With the expansion of the Chola empire, the irrigation activities also increased manifold under the State patronage and there was a greater emphasis on proper maintenance of the tanks. The village assembly had the mandate for this job, and it created some committees for the task.

One of the committees (known as Eri-vāriyam), consisting of six members, was charged with the responsibility of 'Supervision of tanks'. This body held office for 360 days and then the committee was reconstituted. The famous Uttiramerur inscriptions of Parantaka I give an elaborate description of the rules regarding the composition of this Committee. The Eri-vāriyam was mainly concerned with raising resources and their utilisation for the maintenance of irrigation works.

There was another committee called the *Sluice Committee* (Kaliṅgu-Vāriyam) under the village assembly. The duties of this committee were to maintain the sluice gates from wanton damage and other natural causes. The committee made periodic checks to the sluice gates of the irrigation system. This committee existed only in a few village assemblies.

The village assemblies on their part collected various types of land taxes, dues, and cesses. Apart from this, as many as thirty-two tax items connected with the water rates found in the records of the Cholas were ostensibly collected from the tenants and landowners as service charges for the maintenance of the irrigation works.

These aspects point to the foresightedness of the rulers in ancient India and their understanding of the key challenges faced by agriculture in South India. Unlike North India, there are no major river systems that assure the perennial supply of water. The problems like the seasonal drought and consequent famine were checked and brought under manageable control by these well-established practices in irrigation management.

A careful study of this will help current-day planners restore our glory in irrigation management and tackle the current watershed management challenges that we face.

Source: **Srinivasan, T.M. (1997).** "Irrigation and irrigation works", Chapter 7 [7.4] in *History of Technology in India*, Vol. I, A.K. Bag (Ed.), Indian National Science Academy, New Delhi, pp. 555–590.

In Chapter 10, we discussed how a culture of Science, Engineering, and Technology ought to have been part of the ancient Indian society based on archaeological excavations, and remains of metalworking mines and furnaces that were in use long ago. On account of appropriate know-how, it has been possible to extract large quantities of metals from the ores, manufacture corrosion-resistant steels that can stand the test of time, build lofty physical structures of staggering magnitude such as temples, deploy irrigation and water management systems, etc. In the area of health, we find instructions and processes pertaining to surgical practices, preparation of therapeutic formulations involving metallic compounds and their extracts. Ancient Indian literature also points to maritime trade between several parts of peninsular India with the countries in the far east, China, and the Roman Empire. This presupposes knowledge of building ships that are capable of withstanding long voyages across the oceans.

The book on Indian Science and Technology in the 18th century by Dharampal provides interesting insights and suggests that various Indian technologies were in a flourishing state in the 18th century CE. Several contemporary European observers, commented on the sophistication, efficacy and possible adoption in the Europe of the Indian science and technology practices. There are detailed descriptions in the book on widespread practice of inoculation against smallpox, practice of plastic surgery which was transmitted from India to the Europe and the quality of Madras Mortar to name a few[1].

In this chapter, we shall explore the ancient and medieval Indian literature for specific process details and methodologies for other aspects of engineering and technology. It remains an unanswered question as to what happened to these skills and working methodologies that ancient Indians seem to have practiced over several years.

11.1 LITERARY SOURCES FOR SCIENCE AND TECHNOLOGY

As we have seen in some of the earlier chapters (such as Mathematics and Astronomy) there has been a continuous tradition of scientific inquiry in the ancient Indian civilisation for more than 2,000 years. Therefore, it will be possible for us to locate some literary evidences for Science and Technology (S & T) in ancient Indian works, which provides another source for us to make certain conclusions about S & T thinking. Table 11.1 has a list of literary sources that discuss S & T principles and applications. Let us briefly see some areas of applications:

Metallurgy, Material Working, Mining: Fabrication of iron pillars that withstands more than 1000 years of corrosion does not happen by mere accident. It must be preceded by a good knowledge of metallurgy and the availability of experienced ironsmiths. Similar evidence in copper, zinc, brass, and bronze point to the availability of know-how. While references to metals are found in the Vedic corpus, some treatises pertaining to the 5th century CE to 13th century CE discuss these issues in some detail.

The Vedic corpus has multiple references to iron (Ayas), copper (Loha or Lohitāyas), gold (Svarṇa), silver (Rajata), and tin[2]. For instance, the picture of a blacksmith melting iron in his forge is given in Ṛigveda 6.3.4[3]. In Chāndogya-upaniṣad 4.17.7 there is a mention of joining of metals gold, silver, tin, and lead[4]; In Atharvaveda 6.141.2[5], a knife made of 'red metal' has been mentioned. In addition to the word *lohita*, the modern equivalent of copper, viz. *tāmra* occurs in one passage (Atharva, 10.2.11)[6]. Several Indian texts such as Rasaratnākara, Rasāyana-śāstra, and Rasaratna-samuccaya deal extensively on the zinc metallurgy. The Mānasōllāsa, written by the Chalukyan king Someśvara of the 12th century CE describes the process of making metal icons (Idols of Gods for the temples) by the lost wax casting process. This is still practiced in Swamimalai in the Thanjavur district, where there are some world-renowned sthapatis and metalsmiths engaged in idol making. In Caraka-saṃhitā references of zinc, and its extraction from ore is available. Rasaratna-samuccaya of Vāgbhaṭa, is considered to be a metallurgical encyclopaedia of the 13th century CE. There are vivid descriptions and procedures related to mineral purification; extraction of metals; fuels, crucibles, furnaces; metals, alloys, and their properties.

TABLE 11.1 S & T References in Ancient Indian Literature

No.	*Period*	*Source*	*Issues of S & T Discussed in the Text*
1	Up to 1st Century BCE	Ṛgveda	References to Iron, ironsmith and melting process in several passages – no mention of copper
		Atharvaveda, Yajurveda	Mention of six metals (gold, silver, iron, tin, lead, copper), various technical occupational categories
		Chāndogya-upaniṣad, Taittirīya-brāhmaṇa	Mention of black metal and red metal
		Rāmāyaṇa	Description of iron ores, metals and mines
		Manusmṛti	Mention of the use of household utensils made of copper, iron, bronze, brass, tin and lead and enjoins their purification with ashes, acid and water
		Caraka and Suśruta Saṃhitās	Description of surgical instruments, mention of six metals, alloys of copper (brass, zinc), tools for surgical interventions
		Arthaśāstra	Primarily a book on statecraft and public administration, description of various warfare implements such as iron swords, arrows, axes, spades, etc., ores and mines of gold, silver, copper, iron, lead, tin and precious stones, state functionaries such as superintendents of mines, ships and weaving, use of copper in making alloys and in gold and silver coins, goldsmith roles
2	1st Century CE to 10th Century CE	Rasaratnākara of Nāgārjuna	Reduction of metals, recipes for transmutation of base metals, yantras for carrying out several physio-chemical processes, alchemical processes and preparations of mercurial compounds
		Bṛhat-saṃhitā of Varāhamihira	Primarily an astronomy treatise, Geological formations (rain, dust storm, water, etc.) mining, earthquakes, yantras, medicine, botany, architecture (temple, idols, etc.)
		Amara-kośa	Technical terms, vocabulary

(*Contd.*)

TABLE 11.1 S & T References in Ancient Indian Literature

No.	*Period*	*Source*	*Issues of S & T Discussed in the Text*
3	11th Century CE to 14th Century CE	Yukti-kalpataru by Bhojarāja	Architectural principles, ideas of different types of ships and shipbuilding
		Samarāṅgaṇa-sūtradhāra of Mahārājādhirāja Bhojadeva	Architectural principles, different types of yantras (devices), ideas or aerial vehicles
		Mānasollāsa of king Someśvara	Process of making metal icons by the lost wax casting process
		Rasacintāmaṇi of Madanāntadeva	Preparation of medicine, information on chemistry and metallurgy, including the production of zinc
		Rasaratna-samuccaya of Vāgbhaṭa	Contemporary method of production of zinc, different categories of iron, mineral purification; extraction of metals and mercury; fuels, crucibles, furnaces; metals, alloys and their properties, 51 metallurgical implements, 36 types of equipment, 17 types of crucibles, and 9 types of furnaces. It is considered as a metallurgical encyclopaedia of the 13th century, pharmaceutical procedures of metals and minerals, description of many herbo-metallic and mineral formulations

Alchemy: Metallurgy is closely related to alchemy, which is a science of transmutation of metals. Alchemy was practiced in ancient civilisations to find methods to transform other metals into gold. In the Indian tradition, we see similar discussions. Nāgārjuna, a well-known Indian alchemist was a Buddhist monk who authored Rasaratnākara. The Rasaratnākara gives several recipes for the transmutation of base metals into silver and gold, the core subject matter of alchemy. It also describes many chemical processes like the extraction of zinc, mercury, and copper. The text also mentions the names of several yantras (apparatus) for carrying out various physico-chemical processes like distillation, sublimation, extraction, and calcination. Rasendra-cūḍāmaṇi by Somadeva and Rasaprakāśa-sudhākara by Yaśodhara are the other important treatises that deal largely with chemistry and alchemy. A major work, Rasaratna-samuccaya of Vāgbhaṭa extended the application of alchemy into the health and treatment of diseases using heavy metals.

Architecture and Town Planning: The presence of a large number of temples throughout the country points to good knowledge of architecture, iconography, building construction, and construction management practices. Many works address these issues, and we present a list of important literary works in Chapter 12. However, we merely mention two of them here for some other S & T applications.

Samarāṅgana-sūtra-dhāra of Mahārājādhirāja Śrī Bhojadeva of Dhārā, who lived in the 11th century CE is primarily a treatise on architecture (vāstu-śāstra). However, in Chapter 31 of the work, there is a description of different yantras (man-made devices) that can be put to different uses. Further, there is a description of aerial vehicles, which can be considered a good pre-cursor to current-day aerial vehicles. Another interesting work Yukti-kalpataru, by one Bhojarāja, is also considered to be of the same period. There is no clarity as to whether both are by the same author. This work is also mainly on architecture (vāstu-śastra). There are descriptions in this work on types of ships, shipbuilding, etc.

Medical Applications: Caraka-saṃhitā and Suśruta-saṃhitā are the two major medical treatises of ancient India. In Suśruta-saṃhitā, one finds descriptions of surgical instruments for performing several surgical operations. Copper (and its alloy brass and bronze), iron, and other metals including zinc, gold, and silver find mention in both the treatises.

Other Engineering Applications: The Bṛhat-saṃhitā of Varāhamihira is primarily an encyclopaedic work, covering mainly astronomy. However, there are other interesting issues discussed in the work, which include geological formations such as wind patterns, rain formation, earthquakes, and water tables. It also discusses certain aspects related to medicine, yantras, and botany. The lexicographical work of Amarakośa provides the required vocabulary for the engineering aspects, metals, and various physical structures and thus points to the existence of these in the society during that time.

11.2 PHYSICAL STRUCTURES IN INDIA

While archaeological evidence exists throughout the length and breadth of the country, we know very little about the scientific ideas of the people of the ancient civilisation. One possible reason for this could be the reported large-scale burning of over six million manuscripts kept in Nalanda University by the invaders. No civilisation can weather such catastrophic damage inflicted all of a sudden. It is bound to create a certain discontinuity in the knowledge tradition. However, the Indian civilisation is a 'living tradition' and we have a rich heritage of practices and systems that continue till date despite periodic disturbances. There are several ancient physical entities still in use or in a preserved condition. In particular, in India, there are several temples built over the last 2500 years, still either in active use by society or preserved as monuments. These can offer valuable insights into the status of S & T in India. Therefore, one approach to assessing if there were any S & T heritage and knowledge among the people of ancient India is to examine the living specimens of yesteryears. By studying some aspects of ancient buildings and architectural monuments one can make some inferences about the S & T knowledge that ought to have gone behind creating these structures.

The footprints of S & T left behind by ancient Indians in the form of monuments and living physical specimens is too large to be meaningfully analysed and presented in a concise form. It is not even clear if any systematic and scientific study of these has been carried out. Instead, we need to take a slightly different and concise approach. A useful starting point is the UNESCO-approved World heritage sites in India. UNESCO designates certain places as World Heritage as they may have outstanding value to humanity. These could be 'cultural entities' such as ancient ruins, historical structures, buildings, and even cities or 'natural sites' such as rivers, and forests. The cultural sites are often considered to be a creative genius of human beings. We can explore the world heritage sites of the 'cultural' category in India for several reasons. The advantage of this is that these sites represent the developments in culture and social systems, and accumulated knowledge and skills of the society at that time and serve as an outstanding example of architecture, or technology. These sites also portray the ideas, living traditions, and artistic and literary works of the times when they were created. Of the 36 world heritage sites in India, a vast majority of them are cultural sites. Table 11.2 has a list of approved world heritage sites in India up to 13th century CE as of September 2020.

TABLE 11.2 World Heritage Sites in India (up to 13th Century CE)

Sl. No.	*World Heritage Site*	*Period*	*Short Description of the Site and its Importance*
1	Rock Shelters of Bhimbetka, Madhya Pradesh	30,000 years	Five clusters of rock shelters with paintings in some of the shelters dated from 100,000 BCE to 1000 CE amidst dense forest has been discovered. Twenty-one villages surrounding them reflect the traditions displayed in the rock paintings.
2	Champaner-Pavagadh Archaeological Park, Gujarat	Prehistoric and 8th–14th Century CE	Largely unexcavated archaeological, historic and living cultural heritage properties which include prehistoric sites, a hill fortress of an early Hindu capital, and remains of the 16th-century capital of the state of Gujarat.
3	Mahābodhi Temple Complex at Bodh Gaya, Bihar	3rd Century BCE to 6th Century CE	The first temple was built by Emperor Aśoka in 260 BCE around the Bodhi Tree. The oldest temple in the Indian sub-continent built during the Gupta period.
4	Sanchi, Madhya Pradesh	2nd Century BCE to 12th Century CE	A group of Buddhist monuments dated between 200 BCE and 100 BCE. The sanctuary has a plethora of monolithic pillars, palaces, temples and monasteries in different status of preservation.
5	Ajanta Caves, Maharashtra	6th–2nd Century CE	31 rock-cut cave monuments depict richly decorated fresco paintings, which are unique representations of the religious art of Buddhism.
6	Archaeological Site of Nalanda Mahāvihāra at Nalanda, Bihar	5th–12th Century CE	Archaeological remains of a monastic and scholastic institution. It includes stūpas, shrines, and vihāras (residential and educational buildings). The most ancient university of the Indian Subcontinent that functioned uninterrupted for a period of 800 years.
7	Elephanta Caves, Maharashtra	5th–8th Century CE	Consists of two groups of caves – the first is a large group of five Hindu caves containing rock cut stone sculptures, the second, a smaller group of two Buddhist caves.
8	Ellora Caves, Maharashtra	600 CE to 1000 CE	A mix of artistic creation of the ancient civilization of India depicting Buddhism, Hinduism and Jainism. 34 monasteries and temples sculpted contiguously into rock walls.
9	Mahabalipuram, Tamil Nadu	7th and 8th Century CE	About 40 monuments including the largest open-air bas-relief in the world carved out of rock.
10	Hill Forts of Rajasthan	7th–16th Century CE	Rajput forts at Chittor, Kumbhalgarh, Ranthambore, Gagron, Amer and Jaisalmer characterized by its mountain peak settings, utilising the defensive properties of the terrain.
11	Pattadakal, Karnataka	8th Century CE	Nine temples representing a fusion of *Nāgara* and *Drāviḍa* architectural features, as well as a Jain sanctuary. The Virūpākṣa Temple, is an outstanding architectural edifice.

Sl. No.	World Heritage Site	Period	Short Description of the Site and its Importance
12	Khajuraho, Madhya Pradesh	950 CE to 1050 CE	The surviving monuments with striking fusion of sculpture and architecture and unique original artistic creation.
13	Chola Temples, Tamil Nadu	11th and 12th Century CE	Includes three great temples namely, the Bṛhadīśvara Temple, Thanjavur, the Bṛhadīśvara Temple, Gangaikonda-Cholapuram and the Airāvateśvara Temple, Darasuram. The temples testify to the brilliant achievements of the Chola in architecture, sculpture, painting and bronze casting.
14	Rani ki vav (The Queen's Stepwell), Gujarat	11th Century CE	Famous for its size and sculpture. Most of the sculptures are in devotion to Vishnu. It also has an Apsara showcasing 16 different styles of make-up to look more attractive.
15	Qutub Minar, Delhi	Late 12th Century CE	A red sandstone tower of 72.5 metres height with a base of 14.32 metres reducing to 2.75 metres diameter at the top. This also houses the famous Mauryan time iron pillar.
16	Sun Temple, Konark, Odisha	13th Century CE	Built in the form of the chariot of Sūrya with 24 wheels, decorated with symbolic stone carvings.

Source: Adopted from https://en.wikipedia.org/wiki/List_of_World_Heritage_Sites_in_India

The table makes interesting reading. A number of these monuments are places of religious worship cutting across multiple religions (such as Hinduism, Buddhism, and Jainism). There is a variety of things that one sees in this, that can point to several aspects of S & T. For example, iron pillars in the country may point to certain aspects of metallurgy and material science as the archaeological evidence presented in Chapter 10 seem to allude. On the other hand, large granite structures (temples, forts, and step wells) suggest that appropriate stone-cutting tools ought to have been developed. Intricate sculptures and paintings in the caves indicate that highly skilled and well-developed iconography ought to have been in practice. Moreover, chemical technology pertaining to dyes and colours is required to create mural paintings that have stood the test of time.

Temples in India

The long-standing architectural monuments (temples, pillars, and forts) that dot the entire width and breadth of the Indian sub-continent bear testimony to the engineering skills, geometrical accuracy, and judicious use of building materials. They also point to the architectural and project management skills of ancient Indians. While we will merely see some examples of these monuments to highlight the building technology aspects, we will discuss the architectural aspects in Chapter 12.

South Indian Temples

There are several 'great temples' built by Cholas, Pandyas, and Pallavas in Tamil Nadu and others in several parts of South India (such as Vijayanagara temples, and Hoysala temples to name a few) that are living examples of engineering and technology at work. For example, the

Bṛhadīśvara temple (also known as the big temple) built by Raja Raja Chola I between 1003 and 1010 ACE is a UNESCO world heritage site. The sheer size, structural elements, choice of building materials, design choices, and intricate sculptural works make this temple special. Temples in South India follow a distinctive style mostly employing stone structures using hard stone such a granite. The Chola era architects and artisans ought to have innovated the expertise to scale up and build, particularly with a heavy stone, and to accomplish large temple complexes with tall towers. The Thanjavur 'big temple' tower is about 208 feet high. Placing massive stone structures atop the tower is a logistical challenge even today. On the other hand, working with very intricate stone carvings and sculptures running in thousands in the tall towers and the main sanctum sanctorum of the temple requires a high degree of design and execution skills. All these point to an engineering marvel, which rivals modern-day architecture.

A vast majority of South Indian (Dravidian) architecture owe their engineering and building skills to the originator of Pallava architecture and sculpture Mahendravarman I (580–630 CE). The significant contribution of Mahendravarman was the creation of temples for the first time in South India without the use of traditional materials, such as brick, timber, metal, mortar, and plaster. Instead, he created temples using hard rocks like granite. According to an inscription in one of his cave temples at Mandagappattu, "it was excavated in hard rock, a material not tackled either by the contemporary Chalukyas or the earlier dynasties of the Deccan and north India." Following the Pallavas, the Pandyas and the Muttaraiyars also followed the tradition of constructing temples using granite. This tradition reached its pinnacle during the reign of the Cholas, as evident from the grand temples for Bṛhadīśvara at Thanjavur and Gangaikonda-cholapuram. This tradition further continued with Pandyas, Vijayanagar rulers, and Nayaks. Even today this tradition is being followed in South India for temple construction. It is against this background that Mahendravarman's contributions to temple building by tackling the hardest of the rocks and the success he achieved have to be viewed[7].

This paved the way for the introduction of the technique of cutting into rock, which ought to have triggered some associated engineering challenges and innovations. For example, the temples were built in far-off places from the quarry sites. The famous Thanjavur big temple does not have granite sources in about 100 kilometres radius. It is estimated that about 130,000 tonnes of granite were used in the construction of the temple. The śikhara (summit) of the Bṛhadīśvara Temple's vimāna is octagonal and weighs about 25 tons, which is placed on a single block of granite weighing about 80 tons. It is perhaps the largest monolith placed on a towering structure in the world (see Figure 11.1). The questions that beg our attention are, how they brought such large quantities of granite from far-off places to the construction site and how they managed to carry the carved dome atop a 208 feet high tower and place it there safely and firmly. Further, the quantum of granite used is so substantial that one wonders how the logistics and supply chain problem was addressed.

On the other hand, the great emperor Aśoka and his grandson Daśaratha excavated into the hard boulders of quartzose-gneiss near Gaya (Bihar). Their Barabar and the Nāgārjuni caves were excavated by quarrying into the hardest rock and carving and polishing it with a lot of labour. However, this legacy started and ended in the same century. The Mauryan artisans were well trained in the softer sandstone and their tradition continued for centuries. They could easily quarry the material in large blocks, work it to some extent on the quarry-site and transport the semi-finished product over long distances to the destination for final erection and finish.

FIGURE 11.1 The Top Portion of Bṛhadīśvara Temple at Thanjavur
Source: https://www.flickr.com/photos/53319892@N03/24096329400/

Khajuraho Temples

Khajuraho, situated in the Chhatarpur district of Madhya Pradesh has a homogenous group of temples in the region representing a unique style of architecture in North India. These temples were built during the 9th Century CE–12th Century CE. During the early 10th century CE, Chandellas ruled central India with Khajuraho as their capital. On account of their love and patronage significant architectural and literary contributions were made[8]. It is reported that the place had originally 85 temples, but only 25 remain today. These are also at various stages of preservation. Studies reveal that these temples were built between 850 CE and 1150 CE. The earlier tradition of architecture in Khajuraho was one of granite, however, there was a gradual transition to sandstone. Therefore, except for one or two temples, which were built with a large amount of granite, the rest of them were built using a fine-grained variety of sandstone, of varying shades of pink or pale yellow, brought from the quarries of Panna on the east bank of the Ken River.

The architectural elements of the Khajuraho temples show distinctive peculiarities of plan and elevation and point to a degree of sophistication that became the de facto style for central Indian buildings. All the compartments of the temple are interconnected internally as well as externally and are planned in one axis, running east-west and forming a compact unified structure. The details of architectural designs and modelling of sculptures point to a well-established iconographic knowledge and the availability of skilled technicians and sculptors. Decorative motifs and ornaments depicted in the temple sculptures provide evidence to these aspects. The sculptures in Khajuraho are of great iconographical interest. Besides numerous gods and goddesses of the Hindu and Jaina pantheons, different forms of other deities such as Vidyādharas, Gandharvas, Nāgas, Gaṇas, Bhūtas, and Aapsarases are also portrayed on the walls of the temples. With many sculptures of varied themes, the Khajuraho temples vibrate with the exuberance of human warmth, not generally found in other such art forms.

The existence of musical pillars in some of the temples informs us that ancient Indians had a good understanding of the acoustical properties of stones. Above all, building large granite temples (such as the living temples of Cholas in Tamil Nadu) demands a good understanding of sourcing, logistics, and supply chain management. Moreover, knowledge of cutting stones using special tools is inevitable as granite belongs to the category of the hardest substances to work with.

11.3 IRRIGATION AND WATER MANAGEMENT

Hydrology and water management systems are extremely crucial for any agrarian-based economy. Expectedly, there is a long history of developing good irrigation and waste management systems in India. Several archaeological excavations and some of the living dams and anaicuts provide a wealth of information for one to infer these skills of ancient Indians. Several studies in the past have clearly brought out the presence of sophisticated water management systems in Dholavira, an important city in the Sindhu Sarasvati civilisation[9]. The city was surrounded by a series of 16 large reservoirs (7 m deep and 79 m long), with some of them interconnected together, which nearly accounted for about 10% of the area of the city. Recently, a rectangular stepwell was found at Dholavira measuring 73.4 m long, 29.3 m wide, and 10 m deep, making it three times bigger than the Great Bath of Mohenjo-Daro.

There was a survey conducted in an ancient Indian irrigation system at Sanchi, a well-known UNESCO World Heritage site in Madhya Pradesh, and four other known Buddhist sites of Morel-Khurd, Sonari, Satdhara, and Andher, all established between 300 and 200 BCE. The heights of the dams ranged from 1 metre to 6 metres and their lengths from 80 to 1400 metres. Two of the larger dams were equipped with spillways suggesting that flood protection was also taken into account while designing these structures. The Sanchi dams and another well-known Sudarsana dam in Gujarat had several similarities.

Emperor Candragupta Maurya is reported to have constructed the Sudarśana dam in Girnar, Junagadh, Gujarat to which some structural improvements were made during the reign of Aśoka. In an excavation work conducted by the Archaeological Survey of India during 1951- 1955, a canal 45 feet broad and 10 feet deep was traced up to a length of 450 feet in Kumhrar (the site of ancient Pāṭaliputra), a few miles south of Patna, Bihar, which is attributed to the Mauryan period. The canal was linked with the 'Sone River' and also with the 'Ganges'. While excavating the area around the 'Heliodorus' pillar in Vidiśā (present-day Vidisha, Madhya Pradesh), the remains of a 300 BCE canal were discovered.

Mauryans had developed a sophisticated concept for rainwater harvesting and irrigation management, known as the *Ahar–Pyne* system, which is still practiced in the regions of southern Bihar and Chhota Nagpur. The Pynes are constructed channels to utilise the river water flowing through the hilly regions, whereas the Ahars are catchments with embankments to store rainwater and the water released from the Pynes. The Pynes feed many Ahars and several distributaries are then constructed for irrigating the field. The Ahar–Pyne system also works as a flood mitigation system. The Government of Bihar has taken up renovation of the traditional water bodies (Ahar–Pyne system) under the *Jal Jeevan Hariyali* program in 2020. This shows how robust and useful is this ancient system for water harvesting even in modern times.

During the Sangam Period (300 BCE to 300 CE), in the southern parts of India, rainwater-harvesting structures such as tanks were constructed for irrigating the paddy fields. The Grand Anaicut (Kallanai Dam) in Tamil Nadu is the world's oldest 'still in use' dam and is also credited with being the fourth oldest dam in the world and the first in India. The Grand Anaicut was constructed by the Chola King Karikalan during the first century CE on the river Cauvery for the protection of the downstream populations against flood and to provide for irrigation supplies in the Cauvery delta region. In 1804, the British raised the height of the dam by 27 inches. Later the Lower Anaicut was built which replicated the structure of Kallanai. Figure 11.2 is the Grand Anaicut, as it exists today.

FIGURE 11.2 Grand Anicut – World's Oldest 'Still in Use' Dam
Source: https://commons.wikimedia.org/wiki/File:Kallanai.jpg

Another ancient Indian system for water management is the construction of stepwells. Stepwells are primarily water storage systems that were developed in India, primarily to cope with seasonal fluctuations in water availability. Although they are primarily utilitarian, the stepwells depict the finest architectures and integrate culture and religion with social life in an integrated setting as is evident from the existing step-wells. They are commonly found in western India (mainly Gujarat and Rajasthan) and other such arid regions of the Indian subcontinent, extending into Pakistan. The earliest archaeological evidence of wells provided with steps is found at Dholavira. Mohenjo Daro's great bath also has steps in opposite directions, similar to a stepwell construction. Ashokan Inscriptions mention the construction of stepwells along major Indian roads at regular distances for the convenience of travellers. It is interesting to note that Ashoka states that it was a well-established practice even before his time[10]. The Aparājita-pṛccha, a canonical text on art and architecture from Gujarat written in Sanskrit has details on the construction and ornamentation of stepwells[11]. Figure 11.3 is one of the deepest and largest stepwells in India.

FIGURE 11.3 Chand Baori, Rajasthan, One of the Deepest and Largest Stepwells in India
Source: By Doron, https://commons.wikimedia.org/wiki/File:ChandBaori.jpg

11.4 DYES AND PAINTING TECHNOLOGY

Human beings essentially love colours. Adding colour to life both in a figurative and real sense is a long-time pre-occupation of human beings. People in ancient India had knowledge of the properties of different colouring agents and the technique of using dyes. They were using both organic and inorganic substances for these purposes. Dyeing wool or silk is easier since they have both basic and acidic properties and are proteins. On the other hand, dyeing cotton is difficult and it requires substances known as mordants, capable of penetrating and staying in the fibres. Pigments on the other hand are colouring agents. The use of these substances is also found in the ancient Indian society[12].

The spinning of wool and cotton has been common in the Indus valley civilisation. The discovery of a fragment of a dyed fabric at Mohenjo-Daro indicates that the Indus valley people were aware of the dyeing and the need to use mordants on cotton fabrics. Many dyeing vats were also found, confirming the fact that they had the know-how for dyeing fabrics. There are several references in our ancient texts about dyed fabrics with different colours and geometrical designs. These further add credence to the argument that ancient Indians were aware of the use of mordents for dyeing difficult fabrics such as cotton.

In ancient texts there are frequent mentions of certain colours. This includes Indigo (Nīla), Turmeric (Haridrā), Maddar (Mañjiṣṭhā) and Resin (Rajana). For instance, Krishna is supposed to have been draped in a yellow colour dress, while his brother Balarama is fond of blue colour robes. In Buddhist literature, the Vinaya texts describe six sources of dyes: roots, trunks, barks, leaves, flowers, and fruits of trees. Ancient Indians knew the art of selecting colours and mixing them to achieve different shades. Yaśodhara in his Jayamaṅgalā commentary on Vātsyāyana's Kāmasūtra has listed six components of the art of painting amongst which mixing of colours (varṇikā-bhaṇga) is one[13]. According to Viṣṇu-dharmottara-purāṇa, most common colours were white, red, yellow, black and blue. Several hundred other colours and shades may be obtained by a judicious combination of one or more of these basic colours (which a good artist must know)[14]. This suggests that the art of painting was well known in ancient and medieval India and that people knew the technique of preparing pigments, mixing colours to get shades of their choice for painting beautiful pictures.

IKS IN ACTION 11.2

The Art of Making Perfumes

If we read several ancient Indian texts, we find mention of use of perfumes, especially by the royal people. This suggests that ancient Indians knew how to use perfumes and methods of preparing such perfumes. Viṣṇu-dharmottara-purāṇa is one of the Upa-purāṇas and it consists of three kāṇdas. The 64th chapter in the second kāṇḍa has 46 verses in which the issue of perfumes is discussed. The chapter also discusses the technique of preparing different kinds of incenses, perfumed water for bath and scented oil. The process of preparing the perfumes is described in eight steps (known as karmāṣṭakam):

1. Śodhanam–This involves purification of the ingredients
2. Vāsanam–Scenting with the perfumes of flowers
3. Virecanam–The process of cleansing
4. Bhāvanā–Saturation of powder in fluid
5. Pāka–Ripening by preparing a decoction
6. Bodhanam–Revival of scents using certain reviving agents
7. Dhūpanam–Fumigating with perfumed vapours
8. Vedhanam–Further process of revival

Another interesting source for perfumes is Varāhamihira's Bṛhatsaṃhitā (5th century CE).

In Chapter 77, titled Gandha-yukti, he addresses the issue of preparation of perfumes in 37 verses. In the first 12 verses of the chapter, different methods of preparing perfumes using various ingredients are discussed. However, an equally interesting aspect of this chapter is a nice introduction to permutations and combinations. Varāhamihira in the rest of the chapter shows how using 16 ingredients it is possible to blend them in different ways.

If sets of four ingredients are to be made from the 16 ingredients, it results in $16C_4$ (=1,820) combinations. Further if the 16 ingredients are put into 4 sets of four each and in each of them the proportion of the ingredient is 4, 3, 2, and 1 respectively, then there are 96 permutations possible. This means that with the given instructions for perfume making it will be possible to prepare 174,720 (=1,820×96) different perfumes.

Source: **Sastri, V.S. and Bhat, M.R. (1946).** *"Varahamihira's Brihatsamhita"*, Soobbiah & Sons, Bengaluru, pp. 603–612 and **Vyas, K.T. (1997).** *"Cosmetics and Perfumes"*, Chapter 4 [4.2] in *History of Technology in India*, Vol. I, A.K. Bag (Ed.), Indian National Science Academy, New Delhi, pp. 291.

Since all dyes are not soluble in water, they have to be rendered water-soluble by some reduction technique. Indigo has been popular since the pre-Christian era and was an article of export to Egypt and other countries from India. It is an insoluble dye, and some reduction techniques were applied to make it soluble. Thereafter, the cotton fibre was impregnated with the dye and exposed to air for oxidation. Alum, iron sulphate, rind of the lemon, cotton flower, lemon, green dried mangoes, and myrobalans were used as mordants for dyeing.

Dyes were often used in cosmetics. In ancient India, a paste of sandalwood used as cosmetics is often coloured with lac and other dyes, smeared over the body, and applied in different patterns. Collyrium (eye-salve) or añjana made of dark powdered antimony was very popular. Vermilion (sindūra), lac (lākṣā), and a yellow pigment gorocana were used for marking as tilaka on the forehead. Methods of beautifying the body using dyes are mentioned in Nātya-śāstra[15]. In Rasa-ratnākara, a few recipes are given for the preparation of kuṅkuma (used as tilaka). The boiled essence of flowers of the Palāśa tree (also known as the Indian Tulip) should be kept in the Sun in a wide vessel. Finely ground rice powder must be added to this in the proportion of 1:20, making it a fine paste. Further, the lime of half the volume should be mixed and stirred well in bright sunshine. Out of this, pellets can be prepared which will have a bright red colour like kuṅkuma.

In the Ajanta caves, the paintings in the 9th and 10th caves date back to the 1st or 2nd century BCE. Gupta period paintings are found in other caves of Ajanta, Badami, Chittannavasal, and temples of Kanchi and Thanjavur in Tamil Nadu. Ellora mural paintings and Thanjavur paintings of the 11th century CE beautifully depict figures in attractive bright colours and tints. At Ajanta, Badami and other caves, the colours most freely used were white, red, and brown in various tints, dull green, and also blue. These colours were obtained from the available substances. For example, lime sulphate was used for white colour. Using compounds of iron, they obtained red and brown tints. For yellow, the ancient painters used orpiment, a natural arsenic sulphide. The fresco paintings testify to the knowledge and skill of the painters in preparing permanent colours out of vegetable and mineral sources.

The Ajanta paintings employed extracts of plants and minerals in preparing pigments. First, the wall of the caves was given a pre-treatment using a special coating. Viṣṇu-dharmottara-purāṇa discusses different aspects of painting in the third kāṇḍa. In Chapter 40, it gives a formulation for preparing a coating that could be applied first on the surface (wall) where painting has to be done[16]. The ingredients of the coating include three varieties of brick powders mixed with one-third of clay. The other ingredients added to this include gum-resin (guggulu), beeswax (madhūcchiṣṭam), honey (madhu), a type of grass (kundaraka), molasses (guḍam), and safflower (kusumbha) soaked in oil, in equal proportions. Further, powdered lime three-fourths burnt with bel fruit pulp and lampblack should also be added. After adding one-fourth part of the sand, the mixture should be soaked in water for a month to make it lubricous. With this paste, a smooth and firm coat is applied to the wall. It is stated at the end of this procedure that preparing a wall in this manner will ensure that the painting does not get destroyed even after hundred years.

- Ancient Indians knew the art of selecting colours and mixing them to achieve different shades.
- The Ajanta paintings employed extracts of plants and minerals in preparing pigments.
- Viṣṇu-dharmottara-purāṇa discusses different aspects of painting in the third kāṇḍa.

Preparing a Wall for Painting

The process of preparing adamantine for application on a wall as a coating in preparation for painting is described in the text Śivatattva-ratnākara. The process is as follows[17]. Fresh buffalo hide is to be boiled with water until it becomes a sticky butter-like paste. This is cut into small pieces and dried till the pieces become hard. This is the basic formulation for adamantine (known as Vajralepa) and is useful for preparing the initial coating on the wall to make it smooth and amenable for painting pictures. Sometimes it is required to add some pigments so that it can be used as paints. The pieces of Vajralepa can be placed in a mud vessel filled with water and heated so that the pieces can melt. The molten liquid form of Vajralepa can be added to any colouring pigment to get the desired shade of a colour. Soft white sand and conch powder can be added to this Vajralepa and smeared on the surface of a wall three times so that it becomes smooth and glossy. This is followed by a coating using a mixture of Vajralepa and a moon-white mineral called Naga found in Nilgiris. The wall finally is ready for painting pictures.

11.5 SURGICAL TECHNIQUES

Surgical methods, known as Śalya-tantra is an ancient technique known to Indians right from the pre-Christian era[18]. The most dangerous foreign object that required immediate intervention is an arrow, with which the kings and generals were attacked in a battle. The surgical procedures were developed initially to attend to the wounded and remove the foreign objects from the human body thereby saving lives. Foreign objects of every kind are generally referred to as Śalya, but it specifically also means arrow. Śalya-tantra is, therefore, one of the oldest branches of Indian medicine.

In modern surgery, the use of anesthesia, hemostatic, antiseptics, and antibiotics are very important. It is still a matter of speculation as to how ancient Indian surgeons performed surgeries. The knowledge of vital points of the body (known as marma) was known to ancient Indians as was the case probably with Chinese and other older civilisations. This provided some sense of the anatomy of the body, which is required for surgery. Moreover, based on the techniques developed for pre and post-surgery in the Indian system, it is clear that the surgeons were aware of the critical nature of the infections that may develop post-surgery and the need for good post-surgical care. Suśruta is considered the father of Indian surgery. In his work Suśruta-saṃhitā, he has provided a clinical approach to some major surgical operations. Further, the other contributions of Suśruta include an emphasis on the training of surgeons, a meticulous code of personal ethics, and the social conduct of the medical profession in general. These topics will be of interest to today's surgeons and may help serve humanity better.

- Suśruta is considered the father of Indian surgery.
- In Suśruta saṃhitā, he has provided a clinical approach to some major surgical operations.

The status and quality of surgery are best evaluated by the basket of tools and instruments that a surgeon possesses. Suśruta's treatise makes a mention of 101 blunt and 20 sharp instruments. This included varieties of needles (straight, curved, or round body, blunt and cutting), suture materials, splints, and fracture beds for immobilisation of fractures at various sites, fourteen different types of bandages, lion forceps (Siṃha-mukha-yantra), and finger knife (mudrikā-śastra). Figure 11.4 has a sample set of sharp instruments described in Suśruta-saṃhitā. There is also mention of para-surgical measures, i.e. cautery, etc. for checking the bleeding. Suśruta has further mentioned the specific use of some instruments needed for the specific surgical manoeuvre, such as endoscopy to visualise the haemorrhoids and for conducting blind surgical procedures in the case of retropharyngeal abscess or removal of

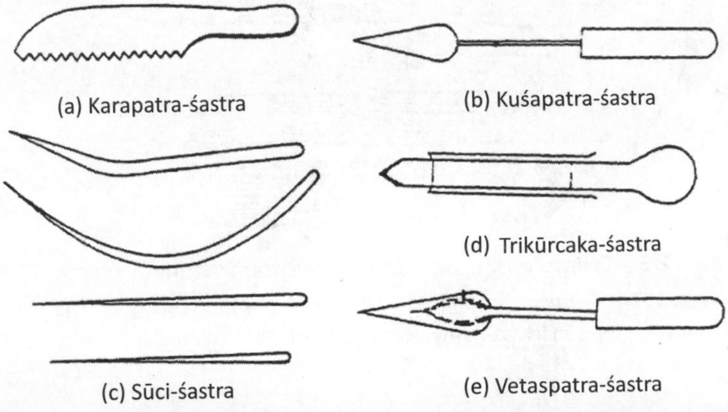

FIGURE 11.4 Some Sharp Instruments Mentioned in Suśruta-saṃhitā

Source: **Sharma, A.N. (1997).** *"Surgical Techniques"*, Chapter 4 [4.4] in *History of Technology in India*, Vol. I, A.K. Bag (Ed.), Indian National Science Academy, New Delhi, p. 335. Reproduced with permission.

the dead foetus through the vagina. These details point to a robust understanding of various aspects of surgery and serve as a good forerunner for modern-day surgeries, which are vast in number, nature, and scope than what we see in Suśruta-saṃhitā.

The clinical approach and surgical treatment of certain disorders mentioned by Suśruta have relevance even today and are being applied in modern surgery. Let us see some examples:

- Suśruta has given a clear description of paracentesis of the abdomen in cases of ascites. After the complete outflow of the fluid, the abdomen should be firmly tied with a many-tailed bandage to prevent further accumulation of fluid.
- Similarly, the ancient surgical approach to remove the urinary calculi appears to be in no way different as worked out in the present era by modern surgeons.
- As per Suśruta-saṃhitā, fumigation of the operation theatre could be done by Guggulu and other similar indigenous drugs to minimise the infection in surgical cases. Modern surgery follows a similar practice of distempering the operation theatres.

On the other hand, some of the surgical practices adopted by Suśruta are unique. For example, certain means of exiting the blood to cure many diseases have been proposed by Suśruta. This includes cupping, the application of leeches, bloodletting (Śirā-vedha) in various other forms. These were used in the treatment of many major diseases such as high blood pressure, sciatica, and other nervous ailments. Suśruta also recommended the use of the barks of several trees such as madhuka, udumbara, aśvattha, palāśa, kakubha, bamboo, sarja, and banyan for the purpose of splints. These barks not only served as an antiseptic in the case of compound fractures but also helped the immobilisation of the affected part. Figure 11.5 has a pictorial representation of bamboo-based splints for the treatment of the fractured hand.

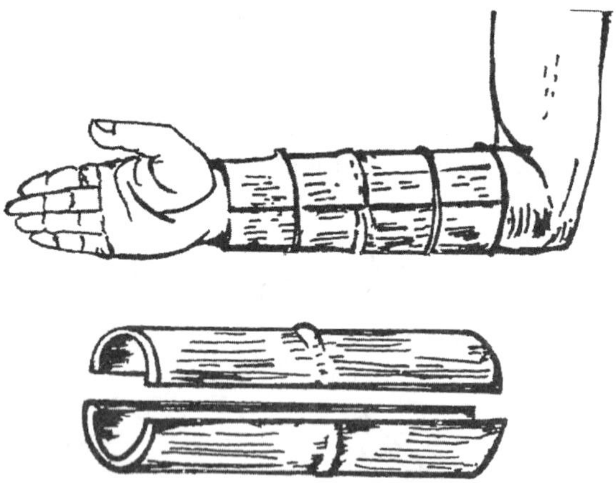

FIGURE 11.5 Example of a Bamboo Splints for Treating a Fractured Hand

Source: **Sharma, A.N. (1997).** *"Surgical Techniques"*, Chapter 4 [4.4] in *History of Technology in India,* Vol. I, A.K. Bag (Ed.), Indian National Science Academy, New Delhi, p. 340. Reproduced with permission.

Rhinoplasty

The list of punishments proposed in dharma śāstras is often executed on the nose or the ear. For example, cutting off the nose and ear was the usual punishment for adultery. This provided an opportunity for the ancient Indian surgeons to perform otoplasty (repair of the ear lobe) and rhinoplasty (surgery to change the shape of the nose). In Suśruta-saṃhitā, Chapter 16 of Śarīrasthāna, a procedure for repairing the nose that has been chopped of has been described in some detail[19]:

- The portion of the nose to be covered should be first measured with a leaf.
- A piece of skin flap of the required size should then be dissected from the cheek or the forehead and turned back to cover the nose.
- The flap of the skin taken from the cheek or forehead must be properly preserved to ensure proper blood circulation in the repaired area.
- The part of the nose is prepared well by making it raw.
- The surgeons should join the two parts quickly but evenly and calmly and keep the skin properly elevated by inserting two tubes inside the nostrils, so that the new nose may look comely.

11.6 SHIPBUILDING

Ancient Indians were very active in the maritime trade. Early Tamil literature of the Saṅgam period (before Common Era), foreign travellers' chronicles, and Roman authors have furnished details of Indian shipping and maritime trade. The Chola kings Raja Raja Chola I (985–1014 CE) and Rajendra Chola I (1014–1042 CE) carried maritime activities with the countries of the Far East and China. Studies suggest that Tamil Pandya embassies were received by Augustus Caesar and Roman historians mention a total of four embassies from the Tamil country.

In Tamil literature, we find references to different names of ships, classes of ships, and their parts, and other relevant information connected with navigation. Although there is no direct reference to the use of monsoon winds as an aid for sailing ships in early literature, the Buddhist Jātaka stories and Jain Canonical texts mention ships moving by force of the wind. Similarly, the Sangam period texts, viz. *Pura-nānūru*, *Aha-nānūru,* and *Madurai-kāñci* delineated different types of seagoing ships as they moved in the seas with the help of wind sails[20]. However, we do not find much information on the art of shipbuilding.

- In Tamil literature we find references to different names of ships, classes of ships, and their parts, and other relevant information connected with navigation.
- The Jātaka stories as well as Buddhist accounts and paintings show that Buddhists were involved in maritime trade.

The Jātaka stories as well as Buddhist accounts and paintings show that Buddhists were involved in maritime trade. The caves of Ajanta, Aurangabad, and Ellora depict Bodhisattva Avalokiteśvara as a saviour of mariners in distressful conditions. The Pāli texts provide some information on Indian shipping. There are multiple references to the number of passengers that the ships carry, thereby giving us some idea of the size of the ships at that time. Prince Vijaya who was banished by his father sailed from Bengal in a fleet of ships carrying more than 700 passengers. Similarly, the ship in which the bride of Vijaya came to Ceylon is supposed to have carried nearly 800 persons. The ship of the Janaka-Jātaka carried 700 persons besides Buddha and that of the Valāhassa-Jātaka carried 500 merchants. The only measurements given of a ship are those of the ship of the Śaṅkha-Jātaka which was 800 cubits (about 360 metres) in length, 600 (about 275 metres) in breadth, 20 fathoms (about 36.6 metres) in depth, and had three masts[21].

IKS IN ACTION 11.3

Surgical Procedure for Piles

Let us consider a large pile mass having a narrow pedicle, moist and projecting, which needs to be treated by surgery. The first step in the pre-operative procedure is to sedate the patient. He should then be seated on a flat plank or on a bed in a covered and clear place on a day with clear sky. The anus should face towards the Sun with the patient in the supine position, and the upper part of his body held in someone else's lap. The waist should be raised a little higher. The neck and the legs should be fixed firmly by a strap and held firmly by assistants. Thereafter, his anus should be lubricated with clarified butter and a sterilized well lubricated rectal speculum should be introduced in the rectum along its passage slowly while the patient is straining.

In this manner, the piles should be visualised and examined properly by the surgeon and the projecting pile mass should be clipped off with a knife and cauterized with fire to prevent bleeding. Thereafter, the pile mass should be mollified by the application of clarified butter mixed with madhuka and the instrument should then be taken out. Following the surgical operation, the patient should be made to get up and sit in warm water, whereas cold water should be sprinkled over him. He should then be taken into a room for post-operative care with proper instructions.

Caustic Application (Kṣāra)

Following the pre-operative procedure, the pile mass should be pressed by a probe, and after wiping with a piece of cotton swab or a piece of cloth, caustic (Kṣāra) should be applied on it. Following this, the surgeon should cover by hand the opening of the instrument until the counting of one hundred (approximately one minute). Then after wiping, the procedure can be repeated for a few times depending on the severity of the disease. The application should be stopped when the piles show the colour of a ripe jambu fruit (Bluish black) and get depressed and shrivelled. Thereafter, the caustic should be washed away by sour gruel, curd, butter milk, vinegar or by the juice of citrous fruits. Then the pile mass should be mollified by the application of clarified butter, or any medicated soothing oil and the instrument should then be taken out.

The post-operative management of the patient should be maintained as described earlier. The remaining pile masses may be treated one by one at intervals of a week each. In case of multiple piles, the right one should be handled first; after the right, the left one and after the left the posterior one; lastly the anterior one should be treated in the same fashion. The post-operative management and the dietary regimen should be according to the instructions mentioned by Suśruta.

Source: **Sharma, A.N. (1997).** *"Surgical Techniques"*, Chapter 4 [4.4] in *History of Technology in India*, Vol. I, A.K. Bag (Ed.), Indian National Science Academy, New Delhi, p. 329.

In the Arthaśāstra, there are several mentions about ships, maritime trade, etc. In book 7, three classes of water-routes were mentioned:

1. the ordinary river routes as well as artificial waterways or canals, called kulyās
2. the routes for coastal traffic carrying on inter-portal communications called kulapatha
3. the ocean-routes called saṃyānapatha

In book 2, different categories of ships (and boats) were mentioned while discussing how appropriate controls must be established in using the waterways. This includes:

1. Saṃyātya-nāva, i.e. ocean-going vessels. It is mentioned that these ships had to pay tolls at the harbours at which they touched

2. Pravahaṇa, which is another name for sea-going vessels
3. Śaṅkha-muktā-grāhiṇya-nāva, which were boats used for pearl-fishing
4. Mahānāva, which were the larger vessels for use in the large rivers that were navigable throughout the year
5. Kṣudraka-nāva, which were smaller boats for use in small and shallow rivers that overflowed in the rains
6. Hiṃsaka, pirate ships, and boats

A French writer, F. B. Solvyns (1811), writes in his Lee Hindous, "In ancient times the Indians excelled in the art of constructing vessels and the present Hindus can in this respect still offer models to Europe so much so that the English, attentive to everything which relates to naval architecture, have borrowed from the Hindus many improvements which they have adapted with success to their own shipping. The Indian vessels unite elegance and utility and are models of patience and fine workmanship[22]."

These references and literary and archaeological evidences point to established maritime trade and inland water transport systems aiding trade and commerce. Information on the details of shipbuilding is scanty in Indian literature. Yukti-kalpataru, a work dated to about 11th century CE, in two chapters, deals with ships and gives details about their construction, varieties, measurements, sizes, decoration, and accommodation[23].

The Yukti-kalpataru deals with certain aspects of shipbuilding in two chapters entitled Niṣpada-yānoddeśa and Jaghanya-jalayānāni[24]. The issues discussed include the characteristics of varieties of woods most suited for the construction of ships, seasons suited for shipbuilding, the classification of vessels for river-going and sea-going and their names and respective measurements, painting and decoration of ships, tying of iron nail to a sea-going vessel, ships with cabins, etc. According to Yukti-kalpataru, there are two broad categories of ships: the ordinary and the special capable of seafaring. In each of these ten different varieties have been specified. These differ in the size of the ships. Table 11.3 has a list of the different varieties of ships and their measurements. Five key attributes of a ship have been mentioned: lightness, hardness, mobility, hollowness, and good proportion.

- The Yukti-kalpataru deals with certain aspects of ship-building in two chapters.
- According to Yukti-kalpataru, there are two broad categories of ships: the ordinary and the special capable of seafaring.

In the Vṛkṣāyurveda four types of wood have been specified. According to Yukti-kalpataru, the wood that is light, hard, and cannot be joined easily is the ideal one for shipbuilding. It cautions against using two different kinds of wood. Such ships built using different kinds of wood bring neither good nor comfort. Moreover, they do not last for a long time, rot, split and sink in the water[25]. It is also mentioned that iron should not be tied to a sea-going vessel using a string because iron may be attracted by magnetic forces in the sea and may cause danger[26].

There are specifications given for building ships with cabins. Cabins could be constructed for the entire length of the ship, at the front or in the middle. Each of them serves a different purpose. For example, a ship with a cabin extending from one end to the other is suitable for the transport of royal treasure, horses, and women. On the other hand, a ship having a cabin in the middle is suitable for use during the rainy season and pleasure trips of kings. A ship having a cabin in the front is convenient for the dry seasons, for long voyages, and naval warfare. Furthermore, a ship having a cabin less than half of its length is likely to be swift.

TABLE 11.3 Different Types of Ships and Their Measurements given in Yukti-kalpataru

Ordinary (Sāmānya)				*Special: Sea-going (Unnata)*			
Name	*Length*	*Breadth*	*Height*	*Name*	*Length*	*Breadth*	*Height*
Kṣudrā (Diminutive)	4.57	1.14	1.14	Dīrghikā (Tall)	9.14	1.14	0.91
Madhyamā (Moderate)	6.86	3.43	3.43	Taraṇī (Moving hither and tither)	13.71	1.71	1.37
Bhīmā (Formidable)	11.43	5.71	5.71	Lolā (Moving hither and tither)	18.28	2.29	1.83
Capalā (Moving to and fro)	13.71	6.86	6.86	Gatvarā (Perishable)	22.85	2.86	2.29
Paṭalā (With covering)	18.28	9.14	9.14	Gāminī (Moving on)	27.42	3.43	2.74
Abhayā (Not dangerous)	20.57	10.28	10.28	Tarī (Running swiftly)	31.99	4.00	3.20
Dīrghā (Tall)	25.14	12.57	12.57	Jaṅghālā (Running Swiftly)	36.56	4.57	3.66
Patraputā (Like a folded leaf)	27.42	13.71	13.71	Plāvinī (Flowing over)	41.13	5.14	4.11
Garbharā (With inner compartments)	31.99	16.00	16.00	Dhāriṇī (Power of possessing)	45.70	5.71	4.57
Mantharā (Curved)	34.28	17.14	17.14	Veginī (Having velocity)	50.27	6.28	5.03

* *Source:* **Mukerji, R. (1977).** "Notes on Ancient Hindu Shipping" in R.G. Bhandarkar Commemoration Volume, Bharatiya Publishing House, Delhi, pp. 447–455.
All values in metres. Conversion: 1 cubit = 0.457 metres

11.7 SIXTY-FOUR ART FORMS (64 KALĀS)

Kalā is defined as that which makes an individual happy and is an art form. The word 'art form' in the context of 'Kalā' has a wider meaning as it does not restrict itself only to performing arts. It includes a variety of specialised skills that are innate to human beings. By getting trained in these skills and by engaging in activities related to these, individuals can become happy and deliver a good job in whatever they do. This formed the bedrock of professional and trade classifications in earlier times. In the ancient Indian tradition, there is a concept of 64 kalās (known as catuḥ-ṣaṣṭi-kalā or vidyā) which captures the essence of various useful skills that individuals can develop. These could form the core of making a living for them. Table 11.4 summarises the 64 kalās organised under eight broad categories. As is evident from the table, this includes certain skills related to artistic pursuits, decoration, beautification of body, cooking, play, and entertainment.

What is however noteworthy is that the list of 64 kalās includes a wide range of engineering and technical skills. An examination of these skills provides further evidence of the use of S & T in ancient Indian society. Weaving and a host of other engineering skills including carpentry, metallurgy, and mineralogy suggest that there was enough application of these ideas in the society and therefore individuals who develop unique skills in these areas can make a living by professing these.

TABLE 11.4 A Summary of 64 Kalās

Type of Skills	*No.*
Artistic Skills: Singing, dancing, painting, theatrics, playing instruments, etc.	8
Skills in Cleaning and Beautifying Body: Dressing, decorating, painting the face, applying aromatics, combing hair, setting ornaments, etc.	9
Skills of Decoration: Making a covering of flowers for a bed, making the groundwork of jewels, covering the bed, applying an admixture of colors, etc.	6
Culinary Skills: Preparing delicious food items and drinks.	3
Play and Entertainment Skills: Playing on music in water, splashing with water, juggling, playing with thread, riddles, gambling, using children's toys, etc.	8
Skills for Intellectual Pursuits: Conversation, practicing language, composing, designing a literary work, lexicography and metres, reciting books, etc.	10
Engineering and Technical Skills: Needle works and weaving, spinning, carpentry, engineering, testing silver and jewels, metallurgy, mineralogy, practicing medicine, mechanics, etc.	14
Special Uncommon Skills: Knowing the mode of fighting of lambs, and birds, knowing conversation between male and female cockatoos, talking with fingers, prediction by heavenly voice, etc.	6

11.8 STATUS OF INDIGENOUS S & T

The available evidence and the scriptural references point to a culture that deployed S & T in various aspects of their day-to-day living. However, it is intriguing that we have no inkling of these today. We have very little awareness of these details. We also do not see the practice of any of these concepts in mainstream activities. Some of the S & T practices still in use have been branded as rural, tribal, and folk practices. They neither receive serious attention nor appreciation. A case in point is the current day water management practices. The trans-Yamuna – Sutlej canal is a multi-crore government initiative to bring water to the desert regions of Rajasthan. While this has not achieved its objectives, the age-old local practices to harvest water have been working well (see the video listed under the second bullet in "Discover IKS" exercise in Chapter 1 of the book for more details on this). In summary, we shall enumerate some broad observations concerning science and technology in ancient times and its current status:

- On the one hand, we find compelling evidence to conclude that ancient Indians had a good understanding of several aspects of science and technology, as we saw in Chapter 10. On the other hand, we are not able to get substantive details in the literature on various aspects of science, engineering and technology. The available information in some cases is broad. A case in point is the shipbuilding industry of ancient India. There

> - There are several indications about the unique aspects of the Indian ships that travelled far and wide and conducted trade with several countries.
> - Literary sources dealing with science, engineering and technology are not very detailed; substantive aspects of knowledge transmission continued in the oral tradition.

are several indications both in the Indian inscriptions and travellers' accounts of foreigners who visited India about the Indian ships and maritime travellers who travelled far and wide and conducted trade with several countries. This clearly indicates that they are adept in shipbuilding without which it may have been a dangerous activity to engage with. There are intricate details and specifications involved in shipbuilding. However, we are unable to extract great details about shipbuilding from the Indian literature.

- As we know, Indian society relied substantially on an oral tradition for the preservation, and dissemination of knowledge across generations. The culture of extensive documentation and relying on written material as the source of knowledge is of fairly recent origin in India, dating back at most to half a millennium. Therefore, we find that literary sources dealing with science, engineering and technology are not very detailed and specific in nature. Perhaps substantive aspects of knowledge transmission continued in the oral tradition. The expertise of the teacher was often a closely guarded secret, available only to the students working with them on the job. Even today we find that there are some traditions in which knowledge is transferred from generation to generation (such as Astrologers, Vaidyas, and Śilpis).
- There was a well-developed caste system that was closely aligned to the skilled craft being practiced by the people. For example, all the metalsmiths and Goldsmiths were of a certain caste and so were the carpenters, the building mechanics, or the Alchemists, to name a few. These practices and know-how were maintained within the family and passed on from generation to generation.
- Large parts of India were subjected to foreign rulers for over 700 years since the beginning of 13th century CE. The priorities of what needs to be encouraged and in what manner were left entirely to the ruling class and this has also contributed substantially to the waning away of indigenous skills, practices, and know-how. For example, much of the knowledge about mining and metalworking that ancient Indians possessed was in grave danger when the British Raj imposed a mining ban in the 18th century. Furthermore, imposing production taxes also contributed to making metalworking practices economically infeasible. This enabled the British Raj to export a large chunk of ores for value addition in England.
- With the advent of modern governance and social systems introduced by the British Raj, many skills and practices went off the mainstream swiftly, thereby making these structures weaker and perhaps irrelevant. Thus, much of ancient art and craft perished over time for lack of institutional support and broad governmental and social acceptance. Moreover, it was not providing enough economic incentives to perpetuate this knowledge across generations. A few remaining skills in the country today suffer from huge survival challenges which are largely economic and institutional in nature.

SUMMARY

- Literary sources dealing with science, engineering and technology are not very detailed and specific in nature. Perhaps substantive aspects of knowledge transmission continued in the oral tradition.
- A study of the archaeological remains and literary works during 5th century CE to 12th century CE points to certain unique aspects of metalworking and metallurgy in India during ancient and medieval times.
- The long-standing architectural monuments (temples, pillars, and forts) that dot the entire width and breadth of the Indian sub-continent bear testimony to the engineering skills, geometrical accuracy, and judicious use of building materials.
- The Grand Anaicut in Tamil Nadu is the world's oldest "still in use" dam. Mauryans had developed a sophisticated concept for rainwater harvesting and irrigation management, known as the "Ahar–Pyne" system, which is still practiced in the regions of southern Bihar and Chhota Nagpur.
- There are several references in our ancient texts about dyed fabrics with different colours and geometrical designs. These indicate that ancient Indians were aware of the use of mordents for dyeing difficult fabrics such as cotton.
- Suśruta is considered as the father of Indian surgery. In his work Suśruta-saṃhitā, he has provided a clinical approach to some major surgical operations.
- In Suśruta-saṃhitā, one finds descriptions of surgical instruments for performing several surgical operations.
- Yukti-kalpataru, a work dated to about 11th century CE, in two chapters, deals with ships and gives details about their construction, varieties, measurements, sizes, decoration, and accommodation.
- In the ancient Indian tradition, there is a concept of 64 kalās (known as catuḥ-ṣaṣti-kalā or vidyā) which captures the essence of various useful skills that individuals can develop.

REVIEW QUESTIONS

1. How many basic colours are there as per the Indian tradition? Enumerate them.
2. Describe a process for the preparation of Kuṅkum.
3. What is the suggested method for preparing a wall before applying paint? Explain its significance.
4. Comment on the statement, "Suśruta is the father of surgery".
5. Compare and contrast the modern surgical practice and that proposed by Suśruta.
6. Explain briefly the surgical practice proposed by Suśruta for plastic surgery and piles.
7. Comment on the statement, "Ancient Indians had a rich heritage of maritime trade".
8. What are the different varieties of ships proposed by Yukti-kalpataru? How do they differ from each other?
9. What are the unique features of South Indian granite temples? Does S & T play any role in building these temples?
10. Did ancient Indians have knowledge of water and irrigation management? Give examples to support your argument.
11. Based on the examination of the world heritage sites in India, is it possible to comment on the S & T heritage of India? Support your argument with examples.
12. What are the major works on Metallurgy and metalworking? What are their main contributions?
13. What do you understand by the term "64 kalās"? Does it have any relationship to S & T culture in India?
14. What is your understanding of the current status of our ancient S & T practices?

DISCOVER IKS

1. The south Indian granite structures offer unique opportunities for one to explore the technological challenges of building such structures. There are two videos given below that discusses issues related to working with granite rocks

   Video 1: Tiger Cave: https://youtu.be/vdkLkAC0RhU

   Video 2: Warangal Fort: https://youtu.be/KBn-JEZhoPE

   After watching the video, prepare a note to answer the following questions:

   (a) What are the challenges in working with granite?
   (b) Does knowledge of engineering technology required to work with granite?
   (c) What do you think about the ability of ancient Indians to work with granite structures in building large structures (such as temples and forts)?

SUGGESTED READINGS

Balasubramaniam, R. (2000). "On the Corrosion Resistance of the Delhi Iron Pillar", *Corrosion Science*, **42**, pp. 2103–2129.

Bag, A.K. (1997). *History of Technology in India*, Vol. I, Indian National Science Academy, New Delhi.

Balasubramaniam, R., Saxena, A., Anantharaman, T.R., Reguer, R., and Dillmann, P. (2004). "A Marvel of Medieval Indian Metallurgy: Thanjavur Forge-Welded Iron Cannon", *JOM*, **56**(1), pp. 17–23.

Biswas, A.K. (1993). "The Primacy of India in Ancient Brass and Zinc Metallurgy", *Indian Journal of History of Science*, **28**(4) pp. 309–330.

Bose, D.M, Sen, S.N. and Subbarayappa, B.V. (2009). "A Concise History of Science in India", *Universities Press*. 2nd ed., Hyderabad.

Chaudhuri, M. (1976). "Shipbuilding in the Yuktikalpataru and Samaraṅgana Sūtradhāra", *Indian Journal of Historical Studies*, **11**(2), pp. 137–147.

Craddock, P.T., Hegde, K.T.M., Gurjar, L.K., and Willies, L. (2017). "Early Indian Metallurgy: The Production of Lead, Silver and Zinc through Three Millennia in North West India", Archetype Publications Ltd., London.

Dharampal (2021). "Indian Science and Technology in the eighteenth century: Some contemporary European Accounts", Dharampal Classic Series 2, Rashtrotthaana Sahitya, Bengaluru.

Gandhi, S.M, (2014). "Copper, Zinc, Lead Ores – Their Exploitation and Metal Extraction by the Ancients in the Northwestern India", *Journal Geological Society of India*, **84** (September 2014), pp. 253–266.

Hegde, K.T.M. (1973). "A Model for Understanding Ancient Indian Iron Metallurgy", *Man, New Series*, **8** (3), pp. 416–421.

Kramrisch, S. (1928). "The Vishnudharmottara – Part III: A Treatise on Indian Painting and Image-making", Kolkata University Press, Kolkata.

Mukerji, R. (1977). "Notes on Ancient Hindu Shipping" in R.G. Bhandarkar Commemoration Volume, Bharatiya Publishing House, Delhi, pp. 447–455.

Neogi, P. (1914). "Iron in Ancient India", Indian Association for the Cultivation of Science, *Bulletin No. 12*, Kolkata.

Neogi, P. (1918). "Copper in Ancient India", Indian Association for the Cultivation of Science, Special Publication No. 1, Kolkata.

Rajput, D.S., Gokarn, R., Jagtap, C.Y., Galib, R., Patgiri, B.J., and Prajapati, P.K. (2016). "Critical Review of *Rasaratna Samuccaya*: A Comprehensive Treatise of Indian Alchemy", *Anc. Sci. Life*, **36**(1) pp. 12–18.

Ray. P. (1956). "History of Chemistry in Ancient and Medeival India", *Indian Chemical Society*, Kolkata.

Srinivasan, S. (2016). "Metallurgy of Zinc, High-tin Bronze and Gold in Indian Antiquity: Methodological Aspects", *Indian Journal of History of Science,* **51**(1), pp. 22–32.

Subbarayapa B.V., (1982). "Glimpses of science and technology in ancient and mediaeval India", *Endeavour, New Series,* **6**(4), pp. 177–182.

Tripati, S. (2011). "Ancient maritime trade of the eastern Indian littoral", *Current Science*, Vol. **100**(7), pp. 1076–1086.

ENDNOTES

1. For a good account of this refer to **Dharampal (2021).** "Indian Science and Technology in the eighteenth century: Some contemporary European Accounts", Dharampal Classic Series 2, Rashtrotthaana Sahitya, Bengaluru.
2. See for example this extract from Camaka Praśna – ... हिरंण्यं च॒ मे ऽयंश्च मे॒ सीसं च मे॒ त्रपुंश्च मे श्या॒मं चं मे लो॒हं चं मे॒ ... hiráṇyaṃ ca̱ me 'yáśca me̱ sīsaṃ́ ca me̱ trapúśca me śyā̱mam̱ cá me lo̱haṃ cá me̱ ...
3. तिग्मं चिदेम महि वर्षो अस्य भसदश्वो न यमसान आसा । विजेहमानः परशुर्न जिह्वां द्रविर्न द्रावयति दारु धक्षत् ॥ tigmaṃ cidema mahi varṣo asya bhasadaśvo na yamasāna āsā | vijehamānaḥ paraśurna jihvāṃ dravirna drāvayati dāru dhakṣat || Ṛgveda Saṃhitā 6.3.4
4. तद्यथा लवणेन सुवर्णँ सन्दध्यात्सुवर्णेन रजतँ रजतेन त्रपु त्रपुणा सीसँ सीसेन लोहं लोहेन दारु दारु चर्मणा ॥ tadyathā lavaṇena suvarṇa̐ sandadhyātsuvarṇena rajata̐ rajatena trapu trapuṇā sīsa̐ sīsena lohaṃ lohena dāru dāru carmaṇā || **Swami Gambhirananda (2003).** "Chāndogya Upaniṣad", Advaita Ashrama, Kolkata, p. 308.
5. लोहितेन स्वधितिना मिथुनं कर्णयोः कृधि । अकर्तामश्विना लक्ष्म तदस्तु प्रजया बहु ॥ lohitena svadhitinā mithunaṃ karṇayoḥ kṛdhi | akartāmaśvinā lakṣma tadastu prajayā bahu || Atharvaveda 6.141.2
6. को अस्मिन्नापो व्यदधात्विषूवृतः पुरूवृतः सिन्धुसृत्याय जाताः । तीव्रा अरुणा लोहिनीस्ताम्रधूम्रा ऊर्ध्वा अवाचीः पुरुषे तिरश्चीः ॥ ko asminnāpo vyadadhātviṣūvṛtaḥ purūvṛtaḥ sindhusṛtyāya jātāḥ | tīvrā aruṇā lohinīstāmradhūmrā ūrdhvā avācīḥ puruṣe tiraścīḥ || Atharvaveda, 10.2.11
7. For a detailed report on this, please see, **Srinivasan, K.R. (1958).** "The Pallava Architecture of South India", *Ancient India*, **14,** pp. 114–138.
8. For more details on this and issues addressed in this section see, **Deva, K. (1959).** "Temples of Khajuraho in Central India", *Ancient India*, **15,** pp. 43–65.
9. For more details on this and the other issues addressed in this section see, **Singh, P.K., Dey, P., Jaina, S.K. and Mujumdar, P.P. (2020).** "Hydrology and water resources management in ancient India", *Hydrol. Earth Syst. Sci.*, **24**, pp. 4691–4707.
10. https://en.wikipedia.org/wiki/Stepwell Last accessed on Oct. 1, 2021.
11. **Jain – Neubauer, J. (1999).** "The stepwells of Gujarat ", *India International Centre Quarterly*, **26**(2), pp. 75–80.
12. **Radha, K. (1997).** "Dyes, Mordantas and Pigments", Chapter 4 [4.3] in *History of Technology in India,* Vol. I, A.K. Bag (Ed.), Indian National Science Academy, New Delhi, pp. 300–323.
13. रूप-भेदः प्रमाणानि भाव-लावण्य-योजनम् । सादृश्यं वर्णिकाभङ्गं षडेते चित्रभागकम् ॥ rūpa-bhedaḥ pramāṇāni bhāva-lāvaṇya-yojanam | sādṛśyaṃ varṇikābhaṅgaṃ ṣaḍete citrabhāgakam || See for details, **Devadutta Shastri (2007).** "Kamasutra of Vatsyayana with Jayamangala commentary", Chaukhambha Sanskrit Sansthan, Varanasi, p. 87.
14. In Viṣṇudharmottara-purāṇa it is mentioned that there are only five basic colours and hundreds of other colours are part of it. मूलरङ्गाः स्मृताः पञ्च श्वेत-पीत-विलोहिताः । कृष्णो नीलश्च राजेन्द्र शतशोऽन्तरः स्मृतः ॥

mūlaraṅgāḥ: smṛtāḥ pañca śveta-pīta-vilohitāḥ | kṛṣṇo nīlaśca rājendra śataśo'ntaraḥ smṛtaḥ || 3.40.16

15. See for example, in Bharata's Nāṭyaśāstra, Chapter 21, the following verse: नेत्रयोरञ्जनं कार्यम् अधरस्य च रञ्जनम् । दन्तानां विविधा रागा: चतुर्णां शुक्लता यथा ॥ netrayorañjanaṃ kāryam adharasya ca rañjanam | dantānāṃ vividhā rāgāḥ caturṇāṃ śuklatā yathā || 21.28

16. For more detail on this see **Kramrisch, S. (1928).** "The Vishnudharmottara–Part III: A treatise on Indian painting and image making", Kolkata University Press, Kolkata, pp. 48–51. Part III of Viṣṇudharmottara Purāṇa has a good discussion on several aspects of Indian painting.

17. See for detail, **Radha, K. (1997).** "Dyes, Mordantas and Pigments", Chapter 4 [4.3] in *History of Technology in India,* Vol. I, A.K. Bag (Ed.), Indian National Science Academy, New Delhi, p. 315.

18. The discussion under this head is lagely drawn from **Sharma, A.N. (1997).** "Surgical Techniques", Chapter 4 [4.4] in *History of Technology in India,* Vol. I, A.K. Bag (Ed.), Indian National Science Academy, New Delhi, pp. 324–343.

19. The relevant verses are given below:
विश्लेषितायास्त्वथ नासिकाया वक्ष्यामि सन्धानविधिं यथावत् ।
नासाप्रमाणं पृथिवीरुहाणां पत्रं गृहीत्वा त्ववलम्बितस्य ॥ 16.27
तेन प्रमाणेन हि गण्डपार्श्वादुत्कृत्य बध्नात्वथ नासिकाग्रम् ।
विलिख्य चाशु प्रतिसन्दधीत तत् साधुबन्धैर्भिषगप्रमत्तः ॥ 16.28
viśleṣitāyāstvatha nāsikāyā vakṣyāmi sandhānavidhiṃ yathāvat |
nāsāpramāṇaṃ pṛthivīruhāṇāṃ patraṃ gṛhītvā tvavalambitasya || 16.27
tena pramāṇena hi gaṇḍapārśvādutkṛtya badhnātvatha nāsikāgram |
vilikhya cāśu pratisandadhīta tat sādhubandhairbhiṣagapramattaḥ || 16.28
For more details see, **Thakral, K.K. (2017).** "Suśruta Saṃhitā by Suśruta", Choukhambha Orientalia, Varanasi, Śarīrasthāna, Chapter 16, p. 6194.

20. **Tripati, S. (2011).** "Ancient maritime trade of the eastern Indian littoral", *Current Science*, Vol. **100**(7), pp. 1076–1086.

21. For details see, **Mukerji, R. (1977).** "Notes on Ancient Hindu Shipping" in R.G. Bhandarkar Commemoration Volume, Bharatiya Publishing House, Delhi, pp. 447–455.

22. *Ibid.*

23. For details see, **Chaudhuri, M. (1975).** "Ship-building in the Yuktikalpataru and Samarangana Sutradhara", *Indian Journal of History of Science*, Vol. **11**(2), pp. 137–147.

24. **Pandit Ishvara Chandra Sastri (1917).** "Yuktikalpataru", Printed by Abhinash Chandra Mandal, pp. 223–230.

25. *Ibid.* pp. 224. विभिन्नजातिद्वयकाष्ठजाता न श्रेयसे नापि सुखाय नौका । नैषा चिरं तिष्ठति पच्यते च विभिद्यते वारिणि मज्जते च ॥ vibhinnajātidvayakāṣṭhajātā na śreyase nāpi sukhāya naukā | naiṣā ciraṃ tiṣṭhati pacyate ca vibhidyate vāriṇi majjate ca ||

26. *Ibid.* pp. 224. न सिन्धुगाद्याहैति लौहवन्धं तल्लोह-कान्तैः ह्रियते हि लौहम् । विपद्यते तेन जलेषु नौका गुणेन बन्धं निजगाद भोजः॥ na sindhugādyāhaiti lauhavandhaṃ talloha-kāntaiḥ hriyate hi lauham | vipadyate tena jaleṣu naukā guṇena bandhaṃ nijagāda bhojaḥ||

CHAPTER

12 Town Planning and Architecture

LEARNING OUTCOMES

After finishing this chapter, you will be able to:

- Develop a historical perspective to town planning and architecture in ancient India
- Get familiarised with the issues addressed in Vāstu-śāstra
- Develop a basic understanding of the limbs of Vāstu-śāstra
- Understand alternative models for town (village) planning
- Understand various aspects of temple architecture

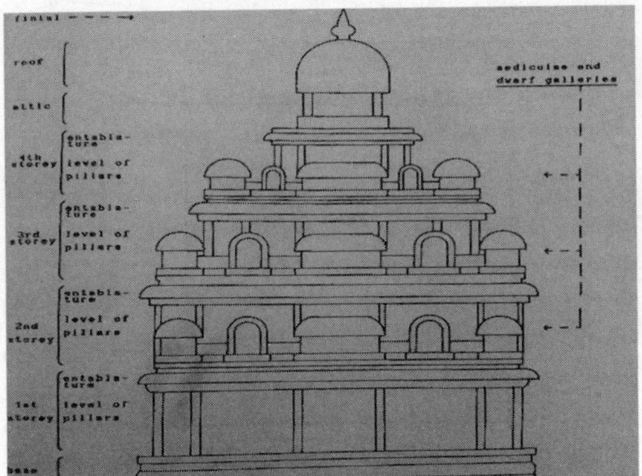

Town planning and architecture is the logical extension of the basic instinct of every living being, to protect self from the weather and other living beings. That also led to the creativity in the field of construction resulting in several incredible temple structures. The figure is the architectural design for a four-storey temple (Subhadraka type) proposed in Mayamata, one of the Vāstu-śāstra treatises.

Source: **Dagens, B. (1994).** "*Mayamatam–Vol. I*", Indira Gandhi National Center for the Arts. p. lxix.

IKS IN ACTION 12.1

Arthaśāstra on Town Planning

Zoning and land use are the concepts used for effective town planning and balanced growth of cities. These issues form part of townplanning in Vāstu-śāstra texts and have been discussed in the Arthaśāstra. Kauṭilya has provided a detailed description of how to plan a fortified city. There are marked areas in a fortified city for specific types of activities. Since a well-articulated varṇa system associated with the type of activities was in vogue, zones are identified with respective varṇas for settlement. The commercial and non-commercial zones are delineated, and appropriate developments permitted within each zone. Figure 12.1 gives a pictorial description of a fortified city.

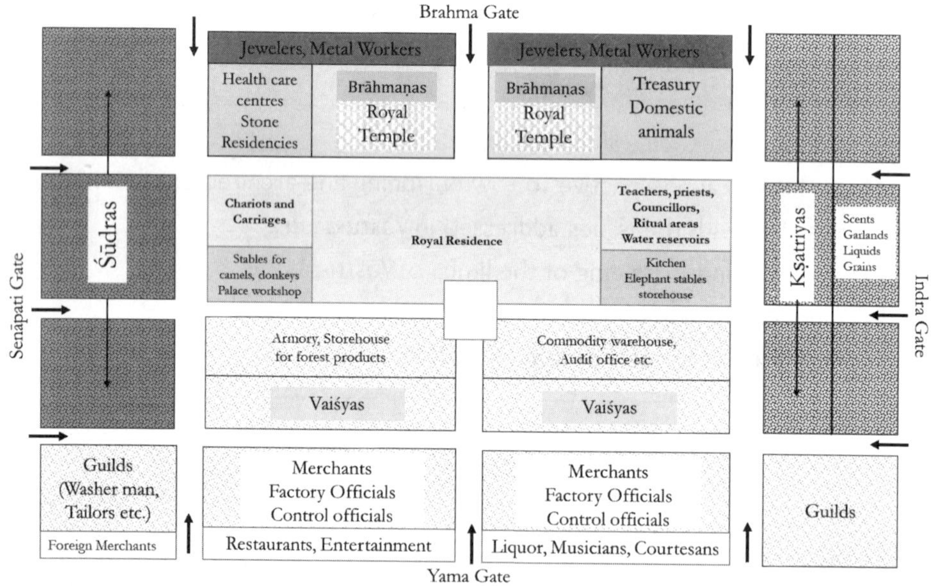

* The crematoriums are outside the city at designated places (not shown in the figure)

FIGURE 12.1 A Schematic Representation of Kauṭilya's Plan for a Fortified City

The demarcated zones and permissible activities in each of the zones are evident from the figure. There are three royal roads (E – W and N – E directions) and 12 gates all around the city. King's palace is supposed to occupy 1/9th area and located a little north of the centre. On the Eastern side, there are merchants of scents, garlands, and liquids. It also houses kṣatriya dwellings. Royal teachers, priests, and ministers are in the north, and artisans, craftsmen, and śūdras are in the west. On the southern side, vaiśyas reside. Moreover, the sale of cooked rice, liquor, and courtesans is located. In the other directions (such as northwest, southwest, etc.) there are other designated offices, storehouses, and manufacturing facilities located as indicated in the figure. In several corners of the city, there will be guilds and corporation workers. At the center of the city, the temples are located. Kauṭilya has provided the specification for the width of the roads to be built in the city. All royal roads, roads to villages and district headquarters, and roads to gardens, groves, and forest are 54 feet wide. On the other hand, roads to military stores, burial grounds, and villages are to be 48 feet wide. Roads to elephant forests are 12 feet wide and roads for chariots are 7.5 feet wide. Finally, roads for the cattle are 6 feet wide and pedestrian walkways are 3 feet wide.

Source: Based on **Shamasastry, R. (1929).** *"Kauṭilya's Arthaśāstra"*, 3rd ed., The Wesleyan Mission Press, Mysuru, pp. 53–57. Figure adopted from **Rangarajan, L.N. (1987).** *"Kautilya–The Arthashastra"*, Penguin Books, New Delhi, p. 192.

Survival instinct is a basic feature of any living entity. To survive the onslaught of weather changes and to protect themselves from other powerful creatures the living beings relied on different types of abodes. For example, birds and insects construct different types of nests and dwellings. Similarly, human beings began with simple temporary structures and dwellings perhaps at the foot of large trees, made of natural resources such as the trunk of trees and branches. Over time, human beings started building residences and more sophisticated structures that provided better utility and aesthetics.

In modern parlance, town planning is a collective set of processes, ideas, and methods using science and technology to deploy the available land for human habitation and other uses in the most efficient way. It is concerned with the development and design of land use and the built environment, including air, water, and infrastructure. Since mobility is a key requirement town planning also deals with the appropriate design of transportation and communication networks and the optimal layout of human settlements. On the other hand, architecture deals with the process of planning, designing, and constructing buildings for human use. This for example includes living places (houses), common meeting places (such as shops, offices, and temples), and other structures (such as a dam, bridge, port, and transport hub). Ancient approaches to town planning and architecture address several of these issues as we shall see in this chapter.

- Archaeological exploration of the Sindhu–Sarasvati civilization points to a well-developed town planning in ancient times.
- Harappa was a planned city with the streets generally oriented along with the cardinal directions.

12.1 INDIAN ARCHITECTURE – A HISTORICAL PERSPECTIVE

The architectural legacy of India can be understood from the archaeological excavations pointing to town planning and a large number of monuments maintained by the Archaeological Survey of India (ASI) pointing to architectural practices of a variety of buildings, most notably temples. The town planning practices of the ancient Indians dating back to 2600 BCE have been studied and well-documented. The extensive archaeological explorations of the Sindhu–Sarasvati civilization and the analysis of sites such as Dholavira, Lothal, Harappa, and Mohenjo-Daro points to a well-developed town planning. In 1921 and 1922, the Indian archeologists Daya Ram Sahni and Rakhaldas Banerjee carried out a systematic excavation of the buried cities of Harappa and Mohanjo-Daro, on the Ravi River in the Indus valley. Geometrically designed and oriented according to the cardinal points, the cities had fortifications, distinct zones, community halls, and manufacturing units including craft workshops and furnaces for the production of copper and bronze tools, weapons, and ornaments. Dholavira, a 48-hectare city located in the Rann of Kachchh where its proportion of rain is very low, had separate drains to harvest rainwater and several dams built across two nearby streams to divert their seasonal waters to a network of large reservoirs within the city[1].

Harappa was a planned city spanning over 25 hectares, with the streets generally oriented along with the cardinal directions. Houses were built with bricks of standardised proportions; some of the larger ones had at least upper storey, roofs consisted of wooden structures covered with grass or leaves. Most houses had individual bathrooms connected to extensive drainage networks. Complex structures, such as Mohenjo-Daro's Great Bath or the so-called Granary, demonstrated advanced planning and construction skills. In the Great Bath's central basin, for example, the floor was made of tightly fitted bricks set on edge and cemented with gypsum plaster to make a watertight surface; it was then covered with a layer of bitumen (natural tar). Structures, such as wells constructed with trapezoid bricks, which prevented inward collapse were also found. Some of these Harappan techniques and concepts were preserved in later Indian architecture: for instance, the general house plan, with rooms organised around a central courtyard, is found in many parts of rural India; the drainage system of the later Ganges civilization appears a manifestation of Harappan legacy. Mohenjo-Daro had houses with two rooms, as well as mansions, underground sewage, drainage systems, public water tanks, granaries, and a great bath.

Based on the archaeological remains of Sindhu–Sarasvati civilisation we can infer that several cities with marked similarities were existing spread over a large geographical area. Town planning and building techniques and construction material used ought to have been standardised. The fire burnt mud bricks used in the sites were of standard geometrical proportions although varying in sizes. The cities had granaries, drains, watercourses, tanks, and wells. This implies a certain level of evolution and maturity in town planning and architecture.

The urban planning and architecture corresponding to the Buddhist and Mauryan times are evident from the inscriptions and reliefs in Sanchi. Other historical counts by visitors also give ample information about the town planning and architecture from 300 BCE to 200 CE. Megasthenes, a visitor around 300 BCE mentions 564 towers and 64 gates in the city walls in the Mauryan capital Pāṭaliputra.

The City of Ayodhyā and Laṅkā

Vālmīki gives a detailed description of the city of Ayodhyā in the fifth sarga of Bālakāṇḍa, in Ramāyaṇa. According to the narrations available in the sarga the highways passed from Ayodhyā to Rājagṛha, the capital of Kekayas. It was mentioned that Ayodhyā was a sprawling city spread over 12 yojanas in length and 3 yojanas in width. The city where many artisans lived had arched outer gateways, well-arranged local markets, and different instruments and weapons. It contained stately edifices decorated with flags and hundreds of śataghnīs (missiles). It had several gardens and mangroves and girdled by sal trees. The city was built based on Aṣṭapada design having a royal palace with highways on four sides by deep and wide ditches and water-filled moats. Its dwellings were constructed on the levelled ground with no space left unutilized. The palaces were perfectly constructed in rows.

Laṅkā is one of the unique cities supposedly built by Viśvakarman, the great architect. As Hanumān enters Laṅkā in search of Sītā, he passes through the city of Laṅkā. Vālmīki gives a good description of Laṅkā in Sargas two to six in Sundarakāṇḍa in varying degrees. According to Vālmīki, Laṅkā was a captivating capital filled with buildings resembling the assembly of planets. It had whitewashed and elevated houses looking like autumnal clouds and well laid out streets decorated with garlands of banners and flag posts, rows of colourful creepers and festoons.

Hanumān saw Laṅkā situated on the top of the mountain, filled with splendid yellow-white palaces, like a city stationed in a sky. There were ramparts, moats filled with water, spiked iron tridents, and Śtaghnī (missile – one which can kill hundreds of people). The city was stretched with rows of buildings all over, with seven or eight storied mansions. The beautiful quadrangles, covered with grains of crystal were as though leaping into the air. It echoed with sounds of krauñca birds, peacocks, and royal swans. Going from one building to another, Hanumān saw mansions of different kinds of structures. The whole city was encircled by a boundary wall. The palace of King Rāvaṇa had spacious places to accommodate big chariots. There were colourful palanquins of several kinds, bowers, picture galleries, spots for sporting, hillocks (artificial) made of wood, apartments for romance, for pleasures and daytime activities. There were storehouses for valuable conches, weapons, bows and arrows.

Although it appears exaggerated on account of the poetic license available for a kavi, it nevertheless provides interesting details on certain aspects of town planning and architecture.

Source: https://www.valmiki.iitk.ac.in/. Last accessed on August 12, 2020.

There are over 3000 ancient monuments and archeological sites and remains of national importance protected by ASI. Overall, there are around 8,000 monuments protected by the government which include temples, mosques, tombs, churches, forts, palaces, stepwells, rock-cut caves, as well as ancient mounds and sites which represent the remains of ancient habitation. In addition to these, there are at least a few lakhs ancient monuments which remain unprotected, and are not maintained using any public funds. The presence of such a large number of monuments in India is an indication of a long history of architecture in the Indian sub-continent. Notably, the temple architecture went through an evolution over time. The earliest known examples of temple architecture belong to cave temples carved out of hard rock found in several places in India. The famous Ajanta and Ellora caves in Maharashtra and the cave temples of Mahabalipuram belong to this category. The temples of Badami, Aihole, and Pattadakal in Karnataka and the Dravidian temples belonging to the Chola and Pandya periods are great examples and living traditions of temple architecture in South India. Another group of temples known as Khajuraho in Madhya Pradesh such as the Kandariya Mahadev Temple and the Orissan type of temples is another example of great temple architecture. Every region in the country has living temples representing outstanding temple architecture in India.

12.2 VĀSTU-ŚĀSTRA – THE SCIENCE OF ARCHITECTURE

Vāstu-śāstra often evokes certain imagination in our minds that it is about building a house or an office complex and locating the rooms in appropriate directions. This is an oversimplified understanding and a misleading perspective. If we study the Vāstu-śāstra texts we notice that it deals with town planning, design of unitary structures including temples and residential houses, sculpture, painting, and furniture. Therefore, Vāstu-śāstra, also known as Vāstu-vidyā or Śilpa-śāstra, taken in the broadest context implies the science related to artistic creations through design, fabrication, and construction. The modern-day equivalent term is architecture. In the Vāstu-śāstra four things are considered as vāstu: the earth or the ground (bhūmi), temple or a palace (prāsāda), conveyance (yāna) and couch (śayana). Of these, bhūmi is considered the primary one and is described first in the vāstu texts.

Vāstu-śāstrā is essentially an art of correct setting whereby one can align the proposed structure to the pañca-bhūtas, and the rotational influence of the sun, moon, and the other planets surrounding the earth. The basic premise of a vāstu is to integrate the design and construction of towns, temples, and other buildings with the natural forces and optimise the whole process of creating the proposed structure so that the intended purpose is achieved. It involves ground preparation, spatial geometry, and space arrangement. In addition to these, the other usual aspects such as measurements, design, and layout aspects of buildings are also considered. Vāstu-śāstra treatises discuss ideas of town planning and design of efficient villages, and integrated temples, water bodies, and gardens within them. Figure 12.2 pictorially lists the topics typically discussed in Vāstu-śāstra. The subject matter of discussion in a Vāstu-śāstra text typically can be organised under five major heads:

Town Planning: Villages are the basic units of human inhabitation. The other elements are towns, forts, the capital city, and military cantonments. Alternative designs for these, choice of an appropriate site for developing these infrastructures, the details of land use patterns (zoning of a village or a town for different uses), etc. are often discussed in detail in all vāstu texts.

Town Planning	Civil Architecture	Temple Architecture	Artistic	Others
• Site selection • Design of towns, villages, capital city • Land use patterns (Zoning)	• Palace • Houses • Forts • Public buildings ▪ Theatre ▪ Library ▪ Other public infrastructure	• Temples • Components of a temple • Iconography (idol making)	• Paintings • Furniture, doors • Sculptures	• Qualification of a Sthapati • Choice of building material, wood, etc. • Site planning (Vāstupuruṣa-maṇḍala)

FIGURE 12.2 Issues Discussed in Vāstu-śāstra

Civil Architecture: Vāstu-śāstra discusses the design and construction of unitary structures such as houses, palaces, temples, and public utilities. These can be segregated into civil and temple architecture. The civil architecture includes the design of a royal palace, houses for different sections of the society, forts, and public buildings such as sabhās (meeting halls), theatre, library, health centre, and other public infrastructures.

Temple Architecture: Elaborate plans and designs for the construction of temples are discussed in all vāstu texts. This also includes descriptions of various components of a temple such as maṇḍapa, garbhagṛha, śikhara, and stūpi and the relative proportions to be adopted. Another related issue discussed in Vāstu-śāstra is iconography. This addresses the issues of making the idols of Gods and Goddesses that occupy the temples.

Artistic Creations: The Vāstu-śāstra also deals with certain aesthetical elements that are part of the above infrastructure. For example, temples and public halls have paintings and different sculptures (other than the Gods and Goddesses). Details on the design of these are part of Vāstu-śāstra. Furthermore, other artistic elements such as furniture, sleeping couches, and doors are also subject matter for discussion in Vāstu-śāstra. For example, in Samarāṅgaṇa-sūtradhāra, a chapter deals with the design of sleeping couches.

Other Issues: In addition to the above topics, we also find other related topics discussed in vāstu texts. For example, the qualifications of a Sthapati (the chief architect), his team of workers, and their roles are discussed in vāstu texts. Another issue that finds discussion relates to the choice of building materials (such as wood). Finally, a generic framework for site planning known as Vāstu-puruṣa-maṇḍala is discussed as the basic design mechanism to layout different components of a proposed structure. This applies to all types of architectural endeavours (Town planning, temple, and civil architecture).

12.2.1 Literary Sources

The subject of vāstu is discussed in several literary sources in the country. References to Sthāpatya are found in the Vedic literature, epics, Purāṇas, Brāhmaṇas, Āgamas, Sūtras, and Śilpa-śāstras. In the Atharvaveda references to different parts of the building such as sitting-room, inner apartment, room for sacred fire, cattle shed, and reception room is found. (Atharvaveda, 9.3). The Śāṅkhāyana-gṛhya-sūtra describes the ceremonies performed for constructing a building. Chapters 252–270 of Matsya-purāṇa deal with issues related to images and buildings. Similarly, in the Agni-purāṇa, Chapters 21–106, 263–272, and 317–326 discuss various aspects of Vāstu-śāstra. The Rāmāyaṇa mentions three types of houses in Ayodhyā: Prāsāda – Palaces for kings, queens, and other royal members, Harmya – Stately mansions and Vimāna – Seven storied buildings. Arthaśāstra of Kauṭilya has several chapters dealing with architecture[2]. Arthaśāstra defines Vāstu as houses, fields, gardens, bridges, lakes, or any place of stay[3]. One of the chapters named Durgavidhānam deals with the construction of different types of forts. Bṛhat-saṃhitā of Varāhamihira (early 6th century CE) also deals with Vāstuvidyā. The text contains a description of temples, Adamantine glue, idols, and installations of idols. Chapter 54 is devoted to Śilpa-śāstra and Chapter 56 to characteristics of temples (Prāsāda-lakṣanādhyāya).

However, there are literary works that deal exclusively with the subject of Vāstu-śāstra. Table 12.1 has a list of major texts and issues covered in them. If we analyse the architectural development in the Indian sub-continent, we notice a certain regional pattern. The literature in the area of architecture and town planning also can be attributed to regional beginnings. With a broad regional classification, we can identify texts with certain regions. For instance, Samarāṅgaṇa-sūtradhāra is typically associated with Malwa region of Madhya Pradesh. In the South Indian texts, the Mayamata and the Mānasāra are associated with the Tamil Nadu region and the Manuṣyālaya-candrika with the Kerala region. Mayamatam comprises about 3300 verses and is divided into 36 chapters. It is considered part of the Śaivite Āgama literature, originating from Tamil Nadu, composed probably during the Chola period[4]. The Mānasāra is a comprehensive treatise on architecture and iconography from the same category as the Mayamata and it consists of about 5400 verses organised in 70 chapters.

- The classification of the architectural texts could be based on the region of origin and their sphere of influence.
- Bṛhat-saṃhitā (6th century CE) deals with Vāstuvidyā (Homebuilding).

TABLE 12.1 Selected Vāstu-śāstra Texts and Issues Covered

Sl. No.	*Text*	*Topics Discussed*
1	Kāśyapa-śilpa	Treatise on architecture and iconography. Prescriptions of the building of the temples, rules for making images of the deities
2	Nārada-śilpa-śāstra	General roads, Water resources, Village and town planning, Fourteen types of towns, Fortification, Palace complex, Interior planning, Superstructure, Residences, Palaces, Furniture, Law courts, an Arts Gallery, Theater, Temple, Iconography, and Paraphernalia
3	Viśvakarma-prakāśa	The orientation of sites, Men and materials to be employed in Vāstu Examination of the different kinds of lands, regions, and soils, Leveling of site, placing the foundation box, The planning of villages, towns, forts, and roads, streets, lanes, Planning of temples, icons of gods, Planning of arsenal, Residential houses
4	Mānasāra	System of measurement, Classification of Architecture, Examination, and Selection of soil, The ground plan (Pāda-vinyāsa), Village Planning, Towns and Ports, Pillars, Entablature, Roof Wood-joinery, One to twelve storied buildings

(Contd.)

TABLE 12.1 Selected Vāstu-śāstra Texts and Issues Covered (*Contd.*)

Sl. No.	*Text*	*Topics Discussed*
5	Mayamata	A treatise on dwelling, deals with all the facets of gods' and men's dwellings, from the choice of the site to the iconography of the temple walls, Descriptions of villages and towns, temples, houses, mansions, and palaces, Selection of proper orientation, right dimensions, and of appropriate materials
6	Samarāṅgaṇa-sūtradhāra	Town planning, house architecture, temple architecture, and sculptural arts, mudrā, the canons of painting, the art of mechanical contrivances – the yantra. Land/Soil examination, Units of measurement, planning of the King's Abode, houses with one, two and four chambers, the definition of the door, pedestal, wall, the art of wood planking and defective or deficient buildings, Definition of temples, etc., Pavilion, Allocation of specific parts to the land
7	Manuṣyālaya-candrikā	Seven chapters covering the following architectural themes: Investigation and selection of land, Site analysis, Houses, Parts of a house, Elements of roof, Ancillary structures

12.3 Vāstu-puruṣa-maṇḍala

The ancient Indian approach to architecture views the site on which the proposed structure is to come as Vāstu-puruṣa. At the outset one may dismiss this as a religious ritual, however, there are deeper implications behind the conceptualisation of Vāstu-puruṣa. The site is not viewed as bare earth but a transformed entity with life. It is, therefore, an artifice in which the ground (bhūmi) is converted into the extent of the manifested universe. According to Garuḍa-purāṇa[5], the vāstu-puruṣa must be placed in the following way: On the north-east the head of the Lord of Vāstu, and his feet towards the south-west and his hands at the south-east and north-west corner of the house. Figure 12.3 is the depiction of the Vāstu-maṇḍala. The center of this maṇḍala is the Brahma-sthala. The eight directions position the natural forces in the site and help the designer align the structure to these forces. Based on the type of planning adopted, these aspects are used to place various facilities. For example, in temple construction, the garbha-gṛha (the sanctum santorum) is placed in the Brahma-sthala. On the other hand, in the design of a fortified capital city, the King's palace is placed at the Brahma-sthala.

From the architectural perspective, vāstu is primarily the planned site of a building or a settlement such as a town and is referred to as vāstu-puruṣa-maṇḍala. Maṇḍala means a polygon, but as a rule, the shape of vāstu-puruṣa-maṇḍala is square. It is the plan of all architectural forms. The site-plan, the ground-plan, the horizontal and vertical sections are regulated by its norm. It is an arrangement by which different parts of an assembly that make up an edifice or a group of edifices or a settlement are positioned with reference to a regular diagram drawn at the time of laying out.

Any site on which a proposed structure needs to be built is conceived in terms of squares. Each square of the vāstu-puruṣa-maṇḍala is attributed to a protecting deity by whose name the square is designated (see Figure 12.3 for a partial representation of these deities in a maṇḍala of 81 squares (9 × 9)). The simple structure is a 1×1 square and at the other extreme, the site

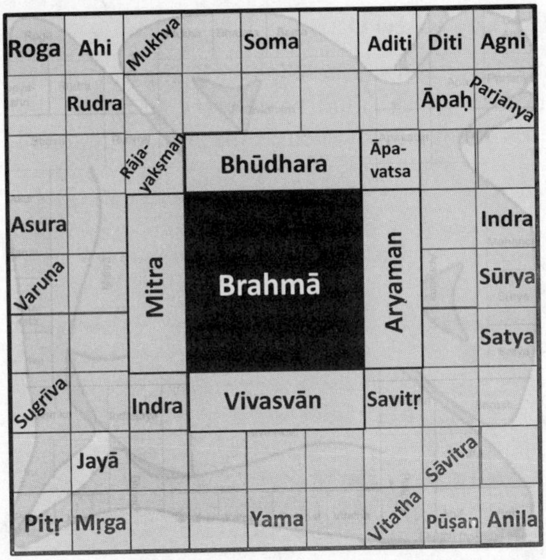

FIGURE 12.3 The Vāstu-puruṣa-Maṇḍala (9 × 9 Grid)
Source: Adopted from **Kramrisch, S. (1946).** "The Hindu Temple", Vol. I, University of Calcutta, p. 32.

could be divided into 1024 squares (32 × 32). Town and village planning may involve larger squares while individual buildings fewer squares. The 64 squares (8 × 8) and 81 squares (9 × 9) are most often employed. Table 12.2 has a representative set of square divisions using vāstu principles. Once the site is conceived in terms of a square structure with a certain number of divisions, then it is viewed using the maṇḍala, which enables integration of the site with natural forces. This guides the entire design process to be adopted.

TABLE 12.2 Division of a Site into Alternative Sizes and Squares

Sl. No.	*Type of Division of the Site*	*No. of Squares*
1	Sakala	One (1×1)
2	Pecaka	Four (2×2)
3	Pīṭha	Nine (3×3)
4	Mahā-pīṭha	Sixteen (4×4)
5	Upa-pīṭha	Twenty-Five (5×5)
6	Ugra-pīṭha	Thirty-Six (6×6)
7	Sthaṇḍila	Forty-Nine (7×7)
8	Manḍūka	Sixty-Four (8×8)
9	Paramasāyika	Eighty-One (9×9)
10	Āsana	Hundred (10×10)

12.4 EIGHT LIMBS OF VĀSTU

The planning and architectural aspects detailed in vāstu can be viewed using an eight-part classification of the concepts. These could be referred to as the limbs of Vāstu. Figure 12.4 depicts the eight limbs. We shall look at each of these in some detail.

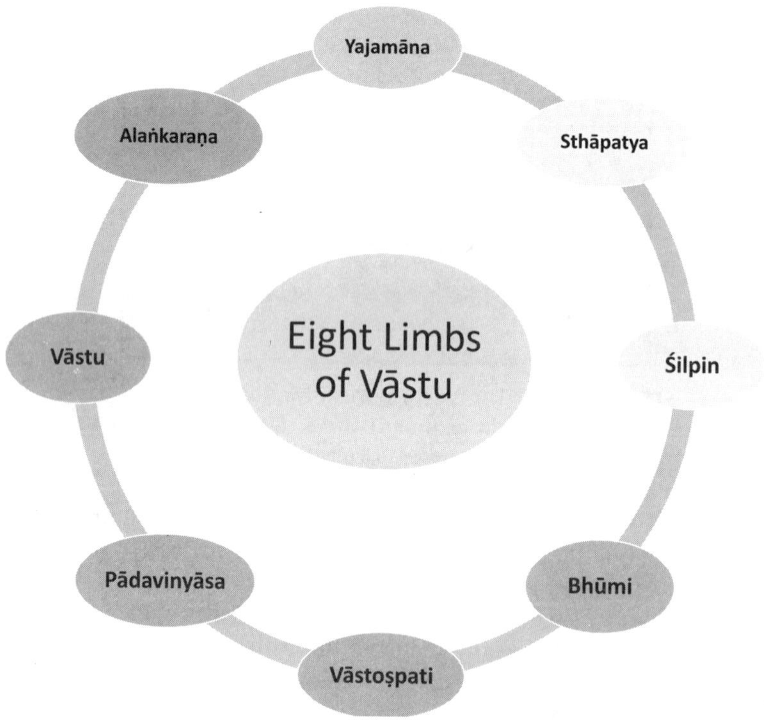

FIGURE 12.4 The Eight Limbs of Vāstu

Yajamāna (Host): The core objective of Vāstu-śāstra is to provide satisfaction, peace of mind, and prosperity for a dweller or the owner. Therefore, the host has a very important role, and the building of any infrastructure depends upon certain characteristics of the yajamāna. This includes horoscope, varṇa, profession, and anthropometrics[6].

Sthāpatya (Architecture): In modern parlance we use the term 'architecture' to denote all aspects of structural design, developing detailed specifications, cost estimation, and unique styling. These issues come under Sthāpatya in vāstu śāstra. Robust structuring is another aspect of consideration. House or any building is built so that the Yajamāna is happy and is prosperous. According to Bhṛgu, the house built must last for at least 200 years and the fort and palaces must have a life span of at least 600 years. The foundation of the building was given the utmost importance in ancient texts[7]. The best style of architecture is none other than the style of God. God has created this whole world. So, a śilpin must try to imitate the style of God according to Vāstu-śāstra texts teachings. One must use wood for the upper part of the house and rocks for the lower part that is what nature teaches. Similarly, the śilpin must be aware of the local issues such as weather, land, water, etc.

Śilpin (Technician): According to Mayamata, four types of technicians (śilpis) work on a building: Sthapati, Sūtragrāhin, Vardhaki and Takṣaka. The master architect is called the *Sthapati* in Vāstu-śāstra. The qualities of Sthapati and the definitions of other technicians are given in Chapter 5 of Mayamata[8]. The Sthapati's job is a creative one, wherein he uses his knowledge of the śāstras, deploys the available material, and directs the skilled labour to create a new structure, much like a poem written by a poet or a painting done by an artist. A Sthapati must have a good knowledge of different sciences while practicing architecture and construction engineering. For example, knowledge of measurements of time, space, area, mass, and motion, and estimation of direction (Dik-sādhana) is important. The Sthapati is required to be well-versed in more than one branch of learning including mathematics, astrology, and the Vedas. Samarāṅgaṇa-sūtradhāra has discussed the desirable qualities of a Sthapati in Chapter 44. At the outset, to be a good Sthapati one must be endowed with four key things[9]:

(a) Śāstra – knowledge of the technical details of the craft

(b) Karma – Practical experience in the application of the knowledge, making him a skilled craftsman

(c) Prajñā – Intuitive insight which endows him with an imaginative mind and artistic orientation

(d) Śīla – Righteous character, which will ensure that he will succeed in all his endeavours with the right attitude and orientation.

The chapter also provides more details on specific skills required of a Sthapati[10]. An intelligent Sthapati should know the limbs of Vāstu-śāstra including signs indicating the nature of a person, mathematics, astronomy, prosody, water channels, arts, and usage of machines. Moreover, he should be able to articulate these according to śāstra and be able to demonstrate them in the application. Specifically, eight types of works have been identified for someone to attain eminence as a Sthapati. These are skills in intricate design, carpentry, building (using bricks, etc), artworks on stone, artificially creating new things out of existing material and gold, and flawless execution of the plan.

Sthapati is the head of the team consisting of Sūtragrāhin, Takṣaka, and Vardhaki. The Sūtragrāhin is the draftsman, who does the layout using the cord (sūtra) for all measurements for the building. The Takṣaka carves stone, wood, and clay while the Vardhaki adds to his work by joining parts and finishing their surfaces. Sūtragrāhin is the disciple of Sthapati and often his son and successor who is considered capable of completing the project if the situation arises. The Vardhaki works with the Takṣaka closely and is subordinate to him. The substantive labor pool of the architectural team will be guilds (*śilpi-sangha*), who are trained in their respective trades. This group is responsible for the construction of the structure, headed by the Sthapati. Sūtragrāhin, Takṣaka, and Vardhaki support Sthapati in the process of construction.

In the selection process of this group, the yajamāna first selects Sthapati depending upon the required scope of knowledge, skills, and experience; past performance and history of Sthapati, and affordability. Once this is done Sthapati and Yajamāna or either of them will select Sūtragrāhin. Sūtragrāhīn will select Vardhaki. Finally, the Takṣaka is selected by Sthapati and Sūtragrāhin.

Bhūmi (Land): The starting point for any architectural endeavour, be it a civil building, a temple, a palace, a fort, or a town is to survey the land on which the proposed structure needs to be established. This involves ascertaining the overall suitability of the site for the proposed construction and an examination of the soil and soil conditions. In Samarāṅgaṇa-sūtradhāra, Chapter 8 (Bhū-parīkṣā) deals with these aspects. The text makes three broad divisions of the lands and the soil (Jāṅgala, Anūpa, and Sādhāraṇa). The chapter also discusses different types of lands suitable for forts, towns, villages, etc. Before selecting land for construction it must be tested thoroughly. Land testing is done for the sake of two purposes; to determine whether the village will be developed and civilised and to ascertain the required quality of the land to construct the building. If the land is wrongly selected then it may lead to an adverse effect on the dweller.

Vāstoṣpati Pūjā (Offerings): The beginning of an architectural endeavour has two activities: Vāstu-pūjana and bali-dāna, in which certain offerings are made and the gods, spirits, and demons are bid to leave. Once the place is emptied of its former contents and occupants, it is ready to assimilate the new ones. Vāstoṣpati deals with various rituals related to Vāstu like homa, śānti, pārāyaṇa (chanting), etc. These are performed by the yajamāna (owner) to receive prosperity, protection, and happiness from the lord Vāstoṣpati also referred to as the master of the area (Kśetrapati). Some of the important rituals to be performed by yajamāna according to Vāstu śāstra include Bhūmi-pūjana – Worshipping the land and finalising the layout, Gṛhārambha – From Śilāsthāpana (Stone laying ceremony) to the completion of the building, Gṛhapraveśa – First entry into the home (Housewarming ceremony), Vāstu-śānti – a process to worship the Vāstu Puruṣa and seek the blessings for positivity, prosperity, and happiness, and Vāstu-pūjana – Periodical rituals.

Pādavinyāsa (Site layout) is the approach to site layout or design for the proposed construction. As we have already discussed the layout of any architectural creation is based on the Vāstu-puruṣa-maṇḍala. Mayamata describes thirty-two designs for site layout[11]. Mānasāra also has a similar description of alternative designs. A 49-cell model is recommended for the renovation of the home, for private homes generally, 64-cell or 81-cell models are recommended. In Chapter 36 of the Mānasāra, layouts for differential areas of a house are specified based on the presiding deities of the grids in the Vāstu-maṇḍala. The suitability of certain grids (such as 49, 64, and 81 squares) for villages depending on the dimensions are indicated.

Vastu (Materials) deals with the materials used and the processing of those materials as required for the construction in detail. Stone was one of the most important materials used. Many temples in India like the Kailāsanātha temple and Bṛhadīśvara temple were built using stones. The stones were obtained from two sources – hills and underground mines. These stones are selected based on colour, defects, age, and gender. The color of the stone should be uniform. Stones with different colours or shades are considered as unsuitable. There is a specific chapter in Bṛhat-saṃhitā, 'Vajra-lepa-lakṣaṇādhyāya[12]' on the preparation of glue which helps in making the building or architecture adamantine (see Chapter 11 for details).

In earlier times, wood was the predominant material used for construction. Therefore, one of the issues discussed in the Vāstu-śāstra texts is bringing wood from the forest for construction. Wood is also used for making images of God and Goddesses and other artefacts. The selection of an appropriate category of material, the suitability of certain trees, and the manner of cutting the trees are discussed in detail in the texts. Before cutting the trees, certain offerings are made to propitiate the trees and their presiding spirits with the mantras quoted in the text. In the cutting operation, if some unfavourable signs are seen, such as the flowing of blood, honey, milk, and clarified butter are noticed, the trees are considered unfit for employment in construction. Using the colour, the oil, and the bark of the tree the age of the trees is ascertained.

Alaṅkaraṇa (Renovations and Decorations) in Vāstu deals with two aspects: interior and exterior design of a building and repairs and modifications. The Interior and exteriors itself is a great subject of the vāstu study. Some important areas of interior and exterior designing discussed in Vāstu-śāstra texts are the decoration of doors and arched decorations over the doorway (Toraṇa), use of beautiful and colorful stones, pedestal designs, wall paintings, carvings and coloring the pillars, and beautiful mural paintings on ceilings. We find several of these aspects in the temple architecture and cave temples in the country. The other aspects of decorations include well-carved furniture, beautification of the main gate of the house, the establishment of a garden, lake, well, fountains, beautification of compound wall, and construction of shrines, statues, etc. inside the compound and outside the home. Similarly, there are many suggestions found in the vāstu-śāstra on the Alaṅkaraṇa of the infrastructure.

TABLE 12.3 Different Categories of Towns Mentioned in Vāstu Texts

Sl. No.	*Category of Town/Village*	*Brief Description*
1	Rājadhānī – Capital city	Primary abode of the king, has a sabhā at the centre of the city
2	Paṭṭana/Puṭabhedana	Second residence of a king, a commercial centre
3	Droṇamukha	Situated on a riverbank, frequented by traders from distant places
4	Durga	A fortified town, 12 types of forts have been discussed
5	Sthānīya	Fortress at the centre of 800 villages
6	Śākhānagara	Subsidiary town
7	Kārvaṭika	Situated at the centre of 200 villages
8	Kheṭa	Smaller town mainly of labour class
9	Nigama	Market mainly of artisans
10	Grāma	Smaller than Nigama
11	Maṭha or Vihārā	A residential University village

Compiled from: **Kumar, P. (1998).** *Bhoja's Samarāṅgaṇa Sūtradhārā*, Vol. I, New Bharatiya Book Corporation, Delhi.

12.5 TOWN PLANNING

Several big cities in developing countries suffer from severe traffic congestion, pollution, and unhealthy living conditions. Business establishments, residential layouts, and manufacturing enterprises have developed in an overlapping fashion adding to the complexities of administering cities and providing safe, healthy, and sustainable environments to society. To respond to these challenges, modern-day urban planning uses concepts such as zoning, which is a process of dividing land in a city or an urban area into zones such as residential, industrial, and commercial. With such classifications, the town planners attempt to specify the dimensions of land area and the form and type of uses it can be put to. Vāstu-śāstra texts have addressed some of these issues and have laid out design principles and methods for town planning and architecture. We have examples of the application of these in the design of Jaipur and Chandigarh. These cities have borrowed ideas from ancient Indian town planning concepts.

Jaipur is built of the Prastara model of the town and Chandigarh resembles the Sarvatobhadra model (see Table 12.4 for a description of these models).

TABLE 12.4 Alternative Designs for Town Planning Found in Mānasāra

Nature of Designs			
Design	*Shape*	*Street and Roads*	*Other Details*
Daṇḍaka	Stick/staff/ phalanx	Parallel set of straight streets (one to five) crossing each other at right angles.	Good for priests, Sages, intellects, etc. No. of houses: 12–300.
Nandyāvarta	Square or oblong	It should have one to five carriage roads together with the surrounding street. Internal roads have one footpath and the outer two footpaths.	
Sarvatobhadra	Square or oblong	Number of car streets varies from 1 to 5.	Town/village is secured by a wall and ditch with four large gates on the sides and as many on the corners.
Padmaka	Lotus shaped	Length and breadth are equal while the enclosing walls can be circular, quadrangular, hexagonal, or octagonal. There are 4 to 8 streets lined with the houses.	According to Mayamata, there are five varieties of this design. Gates are placed in four cardinal directions.
Svastika	Svastika	Two streets passing through the center. Traversing streets are planted in the clockwise direction. The outermost roads should be lined with a single row of houses and other streets with a double row of houses.	Temple at the center, has eight gates (two each on every side).
Prastara	Square or oblong	Space needs to be divided into 4, 9, 16 wards by a network of streets. Roads to be constructed like a chessboard system.	The village needs to have enclosed walls and ditches with four principal gates on the south and subsidiary ones in the corners.
Kārmuka	Semi-circular like a bow	2 car streets, 1 principal street. The number of traverse streets can be from 1 to 5. Houses range on both sides of the streets.	This type of design is best for a seashore. The number of gates is optional.
Caturmukha	Square or oblong	4 car streets on four sides. Two large streets crossing at right angles in the center dividing the whole site into four blocks. Each ward must have four smaller roads.	It is specially meant for traders.

In Vāstu-śāstra, the village is taken as the basic unit of analysis for town planning. A country (rāṣṭra) is made up of a certain number of villages and towns. Arthaśāstra provides a hierarchical system of categorisation of settlements by location, function, and number of inhabitants. The smallest unit is the village; a group of ten villages is under a local administration; two hundred villages are governed by the district headquarters; four hundred villages by a divisional headquarters; and eight hundred villages by a provincial headquarters. Several categories of towns and villages have been proposed in vāstu texts. Typically, the classification of villages and towns is based on the area, location, street plan, types of residences, protective moat, temple, and composition of social groups. For example, in Nārada Śilpaśāstra there is an elaborate discussion on different types of villages and towns. Fourteen different types of towns are described in the text[13]. Table 12.3 is a list of different categories of towns and villages that one can find in ancient Indian texts.

Ancient Indian texts dealing with architecture provide early examples of efforts towards planned urban development. Detailed plans of developing a city using principles of vāstu and the pāda-vinyāsa (creating grids in the vāstumaṇḍala) provide directions for addressing town planning issues. The plans include orderly street systems that are rectilinear and sometimes radial. The city is neatly divided into specialised functional quarters, delineating land use principles (what in modern parlance known as zoning regulations). The centre of the city is typically used for palaces, temples, and civic buildings. There were advanced and well-defined systems of fortification. A good illustration of these aspects of town planning as evident from Kauṭilya's Arthaśāstra is provided in IKS in Action 12.1 (see the box for more details).

Some of the modern towns in India experience severe traffic jams on account of poor planning of arterial roads. Arthaśāstra specifies that the land inside the fort must have six royal roads, three running from east to west, and three from north to south. These forts must have twelve gates. The city must be provided with water, drainage, and covered passages. According to Kauṭilya, the houses in the towns must be constructed on sanitary and regulated principles, violation of which will attract punishment from the state.

12.5.1 Alternative Designs for Town Planning

In Vāstu-śāstra, there is a detailed discussion on village planning. Mānasāra describes eight types of village design, which primarily differ in terms of the shape of the layout and the network of roads established. These are the staff (Daṇḍaka), the quadrangular grid (Sarvatobhadra), quadrangular concentric (Nandyāvarta), lotus (Padmaka), swastika, conch (Prastara), bow (Kārmuka), and four-faced (Caturmukha). Mayamata has also discussed eight types of layouts for the design of the settlements. Table 12.4 has salient features of each of these alternatives. These alternative designs for settlements seek to address different purposes.

- Daṇḍaka type of village is meant for the people who want to spend their retired life (Vānaprastha) peacefully. This type of village is the smallest of all[14].
- Sarvatobhadra, as the name indicates, is a village that is safe from all sides and conducive from different perspectives. So, it can be assumed that such villages should be built where multiple challenges and complexities can arise over time. Chandigarh is an example of the Sarvatobhadra style of the town.
- Kārmuka means a bow. So, the shape of the village must be in a bow formation. These types of villages or towns must be built at the seashore. Poomphar and Kaveripattanam are examples of this model.
- Prastara literally means a couch or a bed. Jaipur city was built on this planning.

Each village was surrounded by a wall of brick or stone and a ditch broad enough to protect the village from some attack. If it is a capital city, the level of fortification is likely to be much more elaborate than what is proposed for a village (see, for example, Chapter 14, Section 14.2.4 for the details of fortification of the capital city). There are four gates at the middle of the four sides, and as many at the four corners. Large streets connect the opposite sides of the main gates. The roads intersect at the centre where there is a temple or a public hall. In the case of a capital city, the royal palace along with the temple complex may occupy the centre of the city. The village is divided into four main blocks and each block further into certain sub-blocks. There are streets laid in a straight line that separates these blocks and sub-blocks. In the case of bigger towns and the capital city, certain zoning conventions are applied, and the city is divided and allocated for residential and non-residential purposes accordingly. The residential zones are allocated as per the caste and the professional background of the people.

An illustration of these elements of village/town planning can be found in Figures 12.5 and 12.6. Figure 12.5 is a Nandyāvarta model for village planning. As we see in the figure, there are 12 exit gates. A network of seven main roads runs across the planned town by connecting the opposite sides. The figure also shows an allocation of blocks of the town to various activities and groups of people. Figure 12.6 is a simplified representation of the swastika model. In this design, there are 8 entry-exit roads to the village. IKS in Action 12.1 (at the beginning of the chapter) has details of the fortified city as spelt out in Arthaśāstra. Despite the differences in the shape and size of these alternatives, we notice similar structural elements and patterns in town planning.

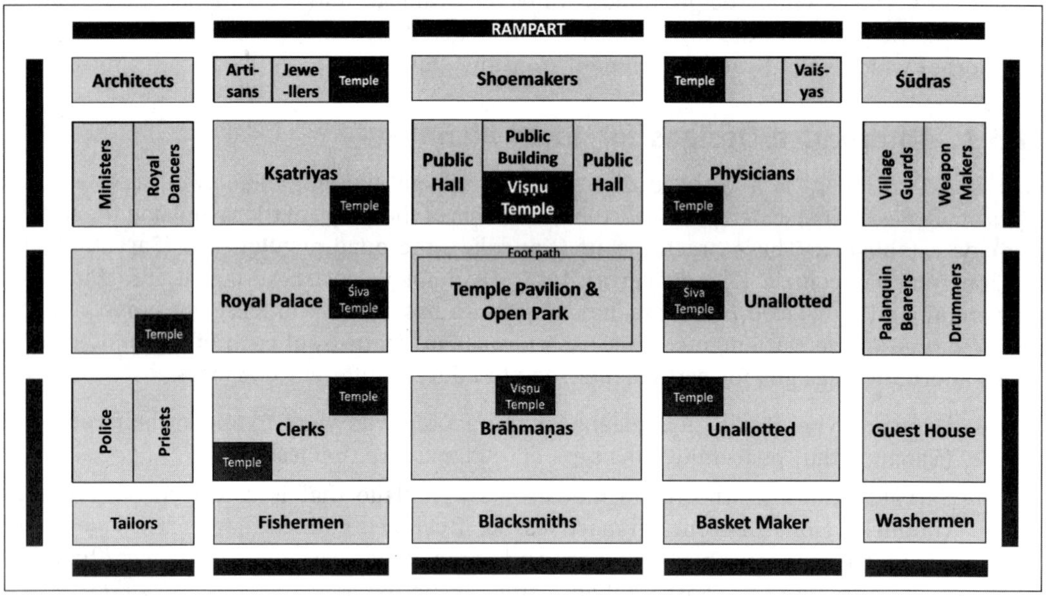

FIGURE 12.5 Village Planning – Nandyāvarta Model

Road Network

One important element of town planning is the determination of different types of roads required, their size, and number. There are several references available in the ancient Indian texts that have addressed this aspect. Viṣṇu Purāṇa (Chapter 38) mentions that vehicular streets (rathyā), avenues (vīthī), and men's roads (nṛṇām mārgaḥ) were constructed separately in the city. Samarāṇgaṇa-sūtradhāra prescribes as many as thirty-four roads in a model town running both from East to West as well from South to North. These roads are so planned as to cover both the interior and exterior planning in the healthiest orientation giving not only the benefit of the sunshine and ventilation but also comfortable residences. Some researchers have studied in detail the artifacts of the excavated Harappan civilization in Dholavira and established the rhythm, patterns, and consistency of proportion with which the town planning activities have been carried out[15].

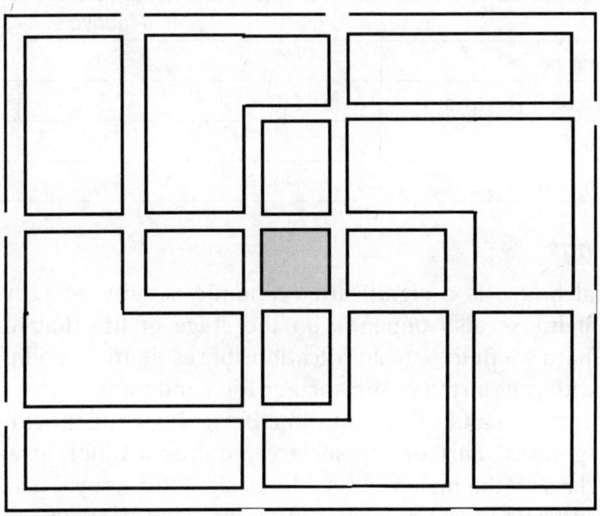

FIGURE 12.6 A Simple Illustration of Svastika Model

Rāja-mārga should be wide enough for free traffic circulation, especially for the fourfold army. It should be made firm with gravel. Vehicle streets must have two-foot paths (Jaṅghā pathas) and are to be located on both sides of it. These roads are laid out in parallel rows cutting one another at right angles. This is the recognised method even in present times. By the time of Arthaśāstra, the measurement of roads has been changed. According to Kauṭilya, royal highways, roads leading to a divisional or a provisional headquarters, roads in the countryside and pasture lands, roads in port towns and cantonments, and roads leading to a village or to a cremation ground are to be fifty-four feet wide. On the other hand, forest roads, roads on reservoir embankments, and roads within the city shall be twenty-seven feet wide[16]. Table 12.5 presents different categories of roads and their suggested widths as prescribed in Vāstu-śāstra texts. A remarkable illustration of the ancient road system is in Jaipur, the city founded by Raja Jai Singh[17].

12.6 UNITARY BUILDINGS

Vāstu-śāstra discusses finer aspects of a variety of individual buildings that make up a town or a village. Broadly three categories of buildings can be identified: Residential buildings,

Temples, and Other public infrastructure. Residential buildings deal with houses for common people, other higher classes, and kings. Public infrastructure comprises a library, health centre, recreation centre, public halls, parks, footpaths, and commercial establishments. Out of its 83 chapters, Samarāṇgaṇa-sūtradhāra has devoted about 20 chapters for house architecture and another 10 chapters to royal architecture. In another 20 chapters, several aspects of temple architecture are discussed.

TABLE 12.5 Different Categories of Roads and Their Widths

Sl. No.	*Type*	*Width of the Road (Feet)*		
		Jyeṣṭha	*Madhya*	*Kaniṣṭha*
1	Rājamārga	36	30	24
2	Mahārathyā (highway)	18	15	12
3	Yānamārga	6	6	6
4	Jaṅghā patha	4½	3¾	3

Residential Buildings

Building a residential house is a significant economic activity. Moreover, as per the Indian tradition, the type of house also depends on the stage of life that one is in (Āśrama). In Samarāṇgaṇa-sūtradhāra we find that classifications of residential buildings are in keeping with the requirements of different strata of the society. Rich and poor cannot have the same quality, size, and richness of material and ornamentation. Designing residential houses for the working class who form the greatest part of the society requires a different approach compared to those in the middle class or the higher class. The design of the royal palace and the residences of the high-ranking officials in the palace is different. Therefore, in Samarāṇgaṇa-sūtradhāra five chapters have been devoted to discussing different types of residential houses. In these chapters, the broad classifications of the ten types of houses (śālā), one-roomed to ten-roomed, are made, and some sub-varieties are described in detail. According to the text, there are 108 one-roomed houses, 52 two-roomed houses, and 72 three-roomed houses[18]. Using the basic configurations of one-roomed, two-roomed, three-roomed, and four-roomed designs, the text proceeds further to make multiple variations of bigger houses (five-roomed up to ten-roomed) in a modular fashion using some permutations. In this manner, a large variety of residential house designs have been proposed. Other related topics such as courtyards, balconies, windows, and other parts of the residential houses that make up for these differences are also discussed.

Royal Palaces

The design of royal residences requires attention to much greater details and aspects other than core living. In addition to the residential quarters for the King and his retinue, other elements including meeting halls (sabhās), other paraphernalia such as the court, the coronation hall, the abodes of ministers, commanders, queens, princes, pleasure-gardens, orchards, and stables for royal animals such as elephants and horses are to be considered. Vāstu-śāstra texts deal with these aspects of royal buildings. Chapter 15 of Samarāṇgaṇa-sūtradhāra provides a quick overview of the design of the royal palace with all its attendant components[19]. The first part of the chapter says that after the town has been planned on the site-plan of sixty-four squares the planning of the royal palace is done. After planning out the fortifications including the ditches, ramparts, and walls and the places of worship allotted to different deities, a piece of land shall be selected at the western side of the centre in the orientation of north for building the king's palace. After these preliminary remarks, the chapter describes in detail the positioning of different elements of the palace suited to the different members of the family of the king and paraphernalia of his establishment.

Sabhā

An assembly hall (Sabhā) is a public hall, often found in palace constructions. These large structures are often characterised by several pillars and fire altars. We find these features even in modern-day equivalent structures (such as in Parliament and Rashtrapati Bhavan). In Samarāṅgaṇa-sūtradhāra, eight different types of designs have been proposed for sabhā construction. The differences are mainly due to the variations in the design of the entrance halls, several pillars, and porches. In the first five of them, several pillars (up to 36) are proposed and in the last three, the design also includes other elements such as corridors[20]. Chapter 25 of Mayamata describes nine types of halls (sabhā) with details of the length, breadth, and the number of main pillars and peripheral pillars. In the same chapter, there is an elaborate description of different types of square-shaped and rectangular-shaped pavilions (maṇḍapa). The details provided include the proportions to be followed, the height and diameter of the pillars, decorations, arrangement of the pillars, and their overall appearance[21].

In addition to these, details of designs pertaining to other public utilities such as a theatre, art gallery, and law court are also found in the vāstu texts[22].

12.7 TEMPLE ARCHITECTURE

Temple (Prāsāda) is an important element of vāstu-śāstra as the building of temples is considered important by the kings and wealthy individuals. Temples present a style that has a precise and harmonious geometry when viewed from all four sides and above. The structure has a square form and grid ground plans, tall towers, and rich and elaborate sculptures decorating the wall and the towers. These depictions present a variety of themes and stories from the Purāṇic lore. It also depicts animals and floral and geometric patterns. Although the temple architecture is found throughout the Indian sub-continent, there are distinctive regional styles found. The north Indian style of temple architecture is known as Nāgara and the South Indian style, Drāviḍa. The vāstu texts have considerably discussed several aspects of temple building of these styles[23]. This includes the components of a temple, the type of temple designs, various structural elements in the temple, iconography for the idols, and the laying out of the temple complex using the Vāstu-puruṣa-maṇḍala. From an architectural perspective the components of the temple are as follows:

Garbhagṛha is the womb or epicentre of the temple, where the presiding deity is to be placed (the sanctum sanctorum). All other structural components of the temple are placed in relation to this. The positioning of the garbhagṛha is established using the Vāstu-puruṣa-maṇḍala.

Maṇḍapa is the pavilion structure in front of garbhagṛha. There could be a few of them beginning with the Mukha-maṇḍapa (entry pavilion), Ardha-maṇḍapa (front pavilion), and Mahā-maṇḍapa (main pavilion). The maṇḍapas are pillared structures of varying numbers and sizes. The maṇḍapas in South Indian temples are classified according to the number of pillars.

IKS IN ACTION 12.2

Temples in India: Marvellous Stone Architecture for Eternity

The temples built in India were mostly stone structures and possessed many unique features that were difficult to fathom today. It required several concepts spanning Mathematics, Astronomy, Building Science, Supply Chain Management, Rock cutting technology, Optics, and Sound engineering. A few examples will help us understand the high level of sophistication with which the architectural aspects of the temples were addressed.

The **Bṛhadīśvara Temple in Thanjavur** was built 1000 years back using granite. The vimāna (tower), a 99-feet structure above the sanctum is one of the tallest in South India. The temple was laid out on a precise plan of 16×16 squares, a design known as Padma-garbha-maṇḍala in the Drāviḍa architecture of southern India. The śikhara, a dome weighing 25 tons rests on a single block of granite, weighing 80 tons, atop the tower. There is no supply of granite in the vicinity of the temple site (about 60 km). It is not clear how the stones were sourced and conveyed to the top of the tower.

There are musical pillars in the maṇḍapa. This requires carving the stones so that the sound frequency is properly established. Several Indian temples have musical pillars. Furthermore, this temple has been a stable structure for 1000 years. In contrast, the leaning tower of Pisa began to lean during its construction in the 12th century, due to soft ground which could not properly support the structure's weight.

The **Sun Temple at Konark** was built of stone in 1250 CE in the form of a gigantic chariot dedicated to the Sun God. The minute and beautiful carvings that characterize the architecture of the temple show its grandeur in iconography. There are 24 elaborately carved stone wheels (symbolizing 24 pakṣas in a year) each of about 12 feet diameter. The seven horses are at the front, symbolically pulling the chariot. The original temple had a *vimāna*, which is estimated to be 229 feet tall. The main *vimāna* fell in 1837.

The wheels of the temple are sundials, which can be used to calculate time accurately to a minute. The placement of the main temple and the Sun God had been aligned in such a way that the first ray of the Sun from the coast would cross the Naṭa Mandir (Dancing Hall) and would fall and reflect from the diamond placed at the crown of the Sun God. Apart from the other challenges of building a massive stone-carved structure, knowledge of astronomy is also required to build such a temple.

One of the most spectacular monuments in the world is the **Kailāsa Temple in Ellora**, which is the largest rock-cut structure anywhere. The cave temples of Ellora extol the spirit of co-existence characteristic of Indian civilization by co-locating temples devoted to Buddhism, Hinduism, and Jainism in one place. The Kailāsa Temple (Cave 16 in Ellora) was built by Kṛṣṇa I, a Rashtrakuta King. It is notable for its vertical excavation–carvers started at the top of the original rock and excavated downward.

The temple was built by digging out a sloping basalt hill with two massive tranches, each 90 m long and joined with a connecting trench 53 m in length. The temple was then carved from the remaining central portion. This resulted in a 32-meter-high structure that seems to come out of the ground. The temple has a three-storey *vimāna* (tower) with an octagonal dome and two huge free-standing columns (*dhvaja-stambhas*) flanking the maṇḍapa entrance hall which has 16 columns set in the groups of four.

According to a UNESCO report, "it is one of the most remarkable of all cave temples in India on account of its striking proportion, elaborate workmanship of architectural members, and sculptural treatment. Furthermore, this temple is decorated with some of the boldest and finest sculpture compositions to be found in India. In the Rang Mahal of this sabhā-maṇḍapa are preserved some beautiful paintings belonging to two different periods."

Granite is difficult to carve even with iron or steel tools. One wonders how very intricate carvings were made in Indian temples built during 600 CE to 1600 CE.

Source: Based on information available on the web and in research reports, https://whc.unesco.org/archive/periodicreporting/apa/cycle01/section2/243.pdf

Prākāra is the open space for circumambulation (pradakṣiṇa) around the garbhagṛha. There could be one or more such parallel structures designed depending on the overall size and span of the temple superstructure.

Viewing the temple from an elevated view, one can identify distinctive elements that make up the structure. These include the following:

Adhiṣṭhāna is the base platform on which the entire superstructure rests. In addition to the foundation deposit, it can have a raised platform (socle) on which the rest of the structure is established. Various designs for the profile of the socle have been proposed in the vāstu texts.

Stambha pillars that support various structures and provide the basis for developing the elevation for the temple.

Prastara is the entablature in the temple structure. Temples are typically multi-storeyed (varying from one to 16 storeys) and the prastara contributes to the substantive part of this. The ground floor is the only habitable storey.

Śikhara is the superstructure, often a tower-like infrastructure that is built above the garbhagṛha. This forms the main elevation to the temple. In the South Indian temples, the superstructure over the garbhagṛha is called Vimāna and the dome-shaped cupola at the top of it is called śikhara.

Stūpi is the finial to the structure.

The South Indian (Drāviḍa) and the North Indian (Nāgara) designs vary in their approach to the design of several aspects of the above components of the temple. The śikhara in a Drāviḍa architecture is a solid dome-shaped roof or cupola. Figure 12.7 depicts various components of a Nāgara style temple. In the vāstu texts, details pertaining to the relative proportions of various components are vividly presented[24]. Moreover, the differences in shapes, numbers, and size of the respective components give rise to a large number of temple designs. Table 12.6 lists some of these variations that contribute to alternative designs of temples.

FIGURE 12.7 Nāgara Temple Architecture (Khajuraho Temple)
Source: https://upload.wikimedia.org/wikipedia/commons/6/60/Architecture_of_the_Khajuraho_temples.jpg

Iconography is a subject discussed in the Vāstu-śāstra texts. The art of idol making requires a good understanding of the relative proportions of different parts of the image (such as eyes, ears, chin, neck, forehead, and limbs). The vāstu texts provide intricate details of these proportions for both a male and a female image so that the idols are good-looking. Varāhamihira in Bṛhat-saṃhitā has laid out five principal heights and lengths for a standard male figure. Similar measurements are also specified for female figures. These have been deployed in Indian idol making in an uninterrupted tradition. The Vāstu-śāstra texts adopt these standards for establishing proportions while discussing iconographic details.

TABLE 12.6 Certain Aspects of Temple Architecture in Ancient India

Sl. No.	*Criterion*	*Remarks*
1	Components of temple architecture	Womb (Garbhagṛha), Front pavilion (Mukha-maṇḍapa), Entry Pavilion (Ardha-maṇḍapa), Base (Adhiṣṭhāna), Pillars (Stambha), Entablature (Prastara), Superstructure (Śikhara), Finial (Stūpi)
2	Types of temples	**Drāvida** – Has tall gopurams and pillared (108 to 1008) halls, huge passageways for pradakṣhiṇa (circumambulation) **Nāgara** – Śikhara gradually inclines inwards in a convex curve, using a concentric rotating-squares and circles
3	No. of storeys	1–12 storeys in South Indian (Drāvida) temples 1–16 storeys in North Indian (Nāgara) temples
4	Shapes of the Vimāna	Square (Vairāja), Circular (Kailāsa), Rectangular (Puṣpaka), Elliptical (Maṇika), Octagonal (Triviṣṭapa)

SUMMARY

- Town planning is a collective set of processes, ideas, and methods using science and technology to deploy the available land for human habitation and other uses in the most efficient way. Architecture addresses the process of planning, designing, and constructing buildings for human use.
- References to Sthāpatya (Architecture and Planning) are mentioned in the Vedic literature, epics, Purāṇas, Brāhmaṇas, Āgamas, Sūtras, and Śilpa-śāstras.
- The extensive archaeological explorations of the Sindhu–Sarasvati civilization and the analysis of sites such as Dholavira, Lothal, Harappa, and Mohenjo-Daro points to a well-developed town planning.
- Vāstu-śāstrā is essentially an art of correct setting whereby one can align the structure to the pañca-bhūtas, and the rotational influence of the sun, moon, and the other planets surrounding the earth.
- At the core of the design using Vāstu principles is the notion of a Maṇḍala. Any site on which a proposed structure needs to be built is conceived in terms of squares.
- The planning and architectural aspects detailed in vāstu can be viewed using an eight-part classification of the concepts. These could be referred to as the limbs of vāstu.
- There is a group of people responsible for the construction of the structure, headed by Sthapati. Sūtragrāhin, Takṣaka, and Vardhaki support Sthapati in the process of construction.
- Mānasāra describes eight kinds of villages, which primarily differ in terms of the shape of the layout and the network of roads established. Vāstu texts have proposed different types of roads required, their size, and number.
- Vāstu-śāstra discusses finer aspects of a variety of individual buildings that make up a town or a village. Broadly three categories of buildings can be identified: Residential buildings, Temples, and other public infrastructure.

- The north Indian style of temple architecture is known as Nāgara and the South Indian style, Drāviḍa. The vāstu texts have considerably discussed several aspects of temple building of these styles.
- Details pertaining to the design of paintings and sculptures in temples and public halls are part of Vāstu-śāstra. Other artistic elements such as furniture, sleeping couches, and doors are also subject matter of discussion in vāstu-śāstra.
- The art of idol making requires a good understanding of the relative proportions of different parts of the image (such as eyes, ears, chin, neck, forehead, and limbs). Iconography is a subject discussed in the vāstu-śāstra texts.

REVIEW QUESTIONS

1. What do you understand by the terms, 'town planning' and 'architecture'?
2. Prepare a one-page note on the status of town planning and architecture in ancient India.
3. Briefly comment on the status of Town Planning in Sindhu–Sarasvati civilization.
4. What do you understand by the term, 'vāstu-śāstra'? What does it deal with?
5. What are the key issues addressed in a typical vāstu text?
6. What are the major architectural texts in ancient India? Outline the issues discussed in these texts.
7. What is a vāstu-puruṣa-maṇḍala? What is the use of it?
8. What are the eight limbs of vāstu? Briefly explain each of these.
9. Who are the key personnel involved in architecture? What are their roles?
10. Enumerate the qualities required of a Sthapati.
11. Prepare a one-page note on the various town (village) planning models suggested by vāstu–śāstra texts.
12. Kauṭilya has addressed several aspects of town planning and architecture in Arthaśāstra. Do you agree with this statement? Prepare a one-page note to support your argument.
13. Are there regional aspects to temple architecture in India? Explain with some examples.
14. Briefly explain how vāstu-śāstra deals with various aspects of temple architecture.

DISCOVER IKS

1. Mahābhārata mentions that the city of Dvāraka, ruled by Krishna was destroyed. While it may be history, the interest to us is if this were true, then it may throw crucial insights into the status of town planning and architecture. While many consider this as mythology, recent deepsea excavation off the coast of Gujarat reveals the submerged city under the sea. Watch the following videos on the city of Dvāraka:
   https://www.youtube.com/watch?v=R1PGp7706HY,
   https://www.youtube.com/watch?v=dcvJY8qSj50.

   After watching the videos develop a note to answer the following questions:
   (a) Why is the exploration of the lost sea of Dvāraka using principles of ocean archaeology important from an international perspective?
   (b) Based on the deep-sea excavation of the lost city of Dvāraka, what are the key inferences one can make about the status of town planning and architecture?
   (c) Based on the interview of Dr. S R Rao, what can we say about the historical and civilizational connectivity between Indus valley civilisation, Mahābhārta, and Dvāraka?

2. The earliest works of temple architecture in India were primarily of rock-cut structures and cave temples. These were found right from the Buddhist times. The video in the link is a documentary on Ellora caves, which houses cave temples and rock-cut structures: https://www.youtube.com/watch?v=k1SE25mURhc.

   After watching the video, prepare a note to answer the following questions:
   (a) What are the main constituents of the Ellorā caves complex?
   (b) List down the major contributions of the ancient Indians to the field of art, architecture.
   (c) What are the unique aspects of the Kailasanātha temple in cave no 16 and the other structures in the Ellora cave complex? How much science and technology knowledge is required to create these structures?

3. Temple architecture in India is very ancient and several temples and cave structures that have stood the test of time and the foreign invaders are living examples. They present a unique set of artifacts that raises several questions in our minds. Watch the video which has a description of Hindu, and Buddhist temples of India: https://www.youtube.com/watch?v=NBG9uHHzxcc. After watching the video, answer the following questions:
   (a) List the structures described in the video and write a paragraph on each of these structures.
   (b) Choose two structures out of the list presented in the video and prepare a detailed note on the unique architectural aspects of these.
   (c) List the complementary skills and knowledge (Astronomy, Mathematics, Optics, Cutting Tool Technology, Logistics, and Supply Chain Management for example) that must have been required to build these temples.

SUGGESTED READINGS

Acarya, P.K. (1996). *Indian Architecture*, Munshiram Manoharlal Publishers, New Delhi.

Acarya, P.K. (1932). *Architecture of Manasara*, The Oxford University Press, Mumbai.

Altekar, R.V. (2010). *Vaastuśāstra: Ancient Indian Architecture and Civil Engineering*, D.K. Printworld, New Delhi.

Bafna, S. (2000). "On the idea of the Mandala as a governing device in Indian architectural tradition", *Journal of Society of Architectural Historians*, **59**(1), pp. 26–46.

Dagens, B. (2007). *Mayamatam*, Indira Gandhi Mational Center for Arts, Motilal Banarasidass, New Delhi.

Ganapathi Sastri, T. (1966). "Samarāṅgaṇa Sūtradhāra of Mahārājādhirāja Bhoja", Oriental Institute, Baroda.

Gangoly, O.C. (1946). *Indian Architecture*. Kutub Publishers, Mumbai.

Gaur, N.A. (2009). *Sthāpatya Ved-Vāstu Śāstra: Ideal Homes, Colony and Town Planning*, New Age Books, New Delhi.

Havell, E.B. (1913). *Indian Architecture*, John Murray, London.

Iyengar, R.N., Kannan, K.S. and Wakankar, S.Y. (2018). *Nārada Śilpaśāstra*, Jain University Press, Bengaluru.

Kashyapa Samhita: https://archive.org/details/KashyapaShilpaShastra/mode/2up

Kramrisch, S. (1946). "The Hindu Temple", University of Calcutta, Kolkata.

Kumar, P. (1998). *Bhoja's Samarāṅgana Sūtradhārā*, New Bharatiya Book Corporation, Delhi.

Nene, A.S. (2012). *Building Materials and Construction Techniques of Ancient India*, Ganga Publications.

Rangarajan, L.N. (2000). *The Arthashastra*, Penguin Random House, Haryana, India.

Ravi Varma, L.A. (1940). *Vastuvidya*, Published under the authority of the government of his highness the Maharaja of Travancore.

Sachdev, V. (2005). "A Vastu Text in the Modern Age", Vishwakarma Darpan, 1969, *Journal of the Royal Asiatic Society*, **15**(2), pp. 165–178.

Sharma, S.K. (2012). *Samarāṅgaṇa Sūtradhāra of Bhojadeva*, Parimal Publications, New Delhi.

Shukla, D.N. (2008). *Vāstu-śāstra: Hindu Science of Architecture*, Munshiram Manoharlal Publishers, New Delhi.

Sinha, A. (1998). "Design of Settlements in the Vaastu Shastras", *Journal of Cultural Geography*, **17**(2), pp. 27–41.

Subrahmanyam, V. (1946). *Varahamihira's Brihat Samhita*, Printed by V.B Subbaiah & Sons, M.B.D. Electric Printing Works, Bengaluru.

ENDNOTES

1. For more details on these archaeological details see, **Danino, M. (2015).** "The invasion that never was", Vivekananda Kendra Prakasan Trust. Chapter 4 – Evidence from Archaeology, pp. 64–72.

2. For more details see, **Kangle, R.P. (2019).** *The Kauṭilīya Arthaśāstra*, Motilal Banarasidass Publishers Private Limited, Delhi. ISBN: 978-81-208-0040-3 (Part II). 2/3/1, 2/4/1,2, pp. 61–68.

3. *Ibid.* गृहं क्षेत्रमारामस्सेतुबन्धस्तटाकमाधारो वा वास्तु । gṛhaṃ kṣetramārāmassetubandhastaṭākamādhāro vā vāstu. (Part II). 3/8/2, p. 216

4. For a good translation of Mayamatam, please refer to **Dagens, B. (1994).** "Mayamatam", Indira Gandhi National Centre for the Arts, New Delhi.

5. वास्तु संक्षेपतो वक्ष्ये गृहादौ विघ्ननाशनम् । ईशानकोणादारभ्य ह्येकाशीतिपदे यजेत् ॥
   ईशाने च शिरः पादौ नैरृतेऽग्न्यनिलौ करौ ।
   vāstu saṅkṣepato vakṣye gṛhādau vighnanāśanam. Īśānakoṇādārbhya hyekāśītipade yajet.
   Īśāne ca śiraḥ pādau nairṛte'gnyanilau karau.
   Pt. Shri Ramtej Pandey (1986). *Garuḍapurāṇa*, The Chowkhamba Vidyabhawan, Varanasi. 1/46/1-2

6. **Altekar, R.V. (2010).** *Vāstuśāstra: Ancient Indian Architecture and Civil Engineering*, D.K. Printworld, New Delhi, p. 89.

7. **Ganapati Shastri, T. (1919).** *The Mayamata of Mayamuni*, Published under the authority of the government of His Holiness The Maharaja of Travancore, Government Press, Travancore. सगर्भं सर्वसम्पत्त्यै विगर्भं सर्वनाशनम् । तस्मात् सर्वप्रयत्नेन गर्भं सम्यग् विनिक्षिपेत् । 8/102 p. 35.

8. Verses 13–25 in Chapter 5 of Mayamata has the details of these. Some relevant extracts are provided here. For more details, see, **Dagens, B. (1994).** *Mayamatam – Vol. I*, Indra Gandhi National Centre for the Arts, New Delhi, pp. 24–27.
   भवन्ति शिल्पिनो लोके चतुर्धा स्वस्वकर्मभिः । स्थपतिः सूत्रग्राही च वर्धकिस्तक्षकस्तथा ॥
   bhavanti śilpino loke caturdhā svasvakarmabhiḥ | sthapatiḥ sūtragrāhī ca vardhakistakṣakastathā ||
   स्थपतेस्तस्य शिष्यो वा सूत्रग्राही सुतोऽथवा । स्थपत्याज्ञानुसारी च सर्वकर्मविशारदः ॥
   सूत्रदण्डप्रपातज्ञो मानोन्मानप्रमाणवित् । शैलदार्विष्टकादीनां सूत्रग्राहिवशानुगः ॥
   तक्षणात् स्थूलसूक्ष्माणां तक्षकः स तु कीर्तितः । मृत्कर्मज्ञो गुणी शक्तः सर्वकर्मस्वतन्त्रकः ॥
   तक्षितानां तक्षकाणामुपर्युपरि युक्तितः । वृद्धिकृद् वर्धकिः प्रोक्तः सूत्रग्राह्यनुगः सदा ॥
   sthapatestasya śiṣyo vā sūtragraāhī suto'thavā | sthapatyājñānusārī ca sarvakarmaviśāradaḥ ||
   sūtradaṇḍaprapātajño mānonmānapramāṇavit | śailadārviṣṭakādīnāṃ sūtragrāhivaśānugaḥ ||
   takṣaṇāt sthūlasūkṣmāṇāṃ takṣakaḥ sa tu kīrtitaḥ | mṛtkarmajño guṇī śaktaḥ sarvakarmasvatantrakaḥ||
   takṣitānāṃ takṣakāṇāmuparyupari yuktitaḥ | vṛddhikṛd vardhakiḥ proktaḥ sūtragraāhyanugaḥ sadā ||

9. स्थापत्यमुच्यतेऽस्माभिरिदानीं प्रक्रमागतम् । ज्ञातेन येन ज्ञायन्ते स्थपतीनां गुणागुणाः ॥ १ ॥
शास्त्रं कर्म तथा प्रज्ञा शीलं च क्रिययान्वितम् । लक्ष्यलक्षणयुक्तार्थशास्त्रनिष्ठो नरो भवेत् ॥ २ ॥
sthāpatyamucyate'smābhiridānīṃ prakramāgatam | jñātena yena jñāyante sthapatīnāṃ guṇāguṇāḥ || 1 ||
śāstraṃ karma tathā prajñā śīlaṃ ca kriyayānvitam | lakṣyalakṣaṇayuktārthaśāstraniṣṭho naro bhavet || 2 ||

For more details see, **Kumar, P. (1998).** *Bhoja's Samarāṅgaṇa Sūtradhārā*, Vol. I, New Bharatiya Book Corporation, Delhi, pp. 246–248.

10. *Ibid.* See for example the following verses from Chapter 44:

सामुद्रं गणितं चैव ज्योतिषं छन्द एव च । सिराज्ञानं तथा शिल्पं यन्त्रकर्मविधिस्तथा ॥ ३ ॥
एतान्यङ्गानि जानीयाद् वास्तुशास्त्रस्य बुद्धिमान् । शास्त्रानुसारेणाभ्युद्य लक्षणानि च लक्षयेत् ॥ ४ ॥
sāmudraṃ gaṇitaṃ caiva jyotiṣaṃ chanda eva ca | sirājñānaṃ tathā śilpaṃ yantrakarmavidhistathā || 3 ||
etānyaṅgāni jānīyād vāstuśāstrasya buddhimān | śāstrānusāreṇābhyudya lakṣaṇāni ca lakṣayet || 4 ||

तथाचाष्टविधं कर्म ज्ञेयं स्थपतिना सदा । आलेख्यं लेख्यजालं च दारुकर्म चयस्तथा ॥ २० ॥
पाषाणसिद्धहेम्नां च शिल्पं कर्म तथैव च । एभिर्गुणैः समायुक्तः स्थपतिर्याति पूज्यताम् ॥ २१ ॥
tathācāṣṭavidhaṃ karma jñeyaṃ sthapatinā sadā | ālekhyaṃ lekhyajālaṃ ca dārukarma cayastathā || 20 ||
pāṣāṇasiddhahemnāṃ ca śilpaṃ karma tathaiva ca | ebhirguṇaiḥ samāyuktaḥ sthapatiryāti pūjyatām || 21 ||

11. Chapter 7 of Mayamatam begins with listing the names of the 32 designs and follows it with description of each design. The 32 designs are mentioned in the verses given below. For more details, see **Dagens, B. (1994).** *Mayamatam – Vol. I*, Indra Gandhi National Centre for the Arts, New Delhi, pp. 37–49.

वक्ष्येऽहं पदविन्यासं सर्ववस्तुसनातनम् । सकलं पेचकं पीठं महापीठमतः परम् ॥ १ ॥
उपपीठमुग्रपीठं स्थण्डिलं नाम चण्डितम् । मण्डूकपदकं चैव पदं परमशायिकम् ॥ २ ॥
तथासनं च स्थानीयं देशीयोभयचण्डितम् । भद्रं महासनं पद्मगर्भं च त्रियुतं पदम् ॥ ३ ॥
व्रतभोगपदं चैव कर्णाष्टकपदं तथा । गणितं पादमित्युक्तं पदं सूर्यविशालकम् ॥ ४ ॥
सुसंहितपदं चैव सुप्रतीकान्तमेव च । विशालं विप्रगर्भं च विश्वेशं च ततः परम् ॥ ५ ॥
तथा विपुलभोगं च पदं विप्रतिकान्तकम् । इन्द्रकान्तपदं चैव द्वात्रिंशत् कथितानि वै ॥ ७ ॥

vakṣye'haṃ padavinyāsaṃ sarvavastusanātanam | sakalaṃ pecakaṃ pīṭhaṃ mahāpīṭhamataḥ param || 1 ||
upapīṭhamugrapīṭhaṃ sthaṇḍilaṃ nāma caṇḍitam | maṇḍūkapadakaṃ caiva padaṃ paramaśāyikam || 2 ||
tathāsanaṃ ca sthānīyaṃ deśīyobhayacaṇḍitam | bhadraṃ mahāsanaṃ padmagarbham ca triyutaṃ padam || 3 ||
vratabhogapadaṃ caiva karṇāṣṭakapadaṃ tathā | gaṇitaṃ pādamityuktaṃ padaṃ sūryaviśālakam || 4 ||
susaṃhitapadaṃ caiva supratīkāntameva ca | viśālaṃ vipragarbhaṃ ca viśvaeśaṃ ca tataḥ param || 5 ||
tathā vipulabhogaṃ ca padaṃ vipratikāntakam | indrakāntapadaṃ caiva dvātriṃśat kathitāni vai || 7 ||

12. **Sastri, V.S. and Bhat, M.R. (1946).** "Varahamihira's Brihat Samhita", V.B Subbaiah & Sons, Bengaluru. Chapter 52, p. 501.

13. For more details see, **Iyengar, R.N., Kannan, K.S. and Wakankar, S.Y. (2018).** *Nārada Śilpaśāstra*, Jain University Press, Bengaluru, pp. 109–135.

14. **Acarya, P.K. (1994).** *Indian Architecture*, Munshiram Manoharlal Publishers, New Delhi, p. 63.

15. There are many published works available. See for example, **Danino, M. (2008).** "New Insights into Harappan Town-planning, Proportions, and Units, with Special Reference to Dholavira", *Man Environment*, **33**, pp. 66–79.

16. **Rangarajan, L.N. (2000).** *The Arthashastra*, Penguin Random House, Haryana, India, p. 157.

17. **Shukla, D.N. (2008).** *Vāstu-Śāstra: Hindu Science of Architecture*, Munshiram Manoharlal Publishers, New Delhi, pp. 268–271.

18. वेश्मनामेकशालानां शतमष्टाधिकं स्मृतम्। द्वापञ्चाशद् द्विशालानां त्रिशालानां द्विसप्ततिः ॥
veśmanām-ekaśālānāṃ śatamaṣṭādhikaṃ smṛtam | dvāpañcāśad dviśālānāṃ triśālānāṃ dvisaptatiḥ || 19.2.
For more details see, **Sharma, S.K. (2012).** *Samarāṅgaṇa Sūtradhāra of Bhojadeva – Vol. I*, Parimal Publications, New Delhi, p. 206.

19. The first seven verses from the chapter are reproduced below:

कृते पुरनिवेशेऽथ चतुःषष्टिपदाश्रये । नियुक्तपरिखासालगोपुराट्टालकेऽपि च ॥ १ ॥
विभक्तरथ्ये परितः प्रविभाजितचत्वरे । क्रमादन्तर्बहिः कॢप्तदेवतायतनस्थितौ ॥ २ ॥
प्रागुदक्प्रवणे देशे प्राग्द्वाराभ्युन्नतेऽथवा । यशःश्रीविजयाधायि मैत्रं पदमधिष्ठितम् ॥ ३ ॥
यथावर्णक्रमायातं चतुरश्रं समं शुभम् । पुरमध्यादपरतोदिक्स्थं कुर्यान्नृपालयम् ॥ ४ ॥
दुर्गेषु भूवशात् कार्यं यद्वा दिक्ष्वपरास्वपि । विवस्वद्भूधरार्यम्णां कार्यमन्यतमे पदे ॥ ५ ॥
त्रिचत्वारिंशता युक्ते ज्येष्ठं स्याद् द्वे धनुःशते । मध्यं शतं तु द्वाषष्टिः शतं साष्टकमन्तिमम् ॥ ६॥
ज्येष्ठे पुरे विधातव्यं ज्येष्ठं राजनिवेशनम् । मध्यमे मध्यमं कार्यं कानिष्ठं च कनीयसि ॥ ७ ॥

kṛte puraniveśe'tha catuḥṣaṣṭipadāśraye | niyuktaparikhāsālagopurāṭṭālake'pi ca || 1 ||
vibhaktarathye paritaḥ pravibhājitacatvare | kramādantarbahiḥ kḷptadevatāyatanasthitau || 2 ||
prāgudakpravaṇe deśe prāradvārābhyunate'thavā | yaśaḥśrīvijayādhāyi maitraṃ padamadhiṣṭhitam || 3 ||
yathāvarṇakramāyātaṃ caturaśraṃ samaṃ śubham | puramadhyādaparatodiksthaṃ kuryānnṛpālayam || 4 ||
durgeṣu bhūvaśāt kāryaṃ yadvā dikṣvaparāsvapi | vivasvadbhūdharāryaṃṇāṃ kāryamanyatame pade || 5 ||
tricatvāriṃśatā yukte jyeṣṭhaṃ syād dve dhanuḥśate | madhyaṃ śataṃ tu dvāṣaṣṭiḥ śataṃ sāṣṭakamantimam || 6||
jyeṣṭhe pure vidhātavyaṃ jyeṣṭhaṃ rājaniveśanam | madhyame madhyamaṃ kāryaṃ kāniṣṭhaṃ ca kanīyasi || 7 ||

For more details see, **Kumar, P. (1998).** *Bhoja's Samarāṅgana Sūtradhārā*, Vol. I, New Bharatiya Book Corporation, Delhi, pp. 62–66.

20. *Ibid.*

नन्दा भद्रा जया पूर्णा सभा स्याद् भाविता तथा । दक्षा च प्रवरा तद्वद् विदुरा चाष्टमी मता ॥ १ ॥
चतुरश्रीकृते क्षेत्रे ततः षोढा विभाजिते । मध्ये पदचतुष्कं स्यात् सीमालिन्दस्तु भागिकः ।। २ ।।
तद्वाद्योऽलिन्दकस्तद्वद् भवेत् प्रतिसरामिधः । प्राग्ग्रीवाख्यस्तृतीयश्च बहिः क्षेत्राच्चतुर्दिशम् ॥ ३ ॥
निसृष्टसौर्ष (धे) यैर्वा स्यादेकस्यां वा यदा दिशि । नन्दा भद्रा जया पूर्णा क्रमेण स्युः सभास्तदा ॥ ४ ॥
षड्भागभाजिते क्षेत्रे कर्णभित्तिं निवेशयेत् । सभा स्याद् भाविता नाम सप्राग्रीवात्र पञ्चमी ॥ ५ ॥
स्तम्भान् षट् त्रिंशदेतासु पञ्चस्वपि निवेशयेत् । स्तम्भान् प्राग्ग्रीवसंबद्धान् पृथगेभ्यो विनिर्दिशेत् ॥ ६ ॥
दक्षेति षष्ठी परितस्तृतीयालिन्दवेष्टिता । प्रवरा सप्तमी द्वारैर्युक्तैषा परिकीर्तिता ॥ ७ ॥
प्राग्ग्रीवद्वारसंयुक्ता विदुरेत्यष्टमी सभा । सभानामिदमष्टानां लक्षणं समुदाहृतम् ॥ ८ ॥

nandā bhadrā jayā pūrṇā sabhā syād bhāvitā tathā | dakṣā ca pravarā tadvad vidurā cāṣṭamī matā || 1 ||
caturaśrīkṛte kṣetre tataḥ ṣoḍhā vibhājite | madhye padacatuṣkaṃ syāt sīmālindastu bhāgikaḥ || 2 ||
tadvādyo'lindakastadvad bhavet pratisarāmidhaḥ | prāggrīvākhyastṛtīyaśca bahiḥ kṣetrāccaturdhiśam || 3 ||
nisṛṣṭasaurṣa (dhe) yairvā syādekasyāṃ vā yadā diśi | nandā bhadrā jayā pūrṇā krameṇa syuḥ sabhāstadā || 4 ||
ṣaḍbhāgabhājite kṣetre karṇabhittiṃ niveśayet | sabhā syād bhāvitā nāma saprāgrīvātra pañcamī || 5 ||
stambhān ṣaṭ triṃśadetāsu pañcasvapi niveśayet | stambhān prāggrīvasaṃbaddhān pṛthagebhyo vinirdiśet || 6 ||
dakṣeti ṣaṣṭhī paritastṛtīyālindaveṣṭitā | pravarā saptamī dvārairyuktaiṣā parikīrtitā || 7 ||
prāggrīvadvārasaṃyuktā viduretyaṣṭamī sabhā | sabhānāmidamaṣṭānāṃ lakṣaṇaṃ samudāhṛtam || 8 ||

For more details see, **Kumar, P. (1998).** *Bhoja's Samarāṅgana Sūtradhārā*, Vol. I, New Bharatiya Book Corporation, Delhi, p. 153.

21. For details see, **Dagens, B. (1994).** *Mayamatam – Vol. II*, Indra Gandhi National Centre for the Arts, New Delhi, pp. 450–513.

22. In Nārada Śilpaśātra the details pertaining to law court, theatre and art gallery are available in three separate chapters. For details see, **Iyengar, R.N., Kannan, K.S. and Wakankar, S.Y. (2018).** "Nārada Śilpaśāstra", Jain University Press, Bengaluru, pp. 211–220.

23. For a thorough analysis of various aspects of the temple architecture please refer to **Kramrisch, S. (1946).** "The Hindu Temple", University of Calcutta, Kolkata.

24. *Ibid.* Chapter 7 discusses in detail the proportionate measurements of the temple leading to alternative designs.

Part 4

HUMANITIES AND SOCIAL SCIENCES IN IKS

CHAPTER

13 Health, Wellness and Psychology

LEARNING OUTCOMES

After finishing this chapter, you will be able to:

- Understand Āyurveda's approach to Health and Wellness
- Get to know the basic framework to health and disease management as spelt out in Āyurveda
- Understand the importance of yoga way of living in maintaining a sound physical and mental health
- Understand the distinctive aspects of Indian Psychology
- Get acquainted with alternative frameworks to understand body–mind–intellect–consciousness that an individual is made of

Yoga way of life was common to all ancient Indian practices including Hindu, Bauddha and Jaina schools. Bauddha school was mainly responsible for its spread to neighbouring countries. The **Gal Vihāra** is a rock temple of the Buddha situated in the ancient city of Polonnaruwa in North Central Province, Sri Lanka. It was built in the 12th century by Parakramabahu I. The large, seated image is 15 feet 2.5 inches (4.636 m) tall and depicts the dhyāna mudrā. Such artefacts depicting Yoga postures are common in several parts of South and Southeast Asia.

IKS IN ACTION 13.1

Impact of Yoga Way of Life on Emotional Intelligence of Managers

The practice of Yoga helps individuals gain control over their minds and lead a stress-free and healthy life. A study was conducted in a unit of a manufacturing company in the state of Gujarat in western India, to experimentally understand the benefits of the yoga way of life. The company had employed about 120 people in the managerial cadre. Those who agreed to participate in the study were divided into two equal groups of 42 each; Group 1 was designated as the yoga group and Group 2 as the physical exercise group.

The yoga group was given 30 hours of yoga practice (75 minutes every day) and 25 hours of theory on the philosophy of yoga spread over six weeks. The theory lectures included topics such as definitions of the yoga way of life, implications of the four types of yoga (Rāja yoga, Karma yoga, Jñāna yoga, and Bhakti yoga) on life, analysis of the aspects of true happiness in life, Aṣṭāṅga yoga steps and the central theme of the universality of consciousness as propounded in these texts. The practice sessions which included āsanas, prāṇāyāma, and relaxation were conducted by a yoga instructor.

The other group was given training in a normal physical workout for an equal number of hours, and lectures on the success factors in life based on modern thought. The topics for theory given to this group included success and happiness, the importance of attitude, self-image, good relationship with others, goal setting, the power of the subconscious mind, communication, motivation, and leadership. The practice given to this group included fast exercises such as spot jogging, bending, body rotation, hand, and leg movements, etc.

Emotional Intelligence (EI) was measured for both the groups, before and after the study, with the help of a standard self-reported questionnaire. In addition, measurements of certain physical parameters such as weight, body mass index, blood pressure, and blood sugar were taken, before and after the study. 80% and 86% of the participants from the yoga group and the control group respectively were from the age group of 21 to 50, while the rest were above 50.

Results and Implications

Statistical analysis of the data was done using SPSS. The average EI score for the yoga group and the control group was 5.50 and 5.61 respectively before the experiment. The average EI score after the experiment for the yoga group and the control group was 5.97 and 5.55, respectively. Statistical tests conducted revealed that EI showed significant enhancement in the yoga group but not in the physical exercise group. The results further showed that while the difference in the average EI between the yoga group and the physical exercise group was not significant before the experiment, the same was statistically significant at the end of the experiment.

The study indicated that a systematic adoption of the yoga way of life can result in better EI among managers, thus paving the way for their better performance as managers. In most organisations, leaders play a pivotal role in driving performance. Several leadership training programmes are being conducted by successful companies. But the yoga way of life is seldom taught systematically as part of these training programmes. Currently, yoga methods are taught more as a means to de-stress individuals and improve personal satisfaction. It would be beneficial to provide systematic exposure to the knowledge enshrined in the texts emphasising the yoga way of life to all managerial cadres of companies.

The study shows that such an initiative would help them personally as well as professionally. They can become more self-aware and self-regulated individuals, with a proper perspective of life and various relationships. It also shows the potential to include a significant component of yoga practices in psychological counselling sessions to improve the efficacy of such interventions.

Source: **Adhia, H., Nagendra, H.R. and Mahadevan, B. (2010).** "Impact of Adoption of Yoga Way of Life on the Emotional Intelligence of Managers". IIMB Management Review, Vol. **22**, pp. 32–41.

Every living being in this Universe feels the need for happiness and makes all-out efforts to achieve this. No living being wants to experience sorrow. In the ancient Indian system, an overall framework was provided for achieving this by setting four life goals (*Dharma, Artha, Kāma,* and *Mokṣa*), known as *puruṣārthas*. To attain these goals in life health plays a major role. Caraka, one of the exponents of health systems in ancient India says, "Health is the best source of right deeds, wealth, desires, and emancipation while diseases are destroyers of this (source), welfare and life itself"[1]. The focus, therefore, is on how to keep oneself hale and healthy.

In the modern systems of medicine such as Allopathy, health is defined at an average level for all human beings. For example, a healthy person's body parameters such as pulse rate, blood pressure, weight, etc. are specified to be a single number, which is at best an average reading. In reality, health is an individual phenomenon as people differ from each other vastly in their constitution, food habits, strength, mentality, age, and adaptability[2]. Therefore, health systems must take these into consideration and the average values can at best serve as broad guidelines. This implies that every individual must know about their nature first in order to ensure that he/she is healthy. The Indian health system plays a significant role in this process and it focuses both on the preventive and curative aspects of health. We shall see in this chapter these concepts of Indian health system, namely Āyurveda. Furthermore, it is increasingly understood that the mental health of a person plays a significant role in overall wellness. Therefore, Psychology is closely linked to health and wellness. We shall also see in this chapter the Indian approach to psychology.

- The Indian health system focuses both on the preventive and curative aspects of health.
- The mental health of a person plays a significant role in overall wellness.

13.1 ĀYURVEDA – DEFINITION OF HEALTH

The word Āyurveda is derived from two words 'āyus' and 'veda'. The word 'āyus' is generally translated as life. The Caraka-saṃhitā defines āyus as the association of body, senses, mind, and Ātman[3]. Āyus is also called as dhāri (one which holds), jīvitam (flow of consciousness), nityaga (one with constant movement) and anubandha (one which binds or connects). The word Veda has already been explained in chapter two. Ancient Indians considered the body to be the first instrument for leading a satisfactory life by performing the right deeds for the attainment of puruṣārthas. Therefore, health was considered an important aspect of life and according to Caraka, Āyurveda must be studied and practiced by every individual in the society[4].

The origin of Āyurveda is traced back to the four vedas. Subsequently three important treatises Caraka-saṃhitā, Suśruta-saṃhitā, and Aṣṭāṅga-hṛdaya serve as the authentic source books. The subsequent three books that are called "Laghu Trayī" are Mādhava-nidāna, Śārṅgdhara-saṃhitā, and Bhāvaprakāśa-nighaṇṭu. These books contain basic concepts of health and disease, disease management, anatomy and physiology, and other related topics. Diseases are classified according to organ systems and function and discussed in detail in Āyurveda. Although there is no record of pharmacological testing during the time period when Āyurvedic texts were written, 50 distinct pharmacological categories of medicinal plants were described.

Āyurveda borrows the philosophical aspects from two major schools of philosophy: Vaiśeṣika and Sāṃkhya. One school follows the Sāṃkhya philosophy which proposed twenty-four elements that make up an individual (see Chapter 3, Section 3.3.1 for more details on this). Āyurveda uses this to understand the anatomy and psychology of a human being in depth. According to Caraka the knowledge of Āyurveda is built on the concepts presented in Vaiśeṣika[5]. As we have seen in Chapter 7, Vaiśeṣika philosophy classifies all existing objects in this world into six padārthas (categories). In Vaiśeṣika, nine categories of substances (Dravyas) have been identified. When these get connected with Indriyas (senses) they become sentient being (cetana-dravyas), and without them, they are called insentient (acetana-dravyas)[6]. All the object in this Universe whether a living being or non-living being is made of a different combination of these dravyas. A good ayurvedic practitioner never overlooks this aspect while treating his patient.

- In 1948, the World Health Organization (WHO) added 'mental wellbeing' as part of heath definition.
- Āyurveda uses tridoṣas to map the symptoms to an appropriate cause by identifying the nature of imbalance among the tridoṣas.

Sāmānya, according to Vaiśeṣika is that which is generally present in many, thereby indicating similarity in characteristics of substances. This padārtha causes an increase in substances with similar qualities. When something with the same guṇa is consumed, the power of disease can increase (as in the case of sugar taken by a diabetic patient). Therefore, with a knowledge of the similar qualities between diseases and the dravya (and the associated guṇa and karma) one can take care of his/her health. On the other hand, Viśeṣa in Vaiśeṣika enables one to distinguish between two substances. Though people may be anatomically similar and may have the same disease, there are specific qualities that make an individual different from the other. Therefore, disease and its cause and impact on individuals can be different. Identifying the distinguishing characteristics of individuals helps in treating an ailment. For example, the specific nature of vāta is roughness, coolness, and lightness. So, when vāta is high in the body it must be treated with food, medicine etc. of a specific nature which is exactly opposite to the nature of the vāta. That exits in the oil. The nature of oil is sticky, hot, and heavy.

The definition of health according to Āyurveda is broad-based and more encompassing. Suśruta's definition of health is as follows:

समदोषः समाग्निश्च समधातुमलक्रियः ।
प्रसन्नात्मेन्द्रियमनाः स्वस्थ इत्यभिधीयते ॥
samadoṣaḥ samāgniśca sama-dhātu-mala-kriyaḥ |
prasanna-ātmendriya-manāḥ svastha ity-abhidhīyate ||

The above definition comprises two parts: The first part states that good health is based on the equilibrium of doṣa (humor), agni (digestive fire), dhātu (seven body tissues), and mala (faeces, urine, and other waste products). The second part states that maintaining physical, mental, and spiritual well-being is also a part of good health. As we can see, the first part relates the health of an individual to the physical aspects (condition of the body), whereas the second part relates health to the psychological aspect (mental well being). Figure 13.1 schematically presents the definition of health. In contrast to the āyurvedic perspective, in the west, early definitions of health focused on the theme of the body's ability to function. When one becomes unhealthy, the effort was to restore the body back to its normal state. It was only in 1948, that the World Health Organisation (WHO) added 'mental wellbeing' as part of the health definition.

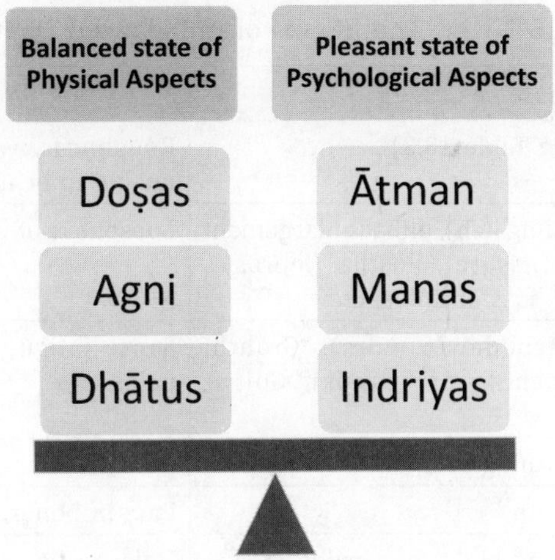

FIGURE 13.1 Health – A Two-dimensional Perspective

Physical Aspects of Health

There are seven dhātus (basic substances) that constitute a human body. Similarly, the doṣas (the humours) and mala (the wastes) are also defined. According to Suśruta a balanced state of the doṣas and agni (healthy digestive fire/juice), vital fluids and tissues of the body being in normal state and quantity and normal processes of secretion constitutes the physical aspect of good health.

Agni plays a vital role[7] and is responsible for health, longevity, and vital breath. When agni extinguishes in the body the person dies and when it is in its normal state the person stays healthy and lives longer. The agni converts the food into energy which helps in the smooth functioning of the body. When Agni is disturbed and becomes low, high, or unbalanced it makes the person ill. The agni in the body is not a flame but is in a liquid form and manifests through the Pitta in the body[8]. The digestive fire metabolises what is eaten. If food is not consumed it metabolises doṣas, then Dhātus, and finally the vital energies[9]. Therefore, an individual must always focus on keeping the agni balanced.

The human body is made up of seven basic substances (dhātus): Plasma (Rasa), Blood (Rakta), Flesh (Māṃsa), Fat (Medas), Bones (Asthi), Bone-marrow (Majjā), and Semen (Śukra). The food that we consume first gets converted into plasma and in a gradual process, it gets converted into flesh and fat. From the fat to bones, bones to bone-marrow, bone-marrow to semen[10]. That semen when accumulated and meditated over gets converted into Ojas, the Vital power. This power leads to a long life without any diseases. An imbalanced state of the dhātus leads to bad health and several diseases. Table 13.1 summarises the effect of imbalanced dhātus in the human body.

Psychological Aspects to Health

The second part of the definition extends the notion of health to the psychological aspect by including a condition that one must also be in a pleasant state of mind and the senses. Mind (Manas) is considered as a subtle internal organ that cannot be seen but its presence can be inferred. For example, when a person eats his/her favourite sweet, he/she enjoys it. This

TABLE 13.1 The Effects of Imbalanced Dhātus

Dhātus	*Increase or Growth*	*Decaying or Loss*
Rasa	Like the Kapha (see Table 13.3)	Roughness, weakness, dryness, fatigue, unable to bear with sounds
Rakta	Visarpa (dry spreading itch), plīha (enlargement of spleen), vidra (fissure), kuṣṭha (leprosy), upakuśa (gumboil), loss of limbs, etc.	Looseness of blood vessels, Roughness
Māṃsa	Gaṇḍa (boils), Arbuda (tumors), Granthi (swelling or hardening of vessels), Goitre, Mumps, etc.	Sense fatigue, Dryness of cheeks, Joint pains
Medas	Weakness and breathing problem	Loss of touch sense, thinness of body
Asthi	Extra bones and teeth	Pain in bones, teeth falls, etc.
Majjā	Heaviness in eyes and body, joint pains	Holes in bones, giddiness, blackouts
Śukra	Over desire for sex, stones in semen	Early ejaculation of semen and blood, pain in testicles, etc.

enjoyment is neither related to the tongue nor the sweet itself. If that would have been the case then the sweet must induce the same enjoyment in everyone, or anything that the tongue consumes must give one happiness. Since this is not the case, it can be inferred that there is some other organ that experiences joy, sorrow, and other such emotions. The mind is an internal sense organ of our awareness of happiness, unhappiness, etc.

Several philosophical and Āyurvedic texts deal with mind in detail. According to our scriptures, the mind is the cause for all the bindings and mokṣa (liberation). If the mind is not controlled, the sense organs will behave in an uncontrolled fashion. As stated in Bhagavadgītā, "The mind is restless, turbulent, strong and unyielding, it is difficult to control as controlling the wind in an open place[11]". According to the Sāṃkhya philosophy, the human mind is a manifestation of Prakṛti (Nature) which is made up of Tri-guṇas. The interplay between the three guṇas decide the psychological and emotional behavior of an individual (see Section 13.6 for some more details on Tri-guṇas). The psychological influence of the mind differs from person to person. In different ratios, it changes the mental and intellectual caliber of an individual. In the framework of Tri-guṇas, all actions, and the behavior of an individual can be classified which further helps us to analyse a person and his activities from a psychological dimension. When rajas and tamas get perturbed, they cause psychological disorders[12]. All the internal enemies of humans like lust, anger, greed, delusion, pride, envy, etc. overwhelm the individual which brings about demonic changes in an individual.

Once we lose control over ourselves the control of buddhi and manas is lost. On account of this, the control over the senses is lost and it leads to the distraction of the senses causing various ill effects. Caraka explains, "When a person performs inappropriate/inauspicious deeds being impaired by the buddhi (ability to comprehend), dhṛti (controlling power), and smṛti (power to recall memories, or reminisce), it leads to an aggravation of all doṣas"[13]. For example, if a person knows that eating ice-cream or cold beverages is not good for his/her health and still consumes those it is an inappropriate deed which leads to an increase in Kapha and

- The tridoṣas have a specific relationship with the time of the day and the type of diet one takes.
- Agni plays a vital role in the digestion process and it is also responsible for strength, health, longevity, and vital breath.

Vāta and causes various diseases related to those doṣas. Caraka relates the psychological state such as buddhi, and indriyas to the cause of diseases. Erroneous use, avoidance, and excessive use of time factor, intellect, and sense objects is the threefold cause of both psychic and somatic disorders[14]. According to Āyurveda, the overall health is determined by physical and psychological environment. To keep these in a balanced and natural state one must follow some specific routines explained in Āyurveda. These address the role of environmental factors, daily routine, seasonal changes, lifestyle, diet, and regular exercise in maintaining health.

Modern systems such as allopathy are primarily oriented towards the treatment of disease. In this approach, drugs are developed based on the concept that the elimination of specific causes of a disease, such as microorganisms, will cure a disease. On the other hand, Āyurveda is oriented toward prevention, health maintenance, and treatment. It is developed on the basic assumption that a disease is the product of an imbalance in the body and mental elements which reduce the body's resistance to diseases. If the imbalance is corrected and the body's defence mechanisms are strengthened by medicine, lifestyle changes, and diet, then the body will resist a disease and eventually eliminate it.

13.2 TRI-DOṢAS – RELATIONSHIP TO HEALTH

One of the cornerstones of health in Āyurveda is the balanced state of the three doṣas (Tri-doṣas). The word doṣa means that which perishes, spoils, or refutes. These doṣas, are Vāta, Pitta, and Kapha. As per the Vaiśeṣika system, the composition of these doṣas is based on the five basic elements. For example, Pitta is formed by the combination of fire and water. Figure 13.2 provides the composition of the three doṣas. According to the Tri-doṣa theory the kapha represents solid material substratum of the human body, pitta the chemical activity and vāta the energy pool of motion and movement. These three doṣas coexist in a predetermined proportion and function in a complementary manner to overall functioning of the body despite their opposite properties and functions. When these are in an equilibrium state the body is said to be in a healthy condition. By contrast, when the doṣas are vitiated they cause disease[15]. To balance or control these doṣas, one must know the nature of the three dośas. Āyurveda gives specific details of the nature of these doṣas[16]. For example, Vāta is considered to be light, cool, dry, mobile, subtle, and rough. Similarly, Pitta is oily, sharp, hot, light, stinking, liquid. On the other hand, Kapha is cool, moist, stable, heavy, slow, and slippery.

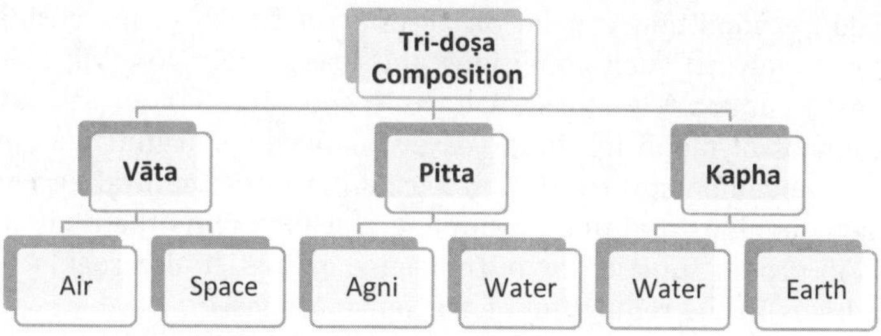

FIGURE 13.2 The Composition of the Tri-Doṣas

The doṣas have a specific relationship with the type of diet one takes. When similar-natured food is consumed, similar natured action is done, and the doṣa increases. For example, the vegetable Okra (Lady's finger) is of Kapha nature. Therefore, if it is consumed the Kapha level in the body gets perturbed. Doṣas are related to the tastes in the food. For instance, sweet, sour, and salty tastes reduce the vāta, whereas pungent, bitter, and astringent tastes exaggerate it. Similarly, astringent, bitter, and sweet tastes reduce pitta, and sour, salty, and pungent tastes exaggerate pitta. In the case of Kapha, pungent, bitter, and astringent tastes reduce it, and sweet, sour, and salty tastes exaggerate it. Therefore, knowing the nature of the body and its current state a person should consume foods of the appropriate taste. A healthy person must have all the six tastes in his food.

Āyurvedic scholars have also identified the organs where these doṣas predominantly exist. All doṣas occupy different parts of the body. Mainly, Kapha occupies the part above the chest of a human body, Pitta occupies the chest to the navel and Vāta occupies the part below the navel. All the Kapha related problems such as cold, and cough are mainly experienced above the chest. The problems which are caused by Pitta such as acidity are experienced between the chest and the navel. Problems related to Vāta such as Arthritis and constipation are experienced mostly below the navel part of the body. These are merely indicative. In reality, the three doṣas, despite their primary location, are there everywhere in the body and are capable of causing disease in any part of the body.

Further, it is also associated with age. Kapha is predominantly present during childhood (till 16th year), morning hours, and the first part of the night. That is the reason we see the children being affected by Kapha often. Pitta dominates during middle age, noon hours, and mid-night hours. Nature of pitta is hot and hot weather or substance increase the level of pitta in the body. It gets cured by the opposite nature, i.e. cold. Vāta is dominant during old age and that is when such people suffer from the disorders related to stomach, pains, arthritis, etc. Vāta gets increased during evening hours and the third part of the night. Table 13.2 summarises these aspects of the tri-doṣas.

TABLE 13.2 Factors Affecting the Tri-doṣas

	Body	*Age*	*Day*	*Night*	*Important Organ*
Kapha	Head to chest	Childhood	Morning hours	First part	Chest
Pitta	Chest to navel	Young age	Noon hours	Mid-night hours	Liver and pancreas
Vāta	Below navel	Old age	Evening hours	Third part	Digestive system

The issue of interest to us is the effect on health when one of the Tri-doṣas goes out of balance. The imbalance condition can be on account of the doṣa increasing or decreasing. Āyurvedic texts have analysed such conditions and listed their possible effects in the body. Table 13.3 provides a summary of these. Vāta is responsible for all the movements in the body like the movement of blood in blood vessels, movement of nutrients, movement of air in lungs, locomotion movements of hands and legs, initiation of natural urges like tears, faces, urination, sneezing, coughing, vomiting yawning, etc. It maintains the dhātus in their normal condition and aids in proper functioning of the sense organs. It also regulates all activities of mind, speech, and helps in the regulation of the sense organs.

One of the major causes of physical illness in a human being is the suppression of urges. There are several urges which when suppressed cause different types of diseases. Table 13.4 has details pertaining to these. On the other hand, Āyurveda recommends the suppression of certain urges to attain psychological health. An intelligent person suppresses the urges of greed, grief, fear, fury, pride, shamelessness, envy, excessive passion as well as covetousness and

TABLE 13.3 The Effects of Doṣas Going Out of Balance

Condition	*Vāta*	*Pitta*	*Kapha*
Balanced	Energy, balanced breath, movement, normality in urges	Digestion, warmth, vision, hunger, thirst, taste, gleam, intelligence, prowess/heroism, delicateness, softness	Stability, moisture, good joints, patience
Exaggerated	Thinness, need of warmth, shivering, constipation, weakness, chattering, dizziness	Yellowness in eyes and excretions, severe hunger and thirst, burning, insomnia, acidity	Indigestion, nausea, laziness, heaviness, frigidity, slack-limbed
Reduced	Exhaustion of limbs, noxious talk, giddiness	Indigestion, coldness, lack of gleam	Giddiness, emptiness in the abode of Kapha, loose joints

TABLE 13.4 Types of Urges that are not to be Suppressed

Urge	*Diseases*
Vāta	Retention of faeces, urine and flatus, distension of abdomen, pain, piles and many other abdominal diseases
Purīṣa (Faeces)	Colic pain, headache, retention of faeces and flatus, cramps in calf muscles and distension of stomach
Mūtra (Urine)	Pain in bladder and genital organ, painful urine, headache, bending of body, distension of lower abdomen
Kṣava (Sneeze)	Spasmodic contraction of the muscles of the neck, headache, facial paralysis, pain in one side of head and weakness of the organs
Tṛṣ (Thirst)	Dryness of throat and mouth, deafness, exhaustion, weakness and cardiac pain
Kṣudh (Hunger)	Abnormal thinness, weakness, change in bodily complexion, bodily discomfort, loss of appetite and giddiness
Nidrā (Sleep)	Yawning, malaise, drowsiness, headache and heaviness in the eyes and many other harmful health issues
Kāsa (Cough)	More cough, disorders related to breath, taste and heart, hiccup, etc.
Śramaśvāsa (Exhaustion)	Chronic enlargement of spleen, heart related problems, fainting, etc.
Jṛmbhā (Yawn)	Bending in body, muscular contractions, numbness, tremor and shaking of the body
Aśru (tears)	Inflammation of nose, eye diseases, heart diseases, inability to eat and giddiness
Vamana (Vomit)	Itching, allergic reaction, loss of appetite, and black pigmentation of face, disease called oedema, anemia, fever, nausea, and deep red inflammation of the skin
Retas (semen)	Pain in genital organ and testicles, general body weakness, cardiac pain and retention of urine are the problems
Hikkā (eructation)	Hiccough, difficulty in breathing, loss of appetite, involuntary shaking of body or limbs (tremor), obstacles in proper functioning of heart and lungs

any urge to speak extremely harshly, critically, falsely, or inappropriately. One should also suppress all the urges which involve causing bodily harm to another person, such as rape, robbery, or injury[17].

IKS IN ACTION 13.2

Dinacaryā: Daily Regimen for Health and Wellness

Āyurveda guides an individual regarding the daily and seasonal routines. In dinacaryā which is the daily regimen for the individual, Āyurveda discusses activities that we can take up right from the time we wake up in the morning till we go to bed at night. Dinacaryā should be followed regularly for ensuring a healthy life. Some of the most important of them are listed below:

Getting up at Brahma-muhūrta

Each muhūrta is of 48 minutes. Brahma-muhūrta which means the muhūrta fit for Brahman is two muhūrtas before the sunrise. Assuming an average sunrise of 6.00 am, it points to 4.24 (or roughly 4.30) am in the morning. Brahman means any intellectual activity like studying, etc. It is considered auspicious as there is peace, purity, freshness, and happiness all around the environment.

Considering the Condition of the Body

Once the individual wakes up, he/she must observe the body and the breath. This helps in planning the day.

Uṣaḥpāna (Drinking Water in the Morning)

One must drink eight handfuls of water before sunrise, even before brushing the teeth in a sitting posture. One who follows this will be able to keep the doṣas balanced. This process also helps to flush out the toxins in the body.

Evacuation of the Bowels

The consumption of water induces natural urges. Proper cleansing of the bowels is very important for a joyful day. These urges happen naturally with autonomic reflex. So, one should not try to evacuate the bowel forcefully. Evacuation of the bowel forcefully leads to various diseases.

Brushing the Teeth

Āyurveda prescribes various stems of trees for this process. Those are Arka, Nyagroda, Khadira, Karañja Kakubha, Karavīra, Sarja, Irimeda, Apāmārga, Mālatī, etc. Ayurveda also advises using Katu (pungent), Tikta (bitter) and Kaṣāya (astringent) tastes for brushing the teeth.

Jihvā Nirlekhana (Tongue Cleaning)

According to Āyurveda cleaning the tongue is a good habit. One can use the leaves of the mango tree or guava tree to clean the tongue. Some metals like copper and brass are also prescribed for this task. Regular cleaning of tongue removes bad taste from the mouth, eliminates the bad odour of the mouth, reduces swelling of the tongue, relieves stiffness of the tongue, and enhances the sense of taste. Thus, it helps in maintaining oral hygiene and stimulating the taste buds.

Applying Sauvīrāñjana to Eyes

The eyes are one of the important human organs. It should be well taken care of. Āyurveda suggests that an individual must apply Sauvīrāñjana (collyrium made from antimony) to the eyes. It balances the oil, water, and mucus required for good eyesight.

Applying Oil to the Body Daily

One must conduct body massage regularly as it is beneficial in delaying aging, relieving fatigue, and mitigating Vāta. Āyurveda suggests that one must atleast apply oil on the head, ears, and foot.

Performing Exercises

Exercise should be a daily activity to keep the body light and in good shape. It also increases the power of digestion. The person suffering from indigestion or any Vāta related problems must not perform exercise. Yoga suggests that one cycle of Sūrya Namaskāra gives exercise to all the limbs of the body. Some breathing exercises provide addtional benefits to both physical and psychological health.

Taking Bath Daily

Taking bath daily makes the body clean and it increases appetite, improves sexual vigor, and enhances the lifespan. It also protects one from bad dreams, cures diseases, and gives peace of mind.

Other activities to be followed as a daily regimen include the following:

- **Udvartana:** Massaging the body with fine powders of various herbs of astringent taste.
- **Karṇapūraṇa:** Filling ears with medicated oil.
- **Dhūmapāna:** Medicated smoking. The powder for this consists of basil leaves, turmeric, etc. It is free from tobacco and nicotine and is good for various Kapha-related problems.
- **Gandhamālya-niveśana:** Wearing an aromatic garland.
- **Pādatrāṇadhāraṇa:** Wearing footwear.
- **Chatradhāraṇa:** Using an umbrella.
- **Tāmbūla sevana:** Chewing betel leaves.
- Avoiding sinful acts, etc.

Some of the above-explained Dinacaryās are to be followed only on specific seasons to balance the tri-doṣas.

13.3 DISEASE MANAGEMENT

Caraka-saṃhitā outlines four components of a disease management scheme: (1) the physician, (2) the drug, (3) the patient, and (4) the attendant[18]. A physician must have proper training, knowledge, and experience. A remedy must be abundantly available, effective, and relatively safe. A patient must provide all information to the physician about the disorder and be compliant and an attendant must have the knowledge of patient care and cleanliness. The treatment framework in Āyurveda consists of three basic steps: Symptoms (Liṅga), Cause (Hetu), and Remedy (Auṣadha). The process of curing a disease invariably begins with the recognition of symptoms. Symptoms merely provide the starting point for understanding the cause. Āyurveda differentiates itself from the modern schools of medicine in the manner in which the symptoms are further pursued to identify the root causes.

13.3.1 Diagnostic Techniques

Āyurveda lays great emphasis on an early and correct diagnosis. In the course of development of the art and science of medicine in ancient India a number of medical techniques were evolved which reflect on the technical skill in this field. These medical techniques can be categorised into three groups: Diagnostic Techniques, Prognostic Techniques and Therapeutic Techniques. In Caraka-saṃhitā, different diagnostic techniques have been discussed in Chapter 8 of Vimāna-sthāna. Broadly, diagnosing the ailment is done using an examination of the patient and an examination of the disease. In each of these, several aspects are studied, as listed in Figure 13.3.

Examination of the Patient	Examination of the Disease
♦ Prakṛti (Constitution)	♦ Darśana-parīkṣā
♦ Vikṛti (Morbidity)	♦ Sparśana-parīkṣā
♦ Sāra (Tissue elements)	♦ Praśna-parīkṣā
♦ Saṃhanana (Organs)	♦ Nāḍī-parīkṣā
♦ Pramāṇa (Body measurements)	♦ Mūtra-parīkṣā
♦ Sātmya (Homologation)	♦ Mala-parīkṣā
♦ Sattva (Psychic condition)	
♦ Āhāra-śakti (Food intake)	
♦ Vyāyāma-śakti (Physical fitness)	
♦ Vayas (Age)	

FIGURE 13.3 Diagnostic Technique Prescribed in Āyurveda Texts

Examination of the patient was given greater emphasis and it essentially involved a study of the personality, strength and vitality of the patient using ten parameters shown in the figure. Since Āyurvedic concepts rest on the concept of Tri-doṣas, doṣa-prakṛti is an important component of the prakṛti analysis. It analyses seven possible constitutions under various combinations of Tri-doṣas. This presents an overall picture of the physical, physiological and psychological state of the patient. In order to make body measurements, the finger breadth (known as aṇguli pramāṇa) is taken as the unit of measurement. The normal measurements of all body parts in terms of aṅguli pramāṇa is described in Caraka and Suśruta saṃhitās.

Various methods to examine the disease have been specified in Āyurveda as listed in Figure 13.3. These are described in some detail below:

- **Darśana-parīkṣā:** This is a visual technique of inspection of the patient, the affected parts etc. The colour, lusture, external appearance, size, shape, etc. are assessed.
- **Sparśana-parīkṣā:** By a physical examination of the patient (palpation) the physician can get knowledge of qualities such as heat, pulsations, etc.
- **Praśna-parīkṣā:** Engaging the patient (and the well-wishers and accompanying people if the patient is unconscious) is a technique to elicit some information which will be helpful in the diagnosis and prognosis of a disease.
- **Nāḍī-parīksā:** This is a common practice adopted by an Āyurvedic practitioner although there is no description of this in the earlier texts. It ought to have been introduced during the medieval period. By observing the pulse, the health of the individual is assessed.
- **Mūtra-parīkṣā:** In addition to the usual urine examination, Āyurveda also prescribes an oil drop test (Taila-binḍu-parīkṣā) in which a drop of oil is left on the surface of the urine and observed. The movement of the oil drop in different directions and its rate of spread and change in colour are used to assess a number of chemical and physical properties.
- **Mala-parīkṣā:** Examination of faeces is a technique for diagnosis of a disease. In Āyurveda, an analysis of the movement of the faecal matter when dropped in water is made to assess the digestive and metabolic conditions of the patient.

13.3.2 Sleep and Food – Impact on Health

Dietary and lifestyle interventions are integral to disease management in Āyurveda. The dietary prescriptions vary according to the overall body constitution on account of the Tri-doṣas and the nature of the disturbed doṣas. We have already seen how the choice of certain foods vitiate the doṣas and therefore need to be avoided. On the other hand, a healthy lifestyle is very important to maximise the effect of treatment. Towards this end, Āyurveda recommends certain practices related to food intake and sleep. There are also daily, and seasonal routines prescribed to lead a healthy life (see IKS in Action 13.2 box for details on the daily routine prescribed in Āyurveda).

Caraka explains that strength, complexion, and growth of the physical body rests on three important pillars: Food, sleep, and celibacy/control over sense organs[19]. A person goes to sleep when the mind is weary, and the tired senses withdraw themselves from their objects. According to Suśruta, the heart which is the place of the Cetanā (consciousness) when it gets covered by tamas one gets sleep[20]. According to Vāgbhaṭa, excess of tamas leads to sleep during the night. Also, when the various channels of the body get enveloped by the kapha and the senses are tired and withdraw themselves from the objects a person gets to sleep[21]. Although we seem to take sleep for granted, sleep is considered an important element in our overall health condition. Sleep is the cause of joy and sorrow, fatness and thinness, energy and weakness, sexual energy and loss of potency, and knowledge and ignorance. When sleep is positive it gives positive results and vice versa.

Six Different Types of Sleep

In the Āyurveda texts, six different types or causes of sleep have been mentioned:

1. **Tamo-bhavā:** When Tamas covers the mind the person tends to fall asleep. This is the reason for a person to feel sleepy after a heavy food. Scholars consider this type of sleep as the root of all inauspiciousness or sins.
2. **Śleṣma-samudbhavā:** Caused by the influence of kapha in the body, the reason why children sleep more compared to elders.
3. **Manaḥ-śarīra-śrama-samudbhavā:** The sleep caused by the exhaustion of the mind and body.
4. **Āgantukī:** Caused by some smell, medicine, massages or smearing of some ointments or oils.
5. **Vyādhy-anuvartinī:** Caused due to some specific diseases like fever caused by Kapha.
6. **Rātri-svabhāva-prabhavā:** The sleep which is natural at night. This is the best of all the mentioned sleeps. It is named as Bhūta-dhātrī which means the supporter of beings. It takes care of a living being like a mother takes care of her child.

The best form of sleep is the one that we get at night. One must sleep according to his nature. It may differ from person to person. Staying awake during the night brings roughness and daytime sleep produces greasiness. Therefore, the best time to sleep is at night. If one feels sleepy at other times, it is prescribed to sleep in a sitting position. Tri-doṣas do not get exaggerated if a person sleeps in a sitting position. For a healthy person, daytime sleep is prescribed during the summer season because the nights in this season are shorter and Vāta gets perturbed which causes roughness. Other than the summer season the day sleep perturbs the Kapha and Pitta[22]. Although untimely sleep is not prescribed there are some specific conditions where daytime sleep is prescribed for a certain category of people. This includes those whose strength is impaired from a lifestyle including singing, studying, drinking alcohol, journeys, walking a long distance, night vigil, carrying heavy loads, old, women, children, patients undergoing therapy, etc.[23].

Sleep problems can be classified into two – insomnia and unwanted/excessive sleep. In this modern world because of various reasons people face the problem of insomnia. Insomnia may lead to other types of diseases such as body pain, heaviness in the head, yawning, laziness, exhaustion, giddiness, indigestion, a state of unconsciousness, and diseases related to vāta[24]. Āyurveda suggests that oil massage, application of medicated powder, taking bath, eating Shali rice with curd, enjoying agreeable aromas and sounds, gentle massaging of the body, pleasing the eyes, application of the cooling ointment, using a clean and smooth bed, and sleeping at an appropriate time will solve the issue of insomnia[25]. Unwanted sleep means sleep at an improper time. Both unwanted and excessive sleep is caused because of perturbed Kapha and Pitta in the individual. Solutions suggested for this are – evacuation therapy, medicated smoking, physical activity, bloodletting, fasting, and subduing the tamas[26].

Food Intake Methods

As diet is critical for the sustenance of life it is one of the pillars of life. An individual must be very cautious about the intake of the food and also of the quantity of food[27]. Most diseases are caused due to wrong and excess intake of food[28]. According to Caraka, the quantity which gets digested easily in time without perturbing the doṣas, dhātus and agni, is appropriate[29]. Consumption of food itself is a great

- Strength, complexion, and growth of the physical body rests on three important sub-pillars: Food, sleep, and control over sense organs.
- Both unwanted and excessive sleep is caused because of perturbed Kapha and Pitta in the individual.

science and needs a careful process. Food is of two types - Guru and Laghu. In simple terms, 'guru' is hard to digest and 'laghu' is easy to digest. Guru causes fatigue, increases the quantity of bodily waste, is strengthening, satiating and growth-promoting and laghu is the opposite and it reduces weight as well as heals the wounds. Food, both heavy as well as light, acquires medicinal value only if taken in the right quantity[30]. Hence, one should always consume in an appropriate quantity as this induces proper action of the digestive fire. Each meal must have a gap of at least three hours and at a maximum of six hours. If the food is consumed within three hours, digestive juices are disturbed. On the other hand, if the food is eaten beyond six hours it leads to a loss of strength. The food to be consumed must not be the one which was cooked before three hours. Such food does not get digested soon and is called tāmasa food as it increases tāmasic tendency in an individual[31].

Role of Water

In Indian culture, water is always considered as nectar or medicine[32]. An individual drinks water to relieve the thirst, to stay healthy and to survive. Consumption of water has many benefits such as, increased energy and relieved fatigue. It promotes weight loss, flushes out toxins, improves skin complexion, maintains regularity, and boosts the immune system. It acts as a natural remedy for headache, prevents cramps and sprains, refreshes the mind, and helps maximize physical performance. Finally, it also relieves constipation and prevents formation of kidney stones. When water is consumed inadequately it causes fatigue, deafness, pain in the heart, dry eyes, dryness in mouth and skin, and weakness.

In modern parlance, we are advised to take two to three litres of water daily to stay healthy. From the basic principles of Āyurveda, we can infer that the quantity of water that one must take daily will depend on the constitution (prakṛti) of a being. It is quite possible that a person can lead a healthy life even after drinking one litre of water and another one may be unhealthy even after drinking three litres. The type of water and the quantity is dependent on one's own prakṛti and the prakṛti of water. According to Āyurvedic texts, there are two types of water; Āntarikṣa and Bhauma. Ānatarikṣa is the type of water which falls directly from the sky. Bhauma is the ground water. The properties of ground water depend upon the type of soil and also on the seasons.

After understanding one's own nature an individual must decide what type of water he/she must drink. However, there are some general instructions given in Āyurveda that can help an individual to keep himself/herself healthy at his/her place of dwelling. One must drink water which is slightly astringent and sweet, subtle, clear, light, neither unctuous nor deliquescent[33]. Water is usually cold. Coldwater relieves intoxication, fatigue, fainting, vomiting, weakness, giddiness, thirst, heat, burning, raktapitta, and poison[34]. But when the cold water is heated up it increases the power of digestion, metabolises undigested food, good for throat diseases, easy to digest, purifies the urinary bladder, indicated in hiccough, cures abdominal distension, vāta and kapha disorders, acute fever and cough[35].

13.3.3 Drugs and Physical Therapy

The drug treatment in Āyurveda primarily consists of herbal formulas, bhasmas, and medicated oils and ghee. A formula made up of several herbs is used rather than a single herb. They are always given with other foods or herbal items, such as honey, ghee, ginger, etc., which help mitigate toxicity and increase absorption of certain ingredients, thus obtaining the desired therapeutic effect. Rasāyanas are equivalent to modern dietary supplements and Āyurvedic physicians determine the appropriate rasāyana required depending on the health needs. Rasāyana refers to compound preparations containing multiple herbs and minerals that improve transportation of nutritional materials to body tissues. Rasāyanas may act in a variety of ways by improving the nutritional value of the food, digestion and absorption of nutrients, transportation of nutrients to tissues, bioavailability of nutrients, metabolism of nutrients in tissues, immunity, and by cleaning the microcirculatory channels or pores (srotas) which improve uptake of micronutrients.

The oxidised form of metal and mineral preparations, called bhasma, is used in Āyurvedic medicine. Bhasmas are metal or mineral powder formulations prepared using instructions and specific procedures provided in Āyurvedic texts. The preparation of these involves several herbs and herbal extracts and subjected to heat treatment and extraction. These procedures detoxify toxic metals like mercury. There are several metal bhasmas described in rasa-śāstra texts such as Rasa-ratna-samuccaya. The common metals and minerals used in making bhasmas for therapeutic use are gold, silver, copper, mercury, iron, and zinc. In Section 10.3.3 we have seen the procedure for extraction of copper for Āyurvedic purposes.

In the 1970s, the government of India appointed a panel of Āyurvedic experts to evaluate the formulas in Ayurvedic texts. The work culminated in compiling a "Ayurvedic Formulary of India"[36], which has over 560 evaluated formulas, 22 bhasmas and 55 rasas.

The management of an illness often starts with cleansing and includes five procedures called pañcakarma, a sub-set of which may be prescribed depending on the condition of the patient. Pañcakarma is designed with the assumption that the major cause of body ailments is the toxic products produced by body metabolism, microorganisms, and other chemicals. If the body organs are not eliminating toxic products from the body, toxic symptoms will develop over a period. Therefore, elimination of toxic products can contribute to overall management of an ailment and its cure. Usually, pañcakarma is a three-step process. In the first step preparatory procedures are done for the body by applying oils and fats, sweating, and suggesting herbs that improve digestion and metabolism. In the second step, purification procedures involving emesis, purgation, enema and bloodletting are undertaken for detoxication. The last step is the post-procedure consisting mainly of recuperative measures in the form of diet and lifestyle change prescriptions.

The points where muscle, cartilage, nerves, and bones join each other are vital and it is believed that the life energies are concentrated at these points. There are 22 vital points on the upper extremities, 22 on the lower extremities, 12 in the abdominal areas, 14 in the back, and 37 in the neck and head. In Āyurveda massage of these points is prescribed to treat diseases and strengthen the body. During massage specially formulated medicinal oils are used. Specific medicated oils and types of strokes are chosen based on the disturbed doṣas, body constitution, injury, and disease condition.

13.4 YOGA WAY OF LIFE – RELEVANCE TO HEALTH AND WELLNESS

Among the many challenges in modern society, work-life balance is one important issue. A growing number of dissatisfied employees complain that they feel deprived of meaning in their workplaces. This introduces considerable stress, physical as well as psychological on working individuals. Workers worldwide face increasing demands to work extended hours and consequently experience considerable work overload. On account of these, in modern corporations, institutions working not-for-profit, and government organisations, there is a concern about the issue of burnout that employees experience at their workplace. Burnout can be defined as the end result of stress experienced but not properly coped with, resulting in symptoms of exhaustion, irritation, ineffectiveness, discounting of self and others, and problems of health (hypertension, ulcers, and heart problems). This has serious consequences including reduced productivity at the workplace, bad health, and mental stress for the employee. If not addressed effectively, this can threaten to manifest as a social problem at a later time. The stress initially experienced at the workplace slowly spills into personal and social life, threatening the health of an individual and his/her family members. The costs of stress are huge and not just financial; there are mental, physical, and social costs as well.

It is well understood today that a psychologically stressed individual suffers from ill-health (both mental health and physical health). If we go to a medical practitioner today for treatment for stress-induced chronic medical conditions such as hypertension and diabetes, in addition to prescribing certain medicines, some advice is also given to make certain changes in the 'lifestyle'. The salient aspects of the changes in lifestyle required to lead a healthy life are found in the yoga way of life. There is a repository of knowledge and accumulated experience in India on the role of yoga as a way of life in enabling individuals to lead a healthy life. Yoga is gaining greater international attention of late. As a holistic science concerned with all aspects of human functioning, the principles and practice of Yoga provide a comprehensive framework by which the psychological and physical aspects of stress can be understood and eliminated.

As we have seen already in Chapter 3, Yoga is one of the six philosophical systems. Patañjali proposed yoga as a mechanism to progressively work on the mental fluctuations to eventually still the mind. Yoga-sūtras outline a method to conduct a life that fosters moderation and promotes harmony. The uniqueness of Yoga way of life lies in its integrated approach to address both the physical and psychological aspects of one's life. According to Yoga, we are unconscious of mental, and emotional processes that habitually create stress. Yoga involves a systematic method by which we can begin to expand our awareness of these processes and eventually gain control over them. In any stress disorder, the para-sympathetic auto-nervous mechanisms fail to function adequately to minimise the impact of stressful stimuli. Among several relaxation practices, Yoga has the potentiality to influence the auto-nervous mechanisms in various ways. This restores the functioning of sympathetic and parasympathetic nervous systems thereby preventing the build-up of stress. In a very practical sense, Yoga gives us the tools and techniques by which we can expand our conscious awareness into the unconscious parts of the mind in order to become aware of the patterns and habits which lead to stress.

The Yoga way of life addresses three aspects concerning one's life. One is a set of physical exercises that help an individual stay physically fit, nimble, and flexible. Another set of practices known as meditation that addresses the issue of calming the mind and providing a stable psychological state. The third set of ideas help reorient one's goals and worldviews so that there is a degree of realism that governs one's living. On account of these practices, an individual experiences stable and balanced physical and psychological states, thereby enabling the individual to reach the health and wellness goals. Figure 13.4 pictorially depicts this idea. Rāja yoga is the path of control of the mind through the practice of Aṣṭāṅga yoga or the eight-fold path given by the sage Patañjali in his Yoga-sūtras. We discussed these eight steps in Chapter 3 (see Section 3.3.2). The eight steps develop and enhance cognitive learning and moral conduct through sustained and continuous physiological and psychological practices.

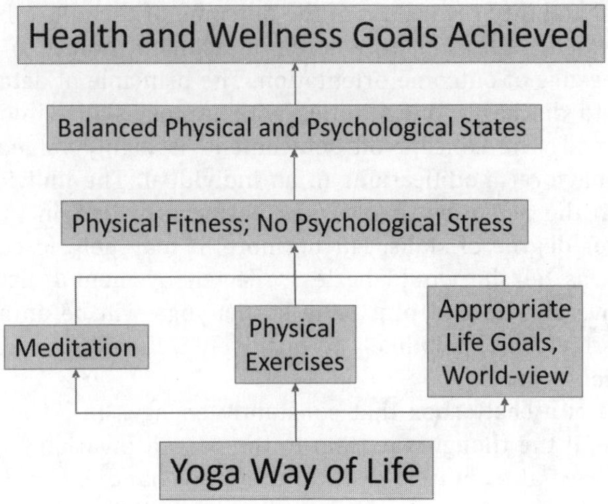

FIGURE 13.4 Yoga Way of Life and Health

Yama and Niyama involve moderating external behavior and thought patterns through mental and physical practices. On the behavioural side, abstention is sought from violence, falsehood, dishonesty, sexual excess, and acquisitive tendencies. On the other hand, the ideals prescribed are purity, contentment, austerity, self-study, and forbearance. Prolonged and diligent practice of this helps the individual remove negative emotions such as anger, jealousy, greed, attachment, ego, and an excessive desire for objects. These are very important in adopting the yoga way of life. Through such a process, the individual attains a state of 'citta-śuddhi' or purity of mind, which is essentially the beginning of the yoga way of life. The stages of 'Āsana and Prāṇāyāma' are meant for disciplining the body and regulating subtle energy flows. Mastery of this and the regularity of these exercises provide a healthy and responsive physical infrastructure to an individual and make the mind subtler, alert, and more observant than before. In the fifth stage of Pratyāhāra, secondary input is regulated so that the mind is not distracted. By acting on the external stimuli using these five stages, it paves the way for control of one's mind.

Yoga works on the subtle aspects of one's mind in the next three stages and thereby makes one more self-aware, and self-regulated. In this manner, 'Dhāraṇā, Dhyāna, and Samādhi' are mainly intended for uplifting one's spiritual self and for heightening consciousness. Practices such as āsanas (postures), prāṇāyāma (breathing control exercises), and kriyās (cleansing regimens) are taught as physical practices. On the other hand, several mediation practices are taught as part of the training for the yoga way of living. While various meditation techniques work at the mental level, all these practices are intended to develop a certain type of awareness within oneself. This in turn is expected to bring about a change in the emotional state of an individual. Furthermore, it will also induce a change in the intellectual and somatic functions of an individual.

- Yoga offers a well-formulated approach to lead a life characterized by health and wellness objectives.
- Yoga works on the subtle aspects of one's mind and thereby makes one more self-aware, and self-regulated.

Karma-yoga is an approach to life situations and activities by focusing on the action on hand without fixation of the mind on outcomes. In such an approach an effort should be made to maintain emotional stability regardless of the outcome of the effort, be it success or failure. This can be called *detached attachment*, wherein attachment points to effort orientation and detachment to the absence of outcome orientation. The principle of detached attachment calls upon the individual to discharge one's duties with utmost skill without desiring the fruits of actions so performed. The concept of 'detachment' is highly valued in Indian culture as it induces several behavioral modifications in an individual. The individual would be able to concentrate better on the activities that he/she performs which, in turn, would lead to the attainment of a higher degree of skills. Furthermore, it may help in conserving the physical and psychic energies as he/she would be less affected by mental distractions, anxiety, and fear of failure. The overall effect of practicing karma-yoga will be an altered perspective on life goals and the worldview. This is likely to positively influence the health and wellbeing of an individual in the long run.

The human mind is a chatterbox that constantly engages in a variety of thoughts about the past or the future. If the thoughts pertain to the past, it invariably gets into an evaluative mode of events that have already unfolded. On the other hand, if the thoughts pertain to the future, it is one of speculation, dreaming, and expectation building on some future course of the event waiting to unfold. Very rarely the mind is able to fix itself to the present. In Bhagavadgītā, Arjuna observed that the mind is fleeting, firm, and strong, and it takes one's attention away at its will making it difficult to control. Patañjali begins the treatise on Yoga with the first sūtra that the very purpose of yoga is to stop the lamentations and block the mounting waves of thoughts emanating out of one's mind. In the set of prescribed methods in yoga, the need for quietening the mind is an important component. This is achieved through some meditative techniques and prāṇāyāma practices. Distractions caused by the fluctuation of the mind were thought to be linked to the fluctuation of the breath. Therefore, meditative techniques often involved regulation of the breath through prāṇāyāma. Moreover, since the mind was thought to be strongly influenced by the body, contemplative practices involved specific postures.

Meditation is a state which enables one to focus on the present moment, leading to a state of thoughtless awareness. While the very purpose of these prescriptions is to help one evolve spiritually and realise one's true nature, these practices also have a useful dimension of health and wellness. Usually, the meditator focuses upon a thought, a vision, or a sound or other sensory experience. There are many types of meditation, but they all have been shown to have a similar effect. Transcendental meditation, essentially a simplified form of yoga, is the skill of effortlessly minimising mental activity so that the body settles into a state of rest deeper than deep sleep while the mind becomes clear and alert. On account of the relaxation obtained through meditation, it has been found to help meditators in the reduction of anger and anxiety.

- Quietening the mind is achieved through meditative techniques and prāṇāyāma practices.
- Meditation is a state which enables one to focus on the present moment, leading to a state of thoughtless awareness.

Several studies point to the beneficial effects of yoga way of life and its relationship to the wellbeing of an individual[37]. In a study done at Swami Vivekananda Yoga Anusandhan Samsthan (SVYASA), Bengaluru, it was found that Yoga training can help people to reduce their heart rate, which has possible therapeutic applications. Another study found Cyclic Meditation Technique, developed by SVYASA, to be more effective in achieving voluntary heart rate variability. A review of studies conducted on Transcendental Meditation (TM) in 200 Universities from 33 countries revealed that TM helps expand consciousness, decrease oxygen intake and stress level, increase basal skin resistance and coherence in EEG, and virtually suspends breathing up to one minute. The yogic lifestyle also resulted in a significant reduction in systolic and diastolic BP, body weight, serum cholesterol, and triglyceride levels. Other studies pointed out that long-term practitioners of Yoga had acquired a remarkable voluntary control over their autonomic processes, which helped them in coping with psychological stress.

13.5 INDIAN APPROACH TO PSYCHOLOGY

Human psychology is a subject matter of great interest in today's society. While we have made significant progress in Science and Technology (S & T) in the last 100 years, compared to the previous two millennia, we cannot say that there has been a commensurate improvement in our 'happiness' quotient. In fact, it seems to have either remained unaffected by the S & T progress or has deteriorated. At the core of human psychology is the basic question of individuals' life goals and purpose, sense of satisfaction, means of achieving them, and challenges one faces in this journey. Therefore, Psychology is a study of the mind and its behavior. Currently, we employ the dominant paradigms from the Western approaches to Psychology, which are of recent origin. It is increasingly felt that the 'positive psychology' paradigm of the West has taken a superficial account of feelings and the behavioral dynamics of an individual. Therefore, issues such as social and family violence, mental health problems, and moral decay faced by society have not been adequately addressed. Moreover, mechanically applying the constructs of Western Psychology in the non-western world has contributed more to the problem.

The Indian approach to psychology has long years of introspection, testing, contemplation, and experiential learning. It draws its frameworks, concepts, tools, and techniques from a rich repository of diverse knowledge traditions starting from the Upaniṣads. The Vedic corpus and the philosophical systems (Darśanas) have carried forward a tradition of discussion on matters of importance to psychology, which serves as the source. Indian psychological tradition is markedly different from the Western approach. In the contemporary discussion in the west, the body is considered as the primary factor that dictates the issue of health and wellness including psychological illness. A more liberal and somewhat less prevalent model of health in the West seeks to include the mind into consideration. In contrast, in the Indian approach, the 'body – mind – consciousness' plays an integrated role in matters of health and wellness. In fact, the role of mind and consciousness is considered very significant when it comes to issues such as wellness, and healing. The goal of health in the Indian tradition is not to find means to merely prolong the longevity and physical aspects of health. On the other hand, it is to promote happiness and a sense of satisfaction and fullness for an individual. The goal of longevity plays a sub-ordinate, but a useful role to this primary goal.

Distinctive Aspects of Indian Psychology

Indian Psychology (**IP**) has certain distinctive elements vis-à-vis the Western counterpart. First of all, the Indian approach to psychology has a much wider canvas to address. Rather than confining itself to merely materialistic and deterministic aspects of human existence, IP considers spiritual dimensions and self-evolution goals. While this has made the problem more complex, it has nevertheless provided a more inclusive agenda for psychological studies.

- The Indian approach to psychology has long years of introspection, testing, contemplation, and experiential learning.
- Indian Psychology is non-religious in its character, simply because the issue addressed is one of human existence, happiness, and contentment.

As already mentioned, IP is firmly rooted in the Indian scriptures and philosophical texts. Despite this it is non-religious in its character, simply because the issue addressed is one of human existence, happiness, and contentment. How an individual should liberate himself/herself from bondage, ignorance, and stress does not have to depend on any religious tradition, nationality, or sex. These are perennial issues of human existence that transcend geographical boundaries and time. Therefore, prescriptions of IP have a wider and international appeal. A case in point is the global acceptance of yoga and meditation, which are considered vital tools for addressing psychological issues in the tradition of IP. Since the issues addressed in IP are one of handling human suffering and alleviation of them leading to final evolution, they do not pose any threat to others. Moreover, there is a multiplicity of views in several matters of psychology and the methods to address them. The six darśanas and Buddha and Jaina schools of philosophical thought offer alternative views on the matter. This provides a wide platter of offerings for every individual to pick and choose what suits the person.

Since the basic premise of IP is to help individuals spiritually grow and liberate themselves from several constraints and limitations of mundane life, the methods adopted are veridical in nature. IP recognises the existence and importance of the 'world within' for every individual as much as the 'world outside' for addressing psychological issues. A vastly developed understanding of the 'world within' is a significant contribution of IP. In IP there is room for both 'objective knowledge' obtained from the 'world outside' and 'subjective knowledge' assimilated through experience, intuition, and deep contemplation. Therefore, every individual has multiple states of existence, a multi-layered understanding of one's own consciousness, and unique methods and possibilities to develop self-awareness. Methods for self-observation and experimentation have been developed to aid this process. Several yogic and meditation practices is one such example belonging to this category.

Basic Tenets of IP

Let us summarize the key features and basic tenets of IP in the light of these discussions.

Construct of a Human Being: According to Vedānta, human beings could be viewed as a composite of three basic ingredients; sat (existence), cit (consciousness), and ānanda (bliss). 'Sat' represents the physical or material infrastructure (substance) of an individual. On the other hand, every person is embodied consciousness (cit), which is the substratum and context to the life experiences. Furthermore, every individual in its purest form is nothing but an expression of bliss. IP derives various frameworks and constructs to articulate each element of this from the repository of ancient Indian knowledge. This representation of an individual is very different from the western conceptualisation and it significantly influences how IP addresses various issues of psychology.

Essential Goal: There is the oneness of humanity, articulated through the notion of ātman in IP. Therefore, the goal of IP is to help individuals come out of ignorance and ultimately experience the oneness and the divinity within. This strips an individual of all limitations and helps him/her realise the true nature and full potential. It requires transforming an individual from a state of conditioned and mechanical thinking to an unconditioned state of freedom.

Nature of Human Beings and Their Pursuit: The true self is the seeker of truth and repository of knowledge. IP begins with a basic premise that every individual is divine, infinite in terms of possibilities, and inherently an embodiment of peace, happiness, and fulfillment. Therefore, it is natural that the pursuit of happiness is the unwavering goal of every individual. However, in an empirical world carrying out a mundane life the individual faces existential constraints that hamper the individual's pursuit of happiness and sense of contentment. Therefore, the goal of IP is to liberate the individual from these constraints and help rediscover one's true nature. This is markedly positivist and grand in its pitch compared to other possible objectives for developing the science of psychology.

Constraints in Life: Nobody wants to either long for unhappiness to descend on them or the happiness that they experience at certain times to fade away. Despite this, our experience of life informs us that the goal of happiness is fleeting. There are several existential constraints that we face, which are responsible for this apparent conflict. We suffer from an ignorance of our true nature and therefore mistake the ego to be the 'self'. This creates a basic fault line in our Body – Mind – Intellect complex and drives several choices that we make in life. Our attachment to impermanent things brings a mixed bag of emotions consisting of attachment, aversion, hate, and envy taking us through a roller coaster of happiness and sadness in life. On account of a distorted understanding and an uncontrolled mind, this results in an unhealthy, stressful, and a defeated life for many. Much of the work in IP addresses the reasons for this apparent conflict between what one's true nature is and what one ends up with in one's life. It provides mechanisms to resolve the conflict in a sustained and systematic manner.

- The oneness of humanity is articulated through the notion of ātman.
- The goal of Indian Psychology is to liberate the individual from life constraints and help rediscover one's true nature.

Mechanisms to Address the Constraints: The underlying principle behind several mechanisms developed to address the constraints is to take the individual through a process of self-realisation. The mechanisms basically address the unsteady mind and help an individual exercise a certain control over it. These include methods for self-observation, recognising higher states of existence, and inquiry into different domains of consciousness. Specific techniques such as yoga, and meditation help in controlling several influences that affect the mind and cloud one's consciousness. The premise is that such a person will experience conflict-free and positive mental health, self-evolve, and passionately engage in societal transformation with a spirit of unselfish work for the common good.

13.6 THE TRI-GUṆA SYSTEM

The Tri-guṇa system provides an overarching framework to understand the physical infrastructure of a human being that forms one of the three basic elements (sat – existence). At an operational level, the tri-guṇa system seeks to answer some of the commonly observed phenomena of human behavior. For example, we always wonder why some people are aggressive while some others are gentle and benign. Similarly, we find some individuals action-oriented, bubbling with energy, and taking leadership roles, while some others play a more intellectual role of calm thinking, guiding, and advising others. The tri-guṇa system provides us some basis to understand these differences and help us make certain behavioral modifications over time to gravitate towards an ideal guṇa composition. As per IP, every individual is a unique construct made of three ingredients, known as tri-guṇas. Tri-guṇa can be summarised in the following manner:

- The tri-guṇa system seeks to answer some of the commonly observed phenomena of human behavior.
- In the Indian system, the triguṇa system is the basic building block of not only human beings but all conceivable forms of physical reality.

- It comprises Sattva–Rajas–Tamas (S-R-T) that serves as the basic ingredients to represent the physical reality
- It characterises something that exists in reality
- It is a mechanism for creating a world of multiplicity including rocks and rivers, plants and trees, and animals and human beings. In short, any manifested reality is supposed to be made of these three basic ingredients.

At the same time, IP also recognises that there is a great degree of commonality, known as divinity among individuals. The tri-guṇa system reconciles this apparent conflict and provides a framework to understand how this gets operationalised. If everything originates from a single source, then there must be a mechanism to 'manufacture' this infinite variety. This is not a new idea for us and there are many examples of this. Let us take a well-known recent example to understand this. Let us say we are preparing a presentation for a meeting using Powerpoint. In our PPT we may include a box with a bulleted list of items and may want to fill the box with a color. If we open a color pallet, it allows us to select any color already available. If we are not happy with the available choices, we can create our own color using a combination of R-G-B. By changing the RGB numbers we can create a very large number of unique shades of color. Figure 13.5 illustrates this for a sample set of colors.

Colour	R	G	B
Black	0	0	0
Dark Blue	25	50	75
Light Grey	208	216	232
Light Green	123	231	132
Dark Grey	100	100	100
Dark Orange	200	100	50
Red	230	81	70
Yellow	225	239	45
Purple	175	170	240
White	255	255	255

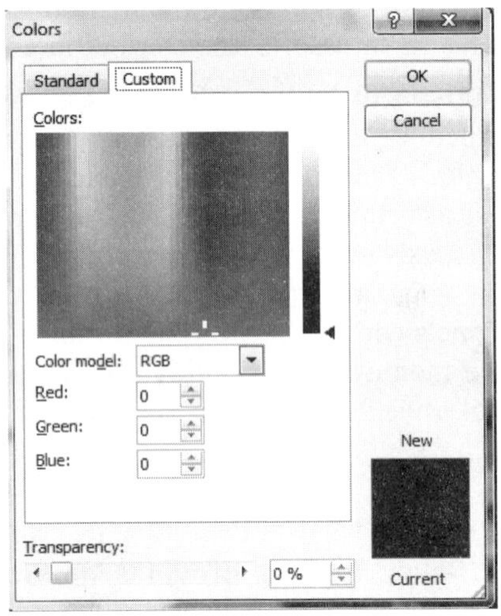

FIGURE 13.5 An Example of Generating Many Colours Using MS Office Colour Palette

This example indicates to us that three ingredients are enough to generate a world of colors using unique combinations of the three. The tri-guṇa system is the equivalent of the color palette. Using three basic ingredients known as Sattva (S), Rajas (R), and Tamas (T) the physical existence can be uniquely determined. While R–G–B can only vary between 0 and 255, the S–R–T has infinite possibilities to choose from. In the Indian system, the Tri-guṇa system is the basic building block of not only human beings but all conceivable forms of physical reality. In a nutshell, it is the R–G–B (or S–R–T) of everything in this universe.

The unique shade of every individual gives rise to the varied patterns setting the stage for psychological analysis of individuals. This basic feature is recognised and taken into consideration by other fields of study in India. For example, as we have already seen in the chapter, Āyurveda recognises the unique nature of every individual. The Sāṃkhya philosophy has elaborately discussed the concept of tri-guṇa and its manifestations. Chapter 14 of Bhagavadgītā concisely introduces the Tri-guṇa system and the characteristics of each ingredient. Chapters 16–18 of Gītā captures the influence of the tri-guṇa on individuals and their behavioral patterns.

- **Sattva** is the source of knowledge and embodiment of happiness that every individual experiences. By its very nature, it binds by attachment to happiness and knowledge. What it essentially implies is that such times when an individual is triggered into deep thinking, contemplation, and focused efforts in search of meaning knowledge, etc. then the sattvic aspect of the individual is manifesting dominantly.
- **Rajas** is known to be the nature of passion, the source of desire, and attachment. It binds fast by the attachment to action. The world of action, desire to achieve name, fame, wealth, and power emanate out of rajas. In other words, when an individual is triggered into action, motivating oneself and others to achieve some goals and targets, the rajasic component is dominantly manifesting in such a person. Expectedly, the behavioral patterns of anger, attachment, aversion, and envy primarily arise out of rajasic tendencies.
- **Tamas**, on the other hand, is born of ignorance, deluding all embodied beings; it binds fast by heedlessness, indolence, and sleep. Tamas represents the gross matter and is the reason for us to rest, sleep, feel mentally dull and lazy. Tamasic tendencies heavily discourage a person from learning, making progress through experimentation, observation, contemplation, and self-reflection.

Table 13.5 summarises these aspects of the three guṇas.

TABLE 13.5 A Summary of the Three Guṇas

Guna	*Binding Force*	*Attributes*	*Effects*
Sattva	Attachment to happiness, knowledge	Stainless, illuminating, flawless	Longing for liberation, divine tendencies
Rajas	Attachment to actions	Source of passion, longing, attachment	Desire, anger, pride, jealousy, egoism, envy, etc. which are modifications of the mind
Tamas	Indolence, sleep, mis-comprehension	Stupefying through ignorance	Absence of right judgment, contrary impression, doubt, etc.

All three guṇas are important for a human being and they play a complementary role in one's life. The only issue of interest is the relative proportions with which the guṇas call the shot while facing several events in one's life. For every individual, the proportion of the three guṇas can keep swinging dynamically from time to time (making us happy at a time and restless at another time). However, in the long run, the guṇas rest in a certain natural balance unique to every individual making the person sāttvika, rājasika, or tāmasika in nature depending on the relative proportion of the three ingredients in long-term equilibrium. From the descriptions of the three guṇas, it follows that increasing the sattva component at the expense of the tamas and to some extent, the rajas is a progressive direction for an individual as it would greatly facilitate the process of self-evolution and growth. Viewed from a psychological perspective, changing the proportion of the three ingredients of tri-guṇa is possible, but is a slow and gradual process. It requires a sustained set of physical and mental practices to do so.

13.7 THE BODY–MIND–INTELLECT–CONSCIOUSNESS COMPLEX

We saw in Chapter 3 that according to the Sāṃkhya school of philosophy every human being is made up of 24 elements (see Chapter 3, Section 3.3.1 for details). Similarly, the Vaiśeṣika defined nine categories of substances (see Chapter 7, Section 7.2.1 for details). Drawing from these, one can identify the physical and psychological components of an individual by defining the gross and subtle body. The gross body corresponds to the physical manifestation of an individual (in the form of the visible organs) which are made of the five bhūtas. The psychological part of an individual constitutes the subtle body. This consists of the five prāṇas, the organs of knowledge and action, and the internal instruments (intellect, memory, mind, and ego). It is normally misunderstood that the external organs such as eyes, ears, etc. are the organs of knowledge and action. In reality, they are mere inert parts of the gross body. The actual organs are internal and indeed subtle to these physical external manifestations. One can understand this from the fact that a blind person has a pair of eyes that look no different from that of a person of normal vision. Similar is the case with other organs as in the case of a dumb or a deaf person. Quite distinct from all these is the true self ('I') which is the consciousness. The person is nothing but embodied consciousness. Figure 13.6 depicts this pictorially.

- Sattva is the source of knowledge and embodiment of happiness that every individual experiences.
- Rajas is known to be the nature of passion, the source of desire, and attachment.

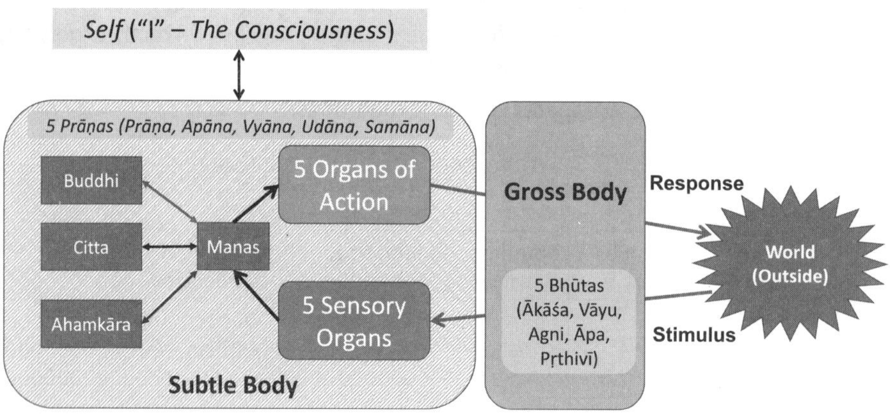

FIGURE 13.6 A Concise Representation of Body–Mind–Intellect–Consciousness

The figure illustrates the process of interaction with the external world by an individual. Stimulus from the external world is gathered primarily through the respective part of the gross body by the sensory organs. For example, the ear picks up the sound waves when somebody speaks to an individual and the pair of eyes read the body language of the person speaking. These signals are fed to the mind, which in some ways plays the crucial linkage between the gross body and the subtle body. The processing of the received signals is done with the help of other antaḥkaraṇas (buddhi, citta, and ahaṃkāra). While buddhi provides a subjective evaluation of the signal, the citta provides a vast repository of accumulated past experiences and memories for analysing the signal. The ahaṃkāra provides a sense of entitlement and agency to the issue on hand. On the basis of these, the signals are sent back to the mind for its final decision on the nature of instruction to be given to the organs of action. The organs of action utilise the physical organs in the gross body and provide a response to the external world. In our example, the statements made by the person is analysed and an appropriate response is given back to the person.

While this appears to be a simple and straightforward process, in reality, complex psychological issues stem on account of the role played by the mind in conjunction with the other antaḥkaraṇas. It is also due to the interface between the antaḥkaraṇas (especially the mind) and the consciousness. The consciousness is the 'sākṣī' and the reference frame for the entire world of transactions an individual goes through in his/her life. It is the 'true self' and the supplier of energy and vitality and therefore is the knower–doer–enjoyer in the process. However, for an untrained mind, the reflected consciousness on the antaḥkaraṇas seems to replace the original and the ahaṃkāra usurps its role. This is at the core of the psychological makeup of individuals and requires a deeper understanding.

- Complex psychological issues stem on account of the role played by the mind in conjunction with the other antaḥkaraṇas.

13.8 CONSCIOUSNESS – THE TRUE NATURE OF AN INDIVIDUAL

For time immemorial, one of the questions that we seek to find an answer for is, "Who am I, and what is my true nature?". The Indian scriptures unequivocally argue that we are nothing, but a bundle of consciousness caged in a physical frame called the body. As already discussed, consciousness is the ultimate essence of a person. In an empirical world, it embodies and expresses itself through the gross and the subtle body. Consciousness is corporeal and therefore one can infer its presence only on the basis of the effects of its existence. Reflected consciousness rather than the consciousness itself is in our domain of understanding and analysis using our antaḥkaraṇas. This is best understood using a contemporary analogy. Think of electrical gadgets such as a fan or a television. When there is electricity passing through the gadget, the gadget works and puts out useful work, in the absence of electricity the gadgets are inert and lifeless. Our gross body and the subtle body similarly get life and energy on account of the presence of consciousness. Just as software can be recognised only by seeing the effects of its working, consciousness also is inferred through its manifold manifestations.

Understanding the manifestations of consciousness is only possible by self-experimentation and deep contemplation of one's experiences. Given the corporeal nature of consciousness, it requires a graded and multi-faceted understanding. Upaniṣadic sages were seekers of this knowledge and were seized of this challenge and developed alternative frameworks to understand consciousness. Expectedly, IP draws from the rich repository of Upaniṣads and other scriptures to conceptualise these multi-fold manifestations of consciousness. We shall look at two such frameworks.

The Pañca-koṣa – A Five-layered Existence

Taittirīya-upaniṣad discusses the notion of an individual using a five-layered concept. It provides a step by step and a graded understanding of consciousness[38]. Koṣa means an envelope or an enclosure. According to this framework, the consciousness is covered by five layers and it radiates through these five layers. This reflected consciousness is often mistaken to be the consciousness although it is a proxy for it. Figure 13.7 illustrates this in simple terms.

- Consciousness is the ultimate essence of a person. Iit embodies and expresses itself through the gross and the subtle body.
- Consciousness is covered by five layers and it radiates through these five layers.

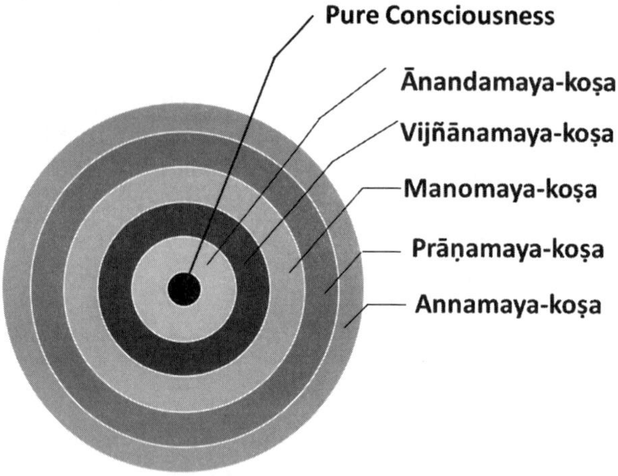

FIGURE 13.7 A Simplified Representation of the Pañca-koṣas

The outermost layer is the annamaya-koṣa, representing the grossest reality of an individual consisting of the physical body. Quite often we tend to mistake our physical body to be ourselves. If that were true, then a person who lost his lower limbs in an accident must have a proportionally reduced sense of 'I' ness, which is never the case. The next layer is known as prāṇamaya-koṣa or the vital energy layer. This is an expression of the life force that we all have, and we feel its manifestation in a physical sense. The third layer is the manomaya-koṣa, the mental layer. This layer is subtler than the other two layers and comprises the mind and its emotions. However subtle it may be, it is still possible to describe an individual in terms of his/her mental processes, feelings, and outcomes. The fourth inner layer subtler than the mental layer is the intellectual layer or the vijñānamaya-koṣa. This describes a person in terms of his/her intellectual makeup. The innermost and the subtlest layer is the layer of bliss. The closest we can get to the pure consciousness and the true identity of an individual is the layer of bliss that is ānandamaya-koṣa. See the box "the five-layered consciousness of an individual" (IKS in Action 13.3) for an illustration of this concept using a real-world example.

The choice of the word 'koṣa' succinctly communicates certain messages about the true identity of an individual. Since koṣa is a cover, it cannot be the original substance itself. It also has a temporary status as it can change with respect to time and place. Finally, if the cover is damaged or has become unfit for use, it can be discarded in preference to a newer, more suitable one[39]. The other aspect to remember is that when an object is covered fully, it can be mistaken to be the object itself. These ideas map directly to the relationship between consciousness and the body (both subtle and gross).

The Four States of Existence

Another framework to describe the consciousness and its multi-faceted manifestations is discussed widely in the Upaniṣads and pertains to four states that a human being undergoes. For example, the Māṇḍūkya-upaniṣad provides a description of the four states of an individual. These are waking state (Jāgrat-avasthā), dreaming state (Svapna-avasthā), deep sleep (Suṣupti-avasthā), and samādhi (Turīyā-avasthā). Western psychologists have not considered the fourth state and have a minimal and a different understanding of the third state. Moreover, analysis of the dream state by the Indian philosophers has yielded a rich understanding of the nature of the association between the mind and the consciousness and the psychological fallout of this on human behavior.

In the normal waking state, the gross body in conjunction with the subtle body is fully engaged with the 'world outside' and it transacts with the outside world. The consciousness is felt by its indirect influence in the matters. The experiences gained are dual, constantly oscillating between pain and pleasure in all its attendant variations. On the other hand, once the consciousness slips to the next level of dreaming state, the gross body becomes inactive. The mind and the antaḥkaraṇas continue to actively engage with the world outside, albeit in an altered framework. The ego 'manufactures' a world within and plays out events and participates in the play both as an active player and as a curious observer. Several rules of the waking world are simply violated in the dream world and the events could be illogical. For example, while sleeping in a bed in a corner room on the outskirts of Bengaluru, the mind will enact an event in the dream in which the same person takes a boat ride to see the beauty of Niagara falls in close quarters. Illogical events and unrelated people will be connected through a chain of events unfolding in the dream. The 'self' if required can create 1000 such 'selfs' and orchestrate a dream involving all of them, that we typically see in a movie. Whether the dream is real, or imaginary is not easy to answer. For example, it is a common experience that in the dream a dog may chase the dreaming person and after running for a long distance the person may sweat and gasp. Disturbed by this event, the dreaming person wakes up with sweat and gaps but there will be neither the dog nor the road on which the dreaming person was running. The dream experience provides a versatile laboratory to deeply analyse certain psychological aspects of human beings, which ancient Indians have extensively carried out.

- In the normal waking state, the gross body in conjunction with the subtle body is fully engaged with the 'world outside'.
- The dream experience provides a versatile laboratory to deeply analyse certain psychological aspects of human beings.

When the consciousness enters the deep sleep state, even the subtle body is completely switched off. In some sense, there is a complete loss of continuity as the gross and subtle bodies are dull and dead. Deep sleep fully charges an individual, relieves him/her of all mental and physical tiredness, and makes the person fresh and energetic. This experience in deep sleep is universal since every living being (not just human beings) will have the same effect after a deep sleep. The consciousness, representing the 'true self' is *alone* in this state and the person will experience bliss without the help of the sense organs but may not be aware of it. On waking up from this state, the 'consciousness' reconnects seamlessly with the gross and the subtle body and transfers this peaceful experience as 'past knowledge'. This is the reason we get up and recount the blissful experience by saying, "I had a very good sleep", in the past tense[40]. This peculiar state of losing continuity and regaining it also provides vital clues to understand the nature of true consciousness, which Indian sages and philosophers have analysed.

The fourth state of consciousness is 'turīyā', which is an enlightened form of consciousness. As per IP, this is the ultimate state that we all need to aspire for. In all other states, there is reflected consciousness but, in this state, an individual directly experiences the true consciousness. In this state of consciousness, the person is fully awake but is

- The fourth state of consciousness is "turīyā', which is an enlightened form of consciousness.
- When the mind withdraws itself from all the sensory objects and empties its content, it will be in an illumined state.

in a state very different from the waking state, where the existence of the body, mind and the intellect are not cognised by the individual and the vibrations (a multitude of thought waves) of the mind cease. On account of this, there is a pure, thoughtless state, where the limitations of an individual are totally absent. In the Indian system, this state is variously described as 'samādhi' and 'nirvāṇa'. With a thought-free mind, one is supposed to realise one's true nature and immense potential. When the mind withdraws itself from all the sensory objects and empties its content, it will be in an illumined state. Yoga and meditation are tools to train and coax the mind into this. Therefore, prolonged and regular use of meditation practices helps an individual enter into this state and stay there for a while. Figure 13.8 summarises these discussions schematically.

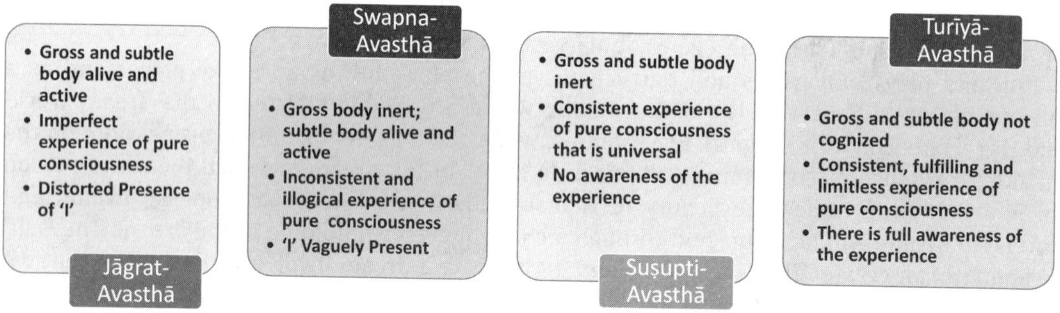

FIGURE 13.8 The Four States of 'Consciousness'

13.9 CONSCIOUSNESS STUDIES – ISSUES

In IP, the body–mind–intellect (antaḥkaraṇas) is considered as the interfacing instrument, connecting the consciousness with the physical body. The reflected consciousness mediates through these internal organs and enables the mind to have knowledge and awareness. For an untrained mind, the consciousness is conspicuous by its absence and is assumed to be fully manifested through the mind. However, in reality, there are significant differences between the mind and the consciousness. Therefore, the manifest and unmanifest components of our existence interact in shaping our response to the unfolding events in our life. In the tradition of IP, the goal of every individual is to explore the inner states of a person finally experiencing the true consciousness. Since the study is about inner states, the method and means of attaining this ought to be different from the normal ones used to deal with the outside world. The methods of self-observation and experimentation required have certain distinctive features:

- There will be a shift in the focus from objects (Viṣaya) to the subject (Viṣayin). When there is a shift in the focus, the means of obtaining the knowledge will be only through experience. The existing means of knowledge and study will be of limited use. The valid means of knowledge (Pramāṇa) available for dealing with the outside world such as *Pratyakṣa* (Perception), *Anumāṇa* (inference), *Upamāṇa* (comparison and analogy) are less useful when it comes to studying the inner nature of an individual.

IKS IN ACTION 13.3

The Five Layered Consciousness of an Individual

Recognising the consciousness inside each one of us is a difficult task and even more difficult is to communicate this idea in a classroom setting. During a lecture on the subject, a video was shown to the students. After watching the video the students participated in a guided discussion, which brought out the essence of the subject.

The video is a self-narrated true story of a person, Mr. Tala Skeels Piggins, who is a professional motorbike racer. Earlier, he was a young and a handsome person undergoing his Navy service in Australia. After he left the navy, he came to the UK in search of a job. He was employed as a Physical Education teacher in a school. Barely after 6 months in the job, he was knocked off a motorcycle that he was riding by a speeding car. He fell on the other side of the road and was runover by a speeding car in the opposite direction, breaking T4 and T5 bones on his back. This left him paralysed chest down. He was told that he had a 30 percent chance of survival and would never walk again if he lived.

As a person who enjoyed and valued physical activity, this was a life changing event. His initial response was one of frustration and a desire to end his life as he saw no possibilities for his future. However, he underwent a dramatic transformation.

Describing what was responsible for this transformation in his own words, "... there is a mechanism within every person that can overcome any trauma that one faces. There is an incredible reserve of strength and resilience inside every person..."

Inspired by this thought, he chose to get back to riding a motorbike and to become a professional racer. After making several modifications to a racing motorbike by adopting newer technologies he entered into professional racing. He continues to strap himself to a motorbike and take on able bodied competitors at one of the most dangerous sports on the planet.

After showing the video, the students were asked to describe Mr. Piggins. The students shared several phrases to describe him. These are summarised in the table below.

The above example demonstrates how we can describe a person in five different ways. In each of these, there is a dominant manifestation of one aspect of the consciousness. The pañca-koṣa concept brings out this aspect of an individual.

Source: Adopted from https://www.youtube.com/watch?v=5VKFIOMrey4

Physical Layer (Annamaya Koṣa)	Vital Energy (Prāṇamaya Koṣa)	Mental Layer (Manomaya Koṣa)	Intellectual Layer (Vijñānamaya Koṣa)	Embodiment of Happiness (Ānandamaya Koṣa)
♦ He is a Man, bike racer ♦ 5 feet 7 inches well-built ♦ A person of fair complexion ♦ A person who is paralysed below his chest in an accident ♦ ...	♦ A person with several months of hard work to go for bike racing ♦ A great fighter against all odds ♦ Energeitic and action oriented ♦ ...	♦ A person with positive attitude ♦ A person of grit and determination ♦ Burning desire to outgrow the limitations arising out of the accident ♦ ...	♦ A great thinker ♦ Constantly thinking to accomplish goals ♦ Intelligent person ♦ Goal oriented task planner ♦ Strategist ♦ ...	♦ A person basking in satisfaction and fulfilment ♦ Happiness and joy of regaining life ♦ Bundle of ecstasy (of winning races) ♦ ...

In the IP tradition, *Śabda* (word, a testimony of past or present reliable experts) is considered as an important means of valid knowledge. Also, valid means of study for the inner world includes śravaṇa (deep listening - sensory understanding), manana (deep internalisation of the ideas - intellectual understanding), and nididhyāsana (deep intuition - experiential assimilation).

- In a study of the inner nature, the 'seer' and the 'seen' are not separate, they seamlessly merge into a single entity. The subject matter of study is not something else, but one's own nature. Also, the object of study is not material but a subtle noncorporeal entity.
- On account of this, there is a distinction made between 'parā-vidyā' and 'aparā-vidya'. Aparā-vidyā constitutes the knowledge of the socio-physical world, known as the mundane and the empirical world that we live in. All science, technology, and management knowledge that we have developed belong to this category. On the other hand, parā-vidyā is about the issues pertaining to our 'inner world'. It will address issues that crop up in our 'inner world' of mental, psychological, and existential nature.
- The normal sensory organs are useful for dealing only with the outside world. They will be of very little or no use in dealing with the internal world. An easy way to understand this is in a large room where 100 people have gathered, if there is sudden power outage nobody needs a torch to know if he/she is in the room. It is a matter of consciousness. On the other hand, we need to use our sensory organs and some artificial source of light to find out if our friend is in the room.
- The study of the inner state will invariably lead to a study of the spiritual nature of an individual. In other words, it is a study of human beings in-depth. In such a study, the questions of what, why, and how become less important. In other words, if Newton's question had been "who made the apple fall?" instead of "why or how did the apple fall from the tree?", it would have led him into the deep nature of human beings and their relationship to other entities in the Universe.

While the mind is considered to be a subtle form of matter, consciousness is non-corporeal. This primarily drives the key differences between the study of the consciousness and the mind in the manner we saw above. On account of these, psychology studies typically involve methods of direct observation (intuition) relying on the blending of the first person (self or the student) and the second person (Guru). Moreover, Indian scriptures highlight the importance of continuous practice (abhāysa) and cultivating a sense of mental dispassion towards the objects and entities in the outside world (vairāgya) to develop a high degree of mental concentration[41]. While at meditation, the mind is able to go beyond mere rational processing of the sensory inputs aided by the brain to accessing consciousness directly. Methods of yoga and mediation help an individual test, experiment, and internally validate the higher mental states. At the core, these have developed practical techniques and rigorous methods of inquiry into the domains of consciousness. In the process, they help an individual discover answers to several existential questions such as the relationship between an individual, the Universe, and the Lord and others pertaining to values, truth, and love.

SUMMARY

- A person who wants to lead a healthy life must show a keen desire to follow the ideas provided in Āyurveda. according to Caraka, Āyurveda must be studied and practiced by every individual in the society.
- Āyurveda borrows the philosophical aspects from two major schools of thought: Vaiśeṣika and Sāṃkhya.
- The treatment framework in Āyurveda consists of three generic steps: Cause (hetu), Symptoms (liṅga), and Remedy (auṣadha).
- When the three doṣas, Vāta, Pitta, and Kapha are in an equilibrium state the body is said to be in a healthy condition. By contrast, when the doṣas are vitiated it causes disease.
- The doṣas have a specific relationship with the time of the day and the type of diet one takes. Moreover, Āyurvedic scholars have also identified the organs where these doṣas predominantly exist.
- Agni plays a vital role in the digestion process and it is also responsible for strength, health, longevity, and vital breath. Agni manifests through the Pitta in the body.
- The food that we consume first gets converted into plasma and in a gradual process, it gets converted into flesh and fat. From the fat to bones, bones to bone-marrow, bone-marrow to semen.
- To keep the internal organs including the mind and intellect in a balanced and natural state one must follow daily routines and seasonal routines prescribed in Āyurveda.
- According to Āyurveda, the quantity of water that one must take daily will depend on the constitution (prakṛti) of a being. An individual must boil and reduce the water before consumption according to the doṣa predominance in the body.
- The principles and practice of Yoga provide a comprehensive framework by which the psychological and physical aspects of stress can be understood and eliminated.
- Several mediation practices are taught as part of the training for the yoga way of living. It will also induce a change in the intellectual and somatic functions of an individual.
- The Indian approach to psychology has long years of introspection, testing, contemplation, and experiential learning. It draws its frameworks, concepts, tools, and techniques from a rich repository of diverse knowledge traditions starting from the Upaniṣads.
- The Tri-guṇa system provides an overarching framework to understand the physical infrastructure of a human being.
- The gross body is the physical manifestation of an individual (in the form of the visible organs) which are made of the five bhūtas. The psychological part of an individual constitutes the subtle body.
- Understanding the manifestations of consciousness is only possible by self-experimentation and deep contemplation of one's experiences.
- A framework to describe the consciousness and its multi-faceted manifestations is discussed widely in the Upaniṣads and pertains to four states that a human being undergoes.
- The body–mind–intellect (antaḥkaraṇas) is considered as the interfacing instrument, connecting the consciousness with the physical body. The reflected consciousness mediates through these internal organs and enables the mind to have knowledge and awareness.
- Indian scriptures highlight the importance of continuous practice (abhyāsa) and cultivating a sense of mental dispassion towards the objects and entities in the outside world (vairāgya) to develop a high degree of mental concentration.

REVIEW QUESTIONS

1. What do you understand by the term 'health' and 'wellness'. Do they mean the same?
2. What do you mean by the term 'Āyurveda'?

3. How does Āyurveda define health? Is it any different from the western definition?
4. Comment on the statement, "Health is both a physical and psychological" issue.
5. Write a note on 'tri-doṣas'. In what way the tri-doṣa influence the health of an individual.
6. What are the prescriptions of Āyurveda to maintain the psychological faculty in a pleasant state?
7. What are the main recommendations of Āyurveda with respect to sleep, food, and drinking water? How will we benefit from following these?
8. Prepare a two page note on the salient aspects of disease management as prescribed in Āyurveda texts.
9. What is the 'yoga way of living'? What benefits accrue to an individual following yoga practice?
10. Comment on the statement, "Rāja-yoga can help an individual maintain a healthy and stress-free life".
11. Does meditation have anything to do with consciousness?
12. How does Indian Psychology differ from the western counterpart?
13. What are the basic tenets of Indian psychology?
14. Enumerate the following as discussed in Indian psychology:
    (a) Tri-guṇa framework
    (b) Four states of consciousness
    (c) Five layers of an individual
    (d) The subtle and Gross body
    (e) Antaḥkaraṇas
    (f) Organs of action and knowledge
15. How does consciousness influence the psychological processes of an individual?
16. What are the challenges of studying human consciousness? Can we use the normal means of knowledge in consciousness studies?

DISCOVER IKS

1. Dr. B M Hegde is known for comparing the Indian and the Western approach to health and wellness. Watch the following video entitled, *Āyurveda over Western Medicine*: https://www.youtube.com/watch?v=HzTvEK1sVi0. After watching the video prepare a two-page note to answer the following questions:
    (a) What is the view of modern science about the human body?
    (b) What do you mean by the Western reductionist approach to medicine?
    (c) How does Āyurveda link the body–mind, and consciousness to health?
2. Watch the following video of Dr. Rammohan, *Āyurveda and Indian Psychology* where he brings out the intricate relationship between Psychology and Āyurveda: https://www.youtube.com/watch?v=DivT7icPvyA. Prepare a note to answer the following questions:
    (a) How does Āyurveda treat the mind and the body?
    (b) How does Āyurveda use the idea of mind and the body in developing treatment?
    (c) What is the relationship between mind and consciousness?
3. Sri Sri Ravishankar has a very large global following. He emphasises the need for a "Yoga way of Life" in this talk he delivered in the European Parliament: https://www.youtube.com/watch?v=S7socwFbYRk. After watching the video, prepare a two-page note to answer the following questions;
    (a) Why do we need yoga, according to Sri Sri Ravishankar?
    (b) From a practical point of view of conducting several day-to-day activities how does yoga help an individual?
    (c) Is Yoga a religious prescription? Why or Why not?

SUGGESTED READINGS

Adhia, H., Nagendra, H.R. and Mahadevan, B. (2010). "Impact of Adoption of Yoga Way of Life on the Reduction of Job Burnout of Managers", *Vikalpa.* **35**(2), pp. 21–33.

Adhia, H., Nagendra, H.R. and Mahadevan, B. (2010). "Impact of Adoption of Yoga Way of Life on the Emotional Intelligence of Managers". *IIMB Management Review.* **22**, pp. 32–41.

Dalal, A.K. and Misra, G. (2010). "The Core and Context of Indian Psychology", *Psychology and Developing Societies,* **22**(1), pp. 121–155.

Dominik, W. (2001). "The Roots of Ayurveda", Penguin Classics, Haryana, India. ISBN: 9780140436808.

Gangadharan, G.G. (2009). "Kṣemakutūhalam composed by Kśemaśarmā", Foundation of Revitalisation of Local Health Traditions (FRLHT), Bengaluru.

Godbole, A., Abhinav, S., Singh, O.P. and Tripathi, J.S. (2017). "Conceptual Study of Agni in Ayurveda", *Pharma Science Monitor: An International Journal of Pharmaceutical Sciences*, pp. 15–20.

Guruprasad, S. (2017). *Sankhya Karika*, Chaukhambha Vidyabharati, Varanasi. ISBN: 978-93-83721-66-5.

Kuppuswami, B. (1985). *Elements of Ancient Indian Psychology*, Vikas Publishing House, New Delhi.

Mishra, L.C. (2004). *Scientific Basis for Ayurvedic Therapies*, CRC Press LLC, Washington D.C.

Pandey, J. (1988). *Psychology in India: The State-of-the-art*, Vols. 1, 2, 3, Sage Publications, New Delhi.

Rao, K.R. (2014). "Positive Psychology and Indian Psychology in Need of Mutual Reinforcement", *Psychological Studies,* **59**(2), pp. 94–102.

Rao, K.R. (2012). "Psychology in the Indian Tradition". *Psychological Studies*, **57**, pp. 323–330.

Rao, K.R., Paranjpe, A.C., and Dalal, A.K. (2008). *Handbook of Indian Psychology*, Cambridge University Press, New Delhi.

Telles, S., Joshi, M., Dash, M., Raghuraj, P., Naveen, K. and Nagendra, H.R. (2004). "An Evaluation of the Ability to Voluntarily Reduce the Heart Rate after a Month of Yoga Practice," *Integrative Physiological and Behavioral Science,* **39**(2), pp. 119–25.

Thakker, J., Chaudhari, S. and Sarkar, P.S. (2011). "Ritucharya: Answers to the Lifestyle Disorders", *An International Quarterly Journal of Research in Ayurveda*, Official Publication of Institute for Post Graduate Teaching & Research in Ayurveda, Jamnagar, Oct-Dec; **32**(4): pp. 466–471. doi: 10.4103/0974-8520.96117. https://www.ncbi.nlm.nih.gov/pmc/articles/PMC3361919/#ref5

Tiwari, P.V. (2018). *Caraka-saṃhitā*, Chaukhambha Vishvabharati, Varanasi.

Valiathan, M.S. (2015). *The Legacy of Caraka*, University Press (India) Private Limited, Hyderabad. ISBN: 9788173716676

Valiathan, M.S. (2015). *The Legacy of Susruta*, University Press (India) Private Limited, Hyderabad.

Valiathan, M.S. (2015). *The Legacy of Vagbhata*, University Press (India) Private Limited, Hyderabad.

Vidyanath, R. (2013). *Illustrated Aṣṭāṅga Hṛdaya of Vāgbhaṭa – Sūtrasthāna*, Chaukhambha Surabharati Prakashan, Varanasi.

ENDNOTES

1. **Pandit Taradatt Pant Shastri (2010).** "Carakasaṃhitā of Agniveśa", Chowkhamba Sanskrit Series Office, Varanasi. धर्मार्थकाममोक्षाणाम् आरोग्यं मूलमुत्तमम् । रोगास्तस्यापहर्तारः श्रेयसो जीवितस्य च ॥ dharmārtha-kāmamokṣāṇām ārogyaṃ mūlam-uttamam | rogāstasyāpahartāraḥ śreyaso jīvitasya ca | Sūtrasthānam Chapter 1, Verse 15, p. 2.
2. **Shukla, V. and Tripathi, R.D. (2017).** Carakasaṃhitā of Agniveśa", Chaukhamba Sanskrit Pratishthan, Delhi. जनपदोद्ध्वंसनमेकेनैव व्याधिना युगपदसमानप्रकृत्याहारदेहबलसात्म्यसत्त्ववयसां मनुष्याणां कस्माद् भवतीति। janapadoddhvaṃsanam-ekenaiva vyādhinā yugapad-asamāna-prakṛty-āhāra-deha-bala-sātmya-sattvavayasāṃ manuṣyāṇāṃ kasmād-bhavatīti | Vimānasthāna Ch. 3, Prose 5, p. 568.
3. *Ibid.* शरीरेन्द्रियसत्त्वात्मसंयोगो धारि जीवितम् । नित्यगश्चानुबन्धश्च पर्यायैरायुरुच्यते ॥ śarīrendriyasattvātmasaṃyogo dhāri jīvitam | nityagaścānubandhaśca paryāyairāyurucyate || Sūtrasthānam Ch. 1, Verse 42, p. 13.
4. *Ibid.* अनुग्रहार्थं प्राणिनां ब्राह्मणैः । आरक्षार्थं राजन्यैः । वृत्त्यर्थं वैश्यैः । सामान्यतः धर्मार्थकामपरिग्रहार्थं सर्वैः । anugrahārthaṃ prāṇināṃ brāhmaṇaiḥ | ārakṣārthaṃ rājanyaiḥ | vṛttyarthaṃ vaiśyaiḥ | sāmānyataḥ dharmārthakāmaparigrahārthaṃ sarvaiḥ | Sūtrasthānam Chapter 30, Prose 29, p. 450.
5. *Ibid.* महर्षयस्ते ददृशुर्यथावज्ज्ञानचक्षुषा । सामान्यं च विशेषं च गुणान् द्रव्याणि कर्म च । समवायं च तज्ज्ञात्वा तन्त्रोक्तं विधिमास्थिताः । लेभिरे परमं शर्म जीवितं चाप्यनित्वरम् ॥ maharṣayaste dadṛśuryathāvajjñānacakṣuṣā | sāmānyaṃ ca viśeṣaṃ ca guṇān dravyāṇi karma ca | samavāyaṃ ca tajjñātvā tantroktaṃ vidhimāsthitāḥ | lebhire paramaṃ śarma jīvitaṃ cāpyanitvaram || Sūtrasthānam, Chapter 1, Verse 28, p. 8.
6. *Ibid,* खादीन्यात्मा मनः कालो दिशश्च द्रव्यसङ्ग्रहः । सेन्द्रियं चेतनं द्रव्यं निरिन्द्रियमचेतनम् । khādīnyātmā manaḥ kālo diśaśca dravyasaṅgrahaḥ | sendriyaṃ cetanaṃ dravyaṃ nirindriyamacetanam | Sūtrasthānam, Chapter 1, Verse 48, p. 18.
7. *Ibid.* आयुर्वर्णो बलं स्वास्थ्यमुत्साहोपचयौ प्रभा । ओजस्तेजोऽग्नयः प्राणाश्चोक्ता देहाग्निहेतुकाः । āyurvarṇo balaṃ svāsthyamutsāhopacayau prabhā | ojastejo'gnayaḥ prāṇāścoktā dehāgnihetukāḥ | Cikitsāsthānam, Chapter 15, Verse 3, p. 358.
8. *Ibid.* अग्निरेव शरीरे पित्तान्तर्गतः । agnireva śarīre pittāntargataḥ | Sūtrasthānam, Chapter 12, Para 11, p. 188.
9. **Gangadharan, G.G. (2009).** "Kṣemakutūhalam Composed by Kśemaśarmā", Foundation of Revitalisation of Local Health Traditions (FRLHT), Bengaluru. आहारं पचति शिखी दोषानाहारवर्जितः पचति । दोषक्षयेऽपि धातून् पचति च धातुक्षये प्राणान् ॥ āhāraṃ pacati śikhī doṣānāhāravarjitaḥ pacati | doṣakṣaye'pi dhātūn pacati ca dhātukṣaye prāṇān || Chapter 3, Verse 29, p. 50.
10. **Kevalakrishna, T. (2016).** "सुश्रुत विरचित सुश्रुत संहिता", चौखम्भा ओरियण्टालिया, वाराणसी। रसाद्रक्तं ततो मांसं मांसान्मेदः प्रजायते । मेदसोऽस्थि ततो मज्जा मज्जायाः शुक्रसम्भवः । rasādraktaṃ tato māṃsaṃ māṃsānmedaḥ prajāyate | medaso'sthi tato majjā majjāyāḥ śukrasambhavaḥ | Sūtrasthānam Chapter 14, Verse 10, p. 144.
11. चञ्चलं हि मनः कृष्ण प्रमाथि बलवद्दृढम् । तस्याहं निग्रहं मन्ये वायोरिव सुदुष्करम् ॥ cañcalaṃ hi manaḥ kṛṣṇa pramāthi balavaddṛḍham | tasyāhaṃ nigrahaṃ manye vāyoriva suduṣkaram || Bhagavadgīta, 6.34. **Swami Chinmayananda (2002).** "The Holy Geeta", Central Chinmaya Mission Trust, Mumbai, pp. 432–434.
12. **Shukla, V. and Tripathi, R.D. (2017).** "Carakasaṃhitā of Agniveśa", Chaukhamba Sanskrit Pratishthan, Delhi. मानसः पुनरुद्दिष्टो रजश्च तम एव च । mānasaḥ punaruddiṣṭo rajaśca tama eva ca | Sūtrasathāna Chapter 1, Verse 57, p. 30.
13. *Ibid.* धीधृतिस्मृतिविभ्रष्टः कर्म यत् कुरुतेऽशुभम् । प्रज्ञापराधं तं विद्यात् सर्वदोषप्रकोपणम् ॥ dhīdhṛtismṛtivibhraṣṭaḥ karma yat kurute'śubham | prajñāparādhaṃ taṃ vidyāt sarvadoṣaprakopaṇam || Sūtrasathāna Chapter 1, Verse 102, p. 693.
14. कालबुद्धीन्द्रियार्थानां योगो मिथ्या न चाति च । द्वयाश्रयाणां व्याधीनां त्रिविधो हेतुसङ्ग्रहः ॥ kālabuddhīndriyārthānāṃ yogo mithyā na cāti ca | dvayāśrayāṇāṃ vyādhīnāṃ trividho hetusaṅgrahaḥ || चरक – १/५४

15. वायुः पित्तं कफश्चेति त्रयो दोषाः समासतः । विकृताऽविकृतं देहं घ्नन्ति ते वर्तयन्ति च ॥ vāyuḥ pittaṃ kaphaśceti trayo doṣāḥ samāsataḥ | vikṛtā'vikṛtaṃ dehaṃ ghnanti te vartayanti ca || (Aṣṭāṅga-hṛdayam – sūtrasthānam 1/6-7)

16. तत्र रूक्षो लघुः शीतः खरः सूक्ष्मश्चलोऽनिलः । पित्तं सस्नेहतीक्ष्णोष्णं लघु विस्रं सरं द्रवम् । स्निग्धः शीतो गुरुर्मन्दः श्लक्ष्णो मृत्स्नः स्थिरः कफः ॥ tatra rūkṣo laghuḥ śītaḥ kharaḥ sūkṣmaścalo'nilaḥ | pittaṃ sasnehatīkṣṇoṣṇaṃ laghu visraṃ saraṃ dravam | snigdhaḥ śīto gururmandaḥ ślakṣṇo mṛtsnaḥ sthiraḥ kaphaḥ || (Aṣṭāṅga-hṛdayam – sūtrasthānam – 1. 11–12)

17. **Shukla, V. and Tripathi, R.D. (2017).** "Carakasaṃhitā of Agniveśa", Chaukhamba Sanskrit Pratishthan, Delhi. लोभशोकभयक्रोधमानवेगान् विधारयेत् । नैर्लज्ज्येर्ष्यातिरागाणामभिध्यायाश्च बुद्धिमान् ॥ परुषस्यातिमात्रस्य सूचकस्यानृतस्य च । वाक्यस्याकालयुक्तस्य धारयेद्वेगमुत्थितम् ॥ देहप्रवृत्तिर्या काचिद्विद्यते परपीडया । स्त्रीभोगस्तेयहिंसाद्यास्तस्या वेगान् विधारयेत् ॥ lobhaśokabhayakrodhamānavegān vidhārayet | nairlajjyerṣyāti rāgāṇāmabhidhyāyāśca buddhimān || paruṣasyātimātrasya sūcakasyānṛtasya ca | vākyasyākālayuktasya dhārayedvegamutthitam || dehapravṛttiryā kācidvidyate parapīḍayā | strībhogasteyahiṃsādyāstasyā vegān vidhārayet || Sūtrasathāna Chapter 7, Verse 27–29, pp. 125–126.

18. This section is mainly based on **Singh, R.H. (1997).** "Medical Techniques", Chapter 4 [4.5] in *History of Technology in India,* Vol. I, A.K. Bag (Ed.), Indian National Science Academy, New Delhi, pp. 344–356 and **Mishra, L.C. (2004).** "Scientific basis for Ayurvedic therapies", CRC Press LLC, Washington D.C.

19. **Tiwari, P.V. (2018).** "Caraka Saṃhitā" Chaukhambha Vishvabharati, Varanasi. त्रय उपस्तम्भा इति- आहारः, स्वप्नो, ब्रह्मचर्यमिति; एभिस्त्रिभिर्युक्तियुक्तैरुपस्तब्धमुपस्तम्भैः शरीरं बलवर्णोपचयोपचितमनुवर्तते यावदायुःसंस्कारात् संस्कारमहितमनुपसेवमानस्य, य इहैवोपदेक्ष्यते । traya upastambhā iti- āhāraḥ, svapno, brahmacaryamiti; ebhistribhiryuktiyuktairupastabdhamupastambhaiḥ śarīraṃ balavarṇopacayopacitamanuvartat e yāvadāyuḥsaṃskārāt saṃskāramahitamanupasevamānasya, ya ihaivopadekṣyate | Sūtrasthāna, Chapter 11, Verse 35, p. 207.

20. **Kevalakrishna, T. (2017).** "Suśruta Saṃhitā by Suśruta", Choukhambha Orientalia, Varanasi. हृदयं चेतनास्थानमुक्तं सुश्रुत देहिनाम् । तमोभिभूते तस्मिंस्तु निद्रा विशति देहिनम् ॥ hṛdayaṃ cetanāsthānamuktaṃ suśruta dehinām | tamobhibhūte tasmiṃstu nidrā viśati dehinam | Śarīrasthāna, Chapter 4, Verse 34, p. 60.

21. https://sa.wikisource.org/wiki/अष्टाङ्गसंग्रहः_सूत्रस्थानम्_अध्याय_१-१०#नवमोऽध्यायः. Last accessed on Oct. 1, 2021. बाहुल्यात्तमसो रात्रौ निद्रा प्रायेण जायते॥ श्लेष्मावृतेषु स्रोतस्सु श्रमादुपरतेषु च। इन्द्रियेषु स्वकर्मभ्यो निद्रा विशति देहिनम् ॥ bāhulyāttamaso rātrau nidrā prāyeṇa jāyate|| śleṣmāvṛteṣu srotassu śramāduparateṣu ca| indriyeṣu svakarmabhyo nidrā viśati dehinam||

22. **Vidyanath, R. (2013).** "Illustrated Aṣṭāṅga Hṛdaya of Vāgbhaṭa – Sūtrasthāna" Chaukhambha Surabharati Prakashan, Varanasi. ग्रीष्मे वायुचयादानरौक्ष्यरात्र्यल्पभावतः । दिवास्वप्नो हितोऽन्यस्मिन् कफपित्तकरो हि सः ॥ grīṣme vāyucayādānaraukṣyarātryalpabhāvataḥ | divāsvapno hito'nyasmin kaphapittakaro hi saḥ || Chapter 7, Verse 56–57, p. 149.

23. *Ibid.* दिवास्वप्नो हितोऽन्यस्मिन् कफपित्तकरो हि सः। मुक्त्वा तु भाष्ययानाध्वमद्यस्त्रीभारकर्मभिः॥ क्रोधशोकभयैः क्लान्तान् श्वासहिध्मातिसारिणः । वृद्धबालाबलक्षीणक्षततृट्शूलपीडितान् ॥ अजीर्ण्यभिहतोन्मत्तान् दिवास्वप्नोचितानपि । धातुसाम्यं तथा ह्येषां श्लेष्मा चाङ्गानि पुष्यति ॥ divāsvapno hito'nyasmin kaphapittakaro hi saḥ| muktvā tu bhāṣyayā nādhvamadyastrībhārakarmabhiḥ|| krodhaśokabhayaiḥ klāntān śvāsahidhmātisāriṇaḥ | vṛddhabālāba lakṣīṇakṣatatṛṭśūlapīḍitān || ajīrṇyabhihatonmattān divāsvapnocitānapi | dhātusāmyaṃ tathā hyeṣāṃ śleṣmā cāṅgāni puṣyati || Chapter 7, Verse 58–59, pp. 149.

24. *Ibid.* निद्रानाशादङ्गमर्दशिरोगौरवजृम्भिकाः । जाड्यग्लानिभ्रमापक्तितन्द्रा रोगाश्च वातजाः ॥ nidrānāśādaṅgamarda śirogauravajṛmbhikāḥ | jāḍyaglānibhramāpaktitandrā rogāśca vātajāḥ || Chapter 7, Verse 64, p. 150.

25. **Tiwari, P.V. (2018).** "Caraka Saṃhitā" Chaukhambha Vishvabharati, Varanasi. अभ्यङ्गोत्सादनं स्नानं ग्राम्यानूपौदका रसाः । शाल्यन्नं सदधि क्षीरं स्नेहो मद्यं मनःसुखम् ॥ मनसोऽनुगुणा गन्धाः शब्दाः संवाहनानि च । चक्षुषोस्तर्पणं लेपः शिरसो वदनस्य च ॥ स्वास्तीर्णं शयनं वेश्म सुखं कालस्तथोचितः । आनयन्त्यचिरान्निद्रां प्रनष्टा या निमित्ततः ॥ abhyaṅgotsādanaṃ snānaṃ grāmyānūpaudakā rasāḥ | śālyannaṃ sadadhi kṣīraṃ sneho madyaṃ manaḥsukham || manaso'nuguṇā gandhāḥ śabdāḥ saṃvāhanāni ca | cakṣuṣostarpaṇaṃ lepaḥ

śiraso vadanasya ca || svāstīrṇaṃ śayanaṃ veśma sukhaṃ kālastathocitaḥ | ānayantyacirānnidrāṃ pranaṣṭā yā nimittataḥ ||Sūtrasthāna, Chapter 21, Verse 52–54, p. 361.

26. **Vidyanath, R. (2013).** "Illustrated Aṣṭāṅga Hṛdaya of Vāgbhaṭa - Sūtrasthāna" Chaukhambha Surabharati Prakashan, Varanasi. योजयेदतिनिद्रायां तीक्ष्णं प्रच्छर्दनाञ्जनम् । नावनं लङ्घनं चिन्तां व्यवायं शोकभीक्रुधः । एभिरेव च निद्रायाः नाशः श्लेष्मातिसंक्षयात् ॥ yojayedatinidrāyāṃ tīkṣṇaṃ pracchardanāñjanam | nāvanaṃ laṅghanaṃ cintāṃ vyavāyaṃ śokabhīkrudhaḥ | ebhireva ca nidrāyāḥ nāśaḥ śleṣmātisaṃkṣayāt || Chapter 7, Verse 62, 63, p. 150.

27. Kṣemakutūhalaṃ summarises the composition of the food that we must take in the following verse:
मधुर-मधुरमादौ मध्यतोऽम्लैकभावः कटुक-कटुकमथान्ते तिक्त-तिक्तं तथैव ।
यदि सुखपरिणामं वाञ्छसि त्वं हि राजन् त्यज-खलजन-सङ्गं भोजनं मा कदाचित् ॥
Madhura-madhauram-ādau madhyato'mlaika-bhāvaḥ kaṭuka-kaṭukam-athānte tikta-tiktaṃ tathaiva |
Yadi sukha-pariṇāmaṃ vāñchasi tvaṃ hi rājan tyaja-khalajana-saṅgaṃ bhojanaṃ mā kadācit ||

The verse advises, "O king if you desire happiness, discard the wicked man's company for it is sweet in the beginning, sour in between and bitter as well as pungent in the end, but never discard your food if this were its nature". See for details, **Gangadharan, G.G. (2009).** "Kṣemakutūhalam composed by Kśemaśarmā", Foundation of Revitalisation of Local Health Traditions (FRLHT), Bengaluru.

28. *Ibid.* According to Kshemendra, those who eat in excess, like beasts, without control, suffer from indigestion which is the root of all disorders. अनात्मवन्तः पशुवद्भुञ्जते येऽप्रमाणतः । रोगानीकस्य ते मूलमजीर्णं प्राप्नुवन्ति हि ॥ anātmavantaḥ paśuvad bhuñjate ye'pramāṇataḥ | rogānīkasya te mūlamajīrṇaṃ prāpnuvanti hi ||

29. **Tiwari, P.V. (2018).** "Caraka Saṃhitā" Chaukhambha Vishvabharati, Varanasi. यावद्ध्यस्याशनमशितमनुपहत्य प्रकृतिं यथाकालं जरां गच्छति तावदस्य मात्राप्रमाणं वेदितव्यं भवति । yāvaddhyasyāśanamaśitamanupahatya prakṛtaiṃ yathākālaṃ jarāṃ gacchati tāvadasya mātrāpramāṇaṃ veditavyaṃ bhavati | Sūtrasthāna, Chapter 5, Prose 4, p. 94.

30. **Gangadharan, G.G. (2009).** "Kṣemakutūhalam composed by Kśemaśarmā", Foundation of Revitalisation of Local Health Traditions (FRLHT), Bengaluru. मात्राशी सर्वकालं स्यात् मात्रा ह्यग्नेः प्रवर्तिका । मात्रा द्रव्याण्यपेक्षन्ते गुरूण्यपि लघून्यपि ॥ mātrāśī sarvakālaṃ syāt mātrā hyagneḥ pravartikā | mātrā dravyāṇyapekṣante gurūṇyapi laghūnyapi ||

31. Accordong to Bhagavadgītā, the food that is cooked three hours ago (one yāma), tasteless, putrid, rotten, refuse and impure, is the food liked by the Tamasic. यातयामं गतरसं पूति पर्युषितं च यत् । उच्छिष्टमपि चामेध्यं भोजनं तामसप्रियम् ॥ yātayāmaṃ gatarasaṃ pūti paryuṣitaṃ ca yat | ucchiṣṭamapi cāmedhyaṃ bhojanaṃ tāmasapriyam || 17.10. **Swami Chinmayananda (2002).** "The Holy Geeta", Central Chinmaya Mission Trust, Mumbai, pp. 1073–1074.

32. Thus, says Vedas – *apsvantaramamṛtam, apsu bheṣajam. Apsu somo'bravīt, antarviśvāni bheṣajā, āpaḥ pṛṇīta bheṣajam,* etc. http://www.vedakosh.com/rig-veda/mandal-1/sukta-023/mantra-rig-01-023-019. Last accessed on Oct. 1, 2021.

33. **Tiwari, P.V. (2018).** "Caraka Saṃhitā" Chaukhambha Vishvabharati, Varanasi. ईषत्कषायमधुरं सुसूक्ष्मं विशदं लघु । अरूक्षमनभिष्यन्दि सर्वं पानीयमुत्तमम् ॥ Īṣatkaṣāyamadhuraṃ susūkṣmaṃ viśadaṃ laghu | arūkṣamanabhiṣyandi sarvaṃ pānīyamuttamam || चरकसंहिता – सूत्रस्थानं – अन्नपानविध्यध्यायः, Sūtrasthāna 27/196-216.

34. **Vidyanath, R. (2013).** "Illustrated Aṣṭāṅga Hṛdaya of Vāgbhaṭa – Sūtrasthāna" Chaukhambha Surabharati Prakashan, Varanasi. शीतं मदात्ययग्लानिमूर्छाछर्दिश्रमभ्रमान् ।तृष्णोष्णदाहपित्तास्रविषाण्यम्बु नियच्छति ॥ śītaṃ madātyayaglānimūrchāchardiśramabhramān |tṛṣṇoṣṇadāhapittāsraviṣāṇyambu niyacchati || Sūtrasthāna, 5/15-16, p. 69.

35. *Ibid.* दीपनं पाचनं कण्ठ्यं लघूष्णं वस्तिशोधनम् । तृष्णोष्मदाहपित्तास्रहिध्माध्मानानिलश्लेष्मसद्यःशुद्धिनवज्वरे । कासामपीनसश्वासपार्श्वरुक्षु च शस्यते ॥ dīpanaṃ pācanaṃ kaṇṭhyaṃ laghūṣṇaṃ vastiśodhanam | tṛṣṇoṣm adāhapittāsrahidhmādhmānānilaśleṣmasadyaḥśuddhinavajvare | kāsāmapīnasaśvāsapārśvarukṣu ca śasyate || Sūtrasthāna, 5/16-17, p. 69.

36. Ayurvedic Formulary of India, 1st ed., Government of India, Ministry of Health and Family Planning, Department of Health, Government of India Press, Delhi, 1978.
37. For some details on the issues discussed here refer, **Adhia, H., Nagendra, H.R. and Mahadevan, B. (2010).** "Impact of Adoption of Yoga Way of Life on the Reduction of Job Burnout of Managers", *Vikalpa.* Vol. **35**(2), pp. 21–33, **Adhia, H., Nagendra, H.R. and Mahadevan, B. (2010).** "Impact of Adoption of Yoga Way of Life on the Emotional Intelligence of Managers". *IIMB Management Review.* Vol. **22**, pp. 32–41 and **Telles, S., Joshi, M., Dash, M., Raghuraj, P., Naveen, K. and Nagendra, H.R. (2004).** "An Evaluation of the Ability to Voluntarily Reduce the Heart Rate after a Month of Yoga Practice," *Integrative Physiological and Behavioral Science*, **39**(2), pp. 119–25.
38. See for more details, **Swami Chinmayananda (2013).** "Taittirīya Upaniṣad" Central Chinmaya Mission Trust, Mumbai. The opening box in Chapter 3 (IKS in Action 3.1) also introduces this aspect.
39. This idea is presented in Verse 22 of Chapter 2 in Bhagavad Gītā. वासांसि जीर्णानि यथा विहाय नवानि गृह्णाति नरोऽपराणि । तथा शरीराणि विहाय जीर्णान्यन्यानि संयाति नवानि देही ॥ vāsāṃsi jīrṇāni yathā vihāya navāni gṛhṇāti naro'parāṇi | tathā śarīrāṇi vihāya jīrṇānyanyāni saṃyāti navāni dehī || 2.22 **Swami Chinmayananda (2002).** "The Holy Geeta", Central Chinmaya Mission Trust, Mumbai, pp. 82–84.
40. This return to waking state is referred to as "प्रत्यभिज्ञा" (pratyabhijñā) and many Indian seekers have discussed this. See for example the verse from Dakṣiṇāmūrty stotram of Ādi Śankarācārya, राहुग्रस्तदिवाकरेन्दुसदृशो मायासमाच्छादनात् सन्मात्रः करणोपसंहरणतो योऽभूत्सुषुप्तः पुमान् । प्रागस्वाप्समिति प्रबोधसमये यः प्रत्यभिज्ञायते तस्मै श्रीगुरुमूर्तये नम इदं श्रीदक्षिणामूर्तये ॥ rāhugrastadivākarendusadṛśo māyāsamācchādanāt sanmātraḥ karaṇopasaṃharaṇato yo'bhūtsuṣuptaḥ pumān | prāgasvāpsamiti prabodhasamaye yaḥ pratyabhijñāyate tasmai śrīgurumūrtaye nama idaṃ śrīdakṣiṇāmūrtaye || 6 **Swami Chinmayananda (1994).** "Hymn to Sri Dakshinamoorthy", Central Chinmaya Mission Trust, Mumbai, pp. 59–66.
41. In Śrimad Bhagavad Gītā, in response to a question from Arjuna about control of the mind, Krishna brings out this aspect, through the following verse: असंशयं महाबाहो मनो दुर्निग्रहं चलम् । अभ्यासेन तु कौन्तेय वैराग्येण च गृह्यते ॥ asaṃśayaṃ mahābāho mano durnigrahaṃ calam | abhyāsena tu kaunteya vairāgyeṇa ca gṛhyate || 6.35. **Swami Chinmayananda (2002).** "The Holy Geeta", Central Chinmaya Mission Trust, Mumbai, pp. 435–438.

CHAPTER

14 Governance and Public Administration

LEARNING OUTCOMES

After finishing this chapter, you will be able to:

- Understand the relevance of Arthaśāstra in Public Administration
- Understand the elements of the Kauṭilyan State
- Get to know administrative set-up envisaged in Arthaśāstra
- Get a glimpse of public administration ideas presented in the epics

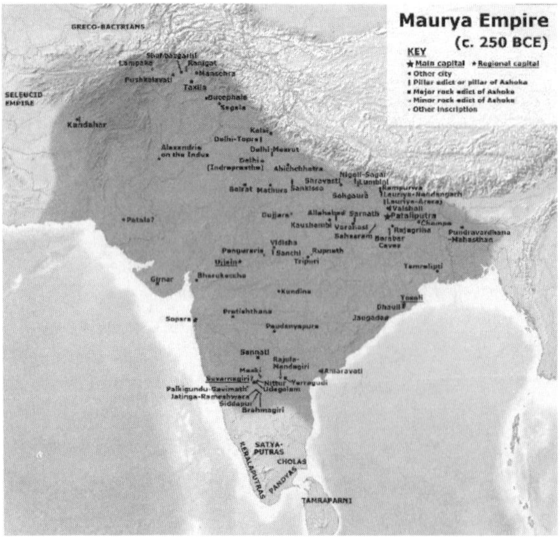

The **Mauryan** dynasty ruled much of the Indian subcontinent from around 321 BCE to 185 BCE. The map shows the vastness of the Mauryan empire. This corresponds to the domain of cakravartin which extends from Himalayas to the seas. Chandragupta, the founder of the dynasty was trained by Cāṇakya. The legend goes that Cāṇakya, also known as Kauṭilya who had come to Pātaliputra to display his knowledge was insulted by Dhana-Nanda, the king of Magadha. Kauṭilya vowed to destroy the Nanda dynasty and found an ideal candidate in Candragupta.

Author: Avantipura7, *Source:* https://commons.wikimedia.org/wiki/File:Maurya_Empire,_c.250_BCE_2.png

IKS IN ACTION 14.1

Rāmāyaṇa on Great Attributes, Dos, and Don'ts of a King

In a monarchy model of public administration and governance, everything revolves around the king. The prosperity of the kingdom, the safety and security of the citizens, and the vulnerability of the kingdom in the eyes of the neighbours are all critically dependent on how the king conducts himself. Therefore, the key attributes required of a king and the dos and don'ts of a king are important aspects of public administration. In the "kaccit sarga" of Rāmāyaṇa (book 2, Chapter 100), we find valuable information pertaining to these issues in verses 65 to 70.

In three verses (65–67) fourteen aspects related to administration have been flagged with a caution to the king to avoid these. This includes atheism, falsehood, anger, inattention, procrastination, evading the wise, indolence, gratification of all five senses, planning alone in the affairs of the kingdom, consultation with people who are proficient in worthless acts, failure to implement decisions, inability to keep the counsel secret, omission of auspicious practices and setting out against all the enemies at a time. These have the propensity to increase wasteful efforts on the part of the king leading to ineffective administration.

There is also a caution that it is not worthwhile to negotiate with 20 different types of monarchs. These monarchs suffer from some inherent limitations or uncontrolled senses leading them to wrong methods of administration. The twenty categories are as follows:

1. A king who is still a child
2. Aged
3. Ailing for some time
4. Ostracized by his own kith and kin
5. Cowardly or nervous
6. Surrounded by cowards
7. Greedy
8. Has greedy associates
9. Has estranged his ministers
10. Extremely voluptuous
11. Confers with fickle minded persons
12. Speaks ill of divine beings & brāhmaṇas
13. Ill-fated
14. Fatalist
15. Afflicted by famine
16. Afflicted by military reverses
17. Mostly remains away from home
18. Has numerous enemies
19. In the clutches of adverse times
20. Not devoted to truth and piety

Some of the attributes a king must possess are also mentioned in this discussion. For example, a list of things whose knowledge the king must fully internalize has been given. This includes:

- The seven limbs of a kingdom,
- The five kinds of fortifications and
- The three Vedas.

These issues have been discussed at some length in the chapter.

The king must also be aware of the eight types of evils born out of anger and ten types of evils born out of desire. Moreover, the king must be well versed with the three puruṣārthas. Furthermore, the king must make special efforts to subjugate the five senses. He must be aware of the six methods adopted for dealing with the neighbouring kingdoms (this issue is discussed in the chapter in some detail), and the adversities caused by divine and human agencies.

Many of the ideas presented here do not apply only to a monarch. The head of any institution such as the Prime Minister of the democratic model of governance, or the CEO of a multi-national corporation can find these guidelines useful. For example, the fourteen aspects related to administration listed here apply to different models of governance.

Artha is one of the three goals for an individual to pursue in his mundane life as per the Indian traditions. Individuals need 'artha' for satisfying all the materialistic requirements and lead a life bereft of livelihood challenges. The material well-being of every individual requires that the security of livelihood be taken care of. In order to ensure the security of livelihood, the acquisition and protection of the earth are important, as the source of livelihood is linked to the manner society is able to meaningfully and sustainably consume the material riches that the earth provides. Viewed in this manner, the meaning of the term 'artha' extends well beyond the goal pursued by individuals to one of the means of ensuring the well-being of society in general. Such a perspective leads us to larger questions on the subject:

- How does society organise itself in such a manner that the earthly resources are properly administered to meet the needs of the society and the individuals?
- How should we establish methods for controlling the consumption of earthly resources?
- What are appropriate governance mechanisms that we need in place to ensure this will be a sustainable process?

In the process of answering the above questions, there is a need to define a structure to the institution ruling a state and the powers and duties of the king. Furthermore, the responsibility of the citizens to contribute to the state exchequer which the king can use to provide security and welfare for the citizens and to protect the state from external aggression. Rāja-dharma is the branch of knowledge that addresses these aspects and provides detailed guidelines and instructions for governance and public administration. In the context of Rāja-dharma the king was viewed as the representative of God or even God himself by the subjects. However, it did not mean that the king had the ultimate authority and the power to act independently. He was required to be the custodian of Dharma and act in accordance with the principles of Dharma. This alone will provide a sustainable mechanism and ensure that the subjects enjoyed peace, security and happiness. Therefore, one finds extensive discussion in all Rāja-dharma texts the need for the king to uphold the Dharma through public administration. Table 14.1 briefly summarises the main texts on Rāja-dharma. While each of them merits greater attention, we draw mainly from Arthaśāstra to discuss various aspects of governance and public administration in this chapter.

In modern parlance, we use phrases such as public administration, governance, and government to denote a structured set of ideas and methodologies to address these issues. The controlling authority is vested in a governance structure that takes care of the citizen welfare by framing policies and ensuring that various constituents of the society abide by these rules. There are different methods of vesting authority. One is a monarchy system in which the king is the sole authority to administer the country. The authority is passed down to generations in a hereditary fashion. On the other hand, democracy is a different method in which the authority is vested in a group of persons elected by the people who administer for a limited time. The process of identifying the next group of persons to administer the country is again exercised by the people. Notwithstanding the type of governance structure in place, the core objectives behind the exercise of securing the earthly resources for sustainable consumption by the citizens remain unchanged.

- Individuals need 'artha' for satisfying all the materialistic requirements and lead a life bereft of livelihood challenges.
- Several earlier teachers including Bṛhaspati and Uśanas are often quoted in Arthaśāstra at several places.

TABLE 14.1 A List of Indian Texts on Rāja-dharma

No.	*Text Name*	*Brief Description of Issues Discussed*
1.	Mahābhārata	The Śānti parva, which presents an elaborate conversation between Yudhiṣṭira and Bhīṣma has extensively covered several aspects of Rāja-dharma. Anuśāsana parva also has details on Rāja-dharma.
2.	Manu-smṛti	Chapters 7 and 9 cover the duties of the king and other elements of Rāja-dharma such as legal and justice.
3.	Arthaśāstra of Kauṭilya	A comprehensive treatise on Rāja-dharma - a composition of earlier thoughts with new perspectives from Kauṭilya. Consists of 6500 verses organised into 15 books.
4.	Nītisāra of Kāmandaka	Inspired mainly by Arthaśāstra - consists of 20 chapters - deals with issues such as obligations of the ruler, governmental organisation, principles and policies of the government, interstate relationships, ethics of envoys and spies, and varieties of battles.
5.	Mānasollāsa or Abhilāṣitārtha-cintāmaṇi of Someśvara III	Encyclopaedic work covering many topics including Rāja-dharma, medicine, horticulture, perfumes, food, architecture and games. The first two books cover issues related to Rāja-dharma in about 40 chapters (about 1600 verses). The issues covered include the qualifications of a king and ministers, their duties and moral characteristics, governance and economic matters.
6.	Rājanīti-ratnākara of Caṇḍeśvara	Deals with Rāja-dharma under various topics organised into 16 chapters, issues discussed include the King and his ministers, fort, army, treasury, administration, appointment of new king, coronation, etc.
7.	Daṇḍanīti of Keśava Paṇḍita	Treatise on criminal justice, topics organised in 6 chapters include the power of the king to punish, nature of offences and punishments, procedure of criminal justice, thefts, adultery, rape, etc., Defamation, libel, slander, Hurt of various kinds; and assaults.

14.1 ARTHAŚĀSTRA – GOVERNANCE AND ADMINISTRATION

The foregoing discussion makes it possible for Arthaśāstra to be regarded as the śāstra concerned with general well-being on earth. Since state activity alone can make such general well-being possible, the protection of the earth and its acquisition, which are an essential part of state activity, are declared to be the province of this śāstra. Arthaśāstra seeks to achieve this in two ways:

- First, it establishes that by able administration, the ruler should protect the territory.
- Secondly, it also proposes methods by which new territory should be acquired by conquest from others, which will invariably lead to the issue of foreign policy.

In other words, Arthaśāstra is the body of knowledge pertaining to state affairs dealing with both the internal issues as well as external matters affecting the state. The main issues discussed in the text include statecraft, economic policy, and military strategy. The Mahābhārata mentions the name Arthaśāstra in two places. It also refers to it by the name Daṇḍanīti. Kauṭilya (also known as Cāṇakya or Viṣṇugupta) wrote Arthaśāstra as it is popularly known now. Kauṭilya's work on Arthaśāstra is not entirely his original contribution. In the very first verse of Arthaśāstra, Kauṭilya acknowledges the work of ancient teachers for the acquisition and protection of the earth. He states that his work is one of bringing those ideas together. Several earlier teachers including Bṛhaspati and Uśanas are often quoted in Arthaśāstra at several places. The significant contribution, however, is in terms of composing them into a cogent piece of literature and offering fresh perspectives. As evident from the text, there are several occasions where Kauṭilya disagrees with the earlier teachers' views and recommendations and provides new perspectives on the issue being discussed.

The work must have been written sometime in the 3rd century BCE. Arthaśāstra had a magic influence on the later writers, who largely followed Kauṭilya's ideas. The prominent among them include the Nītisāra of Kāmandaka and Nīti-vākyāmṛta of Somadeva. After several centuries of lull, it was only in 1905 that Arthaśāstra was rediscovered and brought out by Shamasastry of Mysore by way of an English translation.

The Kauṭilya's Arthaśāstra contains fifteen books or adhikaraṇas. Of these, the first five deal with the internal administration of the state, the next eight with its relations with neighbouring states, and the last two are miscellaneous. The first chapter of the work furnishes a detailed table of contents and enumerates the topics discussed in each adhikaraṇa. Book 2 of Arthaśāstra is the largest and it constitutes nearly one-fifth of the total work. It provides a snapshot of the multi-faceted nature of public administration by describing various activities that the state needs to engage in. It has details on several activities the state administers using many executive officers. It also includes a discussion on other topics such as the settlement of unoccupied land, building a fort, and the layout of the capital city.

The third and the fourth book on the other hand deals with law, crime and punishments. The salaries for different employees of the state are listed in Book 5. Advice to the chief minister on ensuring continuity of rule on the demise of the ruling monarch is also available in Book 5. Although appearing later, in Book 6 the foundational elements of public administration are established. Seven elements that make up a state are described. Various issues of dealing with the neighbouring territories (six aspects of foreign policy, conquest of territories, fortified capital of the enemy territory, etc.) are explained in Books 7 to 13. Book 14 describes certain secret remedies for getting rid of enemies or traitors. Finally, Book 15 defines and illustrates thirty-two tantra-yuktis or methods of treating a subject.

Arthaśāstra at the outset mentions four branches of knowledge that a king must be fully conversant with so that he gains the right knowledge, perspectives, attitude, and strength to discharge his duties as the head of the state. Figure 14.1 pictorially depicts this. The importance of these four can be gauged from the following brief descriptions of each of them:

- **Science of Logic and Argumentation (Ānvīkṣikī):** is considered as the lamp that illuminates all sciences and is fundamental to understand the other branches of knowledge. It serves as the pillar that supports dharma. A proper study of philosophy helps one to distinguish between dharma and adharma expounded in the Vedas, between material gain and loss in the study of economics, and between good and bad policies in the study of politics. It also teaches one the distinction between good and bad use of force.

- **The Vedas (Trayī):** The Ṛgveda, Sāmaveda, and Yajurveda are three Vedas. However, Kauṭilya has also included the Atharvaveda and Itihāsa in his definition the Vedas. A study of the three Vedas is useful because it establishes the social order and the roles and responsibilities of the various sections of the society in practicing dharma.

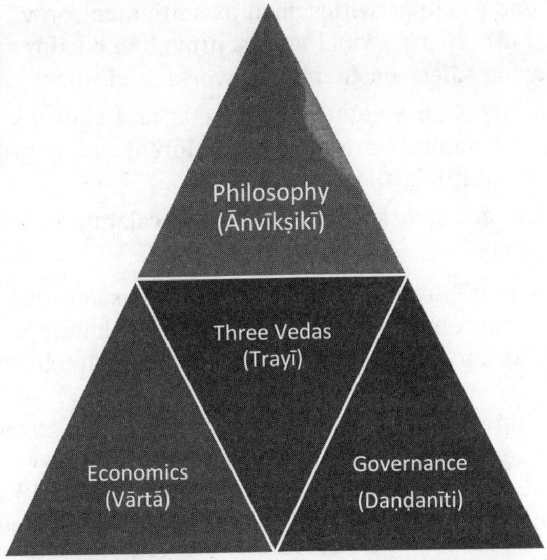

FIGURE 14.1 The Four Branches of Knowledge

- **Economics (Vārtā):** Agriculture, cattle rearing, and trade constitute the economic activity. They are the main sources of wealth for the king to augment his treasury. Only with a healthy treasury the king can maintain a good army and provide security to the citizens and keep the enemies under control.
- **Governance (Daṇḍanīti):** Maintenance of the law and order using a regulatory and a control framework is important. By maintaining order, the king can preserve what he already has and acquire new possessions. Proper functioning of the government and addressing the regulatory and control requirements will ensure that the state makes material progress, the citizens are happy and the weak in the society do not feel oppressed by the stronger forces.

14.2 THE KAUṬILYAN STATE

Arthaśāstra is not about discussing relative advantages or disadvantages of alternative forms of governance or about the origin of the state. Instead, it focuses on practical administration. According to the text, the first and foremost duty of a king is to protect the subjects and their properties. We see in book 4 several issues concerning this aspect such as protection of the people from anti-social elements like deceitful artisans and traders, thieves, dacoits, and murderers, as well as their protection from natural calamities, such as fire, floods, etc. Similarly, in Book 2, Chapter 1 Cāṇakya says that a king must take care of children, old, ill, suffering people, barren woman and woman with child, and those who don't have anybody to look after[1]. In other words, the king must ensure the yoga and kṣema of the subjects. Yoga refers to the successful accomplishment of an object, while kṣema refers to the peaceful, undisturbed enjoyment of that object. Yoga–Kṣema in a way encompasses the idea of welfare, well-being, including the idea of prosperity, and happiness. The state is expected to engage in various kinds of activities that bring welfare to the subjects. Some specific ideas presented in the text includes the following:

- The ruler is to undertake activities such as settlement on virgin land, the building of dams, tanks, and other irrigational works, providing pastures for cattle, opening trade routes and ensuring safety on them, and working of mines[2].
- The sale of commodities, whether indigenous or imported, should be permitted in such a way that the subjects are benefited. Moreover, any profit that may be harmful to the subjects should be avoided.
- When the subjects are struck down by natural calamities, the ruler should take care of them like a father.

On the other hand, Arthaśāstra also cautions that if the ruler does not address the welfare of the citizens, they may become dissatisfied and even take steps to get rid of the ruler. It enumerates several acts on the part of the ruler which are likely to make the subjects dissatisfied with his rule.

A good understanding of the public administration and governance laid out in the Arthaśāstra can be obtained using the framework given in the text that covers all aspects of administration. In Book 6 of Arthaśāstra, seven elements have been listed that make up the state[3]. Figure 14.2 pictorially depicts these elements, known as Prakṛtis. This includes the following:

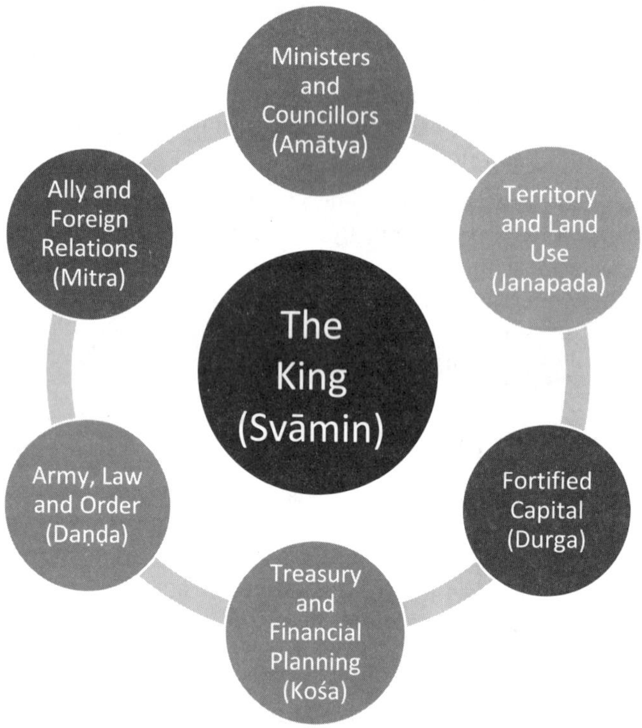

FIGURE 14.2 The Seven Elements of a State

1. The King (Svāmin) – who rules the territory or the protector of the subjects
2. The Minister (Amātya) – by whose counsel the king takes for various decisions
3. The Territory (Janapada) – the populated territory over which the king rules

4. The Fortified Capital (Durga) – the place from where the king rules
5. The Treasury (Kośa) – that provides the wherewithal and stability for the regime
6. The Army (Daṇḍa) – that provides security for the regime as well as establishes law and order in the state
7. The Ally (Mitra) – the other friendly rulers, who form part of the foreign policy of the state

The order of the seven elements needs a special mention. It is proposed that the earlier item in the list is more important than the later ones. Without a doubt, the most important of all the elements is the ruler. However, it is also mentioned that if the adversity in any one of the elements is likely to ruin the other elements, the adversity must be regarded as the most serious, irrespective of the place of enumeration of the constituent. An analysis of each of these components will provide a rich understanding of the various aspects of administration as laid out in Arthaśāstra. In the following sections, we shall analyse each of them in some detail.

14.2.1 The King

According to Kauṭilya, a native of the territory, who follows the teaching of the śāstra and who is free from disease and is strong and of noble birth are the desirable requirements of a ruler. More specifically, the qualities expected in an ideal ruler are enumerated in Chapter 1, Book 6 of Arthaśāstra[4], which includes:

- Inspirational qualities such as the nobility of birth, piety, and truthfulness. These qualities repose confidence in others and make the ruler easily approachable.
- Intellectual acumen such as the ability to understand and think clearly.
- Qualities such as bravery, quickness of decision, and strength of mind.

A ruler may not always be born with all these qualities, therefore, there is an emphasis on the training of the ruler. The training, as proposed in Arthaśāstra consists of four main areas: Philosophical disciplines (Ānvīkṣikī), the Vedic lore (Trayī), Economic principles (Vārtā) and Politics and administration (Daṇḍanīti). These have been discussed in the previous section. Trayī is necessary for developing a good understanding of the social systems and practices and the need to safeguard them. The study of Vārtā would equip the ruler with the right knowledge to regulate the economic life of the state. Knowledge of Daṇḍanīti is necessary for maintaining peace and law and order. Finally, the study of Ānvikṣikī is necessary as it develops a good sense of equanimity in the ruler. With a philosophical orientation, the ruler will be able to develop a wider vision of life and reality and demonstrate greater inner strength and courage to administer.

In Book 1, Chapter 19 of Arthaśāstra there is a suggested schedule of how the king could spend his time in a 24-hour day. Table 14.2 has the details. As evident from the table, the king can spend 10½ hours on his personal needs (including 4½ hours of sleep). The balance time is spent on various aspects of administration.

> - If the ruler does not address the welfare of the citizens, they may become dissatisfied and even take steps to get rid of the ruler.
> - When the subjects are struck down by natural calamities, the ruler should take care of them like a father.

The Manu-smṛiti covers at length the Rāja-dharma in Chapter 7 in about 226 verses, wherein the King's duty vis-à-vis the state is discussed. The topics covered include the establishment of Daṇḍa to protect, duty towards the society, the training required, appointment of ministers, warfare, administrative expenses, treatment of the subjects, foreign affairs, etc.

TABLE 14.2 The Typical 24-hour Schedule for the King

Schedule	*Tasks*
06.00 am – 07.30 am	Review reports on defence, financial matters
07.30 am – 09.00 am	Public audience
09.00 am – 10.30 am	Personal chores
10.30 am – 12.00 pm	Receive revenues, tributes, appoint high officials, allocate tasks
12.00 pm – 01.30 pm	Write letters, confer with councillors, receive information from spies
01.30 pm – 03.00 pm	Personal Time
03.00 pm – 04.30 pm	Inspect and review forces
04.30 pm – 06.00 pm	Consult with chief of defence. End the daytime with prayers.
06.00 pm – 07.30 pm	Interview with secret agents
07.30 pm – 09.00 pm	Personal chores
09.00 pm – 01.30 am	Retire to the bed chamber
01.30 am – 03.00 am	Wake up, meditate on political matters and on work to be done for the day
03.00 am – 04.30 am	Consult with councillors, send out spies
04.30 am – 06.00 am	Religious duties, meeting with the teacher, purohita, astrologer, personal physician, chief cook, etc. At daybreak he shall circum-ambulate a cow, its calf and the bull and then proceed to his court.

IKS IN ACTION 14.2

Vidura-nīti – Advice to a King

Vidura-nīti is a collection of 700 verses of the conversation between king Dhṛtarāṣṭra and Vidura. Dhṛtarāṣṭra with an agitated mind sought the counsel of Vidura and the conversation touched upon several aspects including governance and public administration.

While describing some of the attributes of wise and unwise men, Vidura pointed to a few things about which the king must be alert. Vidura observed that there are seven issues that is best avoided by a king as they could be the source of calamities. Such calamities eventually lead to the destruction of a king, however well he is established.

सप्तदोषाः सदा राज्ञा हातव्या व्यसनोदयाः |
प्रायशो यैर्विनश्यन्ति कृतमूला अपीश्वराः || 1.96

saptadoṣāḥ sadā rājñā hātavyā vyasanodayāḥ |
prāyaśo yairvinaśyanti kṛtamūlā apīśvarāḥ ||

The seven issues that are to be avoided are women, dice, hunting, drinking liquor, harshness of speech, these five, and great severity of punishment and misuse of wealth of the state.

स्त्रियोऽक्षा मृगया पानं वाक्पारुष्यं च पञ्चमम् |
महच्च दण्डपारुष्यम् अर्थदूषणमेव च || 1.97

striyo'kṣā mṛgayā pānaṃ vākpāruṣyaṃ ca pañcamam |
mahacca daṇḍapāruṣyam arthadūṣaṇameva ca ||

By an intelligent use of the numbers from 1 to 7, Vidura suggested a method to handle the issues that a king face. A king must be able to discriminate the two (good and the bad, appropriate, and inappropriate, etc.) by means of one (the intellect) and seek to bring under the control the three (friend, foe and the indifferent) using the four methods (sāma – dāna – bheda – daṇḍa). He should subjugate the five (senses) and know the six (six expedients to be used in foreign policy). Finally, the king must keep away from the seven (women, dice, hunting, drink, harshness of speech, severity of punishment and misuse of wealth) and be happy.

एकया द्वे विनिश्चित्य त्रींश्चतुर्भिर्वशे कुरु |
पञ्च जित्वा विदित्वा षट् सप्त हित्वा सुखी भव || 1.49

ekayā dve viniścitya trīṃś-caturbhir-vaśe kuru |
pañca jitvā viditvā ṣaṭ sapta hitvā sukhī bhava ||

14.2.2 The Amātya

Despite being competent and powerful, a ruler needs a team to accomplish his task. The most important of these are, the ministers, called amātyas (or the Mantrin), which is the second constituent of a state. Amātya has two connotations: one is the chief minister, and in this sense, he is the kingmaker, the other is the group of ministers who represent the highest set of functionaries providing able counselling and administrative support to the king. Constitutionally, the state functionary next in importance to the king is the amātya. The two words amātya and mantrin are used at several places in Arthaśāstra, sometimes interchangeably. However, going by the descriptions available in the text, there is a clear distinction between these two. An amātya is considered to be somewhat lower in terms of qualification and stature than a Mantrin. For example, it is stated in Arthaśāstra that only those who successfully pass all the four secret tests are to be appointed as mantrins, while those who pass only a smaller number of tests are to be appointed as amātyas. Mantrins is the set of counsellors with whom the king will hold secret consulting, whereas amātyas are high-level executive officers. The functions of the amātya include consultations, execution of undertakings, infliction of punishment, settlement and development of new territories, and recovery of fines and taxes, and so on[5].

- The Manusmṛiti covers at length the Rājadharma, wherein the King's duty vis-à-vis the state is discussed.
- Mantrin is the set of counsellors with whom the king will hold secret consulting, whereas amātyas are high-level executive officers.

The most important function of the Mantrin is to advise the king on various matters of public administration. Kauṭilya suggests that the king should appoint three or four counsellors, neither more nor less. According to him, a single mantrin would be difficult to control, and two might quarrel and ruin the state or conspire against the king. On the other hand, a larger number would affect the secrecy of counsel. Apart from the three or four mantrins, there has to be a mantri-pariṣad, consisting of twelve, sixteen, or twenty members (amātyas) depending on the size and power of the state. The functions of the pariṣad include starting work on a new undertaking, continuing an undertaking already begun, improving work, and implementation of orders issued. In case of emergencies, the King summons both the mantrī and the mantri-pariṣad and goes by the majority view of the group.

14.2.3 Settlements and Land Use (Janapada)

The importance of Janapada comes from the fact that a king augments his power by promoting the welfare of his people; for power comes from the countryside which is the source of all economic activity. Therefore, an important dimension of public administration lies in engaging several activities pertaining to the land under the disposal of a kingdom. This includes building forts that provide a haven to the people and the king himself, waterworks to ensure continuous availability of water for agriculture, and establishing trade routes to facilitate trade. Mines are also required to be properly developed as they are a source of war material. Finally, productive forests, elephant forests, and animal herds provide various useful products and animals.

The kingdom invariably has a vast amount of land, and it is important to have a proper understanding of the territory and planning of the land use to achieve the objectives stated above. Arthaśāstra identifies different types of land, forts, irrigation works, forest, mines, and trade routes, and road networks that together constitute the settled territory, known as Janapada.

Organising the kingdom with arable land, pastureland, elephant and timber forests, forts, frontier posts, roads, and trade routes is an integral element of public administration, defence and citizen safety, and sustainable living. The creation of new settlements in the remote villages and inhabiting them with native citizens or foreigners is an activity proposed in Arthaśāstra. The settlers in the villages shall mainly be agriculturists, and the number of families is between 100 and 500. In a developed settlement some lands are granted to Brāhmaṇas, purohitas, teachers, heads of departments, accountants, record keepers, divisional officers, doctors, couriers, and horse trainers. A river, a mountain, a forest, a dry riverbed, a cave, an embankment, or trees like the silk cotton, acacia, and milk tree usually separate two villages.

The administrative divisions of the settlements (Figure 14.3) are indicated in Arthaśāstra as per the following order[6]:

- For each group of ten villages, there shall be a headquarters of a sub-district (Saṃgrahaṇa)
- In the middle of two hundred villages, a district headquarters (Kārvaṭika)
- In the middle of four hundred villages, a divisional headquarters (Droṇamukha) and
- A provincial headquarters (Sthānīya) for every eight hundred villages

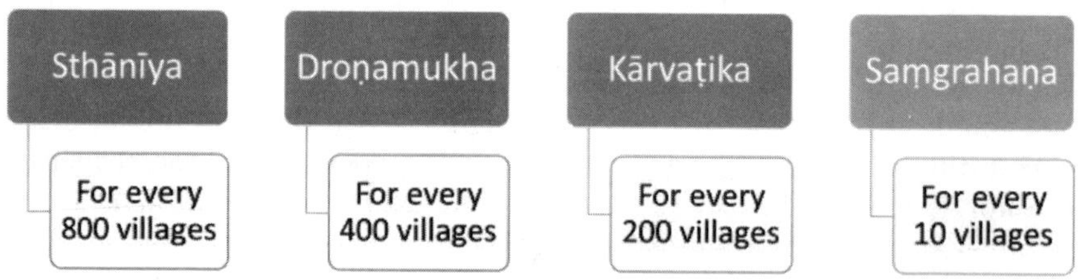

FIGURE 14.3 Administrative Divisions of the Territory

On the frontiers of the country four forts are recommended to be built, one for each quarter, equipped for defence against enemies to guard the entrances to the kingdom. These shall be built in places best suited by nature for defence. Arthaśāstra specifies eight different types of forts with natural defences. These could be on land, forest, water, or desert. A land fort is the easiest to capture, a river fort more difficult, and a mountain fort the most difficult.

14.2.4 Fortified Capital City (Durga)

The defence of the state is based on the fort (Durga) and the army (Daṇḍa). As per the conventions of war in earlier times, the safety and security of the countrymen critically depend on the safety of the king himself. Durga represents the fortified capital of the state, where the king, if hard-pressed by a strong enemy, can entrench himself and withstand a siege over a long period, during which diplomatic moves can be made to ease the enemy's pressure and avert a calamity. Therefore, a good design of the fortified capital city ensures that the king can vouchsafe himself from the enemies, stay for a longer time in the capital city even if there is a siege laid by the invading army and use the borrowed time to wade off the threat and restore normalcy in the kingdom. With a well-designed Durga, there is also a window of opportunity to use the army for offence as well as for defence depending on the strength of the invading army. Since the fort is a place of shelter in which one may have to remain entrenched over a long period during such situations, Kauṭilya states that the fort should have plenty of supplies stored in it and that there should also be secret means of escaping from it in case of need[7]. All these are the key considerations while setting up the Durga. There are several intricate details available in Arthaśāstra on several aspects of capital city planning and fortification of the capital.

In Book 2 of Arthaśāstra details for the layout planning of the capital city is available. There are clear specifications for zoning and land-use patterns. A schematic sketch of the city layout and the relevant descriptions are available in the Chapter 12 on town planning (see IKS in Action 12.1: Arthaśāstra on Town Planning for details). Each household shall be allotted the land necessary and be allowed to store grains and other commodities. Wells shall be provided for every group of ten houses. Since laying of siege by the invading army can extend longer up to a year or more, appropriate quantities of oils, grains, sugar, salt, perfumery, medicines, dried vegetables, fodder, weapons, shields, etc. need to be stored. In the worst case, it can satisfy the needs of the inhabitants of the capital city for more than a year. The armed forces - elephants, chariots, cavalry, and infantry - shall each be under more than one chief. This will inculcate mutual fear among the chiefs and will prevent them from succumbing to the temptations of the enemy.

- A king augments his power from the countryside which is the source of all economic activity.
- In Book 2 of Arthaśāstra details for the layout planning of the capital city is available with clear specifications for zoning and land-use patterns.

The fortification of the capital city is an elaborately planned exercise. According to Arthaśāstra, the capital city needs to be surrounded by three moats, either of a natural water body or artificially made with adequate drainage facilities. Figure 14.4 gives a schematic representation of the arrangement of the three moats. As we see in the picture, the three moats are of increasing width as we approach the fortified capital city. The moats are separated from one another by 6 feet. These moats need to be filled with lotuses and crocodiles. Ramparts are to be erected around the moats, planted with thorny bushes. Parapet walls of 18 or 36 feet high are to be built over the ramparts. Further, the ramparts must have an escape door and exit door as well as paths for the soldiers to go out for attacks. Various defensive mechanisms such as speed-breakers and concealed traps are suggested outside the rampart to retard the advancement of the attacking force. A gateway shall also be built with a tower consisting of a hall, a well, and a border post. The gateway shall have the main door and a side door, and defences against potential ramming by elephants. The bridge over the moat in front of the gateway shall be of removable type.

14.2.5 Treasury and the State Economy (Kośa)

The Arthaśāstra is one of the world's oldest treatises on the economic administration of a state using detailed financial planning. Three main types of economic activity agriculture, cattle rearing, and trade have been recognised and it was expected that these will generate resources in the form of grains, cattle, gold, forest produce, and labor. Using these, the king is supposed to build a treasury and an army. A healthy reserve in the treasury and a strong army are the two pillars using which the king can bring under his control both his people and the enemy. By contrast, a king with

- The Arthaśāstra is one of the world's oldest treatises on the economic administration of a state using detailed financial planning.
- The treasury is more important than the army, for the latter can be raised and maintained only with the help of a well-filled treasury.

a depleted treasury runs a risk of weakening the state by eating into the very vitality of the country. The wealth of the state was the totality of the surplus stored in the king's treasury, the commodity warehouse, the granary, the store for forest produce, and the ordnance depots. All these will constitute the kośa.

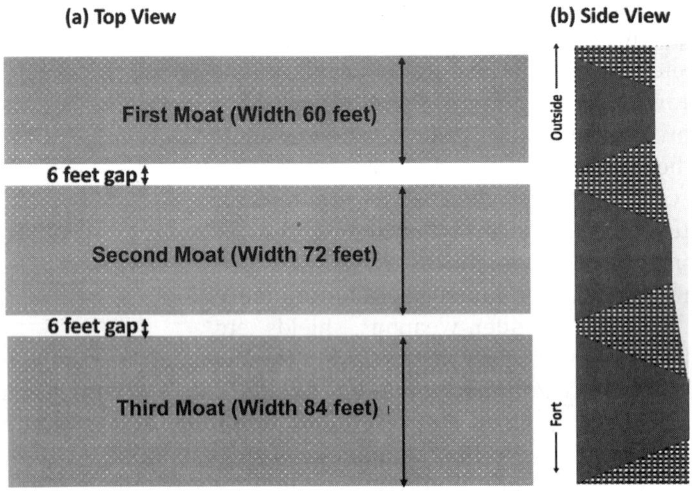

FIGURE 14.4 Arrangement of Moats for the Fortified Capital City

According to Kauṭilya, the treasury is more important than the army, for the latter can be raised and maintained only with the help of a well-filled treasury. The state exercised strict control on economic activity to safeguard the state's income. State income is classified under seven heads: city, country, mines, irrigation works, forests, cattle-herds, and trade-routes[8]. The revenue from these comes in the form of a tax or impost or profit made by state enterprises. Expenditure incurred includes those pertaining to the palace, the armoury, the armed forces, and so on. The sources of income include the following:

1. Mūlya, price realised by the sale of state goods,
2. Bhāga, the share of goods produced by the subjects,
3. Vyājī, a tax imposed on all sales,
4. Parigha, a kind of protective duty for safeguarding state goods,
5. Kḷpta, a fixed levy charged at ports on river-banks,
6. Rūpika, a surcharge on manufactures, and
7. Atyaya, penalties.

A large part of state revenue is received in kind. State stores for such as grains, fats, sugar, salt, wines, spices, vegetables and fruits and so on, also forest produce and manufactured articles, including weapons are available. Part of the stores is made up of goods produced by the state enterprise, the rest being state dues received in kind. The stores would be used for making part payments to state servants and others. What cannot be thus disposed of would be put up for sale. The state engages in buying when there is a glut and selling when there is scarcity. The stores provide a convenient means of creating buffer stocks and preventing a wide fluctuation in prices, thereby exercising a great influence on the state economy.

The state is sustained by the revenue it collects from its subjects, who mainly engage in three principal vocations: agriculture, cattle-tending, and trade. Based on these activities they pay taxes to the state which forms the main revenue for the state. What the state receives from the subjects could be in the form of grains, cattle, money, or various other products. It can also be in the form of supplying labor to the state enterprises. The tax to be remitted under various circumstances has been laid out in great detail in Arthaśāstra.

Agriculture contributes to most of the state revenue. One-sixth share of the agricultural produce is remitted as tax to the state (in the form of the agro-products or grains grown in the land). This is akin to a land tax. Besides the land tax, there is a tax levied for water usage, which ranges from one-fifth to one-third. This is paid over and above the one-sixth levied land tax. Another source of revenue comes from leasing out the pasturelands to herdsmen, who may make a living by cutting and selling the grass.

There are custom houses at city gates with 4–5 officers in charge. All imported goods are subjected to duty. On the other hand, the indigenous products attract excise duty. Excise duties are apparently to be collected at the gates when goods are taken out from the city to the country or brought from the country to the city for sale. The rules stipulate that goods are not to be sold at the place of their production, and fines are prescribed for purchasing metals, etc. directly from mines, or flowers, fruits, etc. directly from the gardens, or grains directly from the fields. In Book 2, Chapter 12 of Arthaśāstra a list of twelve kinds of income derived from the mines is listed out[9]. This includes price charged, share of produce, some sort of sales tax, penalties for violation of state monopoly, compensation for the transfer of state rights of sale to private individuals, inspection fee, and charge for manufacture at 8% of the price. A glossary of terms related to tax, fees, and service charges mentioned in Arthaśāstra is presented in Table 14.3, which is indicative of how elaborate the thinking has been on revenue matters.

TABLE 14.3 A Glossary of Terms Related to Taxes, Fees and Other Charges Found in Arthaśāstra

Sl. No.	*Tax*	*Includes*
1	Customs Duty (शुल्क – śulka)	Import duty (प्रवेश्य – praveśya) Export duty (निष्कर्म्य – niṣkarmya) Octroi and other tolls (द्वारबहिरिकाद्य – dvārabahirikādya)
2	Transaction Tax (व्याजी – vyājī)	Transaction tax for crown goods (मानव्याजी – mānavyājī)
3	Share of Production (भाग – Bhāga)	1/6th share (षड्भाग – ṣaḍbhāga)
4	Tax in Cash (कर – kara)	
5	Taxes in Kind (प्रतिकर – pratikara)	Labour (विष्टि – viṣṭi) Supply of soldiers (आयुधीय – āyudhīya)
6	Others	Countervailing duties or taxes (वैधरण – vaidharaṇa) Road Cess (वर्तनी – vartanī), Royalty (प्रक्रिया – prakriyā) Monopoly tax (परिघ – parigha) Taxes paid in kind by villagers (पिण्डकर – piṇḍakara) Army maintenance tax (सेनाभक्त – senābhakta) Surcharges (पार्श्व – pārśva)
7	Fees and Service Charges	Weights and measures (पौतव – pautava) Passport fees (मुद्रा – mudrā), Ferry fees (तर – tara) Part dues (पत्थन – patthana), Coining fee (रूपिक – rūpika) Land survey charges (रज्जु – rajju) Testing fee (पारीक्षिक – pārīkṣika) Escort charges (अत्वहिक – atvahika) Fixed charges (क्लृप्त – klpta)

On account of some military adventure and adversarial situations, the treasury is likely to be depleted fast. In such situations, special efforts are needed for quick replenishment of the treasury using several methods. These are discussed in Chapter 5.2 of Arthaśāstra,

14.2.6 Law and Order and Security (Daṇḍa)

Arthaśāstra uses the phrase 'Yoga–Kṣema' to describe one of the important duties of the king[10]. The term 'yoga' refers to the successful accomplishment of an objective and 'kṣema' to peaceful enjoyment of prosperity. The 'yoga-kṣema' of the subjects requires that they have the security of person and property. The yoga-kṣema is disturbed by thieves, robbers, anti-social elements like corrupt officials, deceitful merchants, and others. The weeding out of these thorns, is the main task accomplished by Daṇḍa, an important function of the state. The king's three duties concerning internal administration include protection of the state from external aggression, maintenance of the Law and Order in the state, and ensuring the welfare of the citizens. Daṇḍa is a symbol of the ruler's authority and a mechanism to achieve these objectives. The Arthaśāstra devotes the entire book four to this topic.

Daṇḍa may appear to be the coercive power of the state, therefore great care must be taken in deploying Daṇḍa. With the help of Daṇḍa, the ruler should ensure that the strong do not swallow up the weak creating big anarchy in the social system. Arthaśāstra warns a king of the consequences of improper use of Daṇḍa. An unjust or improper use of this power by the ruler might lead to serious consequences, the most serious being a revolt of the subjects against the ruler. The text enumerates several acts on the part of the king which are likely to make the subjects dissatisfied with his rule[11]. If the subjects become dissatisfied, they may join hands with the ruler's enemies or may eventually rise in revolt and themselves slay the king.

- A glossary of terms related to tax, fees, and service charges mentioned in Arthaśāstra is indicative of the elaborate thinking on revenue matters.
- The yoga-kṣema is disturbed by thieves, robbers, anti-social elements like corrupt officials, deceitful merchants, and others.

Figure 14.5 schematically presents the overall structure proposed in Arthaśāstra for the administration of Daṇḍa. As is evident from the figure, the ultimate responsibility for the administration of law and justice rests with the king. This implies that there is an ultimate possibility of appeal to the king to seek redressal of the grievances in the process adopted by the appointed officers. In the daily routine of a king, there is a fixed time when he meets with the citizens and uses the opportunity to hear the appeals and takes a decision on it. The legal system is vertically divided into two: one set addresses the conflicts between two parties and the other crime-related (such as theft, dacoity, cheating, etc.) issues. While the entire conflict resolution system is rested with the Chancellor, the crimes aspect is segregated into the countryside and capital city. The Chancellor is responsible for crimes in the entire countryside, whereas the City Superintendent is responsible for crimes in the capital city. The figure on the right side shows how the Law and Justice system is made accessible to the citizens by placing the courts at different levels of the countryside. At every level, there is a panel of three judges or three magistrates.

The judges are called *Dharmasthas*, a name which apparently refers to the dharma or law, by which they are to be guided in their work. Pradeṣṭṛ, translated as magistrates, is another class of judicial officers. The distinction between judges and magistrates is that the former dealt with all cases concerning transactions between two parties, and the latter was concerned with crimes against society in general. According to Arthaśāstra, a bench of three judges is established at frontier posts, sub-district headquarters, and provincial headquarters. The reference to frontier posts as a seat for a judicial court implies that justice is intended to be made available to the subjects very much nearer to their places of residence than seems possible even today. The qualifications of the judges are equivalent to that of a minister. While adjudicating cases arising from disputes between two parties, the judges are required to discharge their duties objectively and impartially so that they may earn the trust and affection of the people.

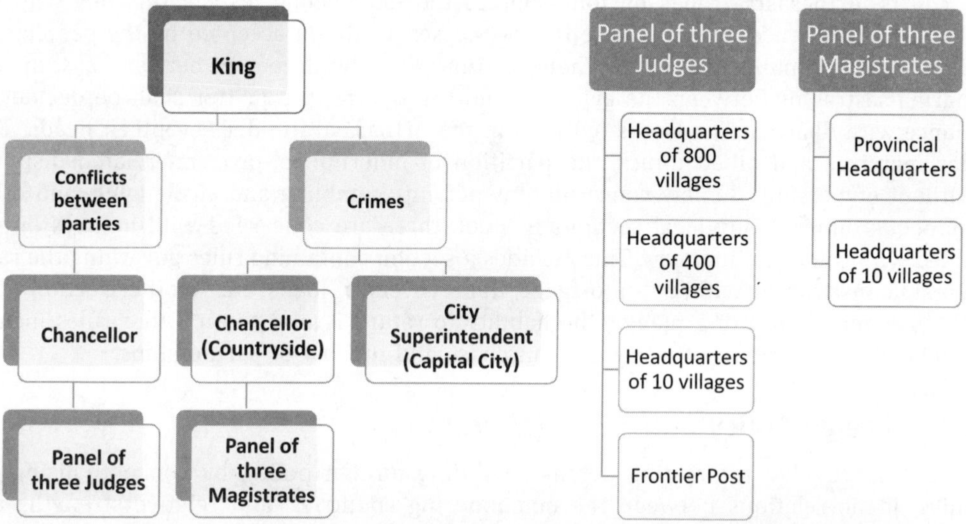

FIGURE 14.5 The Organisational Structure for Daṇḍa

Artisans and craftsmen like weavers, washermen, smiths, etc., are prone to cheat the public with respect to the quality and quantity of the goods produced by them, and by charging exorbitantly for their services. In Arthaśāstra several deficiencies in the professional service offered by artisans have been identified. This includes delay in the delivery of the article, failure to carry out the customer's instructions, appropriation of the material given by the customer for manufacture, and denial that it was ever received by him. Several tricks are described for finding out if the artisan had indeed received the material and if proved he is to be punished as a thief. Certain regulations have been laid out for controlling their activity. For instance it lays down the percentage of increase or decrease in the raw material during the process of manufacture, to which the artisans must conform as well as a scale of wages for the different categories of artisans[12]. Similarly, traders may try to cheat customers by giving them short weight and measure or by adulterating goods or by manipulating prices, and so on. Two officers are mentioned who control the activity of the traders and the markets. In addition to these protections made available to the citizens, the third chapter of book four

- The ultimate responsibility for the administration of law and justice rests with the king.
- A bench of three judges is established at frontier posts, sub-district headquarters, and provincial headquarters.

details measures to be taken for the relief of the subjects when they are overtaken by a natural calamity like floods, fire, epidemics, famine, etc. Arthaśāstra has identified 13 categories of people including the head of departments, those belonging to the judicial administration, etc. who are likely to be corrupt and may resort to accepting bribes or to extorting money. Methods to lay down traps for catching them when they are doing so have been proposed.

The regulatory framework put in place to protect citizens' rights and welfare requires the counterpart in terms of a legal and a judiciary mechanism to address issues arising out of the regulatory mechanisms. In book three of Arthaśāstra, these issues are discussed in some detail, and specific recommendations are made as to how the law needs to be administered in the courts. Arthaśāstra relies on four sources for establishing justice; Dharma, which is based on truth, evidence, which is based on witnesses, tradition accepted by the people, and Royal Edicts, for promulgated law[13]. Whenever there is a disagreement between custom and the dharma-śāstras or between the evidence and the śāstras, the matter shall be decided in accordance with dharma. The law is outlined in the Arthaśāstra under seventeen heads. This includes marriage and allied topics, the partition of inheritance, property-related disputes, sale without ownership, the law concerning ownership, gambling and challenging, and other miscellaneous things. Many legal sections in Book three are concerned with matters largely connected with trade and industry. This includes a section containing rules governing the rates of interest, loans with or without a mortgage, the recovery of loans, etc. Another section deals with the use and misuse of a pledge, the liability to return it, and so on. Other rules include issues related to revocation of a sale or a purchase and the law of partnerships.

14.2.7 Foreign Policy and Allies (Mitra)

Kauṭilya, has provided interesting details regarding foreign policy, by conceptualising the dynamics of the relations between the neighbouring countries, and relative strengths and weaknesses of the kings in a circle of territories around. Books 7, 11, and 12 describe various aspects of foreign policy. However, the guiding principles are laid out in Chapter 2 of Book 6. The basic premise for the foreign policy stems from the status of the other six elements of a state that we have so far discussed, which are internal to the state. The power which a state can bring to bear on promoting its own interests vis-a-vis other neighbouring territories depends on the status of these internal elements. Only when a king ensures that the six internal elements are in a healthy condition, a meaningful foreign policy will be possible. Kauṭilya designates such a king as a 'strong' king. This will determine the nature of allies that a king can have.

Another important perspective that emanates from this is the likely power equation between two kings. Three possibilities exist; the king who initiates a particular action may be equal in strength to the other king, stronger than him or weaker than him. These considerations dictate the nature of foreign policy to be pursued. The conqueror with his allies and vassals, the enemy with a similar circle of kings, other interested parties like the middle and neutral kings are the actors in the drama of diplomacy and war. The tools they use are the six methods of foreign policy, defined in Book 7, Chapter 1[14]. Figure 14.6 briefly describes the six methods to be adopted.

- Arthaśāstra has identified people including the head of departments, those belonging to the judicial administration, etc. who are likely to be corrupt.
- Whenever there is a disagreement between custom and the dharma-śāstras or between the evidence and the śāstras, the matter shall be decided in accordance with dharma.

Sandhi	• Making a treaty containing conditions or terms (Policy of Peace)
Vigraha	• Hostility
Āsana	• Not planning to march on an expedition (Policy of remaining quiet)
Yāna	• Marching on an expedition
Saṃśraya	• Seeking shelter with another king or in a fort
Dvaidhībhāva	• Sandhi with one king and Vigraha with another at the same time (the double policy)

FIGURE 14.6 The Six Methods of Foreign Policy

- **Sandhi** denotes a variety of methods to be adopted to make peace by entering into an agreement with specific conditions. It enables a king to promote the welfare and development of his state without intervening in any conflict in his neighbourhood. Alternatively, the king may use a peace treaty to strengthen alliances.
- **Vigraha** on the other hand, is to take a hostile approach primarily by waging an open or a secret war (taking him by surprise). It can also include diplomatic offensive through available channels.
- **Āsana** amounts to staying quiet (for the moment).
- **Yāna** is preparing (waiting) for an eventual war. The idea behind the strategy is the expectation that the enemy would grow weak or find himself in difficulties or get involved in some war. Alternatively, one can use the time to become stronger.

  Both the above methods represent stages in the transition from peace to war and vice-versa. It is a temporary halt in implementing a policy of peace or war already initiated. The pause may be of short duration waiting for some improvement. Sometimes it may turn out to be a longer pause waiting for the right opportunity. The time borrowed could be used to change the strategy or the target of the proposed action. Deciding to undertake a military campaign involves heavy expenditure related to mobilisation and maintenance of the troops and absence from the capital for a long time, leaving the control in the hands of the Viceroy, with its attendant risks. Therefore, the concept of the pause after declaring war has significant implications.
- **Saṃśraya** is seeking protection. This is adopted when threatened by a stronger king by way of taking refuge in a fort.
- **Dvaidhībhāva** as the name suggests is to simultaneously adopt two policies. One is the policy of making peace (Sandhi) with a neighbouring king, and with his support pursue the policy of hostility (Vigraha) towards another.

Based on the above descriptions one can infer that in general, when one is weaker than the enemy, Sandhi is the policy to be followed, if stronger than him, then vigraha. If both are equal in power, āsana is the right policy, but if one is very strong, yāna should be resorted to. Finally, when one is very weak saṃśraya is necessary,

Alongside the six policies, four means of overcoming opposition - conciliation (sāma), placating with gifts (dāna), sowing dissension (bheda) and use of force (daṇḍa) have also been proposed. The first two could be deployed to subjugate weak kings and the last two for overcoming strong kings[15]. Arthaśāstra also specifies under what conditions these techniques will be useful[16]. Sāma would be best in the case of officers of the enemy whom the latter suspects of treason; Dāna would be most useful in winning over seditious men from the enemy; Bheda would break up confederacies most effectively, while the use of daṇḍa would be most effective in the case of a strong enemy.

The foreign policy measures proposed in Arthaśāstra broadly confines to a situation in which there are several smaller kingdoms, perhaps vying with one another to establish a state of supremacy by conquest, annexing territories, and growing into a bigger kingdom. Nevertheless, many of these ideas merit closer attention and provide greater relevance to craft foreign policy formulations in the contemporary world.

14.3 THE ADMINISTRATIVE SET-UP

The kind of state control over the economy which Arthaśāstra prescribes is not possible without an efficient administration. We, therefore, find a description of elaborate administrative machinery in Arthaśāstra. The entire administrative setup and the hierarchy could be classified under six major heads:

- **King and his Cabinet:** King and his family consisting of the main Queen, the other queens, crown prince (Yuvarāja), other princes, and queen's mother comprise the family circle. The king has his main teacher (Ācārya), the family priest (Purohita), and the officiating priest for doing rituals (Ṛtvik). The others include royal scribe, courtesans, soothsayers, and astrologers. These constitute the personal circle and play an advisory role to the king in formal as well as informal positions. However, in a formal sense, the king has his cabinet of ministers representing various administrative aspects of the state. They provide counselling, participate in key decisions by providing required inputs and offer executional support to the king for implementing all decisions.
- **Palace Administration:** It comprises all administrative aspects pertaining to the palace and the Secretarial aspects of maintaining the king's schedule, personal security, and playing a closer role in the king's daily life. The Chief of Palace Attendants and the Chief of Palace Guards represent the highest position in the hierarchy, and they have several people working under them at various levels.
- **Defence:** It comprises the armed forces concerned with the protection of the state from the external forces, conquest, and expansion of the territory. The Chief of Armed Forces is the Senāpati, who is highly placed in the hierarchy. Defence establishment consists of four categories of armed forces (Infantry, Chariots, Horses, and Elephants) and varying levels of personnel with specific roles and responsibilities. The yuvarāja may be sent instead of the senāpati on an expedition of conquest or to put down a rebellious officer.
- **Treasury:** A significant portion of the taxes and revenues (from agricultural, trade, and other activities) are remitted to the state in kind. Treasury provides the administrative

mechanism to collect a variety of items including gems, grains, edible items, oil, armoury, and other produce due to the state and maintain the storage of these. Moreover, during periods of adversity, the state critically relies on the strength of the treasury to weather the challenges and ensure the welfare of the citizens. Therefore, this is headed by a high-ranking official reporting directly to the king, known as the Treasurer (Sannidhātṛ). He is supported by nation-wide sub-ordinates for the collection, storage, and vouchsafing of the treasury.

- **Revenue and Law and Order:** It focuses on two key elements: financial planning (budgeting), taxation and revenue collection, and the security and welfare of the citizens. Since these are important aspects of public administration, it is headed by a Chancellor (Samāhartṛ), who reports directly to the king. There are elaborate functions performed by the Chancellor and is assisted nationwide by a hierarchy of officials.

  The chief function of the Chancellor is the preparation of the yearly budget and the keeping of accounts. He is to fix the amount of revenue to be collected under each head which makes up the 'body of income'. In this perspective, the Chancellor's role is one of being a finance minister. On the other hand, he is in administrative charge of the entire countryside (except the fortified capital). For administrative purposes, the Chancellor is to divide the country into four divisions, each in charge of an officer called sthānika. Under the sthānika, there are to be junior officers called gopas, each in charge of five or ten villages. With the help of these, the Chancellor will have the required information for preparing the budget.

  He will also have under him another set of servants who will provide him with information on new arrivals in towns and villages and departures from them, on the activity of unsocial elements, on the movements of traders and the goods carried by them, on the loyalty or otherwise of the subjects, on the integrity or its absence among officers and on the activity of robbers and enemy agents to help him maintain law and order in the countryside. The Law-and-order aspects of the fortified capital alone are under the control of another officer, the City Governor-General (Nāgarika) who has similar machinery as that of the Chancellor within the jurisdiction of the fortified capital city.

> - Treasury provides the administrative mechanism to collect a variety of items due to the state.
> - Revenue and Law and Order focus on financial planning (budgeting), taxation and revenue collection, and the security and welfare of the citizens.

- **Civil Administration:** It is the elaborate government administrative machinery to regulate and address all aspects of civilian activities such as agriculture, mining, manufacturing, fisheries, cattle rearing, and trade. Book two of Arthaśāstra has vividly described the entire gamut of activities and the manner the government apparatus must set up efficient administrative units. 33 Heads of the Departments (HoDs), including some for Defence, have been proposed in book two with details on the nature of activities these HoDs need to perform. These are broadly classified into the following: Trade and Industry, Mining, Treasury and Finance, Defence, Land use and Livestock, Citizen Services, and Others.

Table 14.4 has more details on these.

TABLE 14.4 Details of Various Head of Departments Proposed in Arthaśāstra

Trade and Industry	*Mining*
Chief Controller of State Trading	Chief Controller of Mining and Metallurgy
Chief Controller of Weights and Measures	Chief Superintendent of Metals
Chief Textile Commissioner	Chief Superintendent of Mines
Chief Controller of Private Trade	Chief Superintendent of Precious Metals and Jewellery
Treasury and Finance	*Defence*
Chief Superintendent of Treasury	Chief of Ordnance
Chief Master of the Mint	Chief Commander of Cavalry
Chief Salt Commissioner	Chief Commander of Elephant Corps
Chief Superintendent of Warehouses	Chief Commander of Chariot Corps
Chief Controller of Customs and Octroi	Chief Commander of Infantry
Land Use and Livestock	*Citizen Services*
Chief Elephant Forester	Chief Controller of Alcoholic Beverages
Chief Superintendent of Forest Produce	Chief Controller of Entertainers
Chief Surveyor and Timekeeper	Chief Controller of Gambling
Chief Superintendent of Crown Lands	Chief Superintendent of Temples
Chief Superintendent of Crown Herds	*Shipping and Movement*
Chief Controller of Animal Slaughter	Chief Controller of Shipping
Chief Controller of Pasturelands	Chief Controller of Ports and Harbours
Others: Chief Superintendent of Jails	Chief Passport Officer

Source: Based on **Rangarajan, L.N. (1987).** *The Arthashastra*, Penguin Books, pp. 305–307.

An astonishing aspect in Arthaśāstra is the availability of elaborate details on salary to be paid to different government employees belonging to the above six categories has been specified in Book 5, Chapter 3. Normally, one-quarter of the state revenue may be spent on these payments. Using the currency unit paṇa[17], Kauṭilya suggested annual salary levels ranging from 48,000 paṇas to 60 paṇas[18].

14.4 RELEVANCE OF ARTHAŚĀSTRA

Arthaśāstra is a political-economic treatise that presents a public administration and governance framework with an assumed model of a monarchy. Moreover, the broader setting for the text was one of the multiple kingdoms of smaller size, ruled by independent kings, vying with one another to establish their supremacy through conquests and territorial expansion. In the modern-day context, these governance mechanisms and territorial divisions have been replaced with large countries (such as India, China, Russia, the US, etc.) and newer forms of governance such as democracy. Territorial expansion through invasion of the neighbourhood countries is nether feasible not contemplated by most of the countries today as it has catastrophic global implications. The form of government that has been adopted today, a republic with the parliamentary type of democracy was quite inconceivable at the time Arthaśāstra was written down. However, inscriptions in Uttiramerur (near Kanchipuram in Tamil Nadu) describe a form of local democracy prevailing in the villages, with local assemblies and elections being held for various public offices and committees. Notwithstanding these aspects, there are several interesting ideas in Arthaśāstra that have a bearing on modern-day governance and public administration. Let us enlist a few of them below:

- The discussions on various categories of state employees, their qualifications, hierarchical salary structures provide a good understanding of the issues related to public administration. Another interesting aspect pertains to the need for putting in place checks and balances to ensure that those occupying powerful positions do not indulge in corrupt practices and exploit the citizens. These ideas merit closer attention, better understanding, and possible applications in modern-day governance.
- The aspects of criminal investigation and methods required to deploy secret services for keeping a watch over state employees and the political activities of the citizens have relevance for modern applications. The ideas on various methods for criminal investigation are very modern in their approach and appeal.
- The section on the legal aspects of administering the state is largely drawn from the smṛtis and various other commentaries. Our modern-day legal systems are by and large constructed using the same set of resources. Therefore, closer scrutiny of the Books 3 and 4 of Arthaśāstra will be of use even in modern times.

> - The foreign policy ideas presented in Arthaśāstra have useful applications in the wider field of world politics.
> - The epic Rāmāyaṇa has a chapter in Book 2, which provides interesting insights into certain aspects of public administration.

- Arthaśāstra, as we have already seen has taken a position of the control of economic life by the state. Despite alternative governance mechanisms that nations practice today the need for a political-economic perspective to governance is inevitable. Therefore, one can benefit from the elaborate instructions provided in the text concerning various economic policies, sources of revenue for the state, budgeting and planning of government expenditure, etc.
- Despite an altered global scenario of multiple governments and governance structures, we still have certain issues to address. We witness a level of distrust of one nation by another, the pursuit of its own interests by every nation, and the need for securing alliances with like-minded nations in the global order. This calls for a certain approach in terms of developing cross-border intelligence, international negotiations to secure the safety and integrity of one's own nation while deriving certain benefits from the ally. One of the significant contributions of Arthaśāstra is in the area of foreign policy. The ideas presented in the text have useful applications in the wider field of world politics and have the potential to provide new perspectives on this.

14.5 PUBLIC ADMINISTRATION – PERSPECTIVES FROM THE EPICS

Mahābhārata

In the Mahābhārata, at the conclusion of the war of Kurukṣetra between the Pāṇḍavas and the Kauravas Yudhiṣṭhira was anointed as the king of Hastināpura. The devastating post-war effects made Yudhiṣṭhira unhappy, and he was not keen to rule the kingdom. At the suggestion of Kṛṣṇa, Yudhiṣṭhira met Bhīṣma lying in a bed of arrows and took his counsel on governance and administration. The detailed conversation between the two is fully captured in the Śānti-parva (Book 12) of Mahābhārata in about 120 chapters (Chapter 55 to Chapter 173). Several issues related to Rāja-dharma were expounded by Bhīṣma including the origin and purpose of the state, the rule of law, the institution of kingship, the duties and powers of the Rājā, topics such as taxation and the way to collect the taxes, strategies for battle and the method of selecting ministers and counsellors. The way Arthaśāstra has dealt these issues is aligned to the discussion in Śānti-parva. Therefore, there is no need to repeat these details again. However, we shall see some interesting ideas from Śānti-parva in this section.

Importance of Dharma in Governance

Great stress is laid on the personal character and qualities which a Rāja, in whom vast political power is vested, must possess for the proper and effective discharge of his functions. We alluded to the term Rāja-dharma in the opening section of this chapter and remarked that upholding dharma is the supreme goal of a king. We find repeated mention of this idea and extensive discussions around it in Śānti-parva. According to Bhīṣma, the king needs to play a careful role as he leads his subjects by example. If he follows righteousness, then righteousness is honoured everywhere. Whatever acts and things a king like, they are liked by his subjects[19]. That king who considers righteousness as the most effective instrument for accomplishing his objectives and who follows the advice of the righteous indeed shines with righteousness. On the other hand, that king who neglects righteousness and acts with brute force soon falls away from righteousness and loses virtue, wealth and everything else[20].

In Chapter 90, Bhīṣma observes that all creatures prosper in a context nourished by dharma and deteriorate with the decay of Dharma. Further, only in a state where dharma is protected, acquisition and preservation of wealth will be possible. Therefore, a king should follow the dictates of dharma for benefitting his people. A king who has a pious soul and is accomplished is considered to be an embodiment of dharma[21]. In Chapter 68, Bhīṣma observes that the term 'Rāja' originates from the fact that by his reign centred on dharma, the king indeed shines (virājate in Sanskrit means shines). It is the king who establishes peace on the Earth through proper observance of his duties and by establishing a rule of the law founded on the principles of dharma[22]. The rest of the chapter discusses in detail how the king must protect the subjects in the state. According to Bhīṣma, a king of pure soul and devoted to the duty of protecting his subjects by acting righteously acquires merit and fame both here and hereafter[23].

Treasury and Taxation

Treasury is important for public administration as it enables the king to protect the country and the subjects and also sustain the kingdom during periods of distress. Taxation is the main source for filling the treasury. Bhīṣma cautions Yudhiṣṭhira of the care a king must exercise on taxation matters. By levying excessive taxes, not sanctioned in the scriptures, a king may invite the wrath of the subjects. Collecting taxes from the subjects is likened to milking of a cow and it is foolish to cut off the udder in one's anxiety in milking more. Bhīṣma pointed to several just sources of tax collection to augment the treasury. This includes one sixth revenue from the produce of the soil, fines and forfeitures from the offenders, other taxes upon merchants and traders in exchange for safety granted to them[24].

In Chapter 87 of Śānti-parva, Bhīṣma enumerates some details on levying of taxes. He suggests that levying tax on traders must take into consideration the sales and purchases, the food and the dress, the stocks and the profits and the state of the roads (perhaps the quality of infrastructure provided to the trade). Moreover, he observes that no tax should be levied without determining the output and the labour necessary for production. Since nobody will work without sufficient causes the king must tax in such a manner that both he and the producer partake the profits from the output[25].

Rāmāyaṇa

The epic Rāmāyaṇa has a chapter in Book 2, which is a conversation between Bharata and Rāma when Bharata came to meet Rāma in the forest. The events that preceded this are as follows. Bharata was brought back to Ayodhyā from his uncle's place to take the reins as Rāma has already been sent to the forest and Daśaratha passed away suddenly. Bharata, who was unaware of these happenings was not happy with the turn of events and was keen to persuade Rāma to return to Ayodhyā and take the reins. During the meeting between the two, Rāma inquired about the situation in Ayodhyā and asked Bharata a series of questions on various aspects of running the government. This provides interesting insights into certain aspects of public administration. Chapter 100 in Book 2 of Rāmāyaṇa has the details of this conversation. In 76 verses in the chapter, Rāma asked over 70 questions to Bharata. Many of these verses have a repeating word, kaccit (कच्चित्), essentially meaning "hope?", as in, "hope everything is fine at your place". Therefore, this chapter is popularly referred to as "kaccit sarga". The word 'kaccit' is repeated 72 times in the chapter. Rāma displays his administrative prowess and acumen and uses the opportunity to teach Bharata the art and science of governance using a simple methodology of asking questions. The 76 verses cover a wide range of topics on which Rāma perhaps triggered some thoughts in the mind of Bharata on certain aspects of public administration. We shall see a few ideas in this section.

Importance of Wise Counselling

Although the king is the ultimate authority for all decision-making, he cannot discharge all the activities entirely by himself. A competent group of counsellors plays a significant role in the effectiveness of administration. Rāma reminded Bharata that the well-guarded advice of the ministers, learned in scriptures and capable of proper counselling was fundamental for success in the pursuits of a king[26]. As we have already seen, wise counsellors (Amātyas) are one of the seven elements of a state, next only to the king in importance. Rāma posed a series of questions pertaining to this to Bharata: He inquired his brother if he had appointed men, who are brave, learned, of noble birth, exhibiting self-control, and have the skill to guess things from hints[27]. The ministers are in very senior positions advising the king. Therefore, their value system is as important as their competencies. Rāma inquired Bharata if he was able to appoint ministers who were unyielding to bribery and other such temptations and have integrity so that they can discharge superior tasks[28].

Justice and Punishment

Rāma had several pieces of advice to Bharata on the issue of justice and punishment[29]. He cautioned Bharata that if those who ought to be punished are not given the punishment, the king may be slain by the subjects (out of growing dissatisfaction). He also cautioned against punishing an honest, pure-hearted, and venerable person falsely accused of adultery out of avarice without consulting experts in the scriptures. Similarly, he also cautioned against setting free a thief, caught red-handed with sufficient evidence, and interrogated out of greed for money. Taking care of the citizens irrespective of whether they are rich or poor, during periods of hardship is very important. Rāma asked Bharata if his ministers were well-informed of these aspects of governance and addressed these issues without any bias. According to Rāma, the tears falling from the eyes of persons, who are falsely accused and punished for the pleasure of the king will destroy the king's progeny and cattle as well.

A related aspect is one of intelligence gathering on high-ranking officials, who by virtue of their power and position may engage in corrupt practices, coerce the citizens and extract money and other favours from them. Therefore, it is important to put in place counter-balancing forces that can keep a watch on these officials. In enemy territory, these officials are indeed the window to everything happening in the country. Rāma asked Bharata if he was able to collect information (of their secret efforts) intelligently through unrecognisable spies with three of them closely watching each of the eighteen officials (there are eighteen categories of officials in a kingdom) of the enemy's side and fifteen (officials) on his side[30].

Importance of Secrecy

Rāma emphasised another important aspect of public administration, viẓ., maintain the secrecy of information and decisions[31]. In the absence of this, administrative decisions are hard to implement and the threat from traitors and enemies will increase. It can also jeopardise several good initiatives and force the state administration to spend scarce resources in addressing counterattacks. In modern parlance, as we know, an oath of secrecy is an important aspect for the judiciary, legislature, and executive arms of the government. Rāma asked Bharata if he was able to ensure that other kings come to know of his endeavours only when his endeavours succeed or about to succeed. He also inquired if the efforts intended to be made in the future were not revealed to them beforehand. Secrecy applies not only to the king's intentions and initiatives but also to the other highest officials of the state administration. Therefore, Rāma asked Bharata if he was able to also ensure that others are not able to understand his determination or those of his ministers, either by conjecture or by inference or through other means without being explicitly revealed either by him or by his ministers.

Urban Development and Citizen Welfare

Based on a series of questions that Rāma asked Bharata, we get a glimpse of the issues concerning urban development and citizen welfare[32]. Rāma wanted to know from Bharata if the kingdom was adorned with peaceful places, rich in temples, and resting places and sheds where water is stored in tanks for use of the passers-by. He also wanted to know if the agricultural lands were well tilled, fed not just by rains (but also have a supplementary water system to ensure availability of abundant water), studded with mines, and cattle free from cruelties, and rid of fears from beasts. Since agriculture and cattle rearing were perhaps the major economic activities at that time, ensuring that citizens can carry on these activities unhindered will ensure economic prosperity for them. He also wanted to know if men and women were engaged in social festivities happily. These are good indicators of economic prosperity and citizen welfare.

SUMMARY

- Arthaśāstra is the body of knowledge pertaining to state affairs dealing with both the internal issues as well as external matters affecting the state. The main issues discussed in the text include statecraft, economic policy, and military strategy.
- The Kauṭilīya Arthaśāstra contains fifteen books or adhikaraṇas. Of these, the first five deal with the internal administration of the state, the next eight with its relations with neighbouring states, and the last two are miscellaneous.
- Arthaśāstra at the outset establishes four branches of knowledge that a king must be fully conversant with, Philosophy (Ānvīkṣikī), The three Vedas (Trayī), Economics (Vārtā) and Governance (Daṇḍanīti) so that he gains the right knowledge, perspectives, attitude, and strength to discharge his duties as the head of the state.
- A good understanding of the public administration and governance laid out in the Arthaśāstra can be obtained using the framework of seven elements that make up the state.
- According to Kauṭilya, being a native of the territory, following the teaching of the śāstras, being free from diseases and strong and of noble birth are the desirable requirements of a ruler.
- Amātya has two connotations: one is the chief minister, and in this sense, he is the kingmaker, the other is the group of ministers who represent the highest set of functionaries providing able counselling and administrative support to the king.
- Arthaśāstra identifies different types of land, forts, irrigation works, forest, mines, and trade routes, and road networks that together constitute the settled territory, known as Janapada.
- Durga represents the fortified capital of the state, where the king, if hard-pressed by a strong enemy, can entrench himself and withstand a siege over a long period, during which diplomatic moves can be made to ease the enemy's pressure and avert a calamity.
- The wealth of the state was the totality of the surplus stored in the king's treasury, the commodity warehouse, the granary, the store for forest produce, and the ordnance depots. All these will constitute the kośa.
- The state is sustained by the revenue it collects from its subjects, who mainly engage in three principal vocations: agriculture, cattle-tending, and trade. Based on these activities they pay taxes to the state which forms the main revenue for the state.
- The king's three duties concerning internal administration include protection of the state from external aggression, maintenance of the Law and Order in the state, and ensuring the welfare of the citizens. Daṇḍa is a symbol of the ruler's authority and a mechanism to achieve these objectives.
- Arthaśāstra has identified 13 categories of people including the head of departments, those belonging to the judicial administration, etc. who are likely to be corrupt and may resort to accepting bribes or to extorting money. Methods to lay down traps for catching them when they are doing so have been proposed.
- The conqueror with his allies and vassals, the enemy with a similar circle of kings, other interested parties like the middle and neutral kings are the actors in the drama of diplomacy and war. The tools they use are the six methods of foreign policy, defined in Arthaśāstra.

- We find a description of elaborate administrative machinery in Arthaśāstra with details on salary to be paid to different government employees.
- One of the significant contributions of Arthaśāstra is in the area of foreign policy. The ideas presented in the text have useful applications in the wider field of world politics and have the potential to provide new perspectives on this.
- Bhīṣma's counsel to Yudhiṣṭhira on governance and administration is presented in the Śānti-parva of Mahābhārata. Several issues related to Rāja-dharma were expounded by Bhīṣma including the origin and purpose of the state, the rule of law, the institution of kingship, the duties and powers of the Rājā, topics such as taxation and the way to collect the taxes, strategies for battle and the method of selecting ministers and counsellors.
- In the "kaccit sarga" of Rāmāyaṇa, Rāma displays his administrative prowess and acumen and uses the opportunity to teach Bharata the art and science of governance using a simple methodology of asking questions. The 76 verses cover a wide range of topics on which Rāma triggered some thoughts in the mind of Bharata on certain aspects of public administration.

REVIEW QUESTIONS

1. What do you understand by the term 'Arthaśāstra'? How is it related to public administration?
2. What are the four branches of knowledge prescribed in Arthaśāstra? Why is it important for a king?
3. How would you describe the country and state in terms of its elements?
4. What are the qualities expected of a king? What are the dos and don'ts for a king?
5. What do you understand by the term 'amātya'? What role do they play?
6. Briefly explain the territorial divisions of a Kauṭilyan state.
7. What is the importance of a fortified capital city? What considerations that one must bear in mind in designing it?
8. What are the sources of income for a state as per Arthaśāstra? What are the different types of taxes collected?
9. What are the ways by which a state can augment its kośa during periods of adversity?
10. Daṇḍa as stipulated in Arthaśāstra has two dimensions: citizen protection and delivery of fair justice. Comment on this statement.
11. What are the six methods of foreign policy suggested in Arthaśāstra?
12. Ideas of foreign policy specified in Arthaśāstra has contemporary relevance. Comment on this statement.
13. Briefly comment on the administrative setup and hierarchies for public administration suggested in Arthaśāstra.
14. Arthaśāstra has some contemporary relevance. Argue for or against this statement with suitable examples and illustrations.
15. Do the epics contain ideas on public administration and governance? Prepare a one-page note in support of your arguments.
16. Write a one-page note on perspectives of Mahābhārata on the duties and attributes of the king.
17. Prepare a one page note on public administration ideas present in Rāmāyaṇa. Do you find these relevant for contemporary applications?

DISCOVER IKS

1. Given the antiquity of Arthaśāstra, written about 2300 years back contemporary relevance of Arthaśāstra is a question in the minds of the people. Watch the talk by Sriram Balasubramanian on this theme by visiting the URL: https://www.youtube.com/watch?v=Yg_yOUPrB5s. After watching the video prepare a two-page note to answer the following questions:
   (a) Trace the economic indicators of India in the last 1000 years. Could Arthaśāstra have played a role in galvanizing this growth?
   (b) What are the key contributions of Arthaśāstra to the economic development? Prepare specific points on 'Trade and Taxation' and 'Land and Labour'.
2. The author of Arthaśāstra, Kauṭilya has been generally known as the one who installed Candragupta in the throne. Watch this video which has a free-wheeling conversation on some possible aspects of Kauṭilya's personality: https://www.youtube.com/watch?v=25vwv8MglVs. Prepare a note to answer the following questions:
   (a) What is your understanding of Kauṭilya as a personality? Can you list 2 or 3 of them?
   (b) What are some of the unique attributes of Kauṭilya's thinking?

SUGGESTED READINGS

Śrimad Vālmīki-Rāmāyaṇa: Part–I, Gita Press, Gorakhpur, 2014.

Banerjea, P. (1916). *Public Administration in Ancient India*, Macmillan, London.

Bhagat, G. (1990). "Kautilya revisited and re-visioned", *The Indian Journal of Political Science*, Vol. **51**, No. 2 (April–June 1990), pp. 186–212.

Boesche, R. (2003). "Kautilya's Arthasastra on War and Diplomacy in Ancient India", *The Journal of Military History*, Vol. **67**(1), pp. 9–37.

Dutt, M.N. (2008). *Mahābhārata – Vol. VII Śānti Parva*, Parimal Publications, New Delhi.

Gautam, P.K. (2013). "One Hundred Years of Kautilya's Arthasastra", IDSA Monograph Series No. 20. Institute for Defence Studies and Analyses, New Delhi.

Kangle, R.P. (1972). *The Kauṭilya Arthaśāstra – Part I: A Critical Edition with a Glossary*, Motilal Banarsidass, New Delhi.

Kangle, R.P. (1972). *The Kauṭilya Arthaśāstra – Part II: An English Translation with Critical and Explanatory Notes*, Motilal Banarsidass, New Delhi.

Kangle, R.P. (1972). *The Kauṭilya Arthaśāstra – Part III: A Study*, Motilal Banarsidass, New Delhi.

Liebig, M. (2014). "Kauṭilya's Arthaśāstra: A Classic Text of Statecraft and an Untapped Political Science Resource", Working Paper No. 74, Heidelberg Papers in South Asian and Comparative Politics, South Asian Institute, Heidelberg University.

Rangarajan, L.N. (1987). *The Arthashastra*, Penguin Books, New Delhi.

Shamasastry, R. (1929). *Kauṭilya's Arthaśāstra*, 3rd ed., The Wesleyan Mission Press, Mysuru.

ENDNOTES

1. See Arthaśāstra 2.1.26: बालवृद्धव्यसन्यनाथांश्च राजा बिभृयात्, स्त्रियमप्रजातां प्रजातायाश्च पुत्रान् ॥ bālavṛddhavyasanyanāthāṃśca rājā bibhṛyāt, striyamaprajātāṃ prajātāyāśca putrān || 2.1.26. For all references to Arthaśāstra verses quoted in this chapter, refer **Kangle, R.P. (1972).** *The Kauṭilya Arthaśāstra – Part I: A Critical Edition with a Glossary*, Motilal Banarsidass, New Delhi.
2. These activities are specially mentioned in Book 2, Chapter 1 of Arthaśāstra. See for example verses 2.1.1, 2.1.19-20.
3. The seven key elements stated in Arthaśāstra are as follows: स्वामि-अमात्य-जनपद-दुर्ग-कोश-दण्ड-मित्राणि प्रकृतय: । svāmi-amātya-janapada-durga-kośa-daṇḍa-mitrāṇi prakṛtayaḥ: | 6.1.1
4. The qualities expected in an ideal ruler are enumerated in Book 6, Chapter 1 of Arthaśāstra, which includes: तत्र स्वामिसम्पत् । 6.1.2; महाकुलीनो दैवबुद्धिसत्त्वसंपन्नो वृद्धदर्शी धार्मिकः सत्यवागविसंवादकः कृतज्ञः स्थूललक्षो महोत्साहोऽदीर्घसूत्रः शक्यसामन्तो दृढबुद्धिरक्षुद्रपरिषत्को विनयकाम इत्याभिगामिका गुणाः । 6.1.3; शुश्रूषाश्रवणग्रहणधारणविज्ञानोहापोहतत्त्वाभिनिवेशाः प्रज्ञागुणाः । 6.1.4; शौर्यममर्षयः शीघ्रता दाक्ष्यं चोत्साहगुणाः। 6.1.5; वाग्मी प्रगल्भः स्मृतिमतिबलवानुदग्रः स्ववग्रहः कृतशिल्पोऽव्यसनो दण्डनाय्युपकारापकारयोर्दृष्टप्रतीकारी ह्रीमानापत्प्रकृत्योर्विनियोक्ता दीर्घदूरदर्शी देशकालपुरुषकारकार्यप्रधानः सन्धिविक्रमत्यागसंयमपणपरिच्छिद्रविभागी संवृतोऽदीनाभिहास्यजिह्मभ्रुकुटीक्षणः कामक्रोधलोभस्तम्भचापलोभोपतापपैशुन्यहीनः शुक्लः स्मितोदग्राभिभाषी वृद्धोपदेशाचार इत्यात्मसंपत् । 6.1.6.
mahākulīno daivabuddhisattvasaṃpanno vṛddhadarśī dhārmikaḥ satyavāgavisaṃvādakaḥ kṛtajñaḥ sthūlalakṣo mahotsāho'dīrghasūtraḥ śakyasāmanto dṛḍhabuddhirakṣudrapariṣatko vinayakāma ityābhigāmikā guṇāḥ | 6.1.3; śuśrūṣāśravaṇagrahaṇadhāraṇavijñānohāpohatattvābh iniveśāḥ prajñāguṇāḥ | 6.1.4; śauryamamarṣayaḥ śīghratā dākṣyaṃ cotsāhaguṇāḥ | 6.1.5; vāgmī pragalbhaḥ smṛtimatibalavānudagraḥ svavagrahaḥ kṛtaśilpo'vyasano daṇḍanāyyupakārāpakārayor dṛṣṭapratīkārī hrīmānāpatprakṛtyorviniyoktā dīrghadūradarśī deśakālapuruṣakārakāryapradhānaḥ sandhivikramatyāgasaṃyamapaṇaparicchidravibhāgī saṃvṛto'dīnābhihāsyajihmabhrukuṭīkṣaṇaḥ kāmakrodhalobhastambhacāpalobhopatāpapaiśunyahīnaḥ śuklaḥ smitodagrābhibhāṣī vṛddhopadeśācāra ityātmasaṃpat | 6.1.6.
5. See Arthaśāstra 8.1.8: मन्त्रो मन्त्रफलावाप्तिः कर्मानुष्ठानमायव्ययकर्म दण्डप्रणयनममित्राटवीप्रतिषेधो राज्यरक्षणं व्यसनप्रतीकारः कुमाररक्षणमभिषेकश्च कुमाराणामायत्तममात्येषु ॥ mantro mantraphalāvāptiḥ karmānuṣṭhānamāyavyayakarma daṇḍapraṇayanamamitrāṭavīpratiṣedho rājyarakṣaṇaṃ vyasanapratīkāraḥ kumārarakṣaṇamabhiṣekaśca kumārāṇāmāyattamamātyeṣu || Also see 8.1.23: अमात्यमूलाः सर्वारम्भाः जनपदस्य कर्मसिद्धयः स्वतः परतश्च योगक्षेमसाधनं व्यसनप्रतीकारः शून्यनिवेशोपचयौ दण्डकरानुग्रहश्चेति ॥ amātyamūlāḥ sarvārambhāḥ janapadasya karmasiddhayaḥ svataḥ parataśca yogakṣemasādhanaṃ vyasanapratīkāraḥ śūnyaniveśopacayau daṇḍakarānugrahaśceti ||
6. अष्टशतग्राम्या मध्ये स्थानीयं चतुःशतग्राम्या द्रोणमुखं द्विशतग्राम्याः कार्वटिकं दशग्रामीसंग्रहणे संग्रहणं स्थापयेत् ॥ aṣṭaśatagrāmyā madhye sthānīyaṃ catuḥśatagrāmyā droṇamukhaṃ dviśatagrāmyāḥ kārvaṭikaṃ daśagrāmīsaṃgrahaṇe saṃgrahaṇaṃ sthāpayet || Arthaśāstra 2.1.4.
7. For more details on this see, Arthaśāstra, Book 7, Chapter 15.
8. For more details on this see Arthaśāstra Book 2, Chapter 6, Verses 1–8.
9. Verses 35 and 36 in Chapter 12 of Book 2 has the details: एवं मूल्यं च भागं च व्याजीं परिघमत्ययम् । शुल्कं वैधरणं दण्डं रूपं रूपिकमेव च ॥ ३५ ॥ खनिभ्यो द्वादशविधं धातुं पण्यं च संहरेत् । एवं सर्वेषु पण्येषु स्थापयेन्मुखसंग्रहम् ॥ ३६ ॥ evaṃ mūlyaṃ ca bhāgaṃ ca vyājīṃ parighamatyayam | śulkaṃ vaidharaṇaṃ daṇḍaṃ rūpaṃ rūpikameva ca || 35 || khanibhyo dvādaśavidhaṃ dhātuṃ paṇyaṃ ca saṃharet | evaṃ sarveṣu paṇyeṣu sthāpayenmukhasaṃgraham || 36 ||
10. तस्मादरिषड्वर्गत्यागेनेन्द्रियजयं कुर्वीत, वृद्धसंयोगेन प्रज्ञाम्, चारेण चक्षुः, उत्थानेन योगक्षेमसाधनम्, कार्यानुशासनेन स्वधर्मस्थापनम्, विनयं विद्योपदेशेन, लोकप्रियत्वमर्थसंयोगेन, हितेन वृत्तिम् । tasmādariṣaḍvargatyāgenen driyajayaṃ kurvīta, vṛddhasaṃyogena prajñām, cāreṇa cakṣuḥ, utthānena yogakṣemasādhanam, kāryānuśāsanena svadharmasthāpanam, vinayaṃ vidyopadeśena, lokapriyatvamarthasaṃyogena, hitena vṛttim | 1.7.1

11. For more details see 7.5.19–26.

12. These details are available in Book 4, Chapter 1. For more information see, verses 4.1.10, 22, 32, 40, and 42.

13. The four basic sources for establishing justice are as per the following two verses: धर्मश्च व्यवहारश्च चरित्रं राजशासनम् । विवादार्थश्चतुष्पादः पश्चिमः पूर्वबाधकः ॥ 3.1.39. तत्र सत्ये स्थितो धर्मो व्यवहारस्तु साक्षिषु। चरित्रं संग्रहे पुंसां राज्ञामाज्ञा तु शासनम् ॥ 3.1.40. dharmaśca vyavahāraśca caritraṃ rājaśāsanam | vivādārthaścatuṣpādaḥ: paścimaḥ pūrvabādhakaḥ || 3.1.39. tatra satye sthito dharmo vyavahārastu sākṣiṣu | caritraṃ saṃgrahe puṃsāṃ rājñāmājñā tu śāsanam || 3.1.40.

14. The six methods to be adopted in foreign policy are stated in Arthaśāstra Book 7, Chapter 1. They are as follows: षाड्गुण्यमेवैतद् अवस्थाभेदादिति कौटिल्यः । 7.1.5. तत्र पणबन्धः सन्धिः । 7.1.6. अपकारो विग्रहः । 7.1.7. उपेक्षणमासनम् । 7.1.8. अभ्युच्चयो यानम् । 7.1.9. परार्पणं संश्रयः । 7.1.10. संधिविग्रहोपादानं द्वैधीभावः । 7.1.11. इति षड्गुणाः । 7.1.12. ṣāḍguṇyamevaitad avasthābhedāditi kauṭilyaḥ | 7.1.5. tatra paṇabandhaḥ sandhiḥ | 7.1.6. apakāro vigrahaḥ | 7.1.7. upekṣaṇamāsanam | 7.1.8. abhyuccayo yānam | 7.1.9. parārpaṇaṃ saṃśrayaḥ | 7.1.10. saṃdhivigrahopādānaṃ dvaidhībhāvaḥ | 7.1.11. iti ṣaḍguṇāḥ | 7.1.12.

15. The four methods are given in Book 7, Chapter 16: सामदानाभ्यां दुर्बलानुपनमयेत्, भेददण्डाभ्यां बलवतः। 7.16.3. sāmadānābhyāṃ durbalānupanayet, bhedadaṇḍābhyāṃ balavataḥ ||

16. According to Arthaśāstra the four methods are useful under certain conditions. These are given in the following verse: शत्रोः शङ्कितामात्येषु सान्त्वं प्रयुक्तं शेषप्रयोगं निवर्तयति, दूष्यामात्येषु दानं, संघातेषु भेदः, शक्तिमत्सु दण्ड इति । 9.7.72. śatroḥ śaṅkitāmātyeṣu sāntvaṃ prayuktaṃ śeṣaprayogaṃ nivartayati, dūṣyāmātyeṣu dānaṃ, saṃghāteṣu bhedaḥ, śaktimatsu daṇḍa iti | 9.7.72.

17. In Arthaśāstra, there is a description of instructions to minting of a silver coin of 1 paṇa. Based on this, one can conclude that paṇa is a silver coin, with a silver content roughly equivalent to three-quarters of a tola. For details, see, लक्षणाध्यक्षचतुर्भागताम्रं रूप्यरूपं तीक्ष्णत्रपुसीसाञ्जनानामन्यतममाषबीजयुक्तं कारयेत्–पणम् अर्धपणं पादम् अष्टभागमिति, पादाजीवं ताम्ररूपं माषकम् अर्धमाषकं काकणीम् अर्धकाकणीमिति । lakṣaṇādhyakṣacaturbhāgaṃ rūpyarūpaṃ tīkṣṇatrapusīsāñjanānāmanyatamamāṣabījayuktaṃ kārayet – paṇam ardhapaṇaṃ pādam aṣṭabhāgamiti, pādājīvaṃ tāmrarūpaṃ māṣakam ardhamāṣakaṃ kākaṇīm ardhakākaṇīmiti || 2.12.24.

18. For details on the salary levels proposed in Arthaśāstra see, **Rangarajan, L.N. (1987).** *The Arthaśāstra*, Penguin Books, pp. 208–228; 288–293.

19. For details see, **Dutt, M.N. (2008).** *Mahābhārata – Vol. VII Śānti Parva*, Parimal Publications, New Delhi, p. 227.
राज्ञा हि पूजितो धर्मस्ततः सर्वत्र पूज्यते । यद्यदाचरते राजा तत्प्रजानां स्म रोचते ।। rājñā hi pūjito dharmastataḥ sarvatra pūjyate | yadyadācarate rājā tatprajānāṃ sma rocate || 12-75-4

20. *Ibid.* p. 278.
धर्ममेवानुवर्तस्व न धर्माद्विद्यते परम् । धर्मे स्थिता हि राजानो जयन्ति पृथिवीमिमाम् ।। 12-92-6
dharmamevānuvartasva na dharmādvidyate param | dharme sthitā hi rājāno jayanti pṛthivīmimām ||
अर्थसिद्धेः परं धर्मं मन्यते यो महीपतिः । वृद्ध्यां च कुरुते बुद्धिं स धर्मेण विराजते ॥ 12-92-7
arthasiddheḥ paraṃ dharmaṃ manyate yo mahīpatiḥ | vṛddhyāṃ ca kurute buddhiṃ sa dharmeṇa virājate ||
अधर्मदर्शी यो राजा बलादेव प्रवर्तते । क्षिप्रमेवापयातोऽस्मादुभौ प्रथममध्यमौ ।। 12-92-8
adharmadarśī yo rājā balādeva pravartate | kṣipramevāpayāto'smādubhau prathamamadhyamau ||

21. *Ibid.* pp. 270–271.
धर्मे तिष्ठन्ति भूतानि धर्मो राजनि तिष्ठाति । तं राजा साधु यः शास्ति स राजा पृथिवीपतिः ।। 12-90-5
dharme tiṣṭhanti bhūtāni dharmo rājani tiṣṭhāti | taṃ rājā sādhu yaḥ śāsti sa rājā pṛthivīpatiḥ ||
राजा परमधर्मात्मा लक्ष्मीवान् धर्म उच्यते । देवाश्च गर्हां गच्छन्ति धर्मो नास्तीति चोच्यते ।। 12-90-6
rājā paramadharmātmā lakṣmīvān dharma ucyate | devāśca garhāṃ gacchanti dharmo nāstīti cocyate || 12-90-6

यस्मिन्धर्मो विराजेत तं राजानं प्रचक्षते । यस्मिन्विलीयते धर्मस्तं देवा वृषलं विदुः ।। 12-90-15
yasmindharmo virājeta taṃ rājānaṃ pracakṣate | yasminvilīyate dharmastaṃ devā vṛṣalaṃ viduḥ ||
वृषो हि भगवान् धर्मो यस्तस्य कुरुते ह्यलम् । वृषलं तं विदुर्देवास्तस्माद्धर्मं विवर्धयेत् ।। 12-90-16
vṛṣo hi bhagavān dharmo yastasya kurute hyalam | vṛṣalaṃ taṃ vidurdevāstasmāddharmaṃ vivardhayet ||
धर्मे वर्धति वर्धन्ति सर्वभूतानि सर्वदा । तस्मिन् ह्रसति ह्रीयन्ते तस्माद्धर्मं न लोपयेत् ।। 12-90-17
dharme vardhati vardhanti sarvabhūtāni sarvadā | tasmin hrasati hrīyante tasmāddharmaṃ na lopayet ||
धनानि स्त्रवति धर्मो हि धारणाद्वेति निश्चयः। अकार्याणां मनुष्येन्द्र स सीमान्तकरः स्मृतः ।। 12-90-18
dhanāni sravati dharmo hi dhāraṇādveti niścayaḥ| akāryāṇāṃ manuṣyendra sa sīmāntakaraḥ smṛtaḥ ||
प्रभवार्थं हि भूतानां धर्मः सृष्टः स्वयंभुवा । तस्मात्प्रवर्धयेद्धर्मं प्रजानुग्रहकारणात् ।। 12-90-19
prabhavārthaṃ hi bhūtānāṃ dharmaḥ sṛṣṭaḥ svayaṃbhuvā|tasmātpravardhayeddharmaṃ prajānugrahakāraṇāt ||
तस्माद्धि राजशार्दूल धर्मः श्रेष्ठतरः स्मृतः । स राजा यः प्रजाः शास्ति साधुकृत् पुरुषर्षभ ।। 12-90-20
tasmāddhi rājaśārdūla dharmaḥ śreṣṭhataraḥ smṛtaḥ | sa rājā yaḥ prajāḥ śāsti sādhukṛt puruṣarṣabha ||
कामक्रोधावनादृत्य धर्ममेवानुपालयेत् । धर्मः श्रेयस्करतमो राज्ञां भरतसत्तम ।। 12-90-21
kāmakrodhāvanādṛtya dharmamevānupālayet | dharmaḥ śreyaskaratamo rājñāṃ bharatasattama || 12-90-21

22. *Ibid.* p. 203.
राजा ह्येवाखिलं लोकं समुदीर्णं समुत्सुकम् । प्रसादयति धर्मेण प्रसाद्य च विराजते ॥ rājā hyevākhilaṃ lokaṃ samudīrṇaṃ samutsukam | prasādayati dharmeṇa prasādya ca virājate || 12.68.9

23. *Ibid.* p. 256.
भीष्म उवाच । व्यवहारेण शुद्धेन प्रजापालनतत्परः । प्राप्य धर्मं च कीर्तिं च लोकानाप्नोत्यसौ शुचिः ।।12-85-2
bhīṣma uvāca | vyavahāreṇa śuddhena prajāpālanatatparaḥ | prāpya dharmaṃ ca kīrtiṃ ca lokānāpnotyasau śuciḥ||

24. *Ibid.* p. 217.
बलिषष्ठेन शुल्केन दण्डेनाथापराधिनाम् । शास्त्रानीतेन लिप्सेथा वेतनेन धनागमम् ।। 12-71-10
baliṣaṣṭhena śulkena daṇḍenāthāparādhinām | śāstrānītena lipsethā vetanena dhanāgamam ||
दापयित्वा करं धर्म्यं राष्ट्रं नीत्या यथाविधि । तथैतं कल्पयेद्राजा योगक्षेममतन्द्रितः ।। 12-71-11
dāpayitvā karaṃ dharmyaṃ rāṣṭraṃ nītyā yathāvidhi | tathaitaṃ kalpayedrājā yogakṣemamatandritaḥ ||
तस्माद्धर्मेण लाभेन लिप्सेथास्त्वं धनागमम् । धर्मार्थावध्रुवौ तस्य यो न शास्त्रपरो भवेत् ।। 12-71-13
tasmāddharmeṇa lābhena lipsethāstvaṃ dhanāgamam | dharmārthāvadhruvau tasya yo na śāstraparo bhavet ||
अर्थमूलोऽपि हिंसां च कुरुते स्वयमात्मनः । करैरशास्त्रदृष्टैर्हि मोहात् संपीडयन् प्रजाः ।।12-71-15
arthamūlo'pi hiṃsāṃ ca kurute svayamātmanaḥ | karairaśāstradṛṣṭairhi mohāt saṃpīḍayan prajāḥ ||
ऊधश्छिन्द्यात् तु यो धेन्वाः क्षीरार्थी न लभेत् पयः । एवं राष्ट्रमयोगेन पीडितं न विवर्धते ।। 12-71-16
ūdhaśchindyāt tu yo dhenvāḥ kṣīrārthī na labhet payaḥ | evaṃ rāṣṭramayogena pīḍitaṃ na vivardhate ||

25. *Ibid.* p 263.
विक्रयं क्रयमध्वानं भक्तं च परिच्छदम् । योगक्षेमं च संप्रेक्ष्य वणिजां कारयेत् करान् ।।12-87-13
vikrayaṃ krayamadhvānaṃ bhaktaṃ ca paricchadam | yogakṣemaṃ ca saṃprekṣya vaṇijāṃ kārayet karān ||
उत्पत्तिं दानवृत्तिं च शिल्पं संप्रेक्ष्य चासकृत् । शिल्पं प्रति करानेवं शिल्पिनः प्रति कारयेत् ।।12-87-14
utpattiṃ dānavṛttiṃ ca śilpaṃ saṃprekṣya cāsakṛt | śilpaṃ prati karānevaṃ śilpinaḥ prati kārayet ||
फलं कर्म च संप्रेक्ष्य ततः सर्वं प्रकल्पयेत् । फलं कर्म च निर्हेतु न कश्चित् संप्रवर्तते ।। 12-87-16
phalaṃ karma ca saṃprekṣya tataḥ sarvaṃ prakalpayet | phalaṃ karma ca nirhetu na kaścit saṃpravartate ||
यथा राजा च कर्ता च स्यातां कर्मणि भागिनौ । संवेक्ष्य तु तथा राज्ञा प्रणेयाः सततं कराः ।। 12-87-17
yathā rājā ca kartā ca syātāṃ karmaṇi bhāginau | saṃvekṣya tu tathā rājñā praṇeyāḥ satataṃ karāḥ ||

26. For details see, **(2014).** *Śrimad Vālmīki-Rāmāyaṇa: Part-I*, Gita Press, Gorakhpur, pp. 499–505. मन्त्रो विजयमूलं हि राज्ञां भवति राघव । सुसंवृतो मन्त्रधरैरमात्यैः शास्त्रकोविदैः || 2.100.16 mantro vijayamūlaṃ hi rājñāṃ bhavati rāghava | susaṃvṛto mantradharairamātyaiḥ śāstrakovidaiḥ || 2.100.16

27. *Ibid.* कच्चिदात्मसमाः शूराः शृतवन्तो जितेन्द्रियाः । कुलीनाश्चेङ्गितज्ञाश्च कृतास्ते तात मन्त्रिणः || 2.100.15 kaccidātmasamāḥ śūrāḥ śṛtavanto jitendriyāḥ | kulīnāścengitajñāśca kṛtāste tāta mantriṇaḥ || 2.100.15

28. *Ibid.* अमात्यानुपधातीतान् पितृपैतामहाञ्शुचीन् । श्रेष्ठाञ्छ्रेष्ठेषु कच्चित् त्वं नियोजयसि कर्मसु || 2.100.26 amātyānupadhātītān pitṛpaitāmahāñśucīn | śreṣṭhāñchreṣṭheṣu kaccit tvaṃ niyojayasi karmasu || 2.100.26

29. *Ibid.* कच्चिदार्यो विशुद्धात्मा क्षारितश्चापकर्मणा । अपृष्टः शास्त्रकुशलैर्न लोभाद् वध्यते शुचिः || 2.100.56, गृहीतश्चैव पृष्टश्च काले दृष्टः सकारणः । कच्चिन्न मुच्यते चोरो धनलोभान्नरर्षभ ॥ 2.100.57, व्यसने कच्चिदाढ्यस्य दुर्गतस्य च राघव । अर्थं विरागाः पश्यन्ति तवामात्या बहुश्रुताः ॥ 2.100.58, यानि मिथ्याभिशस्तानां पतन्त्यश्रूणि राघव । तानि पुत्रान्पशून् घ्नन्ति प्रीत्यर्थमनुशासतः || 2.100.59 kaccidāryo viśuddhātmā kṣāritaścāpakarmaṇā | apṛṣṭaḥ śāstrakuśalairna lobhād vadhyate śuciḥ || 2.100.56 gṛhītaścaiva pṛṣṭaśca kāle dṛṣṭaḥ sakāraṇaḥ | kaccinna mucyate coro dhanalobhānnararṣabha || 2.100.57, vyasane kaccidāḍhyasya durgatasya ca rāghava | arthaṃ virāgāḥ paśyanti tavāmātyā bahuśrutāḥ || 2.100.58, yāni mithyābhiśastānāṃ patantyaśrūṇi rāghava | tāni putrānpaśūn ghnanti prītyarthamanuśāsataḥ || 2.100.59.

30. *Ibid.* कच्चिदष्टादशान्येषु स्वपक्षे दश पञ्च च । त्रिभिस्त्रिभिरविज्ञातैर्वेत्सि तीर्थानि चारकैः ॥ 2.100.36 kaccidaṣṭādaśānyeṣu svapakṣe daśa pañca ca | tribhistribhiravijñātairvetsi tīrthāni cārakaiḥ || 2.100.36.

31. *Ibid.* कच्चित्तु सुकृतान्येव कृतरूपाणि वा पुनः । विदुस्ते सर्वकार्याणि न कर्तव्यानि पार्थिवाः ॥ 2.100.20, कच्चिन्न तर्कैर्युक्त्या वा ये चाप्यपरिकीर्तिताः । त्वया वा तवामात्यैर्बुध्यते तात मन्त्रितम् ॥ 2.100.21. kaccittu sukṛtānyeva kṛtarūpāṇi vā punaḥ | viduste sarvakāryāṇi na kartavyāni pārthivāḥ || 2.100.20, kaccinna tarkairyuktyā vā ye cāpyaparikīrtitāḥ | tvayā vā tavāmātyairbudhyate tāta mantritam || 2.100.21.

32. *Ibid.* कच्चिच्चैत्यशतैर्जुष्टः सुनिविष्टजनाकुलः । देवस्थानैः प्रपाभिश्च तटाकैश्चोपशोभितः ॥ 2.100. 43, प्रहृष्टनरनारीकः समाजोत्सवशोभितः । सुकृष्टसीमा पशुमान्हिंसाभिः परिवर्जितः ॥ 2.100. 44, अदेवमातृको रम्यः श्वापदैः परिवर्जितः । परित्यक्तो भयैस्सर्वैः खनिभिश्चोपशोभितः ॥ 2.100. 45, विवर्जितो नरैः पापैर्मम पूर्वैः सुरक्षितः । कच्चिज्जनपदस्स्फीतः सुखं वसति राघव || 2.100. 46. kacciccaityaśatairjuṣṭaḥ suniviṣṭajanākulaḥ | devasthānaiḥ prapābhiśca taṭākaiścopaśobhitaḥ || 2.100. 43, prahṛṣṭanaranārīkaḥ samājotsavaśobhitaḥ | sukṛṣṭasīmā paśumānhiṃsābhiḥ parivarjitaḥ || 2.100. 44, adevamātṛko ramyaḥ śvāpadaiḥ parivarjitaḥ | parityakto bhayaissarvaiḥ khanibhiścopaśobhitaḥ || 2.100. 45, vivarjito naraiḥ pāpairmama pūrvaiḥ surakṣitaḥ | kaccijjanapadassphītaḥ sukhaṃ vasati rāghava || 2.100.46

Glossary

अञ्जन (Añjana): Collyrium

अणु (Aṇu): Smallest particle/atom

अदृष्ट (Adṛṣṭa): Invisible

अनुमान (Anumāna): Inference

अन्नमयकोष (Annamayakoṣa): Physical layer

अपौरुषेय (Apauruṣeya): Not composed by a human

अयस् (Ayas): Iron

अयुग्म (Ayugma): Odd number

अर्थ (Artha): Means to fulfil desires

अवस्था (Avasthā): State

अस्थि (Asthi): Bone

आकाश (Ākāśa): Ether

आत्मन् (Ātman): The conscious self

आनन्द (Ānanda): Bliss

आनन्दमयकोष (Ānandamayakoṣa): Embodiment of happiness

आयुर्वेद (Āyurveda): Science of life/health

आयुस् (Āyus): Association of Mind, body, senses and consciousness

आरण्यक (Āraṇyaka): Portion of Vedas which focuses mainly on upāsanā

आसन (Āsana): Conducive sitting posture for yoga

इतिहास (Itihāsa): History

इन्द्रिय (Indriya): Organs (Sensory or Action)

ईश्वर (Īśvara): Supreme being

उत्तरायण (Uttarāyaṇa): Sun's progress towards north

उदाहरण/दृष्टान्त (Udāharaṇa/Dṛṣṭānta): Example (to establish the proposition)

उपनिषद् (Upaniṣad): Portion of Vedas which discusses spiritual knowledge

उपमान (Upamāna): Comparison

उपवेद (Upa-veda): Application oriented knowledge associated with each veda

उपसर्ग (Upasarga): Prefix

उपासनाकाण्ड (Upāsānā-kāṇḍa): Portions of Vedas where techniques of meditation and other concentration are elaborated

ऋषि (Ṛṣi): Seer, one who saw the mantras

औषध (Auṣadha): Medicine

कटपयादि (पद्धति) (Kaṭapayādi (Paddhati)): A system of denoting numbers using Sanskrit alphabets

कर (Kara): Tax

कर्तृ (Kṛtṛ): Performer of an action

कर्म (Karma): Actions

कर्मकाण्ड (Karma-kāṇḍa): Portions of Vedas which deal with ritual and other duties

कल्प (Kalpa): A 'complete user manual' for Vedic life

कान्तलोह (Kāntaloha): Soft iron

कारक (Kāraka): Participant in an action

कालनिर्णय (Kālanirṇaya): Determination of time

कृष्णपक्ष (Kṛṣṇapakṣa): Dark fortnight

कोष्ठी (Koṣṭhī): Furnace

क्रिया (Kriyā): Action

क्षय (Kṣaya): Subtraction

गणित (Gaṇita): Mathematics

गर्भगृह (Garbhagṛha): Sanctum sanctorum

गुण (Guṇa): Quality, Multiplication

गृह्यसूत्र (Gṛhya-sūtra): Guide for household ceremonies

ग्रह (Graha): Planet

ग्राम (Grāma): Village

घन (Ghana): Cube

घनमूल (Ghanamūla): Cube root

चान्द्रमास (Cāndra-māsa): Lunar month

चिति/उपचिति (Citi/Upaciti): Series

चित् (Cit): Consciousness

छन्दस् (Chandas): Meters to which mantras are set

जगत् (Jagat): World

जीव (Jīva): The conscious self

ज्ञानकाण्ड (Jñāna-kāṇḍa): Portions of Vedas which deal with knowledge of the ultimate reality

ज्या (Jyā): Radius

ज्यार्ध (Jyārdha): Half cord

ज्योत्पत्ति (Jyotpatti): Trigonometry

तक्षक (Takṣaka): Carpenter

तर्क (Tarka): Abstract reasoning or Logic

तीक्ष्णलोह (Tīkṣṇaloha): High carbon steel

दक्षिणायन (Dakṣiṇāyana): Sun's progress towards south

दण्ड (Daṇḍa): Punishment

दमन (Damana): Heating and smelting

दर्शन (Darśana): A philosophical world view

दुर्ग (Durga): Fort

द्रव्य (Dravya): Substance

धनुस् (Dhanus): A unit of length, approximately 1.6 meters

धर्म (Dharma): A quality acquired by good actions

धर्मशास्त्र (Dharma-śāstra): Principles of individual, social and political life

धर्मसूत्र (Dharma-sūtras): Guide for social duties

धर्मस्थ (Dharmastha): Judge

धातु (Dhātu): A basic Substance, also a verbal root in grammatical operations

धारणा (Dhāraṇā): Continuous contemplation

ध्यान (Dhyāna): Fixing mind on one object

नक्षत्र (Nakṣatra): Star

निघण्टु (Nighaṇṭu): A collation of synonyms used in Vedas, a thesaurus

नियम (Niyama): Observances

निरुक्त (Nirukta): Etymology of Vedic words

नील (Nīla): Indigo (colour)

पक्ष (Pakṣa): Fortnight

पञ्चकोष (Pañcakoṣa): The five layers of existence

पञ्चलोह (Pañcaloha): An alloy made of five metals: gold, silver, copper, zinc, and iron

पञ्चाङ्ग (Pañcāṅga): Indian calendar consisting of five major components

पट्टन (Paṭṭana): City

पद (Pada): A word

परीक्षा (Parīkṣā): Examination

पादविन्यास (Pādavinyāsa): Site layout

पुराण (Purāṇa): Texts that include historical accounts, mythology and various other things

पुरुष (Puruṣa): Conscious self

पुरुषार्थ (Puruṣārtha): Goals of an individual

प्रकृति (Prakṛti): Root cause of creation, also a base to which a suffix is added in grammatical operations

प्रत्यक्ष (Pratyakṣa): Perception through senses

प्रत्यय (Pratyaya): A suffix used in grammatical operations

प्रत्याहार (Pratyāhāra): Withdrawing cognitive activities

प्रपत्ति (Prapatti): Total submission

प्रमा (Pramā): Valid knowledge

प्रमाण (Pramāṇa): Means to attain valid knowledge

प्रमातृ (Pramātṛ): Seeker of the valid knowledge

प्रमेय (Prameya): Subject of the valid knowledge

प्राण (Prāṇa): Vital energy (in the form of breath)

प्राणमयकोष (Prāṇamayakoṣa): Vital energy Layer

प्राणायाम (Prāṇāyāma): Breath control

प्रासाद (Prāsāda): Temple

बीजगणित (Bījagaṇita): Algebra

ब्रह्मन् (Brahman): The supreme being

ब्राह्मण (Brāhmaṇa): Portion of Vedas which deal with rituals

भद्रगणित (Bhadragaṇita): Construction of magic square

भाग (Bhāga): Division

भूतसंख्या (पद्धति) (Bhūtasaṃkhyā (paddhati)): A system of denoting numbers using the numbers of natural objects

भूमिपूजन (Bhūmipūjana): Worship of the site

मज्जा (Majjā): Marrow

मञ्जिष्ठा (Mañjiṣṭhā): Maddar

मधूच्छिष्टविधान (Madhūcchiṣṭavidhāna): Lost vax casting technique

मनोमयकोष (Manomayakoṣa): Mental layer

मांस (Māṃsa): Flesh

मास (Māsa): Month

मीमांसा (Mīmāṃsā): The discipline that is oriented towards enquiry

मुक्ति (Mukti): Liberation

मुण्डलोह (Muṇḍaloha): Cast iron

मुहूर्त (Muhūrta): A unit of time (approximately 24 minutes)

मूषायन्त्र (Mūṣāyantra): Crucible

मेदस् (Medas): Fat

मोक्ष (Mokṣa): Liberation

यज्ञवेदी (Yajñavedī): Altars made for yajña

यन्त्र (Yantra): Machine or an instrument

यम (Yama): Forbearance

युग्म (Yugma): Even number

युत (Yuta): Addition

योग-क्षेम (Yoga-Kṣema): Accomplishment of an object and its peaceful enjoyment

रक्त (Rakta): Blood

रजत (Rajata): Silver

रजन (Rajana): Resin

रथ्या (Rathyā): Vehicular roads

रस (Rasa): Plasma

राजधानी (Rājadhānī): Capital

राजमार्ग (Rājamārga): Road leading to palace

राशि/राशिचक्र (Rāśi/rāśicakra): Zodiac

लक्षण (Lakṣaṇa): Definition

लाक्षा (Lākṣā): Lac

लिङ्ग (Liṅga): Sign or identifier

लोह/लोहितायस् (Loha/Lohitāyas): Copper

वंश (Vaṃśa): Lineage

वज्रलेप (Vajralepa): Adamantine

वर्ग (Varga): Square

वर्गमूल (Vargamūla): Square root

वर्ण (Varṇa): Smallest unit of sound in Sanskrit language, also denotes a class of people based on occupation

वर्धकि (Vardhaki): One who builds the structure

वस्तु (Vastu): Material

वाद (Vāda): Debate/dialogue

वार (Vāra): Day of a week

वारसंकलित (Vārasaṃkalita): Repeated summation of series

वार्ता (Vārtā): Economy

वास्तुपुरुष (Vāstupuruṣa): Structure personified as man

वास्तुमण्डल (Vāstumaṇḍala): Basic layout of the structure in square

वास्तुशास्त्र (Vāstuśāstra): Science of architecture

विज्ञानमयकोष (Vijñānamayakoṣa): Intellectual layer

विशेष (Viśeṣa): Differentiating factor, speciality

विषुवत् (Viśuvat): Equinoctial day

वीथी (Vīthī): Avenues

वृत्तपरिणाह (Vṛttapariṇāhaḥ): Circumference

वेदाङ्ग (Vedāṅga): Auxiliary works meant for preservation and interpretation of Vedas

व्याकरण (Vyākaraṇa): Vedāṅga that sets out the rules of language, also known as grammar

शङ्कु (Śaṇku): Gnomon

शब्द (प्रमाण) (Śabda (Pramāṇa)): Verbal testimony

शल्यतन्त्र (Śalyatantra): Surgical technique

शाखा (Śākhā): Branch

शाला (Śālā): House

शास्त्र (Śāstra): A discipline

शिक्षा (Śikśā): Vedāṅga that deals with pronunciation

शिखर (Śikhara): Summit

शिरावेध (Śirāvedha): Bloodletting

शिल्पिन् (Śilpin): Technician

शुक्र (Śukra): Semen

शुक्लपक्ष (Śuklapakṣa): Bright fortnight

शुल्बसूत्र (Śulba-sūtras): Rules for measurements and construction of fire altars, sacrificial place, etc.

शोधन (Śodhana): Purification

श्रुति (Śruti): That which is heard, vedic literature

श्रौतसूत्र (Śrauta-sūtras): Guide for Vedic rituals

संकलित (Saṃkalita): Summation of series

संख्या (Saṃkhyā): Actual knowledge, also denotes number

संयोग (Saṃyoga): Conjunction

संवत्सर (Saṃvatsara): Year

संशय (Saṃśaya): Ambiguity

संसार (Saṃsāra): Birth and death cycle

संहिता (Saṃhitā): Portion of the Vedas which consists of mantras

सत् (Sat): Existence

सनातन धर्म (Sanātana dharma): Value systems of ancient Indian society from time immemorial

सभा (Sabhā): Meeting hall

समाधि (Samādhi): Complete cessation of all cognitive activities

समास (Samāsa): Compound word

सर्ग (Sarga): Creation

सर्वतोभद्र (Sarvatobhadra): Pan-diagonal magic square

सामान्य (Sāmānya): Commonality

सिद्धान्त (Siddhānta): Established tenet

सिन्दूर (Sindūra): Vermilion

सुभाषित (Subhāṣita): Well spoken words, verses of wisdom

सूक्त (Sūkta): Group of mantras in the Ṛgveda

सूत्र (Sūtra): Aphorism

सूत्रग्राहिन् (Sūtragrāhin): Draftsman

सौरमास (Sauramāsa): Solar month

स्थपति (Sthapati): Head architect

स्थापत्य (Sthāpatya): Job/skill of an architect

स्वर (Svara): Musical notes (used in recitation of Vedas)

स्वर्ण (Svarṇa): Gold

स्वस्थ (Svastha): Healthy individual

हरिद्रा (Haridrā): Turmeric

हेतु (Hetu): Reasoning of the proposition, cause

हेत्वाभास (Hetvābhāsa): Logical fallacies

Index

Q

R

S